Harmonic generation

$$\cos^2(x) = \frac{1 + \cos 2x}{2} \qquad \sin^2(x) = \frac{1 - \cos 2x}{2}$$

$$THD = \sum_{n=2}^{\infty}$$

$$\int_{-\infty}^{\infty} f(t)\, \delta(\tau - T_0)\, dt = f(t_0) \qquad \int_{-\infty}^{\infty} f(t)\, \delta[a(\tau - T_0)]\, dt = \frac{1}{|a|} f(t_0)$$

$$F(\omega) = \int_{-\infty}^{\infty} f(t)\, e^{j\omega t}\, dt$$

$$f(t) = \frac{1}{2\pi} \int_{-\infty}^{\infty} F(\omega)\, e^{j\omega t}\, dt$$

energy signals $\qquad^{85}$

$$\int_{-\infty}^{\infty} |f_t|^2 dt = \frac{1}{2\pi} \int_{-\infty}^{\infty} |F(\omega)|^2 d\omega$$

fourier Trans
92

$$\int_{-\infty}^{t} \text{impulse response}\, d\tau \quad = \text{unit step response} \quad \boxed{62}$$

Convolution

$$\int_{-\infty}^{\infty} f(\tau)\, h(t - \tau)\, d\tau$$

$$F(\omega)\, H(\omega) = G(\omega)$$
$$f \otimes h = g(t)$$

Amplitude $dB = 10 \log \frac{A_1}{A_0}$

Power $\qquad dB = 20 \log \frac{A_1}{A_0}$

Bessel Table 292

AM suppressed carrier

Double side band

$\phi = f(t) \cos w$ $f_t \cos(w_c t)$

$f(t)$

$F(w)$

$-w \quad w$

w_c
$2w$

w_c
$2w$

Band Pass

$\cos(w_c t)$

$f(t) \longrightarrow \otimes \longrightarrow \phi$

mod

de mod

$\phi \longrightarrow \otimes \longrightarrow \boxed{LPF} \longrightarrow f(t)$

$\cos(w_c t)$

Shape factor $\dfrac{W-60}{W-6}$

PM

$$\Theta(t) = \omega_c t + k_p f(t) + \Theta_0$$

$$\omega_i = \omega_c + k_p \frac{df(t)}{dt}$$

FM

$$\Theta t = \omega_c t + \int_0^t k_f f(\tau) d\tau + \Theta_0$$

$$\omega_i = \omega_c + k_f f t$$

$$\beta = \frac{\Delta \omega}{\omega_m}$$

$$\Delta \omega = k_p f(t)_{max} \omega_m$$

$$\Delta \omega = k_f f(t)_{max}$$

$$\omega_i = \frac{d\Theta(t)}{dt}$$

	Adv	Dis Adv
SSB	Doesn't Double Bandwith	Hard to meet filter requirements
DSB-SC	allows full freq transmition DC efficient	exact freq matching on demod.
DSBLC	Simple demot	inefficient
VSB	conserves spectrum space and Low freq	difficult to get correct phase Response

INTRODUCTION TO COMMUNICATION SYSTEMS

SECOND EDITION

Ferrel G. Stremler
University of Wisconsin, Madison

INTRODUCTION TO COMMUNICATION SYSTEMS

SECOND EDITION

 Addison-Wesley Publishing Company
Reading, Massachusetts
Menlo Park, California • London • Amsterdam
Don Mills, Ontario • Sydney

This book is in the
ADDISON-WESLEY SERIES IN ELECTRICAL ENGINEERING

Library of Congress Cataloging in Publication Data

Stremler, Ferrel G.
    Introduction to communication systems.

    (Addison-Wesley series in electrical engineering)
    Bibliography: p.
    Includes index.
    1. Telecommunication.   2. Signal theory (Telecommuni-
cation)   I. Title.   II. Series.
TK5101.S75   1982   621.38   81-7917
ISBN 0-201-07251-3   AACR2

Reprinted with corrections, December 1982

ISBN 0-201-07251-3
FGHIJ-HA-898765

To Ruth, my wife.

PREFACE

This textbook presents to undergraduates an introductory explanation of communication systems, with the emphasis on signal design and modulation. The approach is therefore tailored to a careful development of the mathematical principles upon which such systems are based, using examples from a wide variety of current communication systems wherever possible. These range from commercial broadcasting and telephone systems to satellite telemetry and radar.

Material added in this second edition is, primarily, about digital methods and reflects the rapidly increasing importance of digital signal transmission and modulation in communication systems. Chapter 7 has been expanded to include Nyquist waveform shaping, asynchronous multiplexing, and pseudonoise sequences. Material in Chapter 9 of the first edition has been divided and expanded into two chapters, one on digital baseband transmission (Chapter 9) and the second on digital modulation (Chapter 10). The presentation in this last chapter includes M-ary digital modulation methods that are currently being investigated for new communication systems designs.

Because this textbook is intended for undergraduates, the material is written in as explicit a manner as possible and is illustrated clearly. Examples and drill problems that, wherever possible, exemplify current practical problems are used frequently, drawing the student in to an active participation in the learning process.

With the example problems worked out in the text, followed by drill problems with answers, the textbook lends itself to self-paced or individualized tutorial instruction. Each chapter ends with a wide selection of problems so that the instructor of a conventional course can adjust the level considerably by assigning problems appropriate to the level of a specific course. Each of the problems is identified by content section number; thus students may refer to the appropriate text sections if they encounter difficulties. A solutions manual is available from the publisher on request.

Basically, the only prerequisites to a course using this textbook are a course in integral calculus and an introductory course in circuit analysis. A course in linear systems analysis would be helpful, but it is not essential.

Although written primarily for undergraduates in an electrical engineering curriculum, this text could also be used by those in other disciplines, in industry, or in telecommunications practice who are interested in learning, reviewing, or up-dating their technical background in communication systems. For these groups the chapter arrangements and frequent examples and drill problems make the text appropriate to independent study. Recommended auxiliary reading lists are included in the summary at the end of each chapter. Books in these lists have been carefully selected, and they should be both accessible and readable to undergraduates. They are listed in approximate order of increasing difficulty. References to specific topics are given as footnotes in the text.

The organization of this textbook is designed to allow maximum flexibility in the choice and presentation of subject matter. If Chapters 2, 3, and 8 constitute review for students in a specific course, there is sufficient remaining material for a one-semester course. If the material in the early chapters is new to the students, some adjustment may be made by deleting the optional material in each chapter and/or not including the material on probability theory. Optional material in each chapter has been designated by a star symbol (★).

Chapter 1 is an introduction to concepts in communication systems and an overview of the book. The Fourier methods of linear systems analysis are reviewed in Chapters 2 and 3, with particular emphasis on what will prove most useful in the succeeding chapters, such as the use of complex notation and interpretations in terms of phasors and spectral representations. Material on the numerical computation of Fourier coefficients and the discrete and fast Fourier transforms are becoming widely used for both computational and signal processing applications, and are included as optional material in Chapters 2 and 3. Problems intended to be solved using numerical methods are designated with a prefix "c" in the margin.

I am making available a set of five video-tapes that accompany the presentation in Chapters 2, 3, and 4. These tapes make use of computer animation and laboratory demonstrations to reenforce the sections about the Fourier series, the Fourier transform, and some of the concepts of linear systems analysis, such as convolution, correlation, and frequency transfer functions. To obtain information about these tapes, write to me at the University of Wisconsin, Department of Electrical and Computer Engineering (Madison, WI 53706).

The material in Chapters 4 through 7 is an introduction to the principles of communication systems. Chapter 4 covers the topics of power spectral density and thermal noise; Chapters 5 through 7 cover the topics of amplitude, angle, and pulse modulation. This organization of material has been influenced by my teaching this material at the undergraduate level. To sustain student interest, the more abstract concepts are interspersed with more practical sections that show how the concepts are being used. Thus the presentation begins with an elementary discussion of noise and then proceeds through amplitude, angle, and pulse modulation. This avoids having one or two chapters devoted entirely to signal-to-noise calculations — a topic that if prolonged fails to retain student interest at the undergraduate level.

The first part of the textbook does not assume a knowledge of probability theory. Presentation of the basic material without probability helps keep the emphasis on signal design and modulation. This treatment ends with Chapter 7, and a course taught from a deterministic point of view could end here also, or could conclude with some of the material in Chapter 10.

For those students for whom the first few chapters are review, there is time in a semester to take up the material of the last three chapters. If students, in addition, have had prior background in probability theory, Chapter 8 can be omitted or used for review. There is ample material in Chapters 9 and 10, in addition to Chapters 4 through 7, for a one-semester course if the optional sections are covered in each chapter.

Chapter 8 is an introduction to the subject of probability and random processes and is presented in such a way that students progress rapidly to the probability-density function and its use in the analysis of communication systems. Chapter 9 builds on this knowledge to introduce such topics as quantization noise and probability of error in baseband transmission. New sections in this edition have been added about partial-response signaling, equalization, M-ary signaling, and the power spectral densities of PCM signals.

Chapter 10 is a fairly complete discussion of digital modulation methods, beginning with amplitude-, frequency-, and phase-shift keying that progresses to modern methods of M-ary digital modulation such as quadrature phase-shift keying, minimum-shift keying, and amplitude-phase keying. The chapter concludes with geometrical representations of digital waveforms and an introduction to maximum likelihood detection. After completing Chapter 10, the student will, it is hoped, be interested in taking an advanced course in communication theory that will employ more statistical concepts.

Most of the appendixes from the first edition are included in this second edition because instructors found they were useful and readily available reference sources. The appendixes on commercial radio and television transmissions have been revised somewhat because of student interest in these topics and, pedagogically, add breadth in background. A new appendix on stereo AM systems has been added to this second edition.

For several semesters, the material in Chapters 4 through 9.7, plus sections 10.1 through 10.4, has been used for a one-semester course at the University of Wisconsin at Madison at the junior/senior level in electrical engineering. Another variation might be to summarize the signal-to-noise sections in Chapters

4 through 7 so that more attention can be given to the material in Chapters 9 and 10. If all material in the textbook is covered, there is ample material for a two-quarter course sequence.

I am indebted to many for their advice and assistance in this second edition. Suggestions and criticisms by reviewers for Addison-Wesley have been most helpful. I appreciate the comments and suggestions made by colleagues and graduate students that have improved the accuracy and clarity of the text. In particular, I wish to thank S. Merchant for his careful reading of the manuscript and his suggestions for changes, and J. Moss and J. Sommers for their helpful comments on Chapter 10. Also, I wish to thank Professor W. P. Birkemeier, Chairman of the Department of Electrical and Computer Engineering at the University of Wisconsin, for his helpful suggestions and criticisms. The encouragement of Dean W. R. Marshall and Dean J. G. Bollinger of the College of Engineering is sincerely appreciated. My appreciation is extended also to those who were most helpful in the initial writing of this textbook and are listed in the preface to the first edition.

Finally, I express my appreciation for the constructive feedback and support of my students who willingly sacrificed a little of their sleep each semester to come to my 7:45 a.m. lectures on the topic of this textbook.

Madison, Wisconsin   F. G. S.
September 1981

CONTENTS

CHAPTER 1

INTRODUCTION

It is difficult to imagine what modern living would be like without ready access to reliable, economical, and efficient means of communication. Communication systems are found wherever information is to be transmitted from one point to another. Telephone, radio, and television are common everyday examples of communication systems. More complicated communication systems guide aircraft, spacecraft, and automated trains; others provide live news coverage around the world, often via satellite—and the list of examples could go on and on. It is hardly an overstatement to say that today communication systems are not only necessary to business, industry, banking, and the dissemination of information to the public, but also essential to the national welfare and defense.

The purpose of this textbook is to present an introductory treatment of communication systems. By ''communication'' we mean the conveying or transmission of information from one place and/or time to another. Admittedly this definition is not very precise, but the subject of communication is very broad. It may, for example, mean anything from a telephone hookup to the use of good gestures, emphasis, and diction in a speech; from the amateur radio operator chatting about the weather to an American Indian sending a smoke signal. Note that the commonality in these examples is that there is information transmitted that is of importance to the recipient.

In this study of communications we shall restrict ourselves to the transmission of information over comparatively long distances. The use of electrical signals (in the broader sense, we may consider light to fall within this class because it is in the electromagnetic spectrum) has almost completely replaced all other forms of information transmission over long distances. This arises mainly because electrical signals are relatively easy to control (compared, for instance, to the fire for the smoke signals) and travel with velocities at or near that of the speed of light. Certainly for long distances, then, a study of communications via electrical signals is appropriate.

What is information transmission? This subject turns out to be more complicated than it might first appear and will, in fact, form a basis for part of our study. From a strictly intuitive viewpoint, we can say that the transmission of information requires that signals vary with time. Consider, for example, a 9-V battery; once the voltage has been established, there is little further information available without changing the voltage. Connection of the battery to a variable resistor or transistor allows such a variation. However, simply the fact that the signal can vary with time is not sufficient. Consider a 120-V, 60-Hz sinusoidal voltage; once the voltage has been established, there is little further information available without changing the amplitude or phase of the sinusoidal voltage, even though the voltage is changing at a 60-Hz rate.

We conclude not only that the transmission of information is related to signals changing with time but also that these changes must be made in an unpredictable way. Thus a necessary requirement is the use of a band of signal frequency content known as "bandwidth" (Chapter 3). Bandwidth is a measure of how rapidly the information-bearing portions of a signal can change, and therefore it is an important parameter in any discussion of communication systems. These intuitive ideas of information transmission will be enough to discuss the basic methods used in electrical communication systems. However, a more complete mathematical description of information transmission also demands some knowledge of probability (Chapter 8) and will be postponed until Chapter 9.

Communication over long distances usually requires that some alterations or other operations be performed on the electrical signal conveying the information in preparation for transmission. Upon reception, known inverse operations are performed to retrieve the information.

In the process of transmission, the signals bearing the information are contaminated by noise. Noise is generated by numerous natural and man-made events and introduces errors in the information transmission. From an engineering point of view, the communication problem consists in designing those portions of the transmission over which one can exercise some control. A criterion for doing this is to keep the information transmission as error-free as possible.

With these desired objectives in mind, we shall consider different communication systems and their basic principles of operation. The emphasis will be on the methods and not on the particular circuits or devices currently employed.

A diagram of the basic units comprising a communication system is shown in Fig. 1.1. Not every communication system makes use of all indicated operations, but each always involves a transmission medium of some kind. The encoder chooses the best form for the signal in order to optimize its detection at the output. The decoder performs the inverse operation to make the best decision, based on the available signals, that a given message was indeed sent. The design of the encoder and decoder must rely on a detailed mathematical description of information transmission. While the subject of coding often carries

Information in   Information out

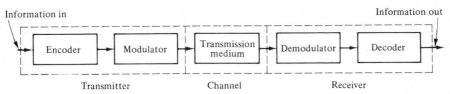

Transmitter   Channel   Receiver

Fig. 1.1 A communication system.

with it an air of secrecy, a more important motive in many modern coding systems is the improved efficiency in conveying information.

The modulator produces a varying signal at its output which is proportional in some way to the signal appearing across its input terminals. For example, a sinusoidal modulator may vary the amplitude, frequency, or phase of a sinusoidal signal in direct proportion to the voltage input. The roles of the encoder and the modulator are similar in that both prepare the signal for more efficient transmission. However, the process of coding is designed to optimize the error-free detection that a given message is being sent, whereas the process of modulation is designed to impress the information signal onto the waveform to be transmitted. The demodulator performs the inverse operation of the modulator to recover the signal in its original form.

The transmission medium is the crucial link in the system. Without it, there would be no communication problem. The transmission medium may include the ionosphere, the troposphere, free space, or simply a transmission line. In any case, attenuation and distortion, as well as noise signals generated in the media and the transmitting and receiving equipment, are introduced. For our purposes, noise signals are any electrical signals (voltages or currents) that interfere with the error-free reception of the message-bearing signal.

Three very basic subsystems of a communication system are indicated by the dashed lines in Fig. 1.1. The central subsystem restricts the flow of information and is called the *channel*. The channel includes the effects of additive noise, interference, propagation, and distortion. It is the limiting factor in the performance of any well-designed communication system. The role of the *transmitter* is to prepare the information to be sent in such a way that it will best cope with the limitations imposed by the channel. The role of the *receiver* is to perform the inverse of the transmitter operations in order to recover the information with the least amount of error possible. Note that, in this broad sense, the transmitter and the receiver as a pair are specifically designed to combat the deleterious effects of the channel on information transmission.

The communication system shown in Fig. 1.1 is capable of one-way transmission and is called a *simplex* (SX) transmission system. In many cases it is desirable to maintain two-way communication, or at least to be able to send a message back to its origin for possible verification, comparison, or control. One method of accomplishing this is to use the same channel alternately for transmission in each direction, as shown in Fig. 1.2. This method of transmission is

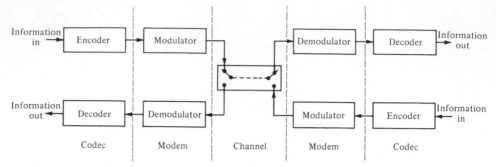

Fig. 1.2 Communication system using half-duplex transmission.

called *half-duplex* (HDX). Although communication flows in both directions in half-duplex transmission, the flow of information is only one-way at any given time.

A third type, *full-duplex* (FDX), is shown in Fig. 1.3. In full-duplex transmission, simultaneous communication is accomplished in both directions. Note that in both HDX and FDX transmission, the modulators and demodulators operate in pairs. This combination of a modulator and a demodulator is called a *modem* (*mo*dulator-*dem*odulator) in data transmission systems. Also, the encoders and decoders operate in pairs, suggesting the term *codec* (*co*der-*dec*oder).

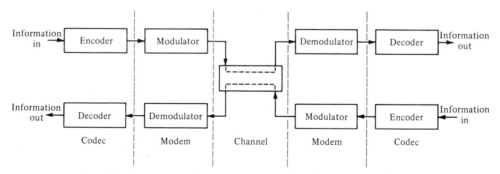

Fig. 1.3 Communication system using full-duplex transmission.

Now consider a channel whose only transmission impairment is that of additive noise. As mentioned previously, a certain minimum bandwidth, B, is necessary for communication. A wider bandwidth would allow more noise to interfere with the information transmission, so it is important to keep the bandwidth in such a channel as small as possible.

We characterize the noise present by its average power, N, and the transmitted signal by its average power, S. If the average noise power is relatively small, then the signal power does not need to be very large for the receiver to determine what information is being sent. (We are, of course, also concerned with efficiency and therefore attempt to minimize the transmitted power required to convey the information to the user.) In contrast, the average signal power required must be relatively large when the noise power is large. We conclude that it is the *ratio* of the average signal power to the average noise power that is important, and not the magnitudes of S and N themselves. This S/N ratio, called the *signal-to-noise ratio*, is an important parameter in the theory and design of communication systems.

All communication systems can be judged in terms of bandwidth, signal-to-noise ratio, and economic (cost) factors. There are various trade-offs between these parameters in any given type of system, and we will examine these later in this book. Some trade-offs between bandwidth and signal-to-noise ratio are emphasized in the following brief discussion of digital systems.

We assume (at least temporarily) that the communication systems are digital and, specifically, that they are binary. Suppose, then, that the information we wish to send at a given instant can be characterized by one sample from n possible input states, all of which are equally likely to occur. For example, this sample could be one of 256 equally likely voltage levels. To send this sample using a binary system, we first generate a digital word composed of m binary symbols. Thus each binary word consists of $m = \log_2 n$ binary digits to represent the one sample out of $n = 2^m$ possibilities. When used in this way, m is called the number of *bits* (*binary digits*) needed to represent the one-of-n possible input states. In this example, for instance, an eight-bit word is necessary to describe one of 256 possible input states.

Next we wish to send this m-bit binary word serially through the channel. In a binary system, these bits are represented by binary symbols (e.g., $+1$ and -1) that are generated at a symbol rate of r symbols per second. The resulting information rate from the transmitter is $R = mr$ bits per second (bps). At the receiver, the transmitted signal is corrupted by the addition of the noise, and as a result the receiver will make some errors. It seems reasonable that this error rate will decrease if S/N is increased. It also seems reasonable that perhaps the error rate could be decreased by a receiver designed to process signals with an increased complexity. Because we are interested in both efficiency and accuracy of communication, a question of great importance to us, then, is the following: For a given channel and a given information transmission rate, is it possible theoretically to make system improvements with the objective of reducing the error rate? The answer to this question, based on the theoretical work of Claude Shannon published in 1949,[†] is in the affirmative *if* the information transmission rate R is such that $R \leq C$, where C is the *channel capacity*. For the type of channel considered here, the channel capacity is given by the Hartley-Shannon law

$$C = B \log_2 (1 + S/N) \quad \text{bps} \tag{1.1}$$

[†] C. E. Shannon, "Communication in the Presence of Noise," *Proceedings of the IRE*, vol. 37 (January 1949): 10–21.

where B is the bandwidth of the channel (in Hz) and S/N is the signal-to-noise ratio. If one tries to send information at too rapid a rate—that is, $R > C$—then the errors begin to increase rapidly and there is no point to trying to design a system to improve the situation. On the other hand, for $R < C$ there is some hope for improvement via good system design. All communication systems to be covered in this textbook, then, will lie within the bound expressed by Eq. (1.1).

Let's pursue our intuitive reasoning farther. Suppose we decide to increase the information transmission rate by increasing the symbol rate r. Because more symbol transitions per second are possible, the required bandwidth must be increased. Thus an increase in information rate can be accommodated by an increase in bandwidth. This merely emphasizes a conclusion we could have obtained directly from Eq. (1.1). What is surprising, however, is that Eq. (1.1) states that bandwidth and signal-to-noise can be exchanged. Thus a smaller S/N ratio may be adequate if the bandwidth is increased, and vice versa. Note that for small S/N the potential trade-off is approximately linear,

$$\log_2 (1 + x) \approx (\log_2 e) \, x \qquad \text{for small } x,$$

but it is exponential for large S/N.

It turns out that the Hartley-Shannon law is applicable to continuous as well as discrete systems, and therefore it is a very powerful and far-reaching result. Its application is restricted, however, to channels with additive noise, and such effects as distortion and interference are not included.

Although the Hartley-Shannon law states that the maximum information transmission rate for a given channel is bounded and suggests that bandwidth can be traded off for S/N, it does not offer a method for the design of a system that will meet these expectations. In other words, it gives us a bound with which we can compare the performance of the systems we design, but it does not give us a procedure for designing systems whose performance will meet that bound. But we should not expect too much. It's fortunate we have the above result.

With this brief introduction to communication systems, we turn now to matters of approach and organization of this textbook.

It is our purpose in this textbook to analyze the principal characteristics of communication systems and to discuss some of their realizations in practice. Our approach will be from a systems viewpoint, in contrast to a study of circuits and devices. This will require mathematical descriptions and representations of the electrical signals (that is, the voltages and currents) that characterize such systems.

The Fourier methods of signal analysis prove to be most useful in our study. Applications of these methods often furnish valuable insights into the signal design aspects of communication systems. These methods are introduced and developed in the first part of this book. An introduction to communication systems, from both an analysis and an applications approach, then follows in the succeeding chapters.

Selected References for Further Reading†

1. A. B. Carlson. *Communication Systems,* Second ed. New York: McGraw-Hill, 1975. Chapter 1 of this text contains a good introductory discussion of communication, communication systems, and modulation.

2. W. D. Gregg. *Analog and Digital Communication.* New York: John Wiley & Sons, 1977.
   Chapter 1 has a good descriptive introduction to historical perspectives, frequency allocations, licensing, and regulations.

3. M. Schwartz. *Information Transmission, Modulation, and Noise,* Third ed. New York: McGraw-Hill, 1980.
   The first chapter of this book provides an introduction to the mathematical formulation of information transmission and communication systems for the interested student.

4. J. C. Hancock and P. A. Wintz. *Signal Detection Theory.* New York: McGraw-Hill, 1966.
   Although primarily a graduate text, Chapter 1 in this book is interesting reading for the undergraduate interested in communication systems.

---

† Selected references for further reading in the material covered are included at the end of each chapter. In general, they are listed in an order of increasing difficulty.

CHAPTER 2

ORTHOGONALITY AND SIGNAL REPRESENTATIONS

We begin this chapter by reviewing some concepts of signal and systems representations that will prove helpful to us later. The concept of orthogonality, for instance, leads to the use of the Fourier series, which in turn is used to introduce the concepts of signal and systems representations in both time and frequency.

2.1 SIGNALS AND SYSTEMS

In a general sense, a *system* is a group of objects that can interact harmoniously and that are combined in a manner intended to achieve a desired objective. A system may in turn be a member (subsystem) of a larger system. A complete hierarchy of systems may be established, each with its identified domain.

A *signal* is an event that serves, or at least is capable, to start some action; i.e., it can incite action. Within energy and power restrictions, we are particularly interested in the concept of a signal and also in the *response* of a system to a given signal. The diagram in Fig. 2.1 illustrates the role of the signal, the system, and the response. Although the concept of a system was introduced first, it will often turn out to be more convenient to use the concepts of the signal and the resulting response to describe the characteristics of a system. Actually, the system will sometimes only be known in terms of its response to given signals.

For our purposes, a signal is defined to be a single-valued function of time; i.e., to each assigned instant of time (the independent variable) there is one unique value of the function (the dependent variable). This value may be a real number, in which case we have a real-valued signal, or it may be complex, and then we can speak of a complex-valued signal. In either case the independent variable (time) is real-valued.

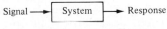

Fig. 2.1 A system diagram.

The complex-valued notation can be used to describe signals in terms of two independent variables, for example $x(t)$ and $y(t)$. Thus the complex-valued notation is convenient for a description of two-dimensional phenomena, such as circular motion, plane wave propagation, etc., as a function of time. The signals used in this book, however, are expressed in terms of time and only one dependent variable (e.g., a voltage versus time). Because we are limited to one dependent variable, we insist that all signals that correspond to physically observable quantities must be real-valued quantities.

But then, why use complex notation at all if our observations are real-valued? In many analyses the mathematical models and calculations are often much simpler, and even more intuitively obvious, if the complex notation is used. For the sake of convenience, the complex notation will prove to be a definite advantage. After all operations have been completed, it is a fairly simple matter to take the real part of the resulting expression. This procedure works whenever superposition holds; we must be cautious in applying it to more general cases.

The preceding remarks can be applied to the description and analysis of physical processes in general. We shall restrict ourselves principally to the description and analysis of *electrical* signals and systems.

An electrical signal may be either a voltage or a current waveform that we are describing mathematically. We are not as interested in "voltage drops," "loop currents," etc., as we are in the time variations of signals, whether they be voltages or currents. It follows that a signal is merely a single-valued function of time that may be used to represent either a voltage or a current for a specific situation. Exceptions to this will sometimes arise, particularly in discussions involving energy and power. In this case, the assumption of a one-ohm resistor is a convenient way to avoid any direct commitment. The value of a particular resistor will just scale the result once the identity of the waveform is established. For all calculations of energy and power a one-ohm resistor is assumed unless otherwise explicitly specified in the given problem.

Sinusoidal signals play a major role in the analysis of communication systems. Such a signal $f(t)$ may be represented as a function of time t by the equation

$$f(t) = A \cos (\omega t + \theta), \tag{2.1}$$

where A is the amplitude, θ is the phase, and ω is the rate of phase change or frequency of the sinusoid in radians per second. It may also be expressed as f in cycles per second (Hz), where $\omega = 2\pi f$.†

The principle of the Fourier methods of signal analysis is to break up all signals into summations of sinusoidal components. This provides a description

---

† The generally accepted unit for frequency in cycles per second (cps) is the *hertz*, abbreviated as Hz. Using prefix multipliers, we then have the following: 1 kHz (one kilohertz) = 10^3 cps; 1 MHz (one megahertz) = 10^6 cps; 1 GHz (one gigahertz) = 10^9 cps.

of a given signal in terms of sinusoidal frequencies. A major objective is a description of how signal (and response) energy and power are distributed in terms of these frequencies. Any description of a response to a given signal will, of course, bring in the characteristics of the system.

2.2 CLASSIFICATION OF SIGNALS

The most useful method of signal representation for any given situation depends upon the type of signal being considered. A few of the classifications most useful to us are discussed here.

Energy Signals, Power Signals.   An energy signal is a pulse-like signal that usually exists for only a finite interval of time or, even if present for an infinite amount of time, at least has a major portion of its energy concentrated in a finite time interval.

For electrical systems, a signal is a voltage or a current. The instantaneous power dissipated by a voltage $e(t)$ in a resistance R is

$$p = |e(t)|^2/R \quad \text{watts} \tag{2.2}$$

and for a current $i(t)$

$$p = |i(t)|^2 R \quad \text{watts.} \tag{2.3}$$

In each instance the instantaneous power is proportional to the squared magnitude of the signal. For a one-ohm resistance, these equations assume the same form. Thus it is customary in signal analysis to speak of the instantaneous power associated with a given signal $f(t)$ as

$$p = |f(t)|^2 \quad \text{watts.} \tag{2.4}$$

Even though the dimensions may not appear to be correct in Eq. (2.4), the multiplication or division by an appropriate resistance is implied in this convention.

Using this convention, the energy dissipated by the signal during a time interval (t_1, t_2) is

$$E = \int_{t_1}^{t_2} |f(t)|^2 \, dt \quad \text{joules.} \tag{2.5}$$

We define an *energy signal* to be one for which Eq. (2.5) is finite even when the time interval becomes infinite; i.e., when

$$\int_{-\infty}^{\infty} |f(t)|^2 \, dt < \infty. \tag{2.6}$$

Examples of several energy signals are shown in Fig. 2.2.

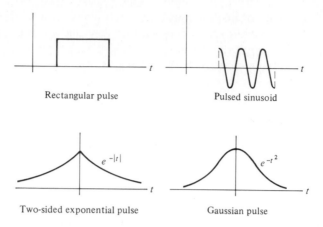

Fig. 2.2 Some energy signals.

The average power dissipated by the signal $f(t)$ during the time interval (t_1, t_2) is

$$P = \frac{1}{t_2 - t_1} \int_{t_1}^{t_2} |f(t)|^2 \, dt. \tag{2.7}$$

If the right-hand side of Eq. (2.7) remains finite but nonzero when the time interval becomes infinite, i.e., if

$$0 < \lim_{T \to \infty} \frac{1}{T} \int_{-T/2}^{T/2} |f(t)|^2 \, dt < \infty, \tag{2.8}$$

then the signal $f(t)$ has finite average power and is called a *power signal*.

Periodic, Nonperiodic. A *periodic signal* is one that repeats itself exactly after a fixed length of time. Thus the signal $f(t)$ is periodic if there is a number T such that:

$$f(t + T) \equiv f(t) \qquad \text{for all} \quad t. \tag{2.9}$$

The smallest positive number T that satisfies Eq. (2.9) is called the *period*. The period defines the duration of one complete cycle of $f(t)$. A periodic signal is a power signal if its energy per cycle is finite, and then the average power need only be calculated over one complete cycle.

Any signal for which there is no value of T satisfying Eq. (2.9) is said to be *nonperiodic*, or *aperiodic*. A borderline case between periodic and nonperiodic signals is that of the "almost periodic signal." This latter type of signal is composed of the sum of two or more periodic signals having incommensurate periods.† An example of such a signal is

$$f(t) = \sin t + \sin \sqrt{2}t. \tag{2.10}$$

This function is "almost periodic" because each term on the right-hand side is periodic, yet there is no period T in which $f(t)$ exactly repeats itself.

Random, Deterministic. A *random signal* is one about which there is some degree of uncertainty before it actually occurs. Such a signal can be thought of as belonging to a collection of signals, each of which is different. If one of these signals is chosen (at random), it may turn out to be quite well defined, as, for example, a sinusoid of fixed frequency but uncertain starting phase. However, if a second sinusoid were chosen from this collection, we could not be certain of the same starting phase. In other cases, future values of the signal may not be predictable even after observation of past values. An example of such a random signal is the output of a radio receiver when tuned off station as it responds to noise arising from disturbances in the atmosphere and its internal circuitry.

A *nonrandom*, or *deterministic*, signal is one about which there is no uncertainty in its values. In almost all cases, an explicit mathematical expression can be written for such a signal. The signals to be discussed in the earlier part of this book are nonrandom. As new methods of analysis are developed, we shall begin to see how certain types of random signals can also be handled.

2.3 CLASSIFICATION OF SYSTEMS

Mathematically, a system is a rule used for assigning a function $g(t)$ (the output) to a function $f(t)$ (the input); that is,

$$g(t) = \mathcal{T}\{f(t)\}, \tag{2.11}$$

where $\mathcal{T}\{\ \}$ is the rule.‡ This rule could be in terms of an algebraic operation, a differential and/or integral equation, etc. For two systems connected in cascade, the output of the first system forms the input to the second, thus forming a new overall system:

$$g(t) = \mathcal{T}_2\{\mathcal{T}_1[f(t)]\} = \mathcal{T}\{f(t)\}. \tag{2.12}$$

---

† Two periods are commensurate if their ratio can be expressed as the ratio of two integers.

‡ The symbol used here is the script "t." Other script letters are used in later chapters to designate specific operators.

As in signal analysis, we find it convenient to classify systems by some of their basic properties. Those most useful to us are discussed below.

Linear, Nonlinear.   If a system is *linear* then superposition applies; that is, if

$$g_1(t) = \mathcal{T}\{f_1(t)\}, \quad \text{and} \quad g_2(t) = \mathcal{T}\{f_2(t)\},$$

then

$$\mathcal{T}\{a_1 f_1(t) + a_2 f_2(t)\} = a_1 g_1(t) + a_2 g_2(t), \tag{2.13}$$

where a_1, a_2 are constants. A system is *linear* if it satisfies Eq. (2.13); any system not meeting these requirements is *nonlinear*.

Time-Invariant or Time-Varying.   A system is *time-invariant* if a time shift in the input results in a corresponding time shift in the output so that

$$g(t - t_0) = \mathcal{T}\{f(t - t_0)\} \quad \text{for any } t_0. \tag{2.14}$$

The output of a time-invariant system depends on time differences and not on absolute values of time. Any system not meeting this requirement is said to be *time-varying*. A system may be linear yet time-varying and vice versa. Two examples of such systems are shown in Fig. 2.3.

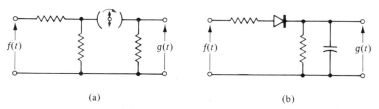

(a)   (b)

Fig. 2.3 Examples of systems: (a) Linear, time-varying; (b) nonlinear, time-invariant.

Realizable, Nonrealizable.   A physically realizable system cannot have an output response before an arbitrary input function is applied. Stated another way, the output of a physical system at $t = t_0$, namely, $g(t_0)$, must depend only on values of the input $f(t)$ for $t \le t_0$. A system having this property is called *physically realizable or causal*.† Any system not meeting this requirement is said to be nonrealizable or noncausal.

---

† Some texts define "realizability" in other ways; this definition will suffice for our purposes.

★ 2.4 SIGNALS AND VECTORS†

Consider the signal $f(t)$ which is defined for all values of time within the interval (t_1, t_2), as shown in Fig. 2.4. This signal is time-limited and has a finite energy, E:

$$E = \int_{t_1}^{t_2} |f(t)|^2 \, dt < \infty.$$

The graph in Fig. 2.4 illustrates one way of specifying $f(t)$; i.e., for each value of the independent variable t we are given the value of the dependent variable, $f(t)$. The closer we take the increments in time, the more exact we know the variations of the signal $f(t)$. Obviously it takes an infinite set of values to completely specify the graph for $f(t)$. From experience, however, we know that as long as we have a few samples for each wiggle on the graph, we can construct a graph that is accurate enough for all practical purposes. We shall place this criterion on a more quantitative basis in Chapter 3 (the sampling theorem).

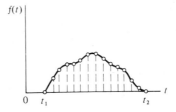

Fig. 2.4 A finite-energy signal.

Alternatively, we may wish to specify $f(t)$ by a countable set of numbers which are not dependent on the explicit choice of the variable t.‡ In other words, we would like to express $f(t)$ as

$$f(t) = \sum_n f_n \phi_n(t), \tag{2.15}$$

where the $\phi_n(t)$ form a set of functions to be specified and the f_n are numbers independent of time.

This idea should not seem particularly surprising to us, for we are quite accustomed to describing a vector in terms of a countable set of numbers. For example, a vector from the origin to the point (1, 1, 1) is that vector which has a length of $\sqrt{3}$ and is inclined at 55° to each of the coordinate axes in the first quadrant. The reason that this vector notation is so convenient, of course, is that each number in the notation represents the projected length of the vector

---

† Optional material in each chapter is designated by a star.

‡ A countable set is a set which can be placed in a one-to-one correspondence to the set of real integers.

upon a given coordinate axis. These coordinate axes are chosen in such a way that they are perpendicular to each other in the vector space. Thus the projection of the vector on each coordinate axis is entirely independent of its projection on each of the other axes. This allows us to write down the equation of the vector immediately in terms of these respective projections and the reference vectors in each coordinate direction.

The analogy between vectors and signals is more than coincidental. In fact, it promises to give us some valuable insights into the description of signals. The use of the more familiar geometric ideas often prove very useful in gaining added perspective into the handling of signal waveforms.

We assume throughout this discussion that all vectors of interest have a finite length. Such vectors can be uniquely specified in space by referring them to a given set of coordinate axes. There must be one coordinate for each dimension of the vector in order to guarantee the uniqueness of the representation. Then we say that the vector space is *complete*.

The scalar ("dot") product between two vectors $\boldsymbol{\phi}_1$, $\boldsymbol{\phi}_2$ is

$$C_{12} = \boldsymbol{\phi}_1 \cdot \boldsymbol{\phi}_2. \tag{2.16}$$

The scalar C_{12} is an indicator of the similarity between the vectors $\boldsymbol{\phi}_1$, $\boldsymbol{\phi}_2$. If C_{12} is zero, we conclude that either (a) $\boldsymbol{\phi}_1$ or $\boldsymbol{\phi}_2$ have zero magnitude, or (b) $\boldsymbol{\phi}_1$ has no component along the vector $\boldsymbol{\phi}_2$. In the latter case the two vectors are mutually perpendicular and are said to be *orthogonal*.

Suppose we generate an orthogonal vector space with the three orthogonal vectors $\boldsymbol{\phi}_1$, $\boldsymbol{\phi}_2$, $\boldsymbol{\phi}_3$. These vectors do not necessarily have unit length; however, we can write

$$\boldsymbol{\phi}_n \cdot \boldsymbol{\phi}_m = \begin{Bmatrix} k_n & n = m \\ 0 & n \neq m \end{Bmatrix}, \tag{2.17}$$

where k_n is the squared length of $\boldsymbol{\phi}_n$. Any vector $\mathbf{A}_1$ in this vector space may be represented in the form

$$\mathbf{A}_1 = A_{11}\boldsymbol{\phi}_1 + A_{12}\boldsymbol{\phi}_2 + A_{13}\boldsymbol{\phi}_3, \tag{2.18}$$

where

$$A_{1n} = \frac{\mathbf{A}_1 \cdot \boldsymbol{\phi}_n}{\boldsymbol{\phi}_n \cdot \boldsymbol{\phi}_n} = \mathbf{A}_1 \cdot \left(\frac{\boldsymbol{\phi}_n}{\boldsymbol{\phi}_n \cdot \boldsymbol{\phi}_n} \right) = \mathbf{A}_1 \cdot \left(\frac{\boldsymbol{\phi}_n}{k_n} \right) \tag{2.19}$$

for $n = 1, 2, 3$. Note that as a result of orthogonality the computation of each scalar product in Eq. (2.19) is not dependent on the computation of any of the others. It also turns out that if the vector space is not complete, the squared length of the error vector remaining is a minimum using the above methods.†

---

† See, e.g., B. P. Lathi, *Signals, Systems and Communications*, New York: John Wiley & Sons, 1965.

We may extend these concepts to a more general N-dimensional space. Such a vector space, of course, does not exist physically in nature. On the other hand, there are many problems which have analogies to an N-dimensional space.

Drill Problem 2.4.1   Three vectors, expressed in the Cartesian coordinate system described by the unit vectors x_1, x_2, x_3, are $A = x_1 - x_2 + 5x_3$; $B = -x_1 + x_2 + x_3$; $C = 3x_1 + x_2 + 2x_3$; $D = x_1 + 5x_2 - 4x_3$. (a) Determine which of these vectors are orthogonal to D. (b) Represent A in terms of the three vectors B, C, D. (c) Compute the squared length of the error vector remaining if A is represented in terms of C, D only.

Answer.   (a) B, C; (b) $A = B + \frac{6}{7}C - \frac{12}{21}D$; (c) 3.

★ 2.5   ORTHOGONAL FUNCTIONS

Returning to the original problem [cf. Eq. (2.15)], recall that we wish to express the function $f(t)$ as a set of numbers (the f_n) which, when expressed in terms of a properly chosen coordinate space [i.e., the $\phi_n(t)$], will specify the function uniquely. It is highly desirable that the set so chosen be a linearly independent set; i.e., that the individual terms are not dependent on each other and that the set is formed by the totality of these terms.†

We have seen that any vector of finite length may be expressed as a sum of its components along n mutually orthogonal vectors provided that these vectors form a complete set of mutually perpendicular coordinates. In addition, this method yields a minimum length (squared) for the error vector when the set is not complete. These ideas naturally form a motivation for choosing a complete set of orthogonal functions, the $\phi_n(t)$. The particular $\phi_n(t)$ chosen are called *basis functions*.

Two complex-valued functions $\phi_1(t)$ and $\phi_2(t)$ are defined to be *orthogonal* over the interval (t_1, t_2) if

$$\int_{t_1}^{t_2} \phi_1(t)\phi_2^*(t)\, dt = \int_{t_1}^{t_2} \phi_1^*(t)\phi_2(t)\, dt = 0. \tag{2.20}$$

Thus if members of a set of complex-valued functions are mutually orthogonal over (t_1, t_2), then

$$\int_{t_1}^{t_2} \phi_n(t)\phi_m^*(t)\, dt = \begin{cases} 0 & n \neq m \\ K_n & n = m \end{cases}. \tag{2.21}$$

---

† A set of functions is linearly independent if no one of the functions can be constructed as a weighted sum of the remaining functions in the set.

The set of basis functions $\phi_n(t)$ is said to be "normalized" if

$$K_n = \int_{t_1}^{t_2} |\phi_n(t)|^2 \, dt = 1 \qquad \text{for all} \quad n.$$

If the set is both orthogonal and normalized, it is called an *orthonormal* set.

The integral of the product of two functions over a given interval is called the *inner product* of the two functions. The square root of the inner product of a function with itself is called the *norm*. Thus Eq. (2.21) is the inner product of $\phi_1(t)$, $\phi_2(t)$ and $K_n^{1/2}$ is the norm. As we shall soon see, the inner product and the norm in signal space are analogous to the dot product and length in vector space.

Now we return to the question of an approximation of a function $f(t)$ in terms of the $\phi_n(t)$ [cf. Eq. (2.15)]; for N terms we have

$$f(t) \cong \sum_{n=1}^{N} f_n \phi_n(t). \tag{2.22}$$

The integral-squared error remaining in this approximation after N terms is

$$\int_{t_1}^{t_2} |\epsilon_N(t)|^2 \, dt = \int_{t_1}^{t_2} \left| f(t) - \sum_{n=1}^{N} f_n \phi_n(t) \right|^2 \, dt. \tag{2.23}$$

The integrand in Eq. (2.23) is nonnegative so that the integral-squared error will go to zero only if the integrand goes to zero.

We wish to minimize Eq. (2.23) by a proper choice of the f_n. Expanding the right-hand side of Eq. (2.23) and allowing an interchange in the order of summation and integration, we have

$$\int_{t_1}^{t_2} |\epsilon_N(t)|^2 \, dt = \int_{t_1}^{t_2} |f(t)|^2 \, dt$$
$$- \sum_{n=1}^{N} \left[f_n^* \int_{t_1}^{t_2} f(t) \phi_n^*(t) \, dt + f_n \int_{t_1}^{t_2} f^*(t) \phi_n(t) \, dt - |f_n|^2 K_n \right], \tag{2.24}$$

where use has also been made of Eq. (2.21). Completing the square within the summation, Eq. (2.24) becomes

$$\int_{t_1}^{t_2} |\epsilon_N(t)|^2 \, dt = \int_{t_1}^{t_2} |f(t)|^2 \, dt$$
$$+ \sum_{n=1}^{N} \left[\left| K_n^{1/2} f_n - \frac{1}{K_n^{1/2}} \int_{t_1}^{t_2} f(t) \phi_n^*(t) \, dt \right|^2 \right.$$
$$\left. - \left| \frac{1}{K_n^{1/2}} \int_{t_1}^{t_2} f(t) \phi_n^*(t) \, dt \right|^2 \right]. \tag{2.25}$$

All three terms on the right-hand side of Eq. (2.25) are nonnegative. Only the second term, however, is dependent on our choice of the f_n and is minimized by

$$f_n = \frac{1}{K_n} \int_{t_1}^{t_2} f(t)\phi_n^*(t)\, dt = \frac{\int_{t_1}^{t_2} f(t)\phi_n^*(t)\, dt}{\int_{t_1}^{t_2} |\phi_n(t)|^2\, dt}. \tag{2.26}$$

Note the close similarity with vector spaces [cf. Eq. (2.19)] and the analogous roles played by the inner product and the dot product.

Using Eq. (2.26) in Eq. (2.25), the minimum integral-squared error in the orthogonal series approximation to $f(t)$ over (t_1, t_2) is

$$\int_{t_1}^{t_2} |\epsilon_N(t)|^2\, dt = \int_{t_1}^{t_2} |f(t)|^2\, dt - \sum_{n=1}^{N} |f_n|^2 K_n. \tag{2.27}$$

The right-hand side of Eq. (2.27) is the difference between two nonnegative quantities (which in turn must be nonnegative). Note that the first term is the energy in $f(t)$. If, for any $f(t)$ having finite energy, i.e., for

$$\int_{t_1}^{t_2} |f(t)|^2\, dt < \infty$$

the $\phi_n(t)$ are such that

$$\lim_{N \to \infty} \int_{t_1}^{t_2} |\epsilon_N(t)|^2\, dt = 0, \tag{2.28}$$

then we say that the orthogonal set $\phi_n(t)$ is *complete* over (t_1, t_2).

For a complete orthogonal set, Eq. (2.15) can now be rewritten as

$$f(t) = \sum_{n=1}^{\infty} f_n \phi_n(t) \tag{2.29}$$

and Eq. (2.27) reduces to

$$\int_{t_1}^{t_2} |f(t)|^2\, dt = \sum_{n=1}^{\infty} |f_n|^2 K_n. \tag{2.30}$$

This relation is known as *Parseval's theorem*.

The representation of a function $f(t)$ by an infinite set of mutually orthogonal functions is called a "generalized Fourier series representation" of $f(t)$. We shall find frequent applications for such representations in our analysis of signals and systems. A possible way of generating the coefficients for such a representation is shown in Fig. 2.5.

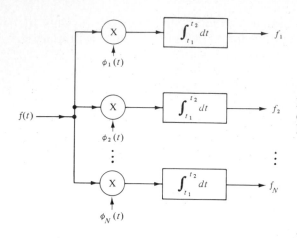

Fig. 2.5 A method of generating the coefficients for the generalized Fourier series representation.

Example 2.5.1   A given rectangular function is defined over (0, 2) by (see Fig. 2.6)

$$f(t) = \begin{cases} 1, & 0 < t < 1 \\ -1, & 1 < t < 2 \end{cases}.$$

We wish to approximate this finite-energy function using a set of functions defined by $\phi_n(t) = \sin n\pi t$, $n > 0$, over the interval (0, 2).

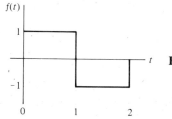

Fig. 2.6 A rectangular function.

Solution.   We can easily show that the functions $\sin n\pi t$ and $\sin m\pi t$ are orthogonal over the interval (0, 2):

$$\int_0^2 \sin n\pi t \sin m\pi t \, dt = \begin{cases} 1 & n = m \\ 0 & n \neq m \end{cases}.$$

[In fact, note that this set is orthonormal over (0, 2).] Then we have the following representation for $f(t)$ over (0, 2):

$$f(t) = \sum_{n=1}^{\infty} f_n \sin n\pi t,$$

where the f_n can be found from Eq. (2.26):

$$f_n = \frac{\int_0^2 f(t) \sin n\pi t \, dt}{\int_0^2 \sin^2 n\pi t \, dt} = \int_0^2 f(t) \sin n\pi t \, dt.$$

For this particular choice of $f(t)$,

$$f_n = \int_0^1 \sin n\pi t \, dt - \int_1^2 \sin n\pi t = \begin{cases} 4/\pi n & \text{for } n \text{ odd} \\ 0 & \text{for } n \text{ even} \end{cases}.$$

Thus the function $f(t)$ is approximated by the series representation

$$f(t) = \frac{4}{\pi} \left(\sin \pi t + \frac{1}{3} \sin 3\pi t + \frac{1}{5} \sin 5\pi t + \frac{1}{7} \sin 7\pi t + \cdots \right)$$

over the interval (0, 2). Figure 2.7 shows the actual function and the approximated

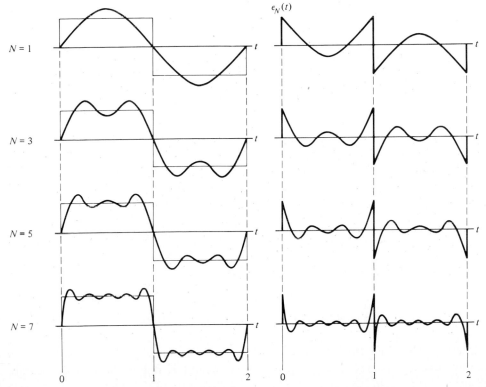

Fig. 2.7 Approximation of a rectangular function by orthogonal functions and the instantaneous (point) error remaining after each successive approximation.

function when the function is approximated with one, two, three, and four terms, respectively. The point error (instantaneous error) is also shown for each approximation.

The error energy (integral-squared error) can be calculated for these approximations by an application of Eq. (2.27):

$$\int_0^2 \epsilon_1^2(t) \, dt = 2 - \left(\frac{4}{\pi}\right)^2 = 0.379,$$

$$\int_0^2 \epsilon_3^2(t) \, dt = 2 - \left(\frac{4}{\pi}\right)^2 - \left(\frac{4}{3\pi}\right)^2 = 0.199,$$

$$\int_0^2 \epsilon_5^2(t) \, dt = 2 - \left(\frac{4}{\pi}\right)^2 - \left(\frac{4}{3\pi}\right)^2 - \left(\frac{4}{5\pi}\right)^2 = 0.134,$$

$$\int_0^2 \epsilon_7^2(t) \, dt = 2 - \left(\frac{4}{\pi}\right)^2 - \left(\frac{4}{3\pi}\right)^2 - \left(\frac{4}{5\pi}\right)^2 - \left(\frac{4}{7\pi}\right)^2 = 0.101.$$

In this case the error energy diminishes rapidly as the number of terms in the approximation is increased. In fact, the calculations above show that about 95% of the energy is contained in the first four terms.

Drill Problem 2.5.1   Find the generalized Fourier representation of the function

$$f(t) = \begin{cases} 1 & -2 < t < 2 \\ 0 & \text{elsewhere} \end{cases}$$

over the interval $(-4, 4)$ using the set of orthogonal functions

$$\phi_n(t) = \cos n \frac{\pi}{4} t, \qquad n = 0, 1, 2, \ldots.$$

Answer

$$f(t) = \frac{2}{\pi} \left(\frac{\pi}{4} + \cos \frac{\pi}{4} t - \frac{1}{3} \cos \frac{3\pi}{4} t + \frac{1}{5} \cos \frac{5\pi}{4} t - \frac{1}{7} \cos \frac{7\pi}{4} t + \cdots \right).$$

Example 2.5.2   Making use of the analogy between signals and vectors, show that, for two finite-energy signals $x(t)$, $y(t)$,

$$\left| \int_{-\infty}^{\infty} x(t) y(t) \, dt \right|^2 \leq \int_{-\infty}^{\infty} |x(t)|^2 \, dt \int_{-\infty}^{\infty} |y(t)|^2 \, dt.$$

Solution.   Consider the vector representation **x** and **y**. The scalar product $\mathbf{x} \cdot \mathbf{y}$ cannot exceed the product of the lengths of **x** and **y**. If θ is the included angle, then

$$|\mathbf{x} \cdot \mathbf{y}|^2 = (|\mathbf{x}| \, |\mathbf{y}| \cos \theta)^2 \leq |\mathbf{x}|^2 \, |\mathbf{y}|^2.$$

The equality holds here if and only if $\cos \theta = \pm 1$, that is, if the vectors are collinear.

Using a signal-space analogy, the above inequality can be rewritten in terms of the inner products of $x(t)$ and $y(t)$:

$$\left| \int_{-\infty}^{\infty} x(t)y^*(t)\, dt \right|^2 \leq \int_{-\infty}^{\infty} |x(t)|^2\, dt \int_{-\infty}^{\infty} |y(t)|^2\, dt.$$

Consistent with x and y being collinear, the equality condition is given by $x(t) = Ky(t)$, $K = $ any real-valued constant.

This result is usually rewritten by letting $z(t) = y^*(t)$ so that

$$\left| \int_{-\infty}^{\infty} x(t)z(t)\, dt \right|^2 \leq \int_{-\infty}^{\infty} |x(t)|^2\, dt \int_{-\infty}^{\infty} |z(t)|^2\, dt,$$

with equality if and only if $x(t) = Kz^*(t)$. This well-known and useful relationship is called the *Schwarz inequality*.

★ 2.6   CHOICE OF A SET OF ORTHOGONAL FUNCTIONS

We have discussed the generalized Fourier series representation of a function of finite energy over a given interval by a linear combination of mutually orthogonal functions. However, many sets of orthogonal functions exist and hence a given function may be expressed in terms of different sets of orthogonal functions. This should not surprise us for, in the analogy to vector spaces, we know that a given vector can be expressed uniquely in different coordinate systems. Often the choice of a specific coordinate system in vector space will make the analysis of a given type of problem easier and a good portion of our experience in vectors is to learn to make certain judicious choices in coordinates and coordinate systems. This same conclusion holds true for signal spaces.

Some examples of sets of orthogonal functions are exponential functions, trigonometric functions, Walsh functions, and Legendre polynomials. We shall not launch into an intensive investigation of each set of orthogonal functions and its properties. (Some are introduced in the problems.) Because we are primarily concerned in applying our results, we ask, Out of all the possible sets of orthogonal functions, which set will be the most convenient one for us to choose? Again, as in vector spaces, the answer to this question depends upon the type of problem to be solved. Therefore we will digress here and, with a little foresight, attempt to make a judicious choice at the outset.

We shall be concerned with systems which are linear and those whose parameters do not vary as a function of time, at least not within given intervals of time. When a signal is applied to such a system, the output can be expressed in terms of linear differential equations with constant coefficients. As an example, consider the *R-L* electrical circuit shown in Fig. 2.8 describing a particular system

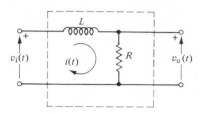

Fig. 2.8 A linear system.

in the above sense. The relationship between the input voltage, $v_i(t)$, and the output voltage, $v_o(t)$, can be expressed as†

$$\frac{d}{dt} v_o(t) + \frac{R}{L} v_o(t) = \frac{R}{L} v_i(t).$$

For a finite-energy input, we could then expect to solve this differential equation to find the output. Unless the input signal is a very simple function of t, however, we have difficulty solving the differential equations encountered and so we turn to writing a representation of the signal in terms of orthogonal functions. However, our choice of a representation for the input signal will affect the difficulty which we may encounter in solving the problem.

Now we entertain an intriguing possibility: if we could choose a function which repeats itself under the operation of differentiation, then such a function at the input will yield the same function at the output multiplied by some algebraic polynomial in terms of the parameters of the function and the differential equation. Furthermore, upon cancelling out the original function from both sides of the differential equation, this algebraic equation will be *only a function of the parameters of the system and the parameters of the signal and not the input signal itself!*

But what type of function demonstrates this type of behavior? The simplest elementary function which repeats itself under the operation of differentiation is that of the complex exponential form $e^{\pm(\sigma+j\omega)t} = e^{\pm st}$, where σ, ω are parameters independent of t. It can easily be verified that such an input signal, when applied to a system which can be described in terms of a differential equation with constant coefficients, will yield an output function of an identical form multiplied by an algebraic equation in $s = \sigma + j\omega$ and the constants of the system, because

$$\frac{d^n}{dt^n} (e^{st}) = s^n e^{st}.$$

---

† Because the current is common to both R and L, we can write

$$\frac{1}{L} \int_0^t [v_i(t) - v_o(t)]\, dt = v_o(t)/R.$$

Taking the derivative of both sides and multiplying by R yields the desired differential equation.

We are interested in expressing finite-energy signals which may exist for negative as well as positive time. Because the real part, σ, of the exponential may cause convergence problems in such cases, we prefer to set $\sigma = 0$ and thus use functions of the form $e^{\pm j\omega t}$. These functions, or linear combinations of them, can exist for all time and analyses using them are usually referred to as "steady-state."

The parameter ω is the rate of phase change or frequency of the complex exponential in radians per second. It may also be expressed as f in cycles per second (Hz), where $\omega = 2\pi f$. Expressions for signals in terms of frequency are said to be in the "frequency domain," whereas expressions in time are said to be in the "time domain."

Drill Problem 2.6.1 Letting $v_i(t) = V_i e^{st}$ and $v_o(t) = V_o e^{st}$, determine the ratio of V_o/V_i in terms of s for the circuit shown in Fig. 2.8.

Answer. $(R/L)/(s + R/L)$.

2.7   THE EXPONENTIAL FOURIER SERIES

We investigate a set of complex-valued exponential functions expressed as

$$\phi_n(t) = e^{jn\omega_0 t}, \tag{2.31}$$

where n is an integer ($n = 0, \pm 1, \pm 2, \ldots$) and ω_0 is a constant to be determined. The value of n is referred to as the harmonic number or *harmonic*.

We consider the following operation on $\phi_n(t)$:

$$
\begin{aligned}
\int_{t_1}^{t_2} \phi_n(t)\phi_m^*(t)\, dt &= \int_{t_1}^{t_2} e^{jn\omega_0 t} e^{-jm\omega_0 t}\, dt \\
&= \frac{1}{j(n-m)\omega_0} [e^{j(n-m)\omega_0 t_2} - e^{j(n-m)\omega_0 t_1}], \qquad n \neq m, \\
&= \frac{1}{j(n-m)\omega_0} e^{j(n-m)\omega_0 t_1} [e^{j(n-m)\omega_0 (t_2 - t_1)} - 1]. \tag{2.32}
\end{aligned}
$$

Except for the trivial case of $t_2 = t_1$, we can force the term within brackets to zero if we choose [since $(n - m)$ is an integer]

$$\omega_0(t_2 - t_1) = 2\pi$$

so that

$$\int_{t_1}^{t_2} e^{jn\omega_0 t} e^{-jm\omega_0 t}\, dt = \begin{cases} (t_2 - t_1) & n = m \\ 0 & n \neq m \end{cases}, \tag{2.33}$$

if

$$\omega_0 = \frac{2\pi}{(t_2 - t_1)}. \tag{2.34}$$

If two functions $\phi_n(t)$, $\phi_m(t)$ meet the condition

$$\int_{t_1}^{t_2} \phi_n(t) \, \phi_m^*(t) \, dt = \begin{cases} \text{constant} & n = m \\ 0 & n \neq m \end{cases},$$

then these two functions are *orthogonal* over the interval (t_1, t_2). Therefore the set of functions

$$\phi_n(t) = e^{jn\omega_0 t}; \qquad n = 0, \pm 1, \pm 2, \ldots$$

forms an orthogonal set over the interval (t_1, t_2) if $\omega_0 = 2\pi/(t_2 - t_1)$.

Next we seek to express an arbitrary signal $f(t)$ in terms of a finite set of complex exponentials by writing

$$f(t) = \sum_{n=-N}^{N} F_n e^{jn\omega_0 t} \qquad (t_1 < t < t_2)$$

where the coefficients F_n are to be determined. It can be shown that the error energy between $f(t)$ and its approximation in terms of a set of complex exponentials decreases to zero as the number of terms taken approaches infinity. When a set of $\phi_n(t)$ meets this condition, it is said to be *complete*. Forming the set, we have

$$f(t) = \sum_{n=-\infty}^{\infty} F_n e^{jn\omega_0 t} \qquad (t_1 < t < t_2). \tag{2.35}$$

Because this $\phi_n(t)$ forms a complete orthogonal set, it is possible to represent any arbitrary complex-valued function $f(t)$ with finite energy by a linear combination of complex exponential functions over an interval (t_1, t_2). The representation of $f(t)$ by the exponential series as demonstrated above in Eq. (2.35) is known as the *exponential Fourier series representation* of $f(t)$ over the interval (t_1, t_2). The coefficients in this series can be found by multiplying both sides of Eq. (2.35) by $\phi_m^*(t) = e^{-jm\omega_0 t}$ and integrating with respect to t over the interval (t_1, t_2). As a result of orthogonality, all terms on the right-hand side of Eq. (2.35) vanish except for the one for $m = n$ and this yields the expression†

$$F_n = \frac{1}{(t_2 - t_1)} \int_{t_1}^{t_2} f(t) e^{-jn\omega_0 t} \, dt. \tag{2.36}$$

In summary, we have found that any given function having finite energy may be expressed as a discrete sum of exponential functions: $\phi_n(t) = e^{jn\omega_0 t}$ over an interval (t_1, t_2), where n is an integer $(n = 0, \pm 1, \pm 2, \ldots)$ and where $\omega_0 = 2\pi/(t_2 - t_1)$.

---

† Note that as a result of orthogonality each coefficient in the series is independent of every other coefficient. The coefficients could be found also by a direct application of Eq. (2.26).

Example 2.7.1 Write a representation of the function $f(t)$ given in Example 2.5.1 in terms of the complex exponential Fourier series.

Solution

$$\omega_0 = \frac{2\pi}{t_2 - t_1} = \frac{2\pi}{2} = \pi,$$

$$F_n = \frac{1}{2}\int_0^2 f(t)e^{-jn\pi t}\, dt$$

$$= \frac{1}{2}\int_0^1 e^{-jn\pi t}\, dt - \frac{1}{2}\int_1^2 e^{-jn\pi t}\, dt$$

$$= \frac{1}{2jn\pi}[-e^{-jn\pi} + 1 + e^{-j2n\pi} - e^{-jn\pi}] = \frac{1}{jn\pi}[1 - e^{-jn\pi}],$$

$$F_n = \begin{cases} 2/jn\pi & n\text{ odd} \\ 0 & n\text{ even}, \end{cases}$$

$$f(t) = \sum_{n=-\infty}^{\infty} F_n e^{jn\omega_0 t},$$

$$f(t) = \frac{2}{j\pi}\left(e^{j\pi t} + \frac{1}{3}e^{j3\pi t} + \frac{1}{5}e^{j5\pi t} + \cdots\right.$$

$$\left. - e^{-j\pi t} - \frac{1}{3}e^{-j3\pi t} - \frac{1}{5}e^{-j5\pi t} - \cdots\right)$$

Drill Problem 2.7.1 Write a representation of the function $f(t)$ given in Drill Problem 2.5.1 in terms of the complex exponential Fourier series over the interval $(-4, 4)$.

Answer

$$f(t) = \frac{1}{\pi}\left(\frac{\pi}{2} + e^{j(\pi/4)t} - \frac{1}{3}e^{j(3\pi/4)t} + \cdots + e^{-j(\pi/4)t} - \frac{1}{3}e^{-j(3\pi/4)t} + \cdots\right).$$

2.8 COMPLEX SIGNALS AND REPRESENTATIONS

For a signal whose instantaneous value is a complex number, the real and imaginary parts form an ordered pair of components given by (suppressing the time dependence for convenience in writing)

$$f = f_r + jf_i. \tag{2.37}$$

The complex conjugate of the signal value is

$$f^* = f_r - jf_i, \tag{2.38}$$

and therefore the real and imaginary parts are

$$f_r = \tfrac{1}{2}(f + f^*) \tag{2.39}$$

$$f_i = \frac{1}{2j}(f - f^*). \tag{2.40}$$

The squared magnitude of the complex-valued signal is equal to the product of the signal and its complex conjugate. Using the fact that $(j)^2 = -1$, we have

$$|f|^2 = ff^* = |f_r|^2 + |f_i|^2, \tag{2.41}$$

so that the square of the signal magnitude is the sum of the squares of the magnitudes of the real and imaginary components.

In the mathematics of linear systems analysis, we are particularly interested in signals of the complex exponential form $f(t) = e^{j\omega_0 t}$. Physically, this particular function may be thought of as describing the motion of a point on the rim of a wheel of unit radius. The wheel revolves (by convention) counterclockwise at an angular rate of ω_0 radians per second. From trigonometry (see Fig. 2.9), we see that the projection of the point on the real axis is $\cos \omega_0 t$, and the projection on the imaginary axis is $\sin \omega_0 t$. Combining, we can write

$$e^{j\omega_0 t} = \cos \omega_0 t + j \sin \omega_0 t. \tag{2.42}$$

An application of Eqs. (2.39) and (2.40) yields the real and imaginary components of the left-hand side of this result. Then, equating the real and imaginary components of both sides of Eq. (2.42), we obtain "Euler's identities":

$$\cos \omega_0 t = \frac{1}{2}(e^{j\omega_0 t} + e^{-j\omega_0 t}), \tag{2.43}$$

$$\sin \omega_0 t = \frac{1}{2j}(e^{j\omega_0 t} - e^{-j\omega_0 t}). \tag{2.44}$$

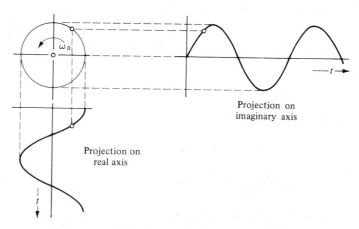

Projection on imaginary axis

Projection on real axis

Fig. 2.9 Complex notation for angular signals.

The addition of several terms, each involving circular motion with differing magnitudes and angular rates, makes the problem a little more complicated. However, we find that if an arrow is drawn from the origin to the reference point on each circle (wheel), the sum can be found by adding the arrows vectorially just as if they were vector quantities. Further, the resultant arrow may be treated just as if it were a vector and translated in coordinates so that its tail is at the origin. The projections on the translated coordinate system describe the real and imaginary components of the resulting summation.

These arrows are called *phasors*. A phasor is specified in the complex plane either by its real and imaginary components or by its magnitude and phase angle. Phasors obey the rules of vectors but they are not vectors; for this reason they are sometimes called pseudovectors.

The exponential Fourier series consists of a summation of complex exponential terms, each with its own magnitude, phase, and angular rate (frequency). Therefore summations of phasors can describe the instantaneous values of complex-valued signals using the exponential Fourier series representation. The usual convention is to let the real part of the phasor represent the real-valued function, although this certainly is not necessary. (We shall interchange the roles in the example below.)

With these added insights, let us reconsider the results of Example 2.7.1. There we obtained the complex exponential Fourier representation of a given function $f(t)$ which resulted in a solution in terms of an infinite series of complex exponentials rotating at integer multiples of π radians per second. Each term in the series is a phasor and the complex coefficients, the F_n, represent the starting angle and magnitude of each phasor. These phasors can be added using the rules of vector addition, and their sum represents the instantaneous complex amplitude and phase of the original function.

For convenience, let us retain only the first three nonzero terms of the series and consider only the interval $0 < t < 0.5$. From the results of Example 2.7.1, the first three terms for positive n are

$$\frac{2}{j\pi} \left(e^{j\pi t} + \frac{1}{3} e^{j3\pi t} + \frac{1}{5} e^{j5\pi t} \right).$$

As a result of the factor of $(-j)$, all three phasors start on the negative imaginary axis so that their projections on the real axis start at zero. For convenience in graphing, let us rotate everything by 90° and take the projections on the imaginary axis.

The phasors are shown in Fig. 2.10. At $t = 0$, all three phasors lie on the axis because the starting phase angles are all zero. As t increases, each of the phasors rotates at different angular rates and the resultant begins to curl in upon itself. The vertical components of the resultant describe the values of the signal representation at the corresponding times marked. Each phasor magnitude is labeled with the particular value of n corresponding to it.

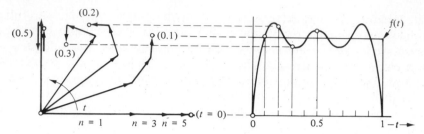

Fig. 2.10 Phasor addition of complex exponential series for function described in Example 2.7.1.†

Figure 2.10 is a graphic illustration of the way in which a vector summation of phasors, spinning at different angular rates, can add up in such a manner that the projection of their sum closely approximates the desired signal. As more and more terms are added to the series, the number of phasors in the sum becomes arbitrarily large. In this way, the tip of the resultant can move very quickly to describe the desired function.

Note that we summed over only positive values of n in the example above and then took the real part. An alternative to this procedure is to use a conjugate pair of phasors. Thus a cosine (real part) or a sine (imaginary part) function may be considered as the resultant of two conjugate phasors, rotating in opposite directions as shown in Fig. 2.11.

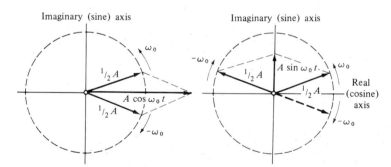

Fig. 2.11 Representation of a cosine and sine by a pair of conjugate phasors.

In both cases the complex conjugate phasors rotate in opposite directions with angular frequencies of $\pm\omega_0$ radians per second to describe real-valued functions of time. Because a *positive frequency* is associated with counterclock-

---

† An excellent demonstration of this type of phasor addition is graphically displayed in the 7-minute computer-animated film, *Harmonic Phasors*, W. H. Huggins and D. D. Weiner (Newton, Mass.: Educational Development Center, 1969).

wise uniform angular motion, this representation suggests the concept of a *negative frequency* to describe the motion of the second phasor. The use of negative frequencies is a convenient way to describe a real-valued function using pairs of complex-valued functions. It is equivalent to taking the real (or imaginary) part of the complex-valued function. The concept of negative frequencies is used throughout this book, largely for mathematical convenience. We must realize, of course, that to represent a real-valued signal the magnitudes of two conjugate phasors must be equal and their instantaneous phase angles must be equal-but-oppositely-directed.

Drill Problem 2.8.1   Using Eqs. (2.37)–(2.40), show that

$$\mathcal{R}e\{f_1 f_2\} = \mathcal{R}e\{f_1\}\mathcal{R}e\{f_2\} - \mathcal{I}m\{f_1\}\mathcal{I}m\{f_2\}.$$

2.9   THE TRIGONOMETRIC FOURIER SERIES REPRESENTATION†

The question may well be asked at this point: If we know that a given function $f(t)$ is real-valued, isn't there an equivalent way of expressing a Fourier series representation of $f(t)$ using a set of real-valued orthogonal functions? Let us investigate by taking the real part of both sides of Eq. (2.35). Making use of the result of Drill Problem 2.8.1, we have

$$f_r(t) = \sum_{n=-\infty}^{\infty} \mathcal{R}e\{F_n\}\mathcal{R}e\{e^{jn\omega_0 t}\} - \sum_{n=-\infty}^{\infty} \mathcal{I}m\{F_n\}\mathcal{I}m\{e^{jn\omega_0 t}\}$$

$$= \sum_{n=-\infty}^{\infty} \mathcal{R}e\{F_n\}\cos n\omega_0 t - \sum_{n=-\infty}^{\infty} \mathcal{I}m\{F_n\}\sin n\omega_0 t. \tag{2.45}$$

The F_n are given by Eq. (2.36). But since we need real-valued functions, we will express new real-valued coefficients in terms of F_n and F_n^*. Therefore we investigate the F_n^* for real-valued signals:

$$F_n^* = \frac{1}{(t_2 - t_1)} \int_{t_1}^{t_2} f(t)e^{jn\omega_0 t}\, dt = F_{-n}. \tag{2.46}$$

Recalling Eqs. (2.39) and (2.40), and using Eq. (2.46), we can write

$$\mathcal{R}e\{F_n\} = \tfrac{1}{2}[F_n + F_n^*] = \tfrac{1}{2}[F_n + F_{-n}], \tag{2.47}$$

$$\mathcal{I}m\{F_n\} = \frac{1}{2j}[F_n - F_n^*] = \frac{1}{2j}[F_n - F_{-n}]. \tag{2.48}$$

---

† J. B. J. Fourier, *Theorie Analytique de la Chaleur*, 1822; republished in English by Dover, N. Y. The first rigorous proof of a version of Fourier's theorem was given by Dirichlet in 1829.

For convenience, let us define new coefficients at this point:

$$a_0 \triangleq F_0, \tag{2.49}$$

$$a_n \triangleq [F_n + F_{-n}] = 2\,\mathcal{R}e\{F_n\}, \qquad n \neq 0, \tag{2.50}$$

$$b_n \triangleq j[F_n - F_{-n}] = -2\,\mathcal{I}m\{F_n\}, \tag{2.51}$$

$$F_n = \tfrac{1}{2}(a_n - jb_n), \qquad\qquad n \neq 0. \tag{2.52}$$

From the definitions of a_n, b_n, it can be seen that a_n is an even function of n while b_n is an odd function of n. Thus everything is completely symmetric in n about $n = 0$, and we could just as well double the coefficients and sum only over positive integer values of n. Following this procedure, Eq. (2.45) can be rewritten as

$$f(t) = a_0 + \sum_{n=1}^{\infty} a_n \cos n\omega_0 t + \sum_{n=1}^{\infty} b_n \sin n\omega_0 t \qquad \text{for } f(t) \text{ real-valued over } (t_1, t_2).$$

$$\tag{2.53}$$

Thus the set of functions $\cos n\omega_0 t$ and $\sin n\omega_0 t$ for $(n = 0, 1, 2, \ldots)$, $\omega_0 = 2\pi/(t_2 - t_1)$, forms a complete orthogonal set over the interval (t_1, t_2). The result given in Eq. (2.53) is called the *trigonometric Fourier series representation of f(t) over the interval* (t_1, t_2). It is capable of describing a real-valued function of finite energy over a given interval. Note that although the functions $\sin n\omega_0 t$, $\sin 2\omega_0 t$, etc., form an orthogonal set over any interval $[t_1, t_1 + (2\pi/\omega_0)]$, this set is not complete. This arises from the fact that we can exhibit a function, namely $\cos n\omega_0 t$, which is orthogonal to $\sin n\omega_0 t$ over the same interval. Hence to complete the set we must include cosine as well as sine functions.

Rather than evaluating the constants a_n and b_n from the F_n, we could have multiplied both sides of Eq. (2.53) by $\cos n\omega_0 t$ and $\sin n\omega_0 t$. Because $\cos n\omega_0 t$, $\cos m\omega_0 t$, $\sin n\omega_0 t$, and $\sin m\omega_0 t$ are all mutually orthogonal, the terms remaining after simplification are†

$$a_n = \frac{\displaystyle\int_{t_1}^{t_2} f(t) \cos n\omega_0 t \, dt}{\displaystyle\int_{t_1}^{t_2} \cos^2 n\omega_0 t \, dt} = \frac{2}{(t_2 - t_1)} \int_{t_1}^{t_2} f(t) \cos n\omega_0 t \, dt, \tag{2.54}$$

$$b_n = \frac{\displaystyle\int_{t_1}^{t_2} f(t) \sin n\omega_0 t \, dt}{\displaystyle\int_{t_1}^{t_2} \sin^2 n\omega_0 t \, dt} = \frac{2}{(t_2 - t_1)} \int_{t_1}^{t_2} f(t) \sin n\omega_0 t \, dt, \tag{2.55}$$

---

† These coefficients could be found also by a direct application of Eq. (2.26).

$$a_0 = \frac{\int_{t_1}^{t_2} f(t)\, dt}{\int_{t_1}^{t_2} dt} = \frac{1}{(t_2 - t_1)} \int_{t_1}^{t_2} f(t)\, dt. \qquad (2.56)$$

The trigonometric Fourier series may be represented in more compact notational form as

$$f(t) = \sum_{n=0}^{\infty} c_n \cos (n\omega_0 t + \phi_n), \qquad (2.57)$$

where

$$c_n = \sqrt{a_n^2 + b_n^2}, \qquad (2.58)$$

$$\phi_n = \tan^{-1}(-b_n/a_n). \qquad (2.59)$$

Substituting Eqs. (2.50) and (2.51) into Eqs. (2.58) and (2.59), we find that this trigonometric representation is related to the complex exponential representation by

$$c_n = 2|F_n| = 2\sqrt{F_n F_n^*}, \qquad n \neq 0, \qquad (2.60)$$

$$\phi_n = \tan^{-1} \frac{\mathscr{I}m\{F_n\}}{\mathscr{R}e\{F_n\}}, \qquad (2.61)$$

and $c_0 = F_0$.

Example 2.9.1   Represent $f(t) = t^2$ in a trigonometric Fourier series over the interval $(0, 2)$.

Solution.   In this case, $t_1 = 0$, $t_2 = 2$, and $\omega_0 = \pi$. The coefficients can be found by a direct application of Eqs. (2.54)–(2.56):†

$$a_0 = \tfrac{1}{2} \int_0^2 t^2\, dt = \tfrac{4}{3},$$

$$a_n = \tfrac{2}{2} \int_0^2 t^2 \cos n\pi t\, dt = 4/(n\pi)^2,$$

$$b_n = \tfrac{2}{2} \int_0^2 t^2 \sin n\pi t\, dt = -4/(n\pi).$$

Thus the trigonometric Fourier series of $f(t) = t^2$ over the interval $(0, 2)$ is

$$f(t) = \frac{4}{3} + \frac{4}{\pi^2} \sum_{n=1}^{\infty} \frac{1}{n^2} \cos n\pi t - \frac{4}{\pi} \sum_{n=1}^{\infty} \frac{1}{n} \sin n\pi t.$$

---

† These integrals can be found in Appendix A.

Any function $f(t)$ can be expressed in terms of a sum of a corresponding even function, $f_e(t)$, and an odd function, $f_o(t)$. These functions can be formed by the following relations:

$$f_e(t) = \tfrac{1}{2}[f(t) + f(-t)], \tag{2.62}$$

$$f_o(t) = \tfrac{1}{2}[f(t) - f(-t)], \tag{2.63}$$

and thus $f_e(t) + f_o(t) = f(t)$.

The properties of even and odd functions are particularly convenient when one considers the trigonometric Fourier series over a symmetric interval $(-T/2, T/2)$. In this case Eq. (2.53) may reduce to a special form depending on whether $f(t)$ is even or odd. If $f(t)$ is even, the product $f(t) \sin n\omega_0 t$ is odd in t, the b_n's are zero, and the result is a cosine series. Similarly, if $f(t)$ is odd, then the product $f(t) \cos n\omega_0 t$ is odd in t, the a_n's are zero, and the result is a sine series. In both cases the nonzero terms can be evaluated by integrating over half the period and then multiplying by two.

Drill Problem 2.9.1   Find the trigonometric Fourier series of the triangular waveform shown in Fig. 2.12 over the interval $(-\pi, \pi)$.

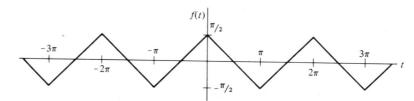

Fig. 2.12 A triangular waveform.

Answer. $f(t) = \dfrac{4}{\pi} \displaystyle\sum_{\substack{n=1 \\ n \text{ odd}}}^{\infty} \dfrac{1}{n^2} \cos nt \qquad -\pi < t < \pi.$

Drill Problem 2.9.2   Find the expansion of the function described by $f(t) = t$ in a trigonometric Fourier series over the interval $(0,2)$.

Answer. $f(t) = 1 - \dfrac{2}{\pi} \displaystyle\sum_{n=1}^{\infty} \dfrac{1}{n} \sin n\pi t \qquad 0 < t < 2.$

2.10   EXTENSION BY PERIODICITY

We have been able to represent a given function with finite energy by a Fourier series over a finite interval (t_1, t_2). Outside this interval the function $f(t)$ and the corresponding Fourier series need not be equal.

Now we would like to extend this representation to periodic signals, i.e., those for which the relation

$$f(t + T) = f(t) \qquad (2.64)$$

holds for all t. We shall assume that $f(t)$ has finite energy over an interval $(t_0, t_0 + T)$. As we consider longer and longer time intervals, the energy accumulates and in the limit the total energy is infinite. However, we also note that the energy in any interval T seconds long is the same as in any other given interval which is T seconds in length. It follows, then, that if we divide by T, the average *rate* of energy, or average power, is a constant.

With this slight change, we can extend all the previous analyses directly over to the case of periodic functions with finite average power simply by (1) taking the interval for finding the coefficients, etc., as the period T of the periodic function; and (2) dividing by this interval in any calculation involving the energy so that we obtain an average energy rate or power. Such representations are said to "converge in mean square" to the periodic function $f(t)$.

It is hardly worth going back and rephrasing all our previous results for the case of the series representation of a periodic function. In fact, for the complex exponential set we chose earlier, it is easy to see that Eq. (2.64) is satisfied if the interval is $T = 2\pi/\omega_0$ because

$$e^{jn\omega_0(t + T)} \equiv e^{jn\omega_0 t} \qquad \text{for} \quad T = 2\pi/\omega_0.$$

It follows that the series representation

$$f(t) = \sum_n F_n e^{jn\omega_0 t}$$

will represent a periodic function over the infinite interval and the representation converges in a mean-square (or average-square) sense.

Graphically, we are saying that since the basic phasor rotates at the angular rate ω_0 rad/sec, the complex exponential Fourier series representation will describe any periodic function whose period is just that time it takes the basic phasor to make one complete revolution. This is shown in Fig. 2.13.

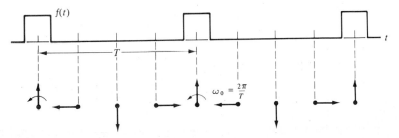

Fig. 2.13 Revolution of a basic phasor in synchronism with a given periodic function.

In general the interval of integration to determine coefficients in the Fourier series is taken over one complete period T to evaluate a given coefficient. Just where the particular interval is taken, however, makes no difference when $f(t)$ is periodic. If the lower limit has an arbitrary value of t_0, then the upper limit will be $(t_0 + T)$. Often it is convenient to take the interval of integration from $-T/2$ to $+T/2$ in order to make use of possible symmetry conditions.

Drill Problem 2.10.1   Determine the trigonometric Fourier series for the symmetric square wave shown in Fig. 2.14.

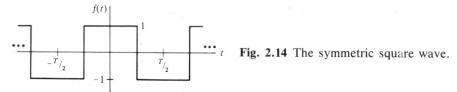

Fig. 2.14 The symmetric square wave.

Answer. $f(t) = \dfrac{4}{\pi}\left(\cos \omega_0 t - \dfrac{1}{3} \cos 3\omega_0 t + \dfrac{1}{5} \cos 5\omega_0 t - \cdots \right)$ where $\omega_0 = 2\pi/T$.

2.11   PARSEVAL'S THEOREM FOR POWER SIGNALS

The average power developed across a one-ohm resistance is [Eq. (2.7)]

$$P = \frac{1}{T} \int_{-T/2}^{T/2} f(t) f^*(t)\, dt \quad \text{watts.} \tag{2.65}$$

Using the exponential Fourier series and substituting in Eq. (2.65),

$$P = \frac{1}{T} \int_{-T/2}^{T/2} \sum_{m=-\infty}^{\infty} F_m e^{jm\omega_0 t} \sum_{n=-\infty}^{\infty} F_n^* e^{-jn\omega_0 t}\, dt \tag{2.66}$$

where $\omega_0 = 2\pi/T$.

Assuming that $f(t)$ is integrable over the interval t_0 to $(t_0 + T)$, we may interchange the order of summation and integration in Eq. (2.66):

$$P = \sum_{m=-\infty}^{\infty} F_m \sum_{n=-\infty}^{\infty} F_n^* \frac{1}{T} \int_{-T/2}^{T/2} e^{j(m-n)\omega_0 t}\, dt. \tag{2.67}$$

But the complex exponential functions are orthogonal over the interval t_0 to $(t_0 + T)$ so that the integral in Eq. (2.67) is zero except for the special case when $m = n$. For this specific condition the double summation reduces to a single summation and we have a new relation for the average power in terms of the magnitudes of the coefficients:

$$P = \sum_{n=-\infty}^{\infty} F_n F_n^* = \sum_{n=-\infty}^{\infty} |F_n|^2. \tag{2.68}$$

Combining Eqs. (2.65) and (2.68), we obtain a relationship which is known as *Parseval's theorem for periodic signals:*

$$P = \frac{1}{T} \int_{-T/2}^{T/2} |f(t)|^2 \, dt = \sum_{n=-\infty}^{\infty} |F_n|^2. \tag{2.69}$$

If we know the time function $f(t)$, we can find the average power. Alternatively, if we know the Fourier coefficients, we can find the average power. The answer obtained in the time domain and in the frequency domain must agree.

Example 2.11.1 Determine the average power of $f(t) = 2 \sin 100t$ using Eqs. (2.65) and (2.68).

Solution. Using Eq. (2.65), we have

$$P = \frac{1}{T} \int_{-T/2}^{T/2} 4 \sin^2 100t \, dt = 2 \text{ W}.$$

The Fourier coefficients of $f(t)$ are[†]

$$F_1 = -j,$$

$$F_{-1} = j,$$

$$F_n = 0 \text{ for all } n \neq \pm 1.$$

Then Eq. (2.68) gives

$$P = \sum_{n=-\infty}^{\infty} |F_n|^2 = F_1 F_1^* + F_{-1} F_{-1}^* = |j|^2 + |-j|^2 = 1 + 1 = 2 \text{ W}.$$

Equation (2.68) illustrates the fact that the power in a periodic function is distributed over discrete frequencies that are harmonically related to one another. The power contained in each frequency component is given by its respective term in Eq. (2.68). Thus if we make a graph of P vs. ω, the power in $f(t)$ is located only at discrete frequencies, as depicted in Fig. 2.15. Such a graph is called the "power spectrum" of the signal $f(t)$.

---

† A shortcut in the determination of these coefficients is to use Euler's identity for 2 sin 100t. The coefficients are found by equating term by term to the corresponding Fourier series.

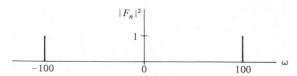

Fig. 2.15 The power spectrum of the function described in Example 2.11.1.

Drill Problem 2.11.1   Repeat Example 2.11.1 for $f(t) = 2 \cos 100t$ and show that its power spectrum is given by Fig. 2.15.

2.12   THE FREQUENCY TRANSFER FUNCTION

In characterizing the behavior of systems, we follow two main plans of analysis, one in the frequency domain and one in the time domain. In both approaches, we use superposition to add up the responses of the system for combinations of elemental functions. Representation in the frequency domain is considered first; the time domain is considered later in Section 2.18. Following the reasoning of the previous sections, we make use of the complex exponential function.

A fundamental property of a linear time-invariant system is that the input and the output are related by linear differential equations with constant coefficients. A typical system could be described by

$$a_0 g(t) + a_1 \frac{dg}{dt} + \cdots = b_0 f(t) + b_1 \frac{df}{dt} + \cdots , \qquad (2.70)$$

where the a's and b's are constants.

Now we use the input signal

$$f(t) = e^{j\omega t} \qquad (2.71)$$

to test the system. A particular solution can be written as

$$g(t) = H(\omega)e^{j\omega t}. \qquad (2.72)$$

Using Eqs. (2.71) and (2.72) in Eq. (2.70), we obtain

$$H(\omega) = \frac{\Sigma_k b_k (j\omega)^k}{\Sigma_m a_m (j\omega)^m}. \qquad (2.73)$$

This important ratio is called the *frequency transfer function* of the system.†
Note that the right-hand side of Eq. (2.73) depends only on the system.

---

† The *transfer function* of a linear time-invariant system is defined as the ratio of the Laplace transform of the output to the Laplace transform of the input for zero initial conditions.

Physically, Eq. (2.73) tells us that a way to test a linear time-invariant system is to apply a sinusoid of known amplitude, frequency, and phase to the input of the system. The output will be another sinusoid *at the same frequency* but the amplitude and phase will, in general, differ from that of the input. Taking the ratio of these two complex coefficients gives the value (in amplitude and phase) of the system transfer function at that frequency. This process may be continued for other frequencies. A plot of the system frequency transfer function may then be obtained by drawing a continuous curve through the points graphed.

We can go one step farther and incorporate the system response characteristics with the components of the system. This is done by multiplying every term giving rise to a derivative by $(j\omega)$ and every term giving rise to an integral by $(1/j\omega)$. This is the method commonly used in ac linear circuit analysis. Use of these methods is demonstrated in Example 2.12.1.

Example 2.12.1   Determine the frequency transfer function of the system shown in Fig. 2.16.

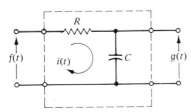

$f(t)$ $i(t)$ $g(t)$ **Fig. 2.16** The RC lowpass filter.

Solution.   The differential equation describing this system is found easily by summing the currents yielding†

$$\frac{1}{R}[f(t) - g(t)] = C\frac{dg}{dt},$$

or

$$f(t) = RC\frac{dg}{dt} + g(t).$$

For the calculation of the system frequency transfer function we let $f(t) = e^{j\omega t}$ and $g(t) = H(\omega)e^{j\omega t}$ so that substitution of the particular solution in the differential equation above becomes

$$e^{j\omega t} = j\omega RCH(\omega)e^{j\omega t} + H(\omega)e^{j\omega t}.$$

The system frequency transfer function is then

$$H(\omega) = \frac{1}{j\omega RC + 1}.$$

---

† Usually a zero source impedance and an infinite load impedance are assumed for defining transfer functions.

To make the calculation even easier, we can replace C by $(j\omega C)^{-1}$ in the circuit and use the property of a voltage divider to obtain the above result by inspection. This is the method commonly used in ac linear circuit analysis.

Drill Problem 2.12.1   A given linear time-invariant system can be described by the differential equation $g(t) + d^4g/dt^4 = d^2f/dt^2$. Determine the system frequency transfer function.

Answer. $H(\omega) = -\omega^2/(1 + \omega^4)$.

In general, $H(\omega)$ is a complex-valued function of frequency and is usually expressed in the polar form

$$H(\omega) = |H(\omega)|e^{j\theta(\omega)} = \left|\frac{G(\omega)}{F(\omega)}\right| e^{j[\theta_g(\omega) - \theta_f(\omega)]}. \tag{2.74}$$

The quantity $|H(\omega)|$ is called the *magnitude response* of the system and $\theta(\omega)$ is called the *phase shift* of the system.

A device available for the rapid measurement of the magnitude response of a system is the sweep generator. Basically, the sweep generator is a voltage-controlled oscillator whose center frequency and frequency sweep can be preset. The amplitude of the resulting sinusoidal waveform is held constant over the range.

The sweep generator is connected to the input of the system under test and a magnitude-responding device is connected to the output and to the vertical deflection on an oscilloscope. A sawtooth voltage derived from the oscilloscope sweep circuit causes the sweep generator to generate sinusoids of constant amplitude but whose frequencies vary in a linear fashion as the oscilloscope trace moves horizontally. The resulting display is a plot of the magnitude of the system frequency transfer function. Because phase shift and the time rate of frequency are related, phase plots are generally not made in this manner.

Note that a transfer function may include a change of units (e.g., from voltage to current, current to voltage, etc.). It may also be used to describe systems containing active elements (gain factors) as long as they are linear and time-invariant.

2.13   STEADY-STATE RESPONSE TO PERIODIC SIGNALS

It follows from Section 2.12 that if the input signal to a linear time-invariant system is described by

$$f(t) = Ae^{j(\omega_1 t + \phi_1)}, \tag{2.75}$$

the output is

$$g(t) = AH(\omega_1)e^{j(\omega_1 t + \phi_1)}. \tag{2.76}$$

Then the system response to a periodic signal becomes very simple if the input is written as an exponential Fourier series,

$$f(t) = \sum_{n=-\infty}^{\infty} F_n e^{jn\omega_0 t},$$

and the resulting output is

$$g(t) = \sum_{n=-\infty}^{\infty} G_n e^{jn\omega_0 t} = \sum_{n=-\infty}^{\infty} H(n\omega_0) F_n e^{jn\omega_0 t}. \tag{2.77}$$

We can also relate the output average power to the input by using Parseval's theorem:

$$P_f = \sum_{n=-\infty}^{\infty} |F_n|^2,$$

$$P_g = \sum_{n=-\infty}^{\infty} |G_n|^2 = \sum_{n=-\infty}^{\infty} |H(n\omega_0)|^2 |F_n|^2. \tag{2.78}$$

Example 2.13.1   Determine the output, $g(t)$, of a linear time-invariant system whose input and frequency transfer function are shown in Fig. 2.17. Calculate the input and output average power.

Fig. 2.17 The input and transfer function for Example 2.13.1.

Solution.   Writing the Fourier series for the input,

$$f(t) = \sum_{n=-\infty}^{\infty} \frac{\sin(n\pi/2)}{n\pi/2} e^{jn2\pi t},$$

and therefore the output is [using Eq. (2.77)]

$$g(t) = \sum_{n=-\infty}^{\infty} H(2\pi n) \frac{\sin(n\pi/2)}{n\pi/2} e^{jn2\pi t} = 1 + \frac{4}{\pi} \cos 2\pi t.$$

The input average power is

$$P_f = \frac{1}{T} \int_{t_0}^{t_0+T} |f(t)|^2 \, dt = \int_{-1/4}^{1/4} 4 \, dt = 2 \text{ W}.$$

The output average power is [using Eq. (2.78)]

$$P_g = \sum_{n=-\infty}^{\infty} |H(n\omega_0)|^2 |F_n|^2,$$

$$P_g = \left(\frac{2}{\pi}\right)^2 + 1 + \left(\frac{2}{\pi}\right)^2 = 1.811 \text{ W}.$$

Note that, as far as energy and power calculations are concerned, it is only the magnitude of $H(\omega)$ that counts.

Drill Problem 2.13.1   The system described in Drill Problem 2.12.1 is used to filter the input signal of Example 2.13.1. Determine a series expression for the output signal; also calculate the output power.

Answer. $g(t) \cong \left(\frac{1}{\pi}\right)^3 \left(-\cos 2\pi t + \frac{1}{27} \cos 6\pi t - \cdots\right)$; 0.52 mW.

2.14   HARMONIC GENERATION

An important practical application of the Fourier series representation is in the measurement of the generation of harmonic content. This harmonic generation may be either intentional or unintentional and we briefly consider both types here.

In some applications it is desirable to generate a sinusoid whose frequency is an exact multiple of that of a given sinusoid. A device with a nonlinear output-input gain characteristic can be used to accomplish this. Suppose, for example, that the output-input gain characteristic is

$$e_o(t) = a_1 e_i(t) + a_2 e_i^2(t), \tag{2.79}$$

where $e_o(t)$, $e_i(t)$ are the output and input signals respectively. If we let $e_i(t) = A \cos \omega_0 t$, then we can write

$$e_o(t) = a_1 A \cos \omega_0 t + a_2 A^2 \cos^2 \omega_0 t$$

$$= \tfrac{1}{2} a_2 A^2 + a_1 A \cos \omega_0 t + \tfrac{1}{2} a_2 A^2 \cos 2\omega_0 t. \tag{2.80}$$

The nonlinear output-input characteristic has therefore resulted in the generation of a second-harmonic term; such a device is called a "frequency doubler." Similarly, a third-order nonlinearity results in generation of third-harmonic content, etc. In practice, the nonlinearities may only be known graphically. In this case, one can use a power series approximation or numerical techniques using a digital computer. The latter approach is illustrated in problems at the end of this chapter.

Because harmonic content is generated by nonlinearities, the above procedure may be reversed. If an amplifier has a linear gain (i.e., output-input) characteristic, a sinusoidal input will result in a sinusoidal output *at the same frequency*. Therefore the presence of harmonic content in the output when only
a single-frequency sinusoid is applied to the input represents distortion resulting
from nonlinearities in the amplifier. A convenient way to measure this distortion
without requiring knowledge of gain or impedance values is to take the ratio of
the mean-square harmonic distortion terms to the mean-square value of the first
harmonic.† This ratio is expressed in percent and is called the *total harmonic
distortion* (THD) of the amplifier. Expressed in terms of the trigonometric Fourier
series, the total harmonic distortion is then

$$\text{THD} = \frac{\sum_{n=2}^{\infty}(a_n^2 + b_n^2)}{a_1^2 + b_1^2}. \qquad (2.81)$$

A computation of the THD of a given waveform is illustrated in the following
example.

Example 2.14.1   A given amplifier is tested with an input 500 Hz sinusoid of
2 mV peak amplitude. The output is found to be composed of the parabolic
sections [t is in msec, $f(t)$ is in V]

$$f(t) = \begin{cases} (1 - 4t^2) & -0.5 < t < 0.5 \\ 4(t - 1)^2 - 1 & 0.5 < t < 1.5 \end{cases}$$

etc. (periodically), as shown in Fig. 2.18. (a) Determine the total harmonic
distortion (THD). (b) What is the linear gain of the amplifier?

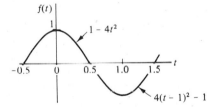

Fig. 2.18 A parabolic waveform.

Solution.   The period is 2 msec; however, because we are only interested in the
relative magnitudes of the Fourier series coefficients, we can scale by 10^3 and
drop the "msec," thereby scaling everything to "sec." By inspection, the average
value of $f(t)$ is zero and the b_n's are zero (as a result of even waveform symmetry).

---

† Zero average value is assumed; otherwise the average value must be subtracted out
before taking the mean-square value.

Computation of the a_n's gives

$$a_n = 32 \frac{\sin(n\pi/2)}{(n\pi)^3}.$$

(a) The mean-square value of $f(t)$ is

$$\overline{f^2} = \frac{1}{T} \int_{-T/2}^{T/2} f^2(t)\, dt = 2 \int_0^{1/2} (1 - 4t^2)^2\, dt = \frac{8}{15}.$$

The total harmonic distortion (THD) is

$$\text{THD} = \frac{\overline{f^2} - \frac{1}{2}a_1^2}{\frac{1}{2}a_1^2} = \frac{16}{15}\left(\frac{\pi^3}{32}\right)^2 - 1 = 0.145\%.$$

(b) The linear gain is $32/2\pi^3 \times 10^3 \cong 516$.

Drill Problem 2.14.1   As an extreme case, suppose the input sinusoid in Example 2.14.1 results in a symmetrical square wave with unit peak amplitude. Calculate the linear gain and THD. [*Hint:* Use the results of Example 2.5.1.]

Answer.   (a) 637; (b) 23.4%.

2.15   THE FOURIER SPECTRUM AND EXAMPLES

The exponential Fourier series is composed of a summation of complex exponentials with the F_n representing the magnitudes and initial phase angles of the harmonically related rotating phasors. The resultant phasor is found by adding the individual phasors vectorially. For all but the simple cases of one or two terms, however, the addition of a series of phasors for each instant of time turns out to be an inconvenient way to describe a signal.

The thought now occurs to us: instead of summing and plotting phasors to describe a signal, why not plot the Fourier coefficients (the F_n) as a function of the angular rate (frequency)? A primary motivation behind this approach is that the resulting diagram will hold for all values of time and we can avoid the summing and plotting for each instant of time. Such a graph of the (complex) Fourier coefficients as a function of frequency for the signal $f(t)$ is called the *Fourier spectrum* of $f(t)$. It is quite common in engineering literature to simply call this the "spectrum" of $f(t)$ with the understanding that it refers to the exponential Fourier representation.

For a periodic signal, we have found that the Fourier coefficients correspond to the relative amplitude and phase weightings of a set of harmonically related phasors. Therefore the Fourier spectrum exists only at $\omega = 0$, $\pm\omega_0$, $\pm 2\omega_0$, $\pm 3\omega_0$, . . . ; i.e., only at discrete values of ω. It is therefore a *discrete spectrum*, sometimes referred to as a *line spectrum*. We may represent this spectrum graphically by drawing vertical lines at the discrete points $\omega = 0$, $\pm\omega_0$, $\pm 2\omega_0$, . . . , with their heights proportional to the amplitudes of the corresponding frequency components (assuming that the coefficients are real-valued). This *amplitude spectrum* then appears on a graph as a series of equally spaced vertical lines with heights proportional to the amplitudes of the respective frequency components.

In general, however, the F_n are complex-valued. To describe the coefficients then requires two graphs, the *magnitude spectrum* and *phase spectrum*. The following example illustrates these different ways of portraying spectra.

Example 2.15.1   Sketch the amplitude spectrum and the magnitude and phase spectrum of the function used in Example 2.11.1: $f(t) = 2 \sin 100t$.

Solution.   The sketches are shown in Fig. 2.19. Note that the magnitude spectrum is symmetrical about $\omega = 0$. This is a general characteristic of magnitude spectra for real-valued $f(t)$, as can be seen from Eq. (2.46).

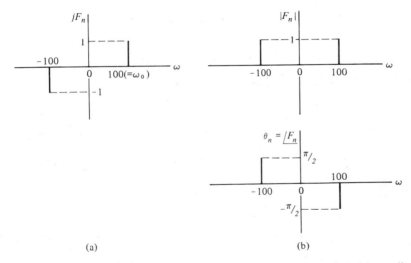

(a)   (b)

Fig. 2.19 Line spectra corresponding to $f(t)$ in Example 2.15.1: (a) amplitude spectrum; (b) magnitude and phase spectra.

We could have used the trigonometric series to describe the spectrum. In this case, we would have one-sided amplitude and phase spectra (i.e., $n \geq 0$ only). The amplitude and phase of each sinusoidal component can be found using Eqs. (2.58) and (2.59). An interesting interpretation of the trigonometric spectrum is illustrated in Fig. 2.20.

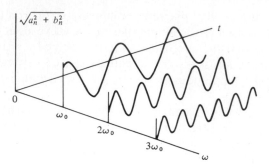

Fig. 2.20 An interpretation of the trigonometric line spectrum.

Now let us find the Fourier spectrum for the periodic gate function shown in Fig. 2.21. The periodic gate function is a generalized case of the square-wave signal in which the width and height of rectangular pulses are variable and one level is at zero amplitude. The gate function and its corresponding spectrum play a major role in the analysis of many systems.

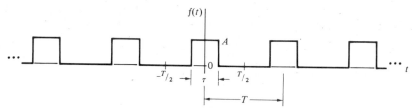

Fig. 2.21 The periodic gate function.

Proceeding, the Fourier coefficients are:

$$F_n = \frac{1}{T} \int_{-T/2}^{T/2} f(t)e^{-jn\omega_0 t} \, dt$$

$$= \frac{1}{T} \int_{-\tau/2}^{\tau/2} Ae^{-jn\omega_0 t} \, dt$$

$$= \frac{-A}{jn\omega_0 T} (e^{-jn\omega_0\tau/2} - e^{jn\omega_0\tau/2}), \qquad n \neq 0,$$

$$= \frac{2A}{n\omega_0 T} \sin(n\omega_0\tau/2), \qquad n \neq 0,$$

$$F_n = \frac{A\tau}{T} \frac{\sin(n\omega_0\tau/2)}{n\omega_0\tau/2}, \qquad n \neq 0.$$

Defining a new normalized variable $x = n\omega_0\tau/2$, we have

$$F_n = \frac{A\tau}{T} \left[\frac{\sin x}{x}\right] \qquad x \neq 0.$$

The function within the brackets occurs so often that we shall assign it the convenient abbreviation "sine over argument"†

$$\text{Sa}(x) \triangleq \frac{\sin x}{x}. \tag{2.82}$$

A graph of Eq. (2.82) appears as Fig. 2.22.

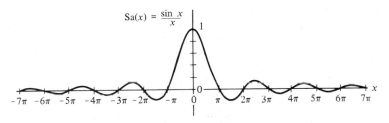

Fig. 2.22 The function Sa(x).

From Fig. 2.22, observe that the amplitude of the function Sa(x) oscillates, decaying in either direction of x and approaching zero as $|x| \to \infty$. The maximum value of this function occurs as x approaches zero, for $[(\sin x)/x] \to 1$ as $x \to 0$. The latter can be seen by expanding sin x in a power series before taking the limit or by using l'Hospital's rule.

Because‡

$$F_0 = \frac{1}{T}\int_{-\tau/2}^{\tau/2} A\,dt = \frac{A\tau}{T},$$

the exponential Fourier representation of the periodic gate function is given by

$$f(t) = \frac{A\tau}{T}\sum_{n=-\infty}^{\infty} \text{Sa}(n\pi\tau/T)e^{jn\omega_0 t}, \tag{2.83}$$

where

$$\omega_0 = 2\pi/T.$$

Because the F_n are real, we need only display the amplitude spectrum. Aside from the scaling constant A, this spectrum is dependent only on the parameter choices τ and T.

---

† This abbreviation is not used by all authors. Another common abbreviation is

$$\text{sinc }(x) \triangleq \frac{\sin \pi x}{\pi x}.$$

‡ The coefficient F_0 represents the average value of the signal $f(t)$. Note that this term can be changed without affecting any of the other harmonic components in the exponential Fourier series.

It is instructive first to keep τ fixed and vary the period T. As T increases, two effects are noticed: (1) the amplitude of the spectrum decreases as $1/T$, and (2) the spacing between lines decreases as $2\pi/T$. These effects are shown in the drawings of Fig. 2.23. Note that the envelope of the spectra remains fixed except for the amplitude scale factor. As the period T increases, the spacing between components, $\Delta\omega = 2\pi/T$, becomes smaller and there are more and more frequency components in a given range of frequency. The amplitudes of these components decrease as T is increased.† Although the amplitude of the spectrum goes to zero, the *shape* of the spectrum does not change. We conclude that *the shape, or envelope, of the spectrum is dependent only upon the pulse shape and not on the repetition period T*.

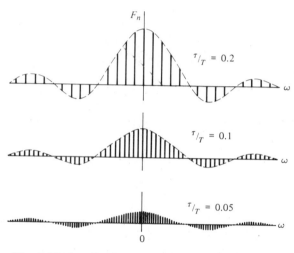

Fig. 2.23 Amplitude spectra for various values of τ/T, τ fixed.

It is also instructive to keep T fixed and vary τ (with the restriction that $\tau < T$). As τ increases, (1) the amplitude of the spectrum increases proportional to τ, and (2) the frequency content of the signal is compressed within an increasingly narrower range of frequencies. This trend is shown in Fig. 2.24. Thus there is *an inverse relationship between pulse width in time and the frequency "spread" of the spectrum*. A convenient measure of the frequency spread is the distance to the first zero crossing of the Sa(x) function.

Summarizing, the spectrum of a periodic function is a line spectrum with spectral components located at multiples of the basic repetition frequency. The spacing between frequency components is inversely proportional to the repetition

---

† In the limit as $T \to \infty$, only a single pulse remains because its nearest neighbors have been removed to infinity. In this limiting case, the frequency spacings approach zero and the spectrum becomes continuous (see Chapter 3).

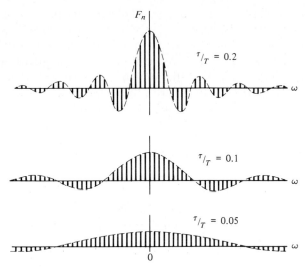

Fig. 2.24 Amplitude spectra for various values of τ/T, T fixed.

period T. The frequency spread of the envelope of the spectrum is inversely proportional to the pulse width τ. The shape of the spectrum depends only on the pulse shape and not on the repetition period T. Finally, even though these effects are illustrated here using the periodic gate function, the general conclusions hold for all periodic functions.

    A device available for laboratory measurements of spectra of power signals is the scanning spectrum analyzer. Basically, the scanning spectrum analyzer is designed to accept only a very narrow spread (band) of frequencies, measure the power, and display the square root of the power as the vertical deflection on an oscilloscope. The placement of the frequencies accepted is movable electrically.

    A simplified block diagram of a scanning spectrum analyzer is shown in Fig. 2.25. A sawtooth voltage derived from the oscilloscope sweep circuit causes the

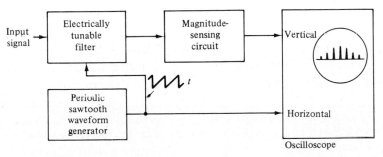

Fig. 2.25 Block diagram of a scanning spectrum analyzer.

analyzer to sweep across the frequencies of interest in a linear fashion as the oscilloscope trace moves horizontally. Any nonzero spectral component present results in a vertical displacement of the oscilloscope trace as its frequency is passed. This can give a close approximation to the magnitude spectrum of the input waveform. The accuracy of the approximation is governed by how narrow a band of frequencies the analyzer uses to make the measurement and by how slowly it sweeps over the desired range of frequencies.

★ 2.16   NUMERICAL COMPUTATION OF FOURIER COEFFICIENTS

Calculation of the Fourier series coefficients may be approximated numerically. This becomes advantageous when an analytical expression for $f(t)$ is not known and $f(t)$ is available as numerical data points, or when the integration is difficult to perform or must be repeated often with parameter changes. We shall outline the approach here using only the most straightforward methods of numerical approximation.

Equations for the calculation of trigonometric Fourier series coefficients are given in Eqs. (2.54)–(2.56). Choosing the cosine term as an example, we have

$$a_n = \frac{2}{T} \int_0^T f(t) \cos n\omega_0 t \, dt, \tag{2.84}$$

where $\omega_0 = 2\pi/T$. We now assume that M equally spaced data points representing $f(t)$ over $(0, T)$ are either known or available to us, and that the interval between each such data point is Δt.[†] Approximating the integration in Eq. (2.84) with a summation of rectangular strips each Δt in width, we have[‡]

$$a_n \cong \frac{2}{T} \sum_{m=1}^{M} f(m \cdot \Delta t) \cos \left[n(2\pi/T)(m \cdot \Delta t) \right] \Delta t, \tag{2.85}$$

where

$$\Delta t = T/M. \tag{2.86}$$

Using Eq. (2.86) in Eq. (2.85), we obtain our desired result:

$$a_n \cong \frac{2}{M} \sum_{m=1}^{M} f(m \cdot \Delta t) \cos (2\pi mn/M). \tag{2.87}$$

Thus once the M equally spaced data points have been determined, the computation is expressed in terms of the integers m (the integer time scale) and n (the integer frequency scale).

---

† If the data points are not equally spaced, a subroutine can be employed to make a "best fit" numerically to the available data and then interpolate to obtain the equally spaced data samples.

‡ Equivalently, the summation may go from 0 to $(M - 1)$.

Computation of the other Fourier coefficients follows in the same manner so that

$$b_n \cong \frac{2}{M} \sum_{m=1}^{M} f(m \cdot \Delta t) \sin (2\pi mn/M), \qquad (2.88)$$

$$a_0 \cong \frac{1}{M} \sum_{m=1}^{M} f(m \cdot \Delta t). \qquad (2.89)$$

An example of such a calculation for a known function demonstrates the procedure.

Example 2.16.1 Using 100 equally spaced sample points per period, compute the coefficients of the first 10 harmonic terms of the trigonometric Fourier series for the triangular waveform of Drill Problem 2.9.1.

Solution. Over the interval $(0, 2\pi)$,

$$f(t) = \begin{cases} \pi/2 - t & 0 \le t \le \pi/2 \\ t - 3\pi/2 & \pi/2 \le t < 3\pi/2 \end{cases}$$

and thus we have

$$T = 2\pi,$$
$$M = 100,$$
$$\Delta t = T/M = \pi/50.$$

Using Eqs. (2.87)–(2.89), we find

$$a_n \cong \frac{2}{100} \sum_{m=1}^{100} f(m\pi/50) \cos (\pi mn/50), \text{ etc.}$$

A sample program and corresponding numerical solution for a typical programmable calculator is shown in Table 2.1.† [*Note:* The choice of variables in the program displayed is $a_n \rightarrow A$; $b_n \rightarrow B$; $n \rightarrow C$; $m \rightarrow X$.] One look at the solution printout reveals that most of the computation time is being spent to compute coefficients which are zero! Actually, we should have known this as a result of the symmetry in the waveform.

It is interesting to compare the above results determined by numerical integration to the exact results determined in Drill Problem 2.9.1. A comparison is shown below.

n	Approximate b_n	"Exact" b_n	Difference
1	1.273659	1.273240	+0.000419
3	0.141891	0.141471	+0.000420
5	0.051351	0.050930	+0.000421
7	0.026407	0.025984	+0.000423
9	0.016145	0.015719	+0.000426

---

† This type of program is chosen here because it is easy to read and follow.

Table 2.1   Program and Solution Printout for Example 2.16.1

Program	Solution printout
1: 0→C	0
2: 0→A;0→B;1→X	−.000000
	1
3: IF X> 50;GTO 5	1.273659
4: .02(π/2−πX/50)COS(πCX/50)→Z;GTO 6	.000000
	2
5: .02(πX/50−3π/2)COS(πCX/50)→Z	.000000
6: Z+A→A	.000000
	3
7: IF X> 50;GTO 9	.141891
8: .02(π/2−πX/50)SIN(πCX/50)→Z;GTO 10	.000000
	4
9: .02(πX/50−3π/2)SIN(πCX/50)→Z	.000000
10: Z+B→B	.000000
	5
11: IF X=100;GTO 13	.051351
12: X+1→X;GTO 3	.000000
	6
13: IF C> 0;GTO 15	.000000
14: FXD 0;PRT C;FXD 6;PRT A/2;GTO 16	.000000
	7
15: FXD 0;PRT C;FXD 6;PRT A,B	.026407
16: IF C=10;GTO 18	.000000
	8
17: C+1→C;GTO 2	.000000
18: END	.000000
	9
	.016145
	.000000
	10
	.000000
	.000000

Note that agreement between the two is good, but not perfect. One who is acquainted with numerical integration procedures may be tempted to try to improve the numerical accuracy by more sophisticated integration procedures. However, this is not the main source of error here; we shall devote some attention to this shortly.

Drill Problem 2.16.1   (a) Write a program in Fortran for Example 2.16.1. (b) Run your program to investigate the effects on the accuracy of the coefficients for the following number of equally spaced sample points per period: 20; 40; 60; 80.

Answer for a_1. $+0.010523$; $+0.002621$; $+0.001164$; $+0.000654$.

When M equally spaced data points are used to represent a given waveform for computation of Fourier coefficients, that waveform is not only made periodic with period M but the data points themselves also exhibit a periodicity (with period $1/M$). The net result is that orthogonality between terms with differing harmonic numbers is guaranteed only if that difference is less than $M/2$. Taking the harmonic numbers (n) as the consecutive integers starting with zero, we have

$$n \le M/2. \tag{2.90}$$

In other words, *the highest harmonic coefficient which can be determined uniquely from M sampled data points is that one for which the harmonic number is less than or equal to one-half the number of sampled data points.*

An illustration of a typical line spectrum of a periodic function and a line spectrum of its numerical approximation is shown in Fig. 2.26. Note that there is some approximation error even for those coefficients for which $n < M/2$ but that this approximation error becomes proportionally very large for $n > M/2$. The error arising in those terms for which $n < M/2$ is called "aliasing error" and is usually the main source of error in the numerical approximation. Aliasing error is the subject of the next section. Using more sampled data points for a fixed T (i.e., increasing M and thus decreasing Δt) aids in decreasing the amount of aliasing error present.

★ 2.17   EFFECTS OF ALIAS TERMS

If all frequency components in a given waveform are within the harmonic range given by Eq. (2.90), the Fourier coefficients can be determined within the accuracy of the numerical integration procedures. However, the magnitudes of the frequency components in a given waveform do not always lie within a finite range but rather may decrease gradually as n becomes large. Because orthogonality is no longer guaranteed, the numerical evaluation of a coefficient at a given harmonic within the range given by Eq. (2.90) may actually include waveform components which otherwise would be assigned to higher coefficients that lie outside this range (see Fig. 2.26). This type of error is called *aliasing error*.

It turns out that only certain harmonic terms contribute (erroneously) to the result.† The net effect on the computation can be summarized as follows. Designating the true values of the Fourier coefficients by a_n', b_n', the coefficient values actually computed from the M equally spaced data points are given by

$$a_n = a_n' + \sum_{k=1}^{\infty} (a_{kM+n}' + a_{kM-n}'), \tag{2.91}$$

$$b_n = b_n' + \sum_{k=1}^{\infty} (b_{kM+n}' - b_{kM-n}'). \tag{2.92}$$

Thus if we restrict the range of computation to $n \le M/2$, Eqs. (2.91) and (2.92)

---

† The proof of these statements is postponed until Chapter 3.

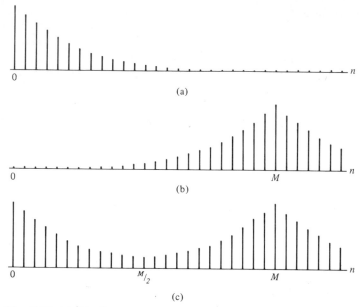

Fig. 2.26 (a) The line spectrum of a periodic function; (b) spectral replica produced by sampling; (c) composite numerical approximation to the line spectrum.

show that the numerical accuracy of the computed Fourier coefficients a_n, b_n is dependent on how fast the true coefficients decrease for $n \leq M/2$. The terms within the summations in Eqs. (2.91) and (2.92) are called alias terms and arise as a result of the sampled nature of the data. The presence of the alias terms is usually the main source of inaccuracy in the numerical computation of the Fourier coefficients. Their effect can be minimized for a given waveform with T fixed by decreasing Δt and thereby increasing the number of sample points, M.

To emphasize the effect of the alias terms on numerical accuracy, let us attempt to estimate the accuracy of the first Fourier coefficient a_1 in Example 2.16.1. We define the error in the numerical approximation by

$$\epsilon_n = a_n - a_n'.$$

Using Eq. (2.91) and $M = 100$, the error in a_1 is

$$\epsilon_1 = \sum_{k=1}^{\infty} (a_{100k+n}' + a_{100k-n}').$$

From the results of Drill Problem 2.9.1, we know that the true Fourier coefficients for this waveform are

$$a'_n = \begin{cases} 4/(\pi n^2) & n \text{ odd} \\ 0 & n \text{ even} \end{cases}.$$

Therefore the aliasing error in the computation of the first Fourier coefficient is

$$\epsilon_1 = \frac{4}{\pi} \left[\frac{1}{(101)^2} + \frac{1}{(201)^2} + \frac{1}{(301)^2} + \frac{1}{(401)^2} + \cdots \right.$$
$$\left. + \frac{1}{(99)^2} + \frac{1}{(199)^2} + \frac{1}{(299)^2} + \frac{1}{(399)^2} + \cdots \right],$$

$$\epsilon_1 \cong \frac{8}{\pi} \frac{1}{(100)^2} \left(1 + \frac{1}{2^2} + \frac{1}{3^2} + \frac{1}{4^2} + \cdots \right).$$

However, we can also write†

$$\sum_{n=1}^{\infty} \frac{1}{n^2} = \frac{\pi^2}{6},$$

so that

$$\epsilon_1 \cong \frac{8}{\pi} \frac{1}{(100)^2} \frac{\pi^2}{6} = 0.000419.$$

A comparison with the results of Example 2.16.1 reveals that all of the approximation error (to six places) is due to aliasing, despite the fact that we used 100 data samples to compute the first harmonic coefficient! Thus the effects of aliasing are not always negligible. Also, as the harmonic number n is increased, some terms in the summation in Eq. (2.91) or Eq. (2.92) become more predominant so that the aliasing error generally increases with increasing n. This trend is evident in Example 2.16.1.

For a given number of data points, the effect of aliasing depends on the rate of convergence of the Fourier coefficients as n becomes large. In many cases the true coefficient behavior may not be known. However, the rate of convergence of the coefficients can often be estimated either from some knowledge of the physical nature of the source of the data or from the characteristics of the waveform itself. We shall take up the latter topic again in a later section of this chapter.

What happens if the computation of a coefficient is attempted for $n > M/2$? In this case one of the alias terms in Eq. (2.91) or (2.92) will predominate. For

---

† At points of finite discontinuity, the Fourier series converges to the arithmetic average of the values of the function to either side of the discontinuity. Thus this result can be obtained by evaluating the result of Example 2.9.1 at $t = 2$.

instance, if the computation of a_{99} in Example 2.16.1 were attempted, Eq. (2.91) gives

$$a_{99} = a'_{99} + (a'_{199} + a'_{299} + \cdots) + (a'_1 + a'_{101} + \cdots).$$

The term a'_1 will, of course, predominate if the coefficients in the series are convergent. In fact, if one writes out these results for a few different values of n, it becomes evident that all the computed a_n are symmetric about $n = M/2$ and the b_n are antisymmetric about $n = M/2$. This is shown in the following example.

Example 2.17.1 Numerically evaluate the coefficients of the first 9 harmonic terms of the trigonometric Fourier series for the waveform defined by

$$f(t) = \begin{cases} t & 0 < t < \pi \\ 0 & \pi < t < 2\pi \end{cases}.$$

Use 10 sample points per period. At points of finite discontinuity in the waveform, use the average of the values to either side of the discontinuity.

Solution. $T = 2\pi$; $M = 10$; $\Delta t = 2\pi/10 = \pi/5$;

$$a_n \cong \frac{2}{10} \sum_{m=1}^{10} f(m\pi/5) \cos(2\pi mn/10), \text{ etc.}$$

A sample program and solution printout is shown in Table 2.2. Note the symmetry properties of the coefficients about the harmonic number $n = M/2 = 5$.

Drill Problem 2.17.1 Write a program for computation of the trigonometric Fourier coefficients for the waveform of Drill Problem 2.9.2. Use 10 sample points over the interval $(0,2)$; assign the average value of the waveform to the sample point at $t = 2$. (a) Determine the numerical error in b_1. (b) What is the relationship b_9 to b_1?

Answer.   (a) $+0.021083$; (b) equal but reversed sign.

Computation of coefficients in the exponential Fourier series follows in a similar manner through use of Euler's identities. Thus we can rewrite Eq. (2.36) as

$$F_n = \frac{1}{T} \int_0^T f(t) \cos n\omega_0 t \, dt - j\frac{1}{T} \int_0^T f(t) \sin n\omega_0 t \, dt \qquad (2.93)$$

$$= \frac{a_n}{2} - j\frac{b_n}{2}, \qquad (2.94)$$

and the numerical computation of the a_n, b_n has already been discussed.

While computation of coefficients for the exponential Fourier series is essentially the same as for the trigonometric series, interpretation of the results

Table 2.2   Program and Solution Printout for Example 2.17.1

Program	Solution printout
	0
1: 0→C	.785398
	1
2: 0→A;0→B;1→X;1→Y	−.657984
	.966833
3: IF X=5;.5→Y	2
4: .2(πX/5)Y COS(πCX/5)→Z	.000000
	−.432403
5: Z+A→A	3
6: .2(πX/5)Y SIN(πCX/5)→Z	−.095999
	.228250
7: Z+B→B	4
8: IF X=5;GTO 10	.000000
	−.102077
9: X+1→X;GTO 3	5
10: FXD 0;PRT C	−.062832
	.000000
11: IF C=0;FXD 6;PRT A/2;GTO 13	6
12: FXD 6;PRT A,B	.000000
	.102077
13: IF C=10;GTO 15	7
14: C+1→C;GTO 2	−.095999
	−.228250
15: END	8
	.000000
	.432403
	9
	−.657984
	−.966883

is, in fact, slightly different. Recall that there are both positive and negative frequencies in the exponential series, but only positive frequencies in the trigonometric series. For real-valued $f(t)$, the relation between the positive and negative frequency components in the exponential series is that of a complex conjugate; i.e., $F_{-n} = F_n^*$ [cf. Eq. (2.46)]. Suppose we numerically compute the coefficients for the exponential series using Eqs. (2.87), (2.88), and (2.94). The computed coefficients for $0 \le n \le M/2$ yield the positive frequency components, just as they did for the trigonometric series. But those coefficients computed for $M/2 \le n \le M$ are the complex conjugates of those for $0 \le n \le M/2$ [cf. Eqs.

(2.91), (2.92), and (2.94)]. Thus we conclude that *the positive frequency components in the numerical computation of the exponential Fourier series coefficients are given over the harmonic range* $0 \le n \le M/2$ *and the negative frequency components are over* $M/2 \le n \le M$. The numerical accuracy of the exponential series is affected by aliasing in the same way that it is for the trigonometric series.

This introduction will enable the reader to proceed to use the digital computer to numerically evaluate the Fourier series. A more theoretical basis for understanding the aliasing effects must be postponed until Chapter 3.

Example 2.17.2   Compute and plot the magnitude spectrum of the exponential Fourier series for the periodic waveform shown in Fig. 2.27(a). Use 50 sample points per period and plot the spectrum over the range $0 \le n < 50$. At points of a finite discontinuity in the waveform, assume continuity to the left.

Solution.   See Fig. 2.27(b).

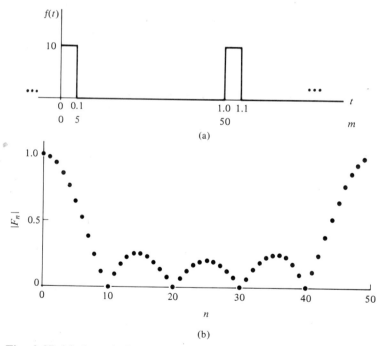

(a)

(b)

Fig. 2.27 (a) A periodic rectangular function; (b) its computed magnitude spectrum using the discrete exponential Fourier series (50 sample points).

Drill Problem 2.17.2   Determine the common characteristic(s) which are true for the following coefficients in the exponential Fourier series ($k = 0, 1, 2, \ldots$): (a) $n = kM$; (b) $n = kM/2$.

Answer.   (a) All represent the average value of the waveform; (b) All must be real-valued.

2.18   SINGULARITY FUNCTIONS

There is a particular class of functions which plays a role of such importance in signal analysis that we treat it separately here. Members of this class have simple mathematical forms but they are either not finite everywhere or they do not have finite derivatives of all orders everywhere. For this reason they are called *singularity functions.*

Singularity functions are mathematical idealizations and, strictly speaking, do not occur in physical systems. They are useful in signal analysis because they serve as good approximations to certain limiting conditions in physical systems. This, together with their simple mathematical forms and properties, makes it possible often to evaluate complicated expressions which might otherwise be impossible—or at least very difficult—to solve.

A singularity function of great importance to us is the unit impulse function (also called the Dirac delta or impulse function), $\delta(t)$. This function has the property exhibited by the following integral:

$$\int_a^b f(t)\, \delta(t - t_0)\, dt = \begin{cases} f(t_0) & a < t_0 < b \\ 0 & \text{elsewhere} \end{cases} \qquad (2.95)$$

for any $f(t)$ continuous at $t = t_0$, t_0 finite. The impulse function selects or sifts out a particular value of the function $f(t)$, namely, the value at $t = t_0$, in the integration process. Equation (2.95) is often referred to as the *sifting property* of the impulse function. An example illustrating the use of the sifting property follows.

Example 2.18.1   Evaluate the definite integral $\int_{-\infty}^{\infty} e^{\cos t} \delta(t - \pi)\, dt$.

Solution. $\int_{-\infty}^{\infty} e^{\cos t} \delta(t - \pi)\, dt = e^{\cos \pi} = e^{-1} = 0.368$.

Drill Problem 2.18.1   Evaluate:

$$\text{(a)} \int_1^{\infty} e^{-x^2} \delta(x)\, dx, \qquad \text{(b)} \int_1^{100} \log t\, \delta(t - 10)\, dt.$$

Answer.   (a) 0; (b) 1.

As seen from Eq. (2.95), the impulse function is no ordinary function. However, we can treat $\delta(t)$ as a function formally obeying the rules of integration provided that all conclusions are based on Eq. (2.95) and not on any point properties of $\delta(t)$. Using Eq. (2.95) we can demonstrate formally the following properties of the impulse function.

Area (strength).   If $f(t) = 1$, Eq. (2.95) becomes

$$\int_a^b \delta(t - t_0)\, dt = 1, \qquad a < t_0 < b. \tag{2.96}$$

Therefore $\delta(t)$ has unit area.† In an entirely analogous manner, $A\delta(t)$ has an area of A units.

Amplitude.   We have

$$\delta(t - t_0) = 0 \quad \text{for all} \quad t \ne t_0. \tag{2.97}$$

That this must be true follows directly from a consideration of Eq. (2.95); i.e., all values of $f(t)$ for $t \ne t_0$ are neglected (given zero weight) in the integration process. *The amplitude at the point $t = t_0$ is undefined.*

Graphical Representation.   It is fairly obvious from the above properties that we are going to encounter some difficulty in graphing the impulse function. To avoid any attempt to display the amplitude at $t = t_0$, we shall draw an arrow at the point $t = t_0$ as an indicator of the impulse function. The *area* or *strength* of the impulse is designated by a quantity in parentheses beside the arrow or by the height of the arrow (or both), as shown in Fig. 2.28(a). An arrow pointing downward indicates negative area.

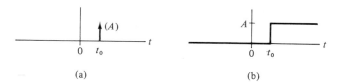

(a)   (b)

Fig. 2.28 Graphical representations for (a) $A\ \delta(t - t_0)$; (b) $Au(t - t_0)$.

Symmetry.   We define $\delta(t)$ to be an even function; that is,

$$\delta(t) = \delta(-t). \tag{2.98}$$

This follows from the fact that it makes no difference in Eq. (2.95) whether the point $t = t_0$ is approached from the left or from the right in the integration.

Time Scaling.   A scaling in the argument of the impulse function can be converted as follows:

$$\delta(at) = \frac{1}{|a|}\delta(t). \tag{2.99}$$

---

† Area in this sense denotes "strength" or "weight" of the impulse function.

This can be shown using the change of variable $x = at$; for $a > 0$,

$$\int_{-\infty}^{\infty} f(t)\, \delta[a(t - t_0)]\, dt = \int_{-\infty}^{\infty} f\left(\frac{x}{a}\right) \delta(x - at_0)\, \frac{dx}{a} = \frac{1}{a} f(t_0).$$

Repeating for $a < 0$ and letting the change of variable be $x = at$, we have

$$\int_{-\infty}^{\infty} f(t)\, \delta[a(t - t_0)]\, dt = \int_{\infty}^{-\infty} f\left(\frac{x}{a}\right) \delta(x - at_0)\, \frac{dx}{a} = -\frac{1}{a} f(t_0)$$

Combining these two, we can write

$$\int_{-\infty}^{\infty} f(t)\, \delta[a(t - t_0)]\, dt = \int_{-\infty}^{\infty} f(t) \frac{1}{|a|} \delta(t - t_0)\, dt = \frac{1}{|a|} f(t_0).$$

Graphically, the introduction of this scale factor is necessary to retain unity area in the definition of the impulse function.

Example 2.18.2   Evaluate the definite integral

$$\int_{-\infty}^{\infty} t^2 e^{-\sin t} \cos 2t\, \delta(2t - 2\pi)\, dt.$$

Solution.   Using the time scaling property and then the sifting property in succession, we have

$$\int_{-\infty}^{\infty} t^2 e^{-\sin t} \cos 2t\, \delta(2t - 2\pi)\, dt = \frac{1}{2}\int_{-\infty}^{\infty} t^2 e^{-\sin t} \cos 2t\, \delta(t - \pi)\, dt = \frac{1}{2}\pi^2.$$

Drill Problem 2.18.2   Evaluate the definite integral

$$\int_{-\infty}^{\infty} \delta(1 - \pi t) \cos(1/t)\, dt.$$

Answer. $-1/\pi$.

Multiplication by a Time Function.   Sometimes we wish to use $\delta(t)$ without the defining integration. Such operations are formal only and result in the following relation:

$$f(t)\, \delta(t - t_0) = f(t_0)\, \delta(t - t_0), \qquad f(t) \text{ continuous at } t_0. \qquad (2.100)$$

This can be shown easily by integrating both sides.

Relation to the Unit Step Function.   The unit step function is that function defined by

$$u(t - t_0) \triangleq \begin{cases} 1 & t > t_0 \\ 0 & t < t_0 \end{cases}. \qquad (2.101)$$

A graphical representation of $Au(t)$ is shown in Fig. 2.28(b). Using Eq. (2.95) and letting $a = -\infty$, $b = t$, and $f(t) = 1$,

$$\int_a^b f(\tau)\,\delta(\tau - t_0)\,d\tau = \int_{-\infty}^t \delta(\tau - t_0)\,d\tau = \begin{cases} 1 & t > t_0 \\ 0 & t < t_0 \end{cases} = u(t - t_0),$$

or

$$\int_{-\infty}^t \delta(\tau - t_0)\,d\tau = u(t - t_0). \qquad (2.102)$$

Therefore the integral of the unit impulse function is the unit step function. The converse can also be shown by formally differentiating both sides of Eq. (2.102):

$$\frac{d}{dt}\int_{-\infty}^t \delta(\tau - t_0)\,d\tau = \frac{d}{dt} u(t - t_0),$$

$$\delta(t - t_0) = \frac{d}{dt} u(t - t_0).$$

The derivative of the unit step function at $t = t_0$ is the unit impulse occurring at $t = t_0$.

Example 2.18.3 Compute and graph the derivative of the rectangular pulse shown in Fig. 2.29(a).

Solution.   Writing an equation for the pulse in terms of step functions, we have

$$f(t) = Au(t) - Au(t - t_0),$$

$$\frac{df}{dt} = A\,\delta(t) - A\,\delta(t - t_0).$$

We can check our result by integrating

$$\int_{-\infty}^t \frac{df}{d\tau}\,d\tau = \int_{-\infty}^t [A\,\delta(\tau) - A\,\delta(\tau - t_0)]\,d\tau = Au(t) - Au(t - t_0),$$

confirming our result. The corresponding graphical solution is shown in Fig. 2.29(b).

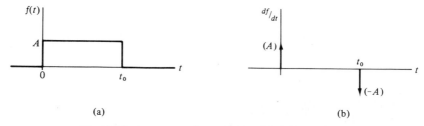

(a)   (b)

Fig. 2.29 A rectangular pulse and its derivative.

Drill Problem 2.18.3 It is given that $f(t) = u(t) - u(t - 3) - K \delta(t - 4)$. Determine the required numerical value of K such that $\int_{-\infty}^{\infty} f(t) \, dt = 0$.

Answer. 3.

The operations indicated involving the impulse function all arose formally from its integral definition. We have observed that the impulse function is not really a true function in a mathematical sense. The impulse function has, however, been justified mathematically using a theory of "generalized functions."[†] In this approach, the impulse function is defined as the limit of a sequence of regular well-behaved functions having the required property that the area remains a constant (unity) as the width is reduced. All operations are performed on this sequence. Finally the limit of this sequence is taken to define the impulse function as the width is reduced toward zero.

The defining sequence of pulses is not unique and many pulse shapes can be chosen. In fact, the shape of the particular pulse is relatively unimportant as long as the sequence satisfies the conditions that (1) the sequence formed describes a function which becomes infinitely high and infinitesimally narrow in such a way that (2) the enclosed area is a constant (unity). For example, the following sequences all satisfy these conditions (see Fig. 2.30):

a) Rectangular pulse: $\delta(t) = \lim_{\tau \to 0} \frac{1}{\tau} \left[u\left(t + \frac{\tau}{2}\right) - u\left(t - \frac{\tau}{2}\right) \right].$ (2.103)

b) Triangular pulse: $\delta(t) = \lim_{\tau \to 0} \frac{1}{\tau} \left[1 - \frac{|t|}{\tau} \right], |t| < \tau.$ (2.104)

c) Two-sided exponential: $\delta(t) = \lim_{\tau \to 0} \frac{1}{\tau} e^{-|2t|/\tau}.$ (2.105)

d) Gaussian pulse: $\delta(t) = \lim_{\tau \to 0} \frac{1}{\tau} e^{-\pi(t/\tau)^2}.$ (2.106)

e) Sa(t) function: $\delta(t) = \lim_{\tau \to 0} \frac{1}{\tau} \text{Sa}(\pi t/\tau).$ (2.107)

f) Sa$^2(t)$ function: $\delta(t) = \lim_{\tau \to 0} \frac{1}{\tau} \text{Sa}^2(\pi t/\tau).$ (2.108)

In Eq. (2.107), it is helpful to note that $1/\pi \int_{-\infty}^{\infty} \text{Sa}(x) \, dx = 1$. This sequence can also be rewritten in slightly different form by letting $\pi/\tau = k$ so that

$$\delta(t) = \lim_{k \to \infty} \frac{k}{\pi} \text{Sa}(kt).$$ (2.109)

[†] M. J. Lighthill, *Fourier Analysis and Generalized Functions*, Cambridge University Press, 1959.

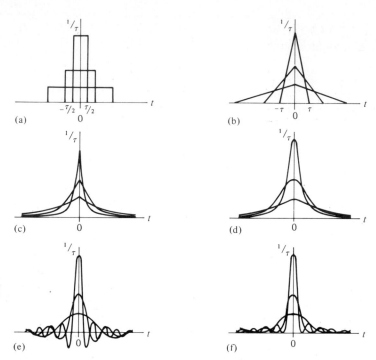

Fig. 2.30 Function sequence definition of the impulse function:
(a) rectangular pulse; (b) triangular pulse; (c) two-sided exponential;
(d) gaussian pulse; (e) Sa(t) function; (f) Sa²(t) function.

In a similar manner, Eq. (2.108) can be rewritten as

$$\delta(t) = \lim_{k \to \infty} \frac{k}{\pi} \, Sa^2(kt). \tag{2.110}$$

2.19 IMPULSE RESPONSE

To describe the response of a system in the time domain, we shall use reasoning analogous to that used in the frequency domain. The time function whose occurrence is the most definite of the elementary functions at our disposal (i.e., the one having the least time spread) is the impulse function. If we test the system with a unit impulse, the resulting output will describe the behavior of a system over all time in response to the unit impulse function occurring at one point in time. The resulting output is called the *impulse response* of the system.

This can be expressed mathematically in the following manner. If the input to the system is $f(t)$ and the corresponding output is $g(t)$, then, from Eq. (2.11), we have

$$g(t) = \mathcal{T}\{f(t)\}. \tag{2.111}$$

Specifically, we choose $f(t) = \delta(t)$, set all initial conditions equal to zero, and define the resulting output as the impulse response:

$$h(t) \triangleq \mathcal{T}\{\delta(t)\}. \tag{2.112}$$

For simple systems, Eq. (2.112) may be applied directly. For more complicated systems, a sequence such as one of those shown in Fig. 2.30 may be used for the input. After obtaining a solution for the corresponding output, the impulse response may be found by taking the limit.

Example 2.19.1   Determine the impulse response of the system shown in Fig. 2.31(a).

Solution.   To find the impulse reponse, let $f(t) = \delta(t)$. Then (see Fig. 2.31a)

$$y(t) = \delta(t) - \delta(t - t_0),$$

and

$$h(t) = \int_{-\infty}^{t} [\delta(\tau) - \delta(\tau - t_0)] \, d\tau = u(t) - u(t - t_0).$$

This is an example of a "holding circuit" commonly used in pulse systems. Its impulse response is sketched in Fig. 2.31(b).

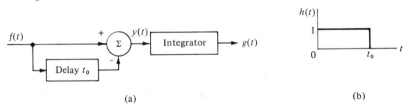

(a)   (b)

Fig. 2.31 A holding circuit and its impulse response function.

Drill Problem 2.19.1   Determine the impulse response of the system shown in Fig. 2.32 for (a) $K = 1$ and (b) $K < 1$. This type of system is useful in the detection of periodic signals if the period is known.

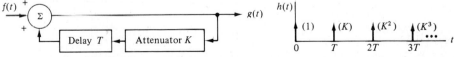

Fig. 2.32 A recirculating-type filter and its impulse response.

Answer.   (a) $h(t) = \sum\limits_{n=0}^{\infty} \delta(t - nT)$; (b) $h(t) = \sum\limits_{n=0}^{\infty} (K)^n \, \delta(t - nT)$.

★2.20   CONVERGENCE OF THE FOURIER SERIES

Finally, with our added perspectives, we can now state some rather general results for the Fourier series. If a signal $f(t)$ meets the conditions, which are known as the "Dirichlet conditions," over the interval $(0, T)$:

1. $f(t)$ has only a finite number of maxima and minima in the interval T;
2. $f(t)$ has only a finite number of finite discontinuities (i.e., $f(t)$ is "piecewise continuous") in the interval T;
3. $f(t)$ satisfies the inequality

$$\int_0^T |f(t)|\, dt < \infty;$$

then a Fourier series representation can be written which converges to $f(t)$ at all points of continuity. These conditions are sufficient but not necessary, and we shall encounter some mathematical functions which do not satisfy the Dirichlet conditions and yet have a Fourier series representation.

The Dirichlet conditions are satisfied for signals which occur in the laboratory because we insist that the somewhat weaker condition

$$\int_0^T |f(t)|^2\, dt < \infty$$

always holds. Stated another way, a Fourier series can always be used to represent any signal which has finite energy over $(0, T)$.

For any function satisfying the Dirichlet conditions, the magnitude of the coefficients decreases as n becomes large at least as fast as n^{-1}; that is,†

$$|F_n| \le \frac{M_N}{n} \quad \text{for} \quad n > N, \tag{2.113}$$

where M is some positive number which is independent of n. If the given function has a derivative satisfying the Dirichlet conditions, the Fourier coefficients decrease at least as fast as n^{-2}. Likewise, if the given function has a second derivative satisfying the Dirichlet conditions, the Fourier coefficients decrease at least as fast as n^{-3}. In general, if the kth derivative satisfies the Dirichlet conditions, then we know that the magnitude of the Fourier coefficients will decrease at least as fast as $n^{-(k+1)}$ as n becomes very large.

---

† The analogous relations for the trigonometric Fourier series are

$$|a_n| \le \frac{M_N}{n}\,, \ |b_n| \le \frac{M_N}{n} \quad \text{for} \quad n > N.$$

The unit impulse function does not satisfy the Dirichlet conditions. However, the integral of a periodic sequence of impulse functions satisfies the Dirichlet conditions as long as the net area per period is zero. Also, the unit impulse function arises whenever a sufficient number of derivatives are taken of a piecewise continuous function. Therefore the order of the derivative of a given function that first exhibits impulse functions represents that negative order of n which bounds the rate of decrease of the Fourier coefficients. For example, the magnitude of the Fourier coefficients for a square wave must decrease at least as fast as $1/n$ for large n, those for a triangular waveform as $1/n^2$, those for a waveform composed of parabolic sections as $1/n^3$, etc. Reasons why this is true will become evident when derivative properties of the Fourier transform are discussed in Chapter 3. For the present, these observations can serve as useful checks in our calculations of Fourier coefficients.

The Fourier series representation converges in mean-square (i.e., the mean-square error approaches zero uniformly and absolutely as $N \to \infty$) to $f(t)$ for every $f(t)$ satisfying the Dirichlet conditions.† For many functions this convergence is rapid enough that a truncation of the series after only a few terms gives a fairly good approximation to the original function.

At points of finite discontinuity in the function, the Fourier series representation converges to the arithmetic average value of the function to either side of the point of finite discontinuity. As the number of terms (N) in the Fourier series is increased, the mean-square error between the representation and the given function decreases and the approximation to the given function improves everywhere except in the immediate vicinity of a finite discontinuity. However, in the neighborhood of points of discontinuity in $f(t)$ the Fourier series representation fails to converge even though the mean-square error in the representation approaches zero. Hence even though the number of terms is increased, the representation exhibits overshoot near points of discontinuity of $f(t)$. This behavior is known as the Gibbs phenomenon, after its discoverer, and is shown in Fig. 2.33 for a periodic sawtooth waveform. Note that the overshoot peak moves closer to the point of discontinuity as more terms in the series are added. Even though the mean-square error approaches zero in the limit for large n, this amplitude overshoot approaches a value approximately 9 percent greater than the amplitude of the sawtooth waveform. Therefore "convergence in mean-square" implies convergence in amplitude in the neighborhood of points of continuity of $f(t)$ only.

---

† Let $f^N(t) = \sum_{n=1}^{N} f_n(t) = \lim_{N \to \infty} f^N(t)$ over $(0, T)$. If for each $\epsilon > 0$ there exists some integer M such that for every t in $(0, T)$, $|f^N(t) - f(t)| < \epsilon$, $M \le N$, we say that $f^N(t)$ converges *uniformly* to $f(t)$ in $(0, T)$. If $\sum_{n=1}^{N} |f_n(t)|$ converges uniformly, then $f^N(t)$ converges *absolutely and uniformly* to $f(t)$ in $(0, T)$.

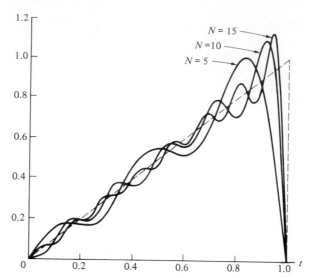

Fig. 2.33 Gibbs phenomenon for a periodic sawtooth waveform.

2.21 SUMMARY

In a systems approach, an input signal and the resulting response are used to characterize the behavior of a given system. For our purposes a signal is defined to be a single-valued function of time. This function may be complex-valued for mathematical convenience but our observations in practice are real-valued.

Signals may be classified into basic categories by their finite energy or power. Other useful categories include those of periodic or aperiodic signals and random or deterministic signals.

Mathematically, a system is a rule for mapping inputs into outputs. If the system is linear, superposition applies. If the system response is not dependent on absolute values of time, but only on time differences, the system is time-invariant.

A useful analogy may be made between vectors and signals. Use of orthogonality allows the representation of a given vector or signal by a linearly independent set of known vectors or signals. Only the coefficients need to be determined for a given case. Many different sets of orthogonal functions or vectors may be chosen to represent a given signal or vector.

Operations of particular interest are those which can be described by linear differential equations with constant coefficients. The Fourier series proves very useful because the exponential form repeats under differentiation. The complex exponential and trigonometric functions yield coefficients in terms of frequency which retain a correct measure of energy and power. Parseval's theorem equates the energy and power in terms of frequency to that expressed in terms of time.

The Fourier series representation over an interval converges in an integral-square (energy) sense at points of continuity in the signal. Signals with a finite energy rate (power) may be handled by dividing by the interval. This is useful in representing periodic signals if the interval is equated to the period. Such representations converge in a mean-square sense at points of continuity in the signal.

A plot of the Fourier coefficients as a function of frequency is called the spectrum of the signal. The spectrum of a periodic signal consists of a set of equally spaced lines whose heights indicate the relative weights of each frequency component. A signal meeting certain conditions, known as the Dirichlet conditions, can always be described uniquely in terms of both time and frequency.

Singularity functions are simple mathematical functions which may not be finite everywhere or do not have finite derivatives of all orders. The unit impulse function is of particular importance in this class of functions and can be defined in terms of its integral properties. The unit impulse function allows us to handle functions which exist only at a point and yet exert a dominant influence on the integration process. The unit step function is the indefinite integral of the unit impulse function.

A linear time-invariant system is determined uniquely by its transfer function or by its impulse response. The transfer function is a complex-valued function which can be found at a frequency by applying a complex exponential of fixed frequency to the input and dividing the steady state output by this input. The impulse response can be found by applying a unit impulse to the input and recording the output waveform. By use of linearity, it may also be found by applying a unit step and differentiating the output waveform.

The output of a linear time-invariant system has only those frequencies which are present at the input. The magnitude and phase of each frequency term is modified by the system transfer function. Systems which are not linear or time-invariant can generate new frequency (harmonic) content. The Fourier spectrum is a very sensitive indicator of harmonic content.

Computation of Fourier coefficients from M numerical data points limits the computation to $M/2$ harmonic terms. Higher harmonic terms are reflected about $M/2$ with conjugate symmetry and affect the numerical accuracy of the computed coefficients. This is called aliasing. Effects of aliasing are minimized by a rapid decrease in Fourier coefficient magnitudes with increasing frequency and by choosing more sample points per interval. A bound on the rate of decrease of Fourier coefficients with increasing frequency can often be estimated from a knowledge of the signal waveform.

Selected References for Further Reading

1. A. Papoulis. *Circuits and Systems: A Modern Approach.* New York: Holt, Rinehart and Winston, 1980.
   A terse introduction to the Fourier series with added emphasis on the discrete series and aliasing.

2. B. P. Lathi. *Signals, Systems and Communication*. New York: John Wiley, 1965. Chapter 3 is a treatment of the analogy between signal and vector representations using a minimization of mean-square error.

3. D. K. Cheng. *Analysis of Linear Systems*. Reading, Mass.: Addison-Wesley, 1959. Chapter 5 has a good discussion of the Fourier series, minimization of mean-square error, and convergence properties.

4. C. M. Close. *The Analysis of Linear Circuits*. New York: Harcourt, Brace & World, 1966.
   Chapter 9 has a good discussion of the Fourier series with examples and applications in linear circuits and the use of the frequency transfer function for periodic inputs.

5. H. S. Carslaw. *Theory of Fourier's Series and Integrals*. London: Cambridge University Press, 1930. (Reprinted by Dover Publications, New York, 1952.)
   A classic treatment, at a more advanced level, of the trigonometric Fourier series and its application to problems in heat transfer.

6. E. A. Guillemin. *The Mathematics of Circuit Analysis*. New York: John Wiley, 1949. A more detailed and advanced treatment of the Fourier series, particularly on the topics of convergence and the Gibbs phenomenon.

Problems

2.2.1 Classify the following signals as energy signals or power signals, and find the normalized energy or normalized power of each. (All signals are defined over $-\infty < t < \infty$.)

a) $\cos t + 2 \cos 2t$
b) $\exp(-|t|)$
c) $\exp(j2\pi t)$
d) $\exp(-|t|) \cos 2t$

2.3.1 Let $f(t)$ be the input to a given system and $g(t)$ be the corresponding output. The input-output relationships of several systems are given below. Classify the systems into one or more of the categories discussed in Section 2.3.

a) $g(t) = 1 + f(t + 1)$
b) $g(t) = 2tf(t)$
c) $g(t) = 2tf^2(t)$
d) $g(t) = \int_0^t f(\tau)\, d\tau$.

★ **2.4.1** Two vectors, expressed in the Cartesian coordinate system, are $\mathbf{A} = 3\mathbf{x}_1 + 4\mathbf{x}_2 + 5\mathbf{x}_3$; $\mathbf{B} = \mathbf{x}_1 + 3\mathbf{x}_2 - 3\mathbf{x}_3$.

a) Show that $\mathbf{A}, \mathbf{B}$ are mutually orthogonal.
b) Calculate the length of each vector.

★ **2.4.2** A three-dimensional coordinate system is defined by the following three mutually orthogonal vectors: $\boldsymbol{\phi}_1, \boldsymbol{\phi}_2, \boldsymbol{\phi}_3$, where: $\boldsymbol{\phi}_1 = \mathbf{x}_1 + \mathbf{x}_2$; $\boldsymbol{\phi}_2 = \mathbf{x}_1 - \mathbf{x}_2$; $\boldsymbol{\phi}_3 = \mathbf{x}_3$.

a) Show that $\boldsymbol{\phi}_1, \boldsymbol{\phi}_2, \boldsymbol{\phi}_3$ are mutually orthogonal.
b) Represent the vectors $\mathbf{A}$ and $\mathbf{B}$ of Problem 2.4.1 in terms of this set of orthogonal vectors.

★ **2.5.1** Two signals are orthogonal over an interval (t_1, t_2).

    a) Show that the energy of the sum of these two signals is equal to the sum of their energies.

    b) Extend your result to the case of n mutually orthogonal signals.

★ **2.5.2** The first term in the representation of $f(t)$ in Example 2.5.1 is $(4/\pi) \sin \pi t$. Retaining only this term, the error is

$$\epsilon_1(t) = f(t) - (4/\pi) \sin \pi t.$$

    a) Show that this error is orthogonal to the function $\sin \pi t$ over the interval $(0, 2)$.

    b) Show that the error energy is given by the difference in energies of $f(t)$ and of $(4/\pi) \sin \pi t$.

★ **2.5.3** A given set of functions, $\phi_n(t)$, is shown in Fig. P–2.5.3.†

    a) Show that these functions form an orthogonal set over the interval $(0, 1)$. Is the set an orthonormal set?

    b) Represent the given signal $f(t) = 2t$ over the interval $(0, 1)$ using this set of orthogonal functions.

    c) Sketch $f(t)$ and the representation of $f(t)$ on the same graph and compare.

    d) Compute the energy in each term of the series and the error energy remaining after each term is included.

$$\phi_0(t)$$

Fig. P–2.5.3.

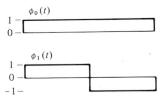

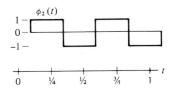

★ **2.5.4** A given set of three functions is‡

$$\phi_0(t) = 1; \qquad \phi_1(t) = t; \qquad \phi_2(t) = \tfrac{3}{2}t^2 - \tfrac{1}{2}.$$

    a) Show that these functions are mutually orthogonal over the interval $(-1, 1)$.

    b) Represent the signal $f(t) = |t|$ over the interval $(-1, 1)$ using this set of functions.

---

† These are selected functions from a larger set of orthogonal functions called Walsh functions. Designating members of the set by $w_n(t)$, where n is the number of zero crossings in the open interval, we find the above set is formed by $w_0(t)$, $w_1(t)$, $w_3(t)$.

‡ These functions are the first three members of a set of orthogonal functions called Legendre polynomials.

★ **2.5.5** Two given functions are:

$$\phi_1(t) = e^{-|t|}; \qquad \phi_2(t) = 1 - Ae^{-2|t|}.$$

a) Determine the constant A such that $\phi_1(t)$ and $\phi_2(t)$ are orthogonal over the interval $-\infty$ to $+\infty$.

b) Repeat (a) for the interval $(-1, 1)$.

2.7.1 The complex exponential Fourier series of a given signal $f(t)$ over an interval $(0, T)$ is

$$f(t) = \sum_{n=-\infty}^{\infty} \frac{3}{4 + (\pi n)^2} e^{jn\pi t}.$$

a) Determine the numerical value of T.

b) What is the mean (average) value of $f(t)$ over the given interval?

c) The component of $f(t)$ at a certain frequency can be expressed as $a \cos 3\pi t$. Determine the numerical value of the constant a.

2.7.2 Find the exponential Fourier series representation for the signal $f(t) = 2t$ over the interval $(0, 1)$.

2.7.3 The expression for the coefficients in the exponential Fourier series for a given $f(t)$ over (t_1, t_2) is given by Eq. (2.36). A new function, $g(t)$, is formed from $f(t)$ using the operations below. The series coefficients for $g(t)$ are designated by G_n. Determine the relations for the G_n in terms of the F_n for each of the following:

a) $g(t) = f(2t)$,

b) $g(t) = f(t - 2)$,

c) $g(t) = e^{j\omega_0 t}f(t)$.

2.8.1 A transformer with a rotatable primary and two fixed secondaries is shown in Fig. P–2.8.1.

a) Assuming that the voltage induced in the secondaries is proportional to the projection of the primary on the secondaries, show that the voltage appearing at aa' can be expressed as $A\,\mathcal{R}e\{e^{j\theta}\} \cos \omega_0 t$, and that the voltage appearing at bb' can be expressed as $B\,\mathcal{I}m\{e^{j\theta}\} \cos \omega_0 t$.

b) If (a) above is true, and $A = B$, then the voltages aa', bb' form an ordered pair of signals which can represent the circular position described by $e^{j\theta}$. (This principle is used in ac angular positioning systems.) Demonstrate that an exact replica of the system shown in Fig. P–2.8.1 will position itself uniquely to the identical angular position as the original shaft angle. What happens if one pair of wires is reversed? (Examine all possible combinations.)

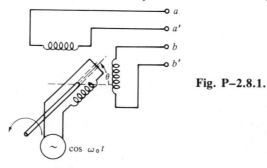

Fig. P–2.8.1.

2.9.1 Represent the signal $f(t) = e^{-t}$ over the interval $(0, 1)$ using:

  a) the exponential Fourier series.
  b) the trigonometric Fourier series.
  c) Compare your results using Eqs. (2.49)–(2.51).

2.9.2 a) Determine the trigonometric Fourier series representation for the signal $f(t) = 2t$ over the interval $(0, 1)$.
  b) Compare your result with that of Drill Problem 2.9.2.

2.10.1 A sine-wave oscillator, intended to generate a one-volt peak sine wave at ω_0 radians per second, is found to generate the following waveform:

$$f(t) = 0.900 \sin \omega_0 t + 0.300 \sin 3\omega_0 t.$$

  a) Determine the error energy present in the oscillator waveform over one period if $\omega_0 = 4\pi$ radians per second.
  b) Dividing this energy by the period gives an averaged error energy or "mean-square error." Compute the mean-square error for the oscillator.

2.10.2 Calculate the mean-square error resulting in the approximation of a two-volt peak-to-peak square wave by only its fundamental sinusoidal component. Assume that the average value is zero. Does your result depend on the period of the square wave? (*Hint:* Use results of Example 2.5.1 and Problem 2.10.1.)

2.10.3 A common "rule of thumb" used in the design of amplifiers for square waves is that the amplifier should pass all harmonics up to the tenth in order to obtain good reproduction of the waveform. Compute the corresponding mean-square error as a percentage of the average power in the waveform. (Assume zero average value.)

2.10.4 Show that if a periodic signal satisfies the half-wave symmetry condition $f(t) = -f(t \pm T/2)$, all even harmonic amplitudes in the Fourier series will vanish, with the possible exception of the $n = 0$ term.

c† **2.11.1** Determine the minimum number of harmonic terms that must be retained in Problem 2.9.1(a) to include 99% of the energy of $f(t)$ in the given interval.

2.11.2 Show that Parseval's theorem for the trigonometric Fourier series is

$$\frac{1}{T} \int_0^T |f(t)|^2 \, dt = a_0^2 + \frac{1}{2} \sum_{n=1}^{\infty} (a_n^2 + b_n^2).$$

2.12.1 a) Determine the frequency transfer function of the system shown in Fig. 2.8.
  b) Sketch the magnitude and phase as a function of frequency and label the points at which the magnitude is down to $1/\sqrt{2}$ its highest value.
  c) Compare the results of this system to that of Example 2.12.1.

2.12.2 The differential equation relating the output voltage, $g(t)$, to the input voltage, $f(t)$, of the system shown in Fig. P–2.12.2 is

$$LC\frac{d^2g}{dt^2} + g(t) = f(t).$$

Find the frequency transfer function of this system.

---

† Problems intended to be solved using numerical methods are designated with c in the margin.

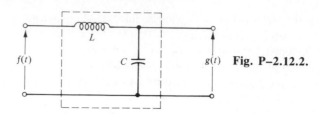

Fig. P–2.12.2.

2.13.1 A periodic gate signal is applied to the input of the system whose transfer function is shown in Fig. P–2.13.1. It is given that

$$f(t) = \sum_{n=-\infty}^{\infty} \frac{A\tau}{T} \frac{\sin (n\omega_0\tau/2)}{n\omega_0\tau/2} e^{jn\omega_0 t},$$

where τ is the pulse width and $T = 2\pi/\omega_0$ is the period. Determine the output of the system.

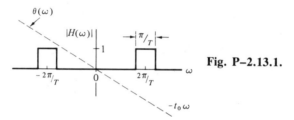

Fig. P–2.13.1.

2.13.2 Let $R = 100$ kΩ, $C = (5/2\pi)$ μF in the system shown in Fig. 2.16 and let the input $f(t)$ be the signal of Example 2.13.1.

a) Determine the output, $g(t)$. Express your answer in terms of cosines and truncate the series at 4 Hz.

b) Compute the percentage of the mean-square value of the input remaining at the output of the system.

2.14.1 A periodic signal $f(t)$ is shown in Fig. P–2.14.1. For $a = 1$, this signal is a periodic square wave.

a) Find the trigonometric Fourier series of $f(t)$ for arbitrary values of a.

b) Determine the effect on the amplitude of the second harmonic of $f(t)$ when there is a very small departure from perfect square-wave symmetry. To do this, let $a = 1 - \epsilon$, where $\epsilon \ll 1$, and find the second harmonic dependence on ϵ. The parameter a is sometimes called the signal symmetry factor (where $0 \leq a \leq 1$).

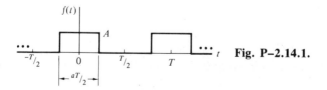

Fig. P–2.14.1.

2.14.2 Sometimes it is advantageous to truncate or "clip" a sinusoidal waveform. This generates harmonic content and can be used, for example, to triple the frequency of a given sinusoidal signal. A truncated sinusoidal waveform is shown in Fig. P–2.14.2. Note that $B < A$, and $\theta_0 = \sin^{-1}(B/A)$.

a) Determine the amplitude of the third harmonic of the truncated waveform.

b) Calculate the percentage of the amplitude of the original sinusoid present in the third harmonic for $B = A/2$.

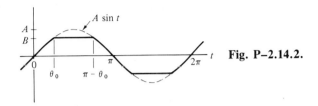

Fig. P–2.14.2.

2.14.3 Another possibility for generation of harmonic content is to use a nonlinear device which follows the characteristic shown in Fig. P–2.14.3. For this device,

a) Determine the magnitude of the third harmonic of the resultant waveform.

b) Solve for the optimum angle θ_0 which yields maximum third harmonic magnitude.

c) Compare the resulting maximum third harmonic magnitude with the results of Problem 2.14.2(b).

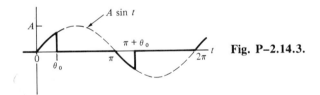

Fig. P–2.14.3.

2.14.4 a) Solve for the optimum clipping angle, θ_0, in Problem 2.14.2 which yields maximum third harmonic magnitude for a given input sinusoidal amplitude A.

b) Calculate the total harmonic distortion (THD) for Problem 2.14.2(b).

2.15.1 Sketch the Fourier spectrum of the exponential and trigonometric series for the waveform in Problem 2.7.1.

2.15.2 A given periodic signal (in volts) $f(t) = 4 \sin 8\pi t + 2 \cos 12\pi t$ is developed across a one-ohm resistor.

a) Determine the highest fundamental frequency possible for this signal.

b) Find the total mean (average) power in $f(t)$.

c) Calculate the percentage of the total mean power contained in each harmonic up to the fifth harmonic using the fundamental of part (a).

d) Sketch the complex Fourier spectrum for $f(t)$.

2.15.3 The output of a simple rectifier circuit with a smoothing capacitor and a resistive load is shown in Fig. P–2.15.3.

a) Find the exponential Fourier series for this waveform.
b) Sketch the Fourier magnitude line spectrum.
c) Calculate the percentage of the total average power contained in the fundamental.

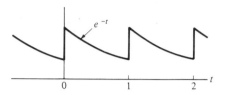

Fig. P–2.15.3.

c★ **2.16.1** Using numerical methods and dividing the period into $5°$ increments, compute and plot the magnitude line spectrum of the trigonometric Fourier series of Problem 2.14.2(b) for the first 20 harmonics.

c★ **2.16.2** Compute and plot the magnitude line spectrum of the waveform of Problem 2.14.3(b) using the directions of Problem 2.16.1. At points of finite discontinuity, use the average of the values to either side of the discontinuity.

c★ **2.16.3** Compute and plot the magnitude line spectrum of the trigonometric Fourier series for the waveform

$$f(t) = \begin{cases} 4 \sin 20\pi t & 0.0 \le t \le 0.5, \\ 0 & 0.5 \le t \le 2.0, \end{cases}$$

when repeated periodically with $T = 2$. Choose 100 sample points per period and plot for the first 50 harmonic terms.

c★ **2.17.1** Compute the exponential Fourier series coefficients for the first ten positive harmonic frequencies and for the first ten negative harmonic frequencies for the waveform described by

$$f(t) = 5 \exp(-t), \qquad 0 \le t \le 2\pi.$$

Use twenty sample points over the period $T = 2\pi$ and follow the instruction of Problem 2.16.2 at the finite discontinuities.

c★ **2.17.2** Compute and plot the magnitude and angle spectra of

$$f(t) = (1/t) - (1/4), \qquad 0 < t \le 5,$$

using the exponential Fourier series. Use 50 equally spaced sample points per period and plot over the integer frequency increments: 0, 1, 2, . . . , 50.

2.18.1 Show that a periodic train of unit impulse functions, spaced T seconds apart, may be represented by the following trigonometric Fourier series:

$$\frac{1}{T} + \frac{2}{T} \sum_{n=1}^{\infty} \cos n\omega_0 t, \quad \text{where} \quad \omega_0 = 2\pi/T,$$

if one of the impulse functions is at $t = 0$.

2.18.2 Evaluate the following integrals:

a) $\int_{-\infty}^{\infty} (t^3 + t^2 + t + 1) \, \delta(t - 3) \, dt$

b) $\int_{-\infty}^{\infty} \delta(t + 3) e^{-t} dt$

c) $\int_{-\infty}^{\infty} \delta(t - 2) \cos \pi(t - 3) \, dt$

d) $\int_{-\infty}^{\infty} (t^3 + 4) \, \delta(1 - t) \, dt$

e) $\int_{-\infty}^{\infty} (t^2 + 3) \, \delta(3t - 9) \, dt$

f) $\int_{-\infty}^{\infty} (t^2 + 2) \, \delta(\tfrac{1}{2}t - 1) \, dt$

2.18.3 Evaluate the following integrals:

a) $\int_{-\infty}^{\infty} tu(2 - t) \, u(t) \, dt$

b) $\int_{-\infty}^{\infty} [\delta(t) + u(t) - u(t - 2)] \, dt$

c) $\int_{-\infty}^{\infty} \delta(t - t_0) \, u(t - t_1) \, dt$

d) $\int_{-\infty}^{t} u(\tau - 1) \, d\tau$

2.18.4 A given function $f(t)$ is described by

$$f(t) = \begin{cases} K(t - \epsilon)^2/\epsilon^3 & -\epsilon < t < \epsilon \\ 0 & \text{elsewhere} \end{cases}.$$

Show that as $\epsilon \to 0$, $f(t)$ can be used as a limiting sequence to define a unit impulse at $t = 0$ if $K = 3/8$.

2.19.1 A rectangular pulse of height $1/\Delta\tau$ and of width $\Delta\tau$ is applied to the input terminals of a RC low-pass filter (cf. Fig. 2.16) at $t = 0$.

a) Write an expression for the output signal.

b) Take the limit as $\Delta\tau \to 0$ and show that the impulse response of the filter is

$$h(t) = \frac{1}{RC} e^{-t/RC} u(t).$$

c) Determine the dimensions of the voltage transfer function, $H(\omega)$, and the impulse response function, $h(t)$.

2.19.2 The impulse response of a given system is $h(t) = 0.2e^{-0.2t} u(t)$.

a) Determine the unit step response of the system.

b) A low-frequency periodic (symmetric) square wave with a 2-volt P/P amplitude is applied to the system. Compute the mean-square error between the input and the output of the system when the period of the square wave is 100 sec.

c) Repeat part (b) when the period is 1 sec.

★ **2.20.1** Estimate the minimum rate of convergence (for large n) of the Fourier coefficients for the periodic waveforms shown in Fig. P–2.20.1.

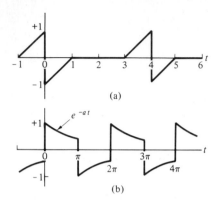

(a)

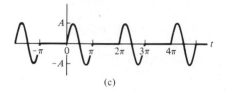

(c)

(b)

Fig. P–2.20.1.

★ **2.20.2** Estimate the minimum rate of convergence (for large n) of the Fourier coefficients for the waveforms described in

a) Problem 2.14.1;
b) Problem 2.14.2.

CHAPTER 3

THE FOURIER TRANSFORM AND APPLICATIONS

In Chapter 2 we found how to represent a given function in terms of a complex exponential (or trigonometric) series over a finite interval. For the particular case in which the function is periodic, this representation can be extended over the entire interval $(-\infty, \infty)$.

Now let us consider another limiting case: the frequency domain description of one signal pulse. Our approach will be to consider a periodic signal and represent it in terms of complex exponentials with a period approaching infinity to yield, at least formally, a representation for the aperiodic signal. This representation will completely describe the given aperiodic signal *for all values of time*.

3.1 REPRESENTATION OF AN APERIODIC FUNCTION OVER THE ENTIRE REAL LINE

In Chapter 2 we discussed the Fourier exponential series representation of a given time function. It was observed that as the period T is made larger, the fundamental frequency becomes smaller and the frequency spectrum becomes more dense while the amplitude of each frequency component decreases. The *shape* of the spectrum, however, remains unchanged with varying T.

Suppose we are given an aperiodic function $f(t)$, as shown in Fig. 3.1(a). We wish to represent this function as a sum of exponential functions over the entire interval $(-\infty, \infty)$. For this purpose, we construct a new periodic function $f_T(t)$ with period T so that the function $f(t)$ is forced to repeat itself completely every T seconds.

The original function can be obtained back again by letting $T \to \infty$, that is,

$$\lim_{T \to \infty} f_T(t) = f(t). \qquad (3.1)$$

In other words, taking the limit as the period approaches infinity effectively removes all the nearest neighbors of the pulse $f(t)$ to infinity. To make this

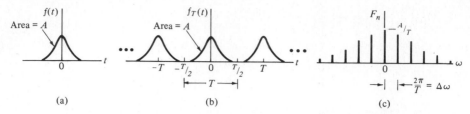

Fig. 3.1 Generation of a periodic function and its line spectrum.

method effective, we adopt the following procedure: (1) the desired pulse is located at the origin, and (2) the limit as $T \to \infty$ is taken symmetrically, measured from the origin.

The new function $f_T(t)$ is a periodic function and consequently can be represented by an exponential Fourier series. This exponential series is written as

$$f_T(t) = \sum_{n=-\infty}^{\infty} F_n e^{jn\omega_0 t}, \tag{3.2}$$

where

$$F_n = \frac{1}{T} \int_{-T/2}^{T/2} f_T(t) e^{-jn\omega_0 t} \, dt \tag{3.3}$$

and

$$\omega_0 = 2\pi/T. \tag{3.4}$$

Before taking any limiting operation, however, we have to adjust things a little so that the magnitude components of the F_n don't all go to zero as the period is increased. We make the following changes:

$$\omega_n \overset{\Delta}{=} n\omega_0, \tag{3.5}$$

$$F(\omega_n) \overset{\Delta}{=} TF_n. \tag{3.6}$$

When we use these definitions, Eqs. (3.2) and (3.3) become

$$f_T(t) = \sum_{n=-\infty}^{\infty} \frac{1}{T} F(\omega_n) e^{j\omega_n t}, \tag{3.7}$$

$$F(\omega_n) = \int_{-T/2}^{T/2} f_T(t) e^{-j\omega_n t} \, dt. \tag{3.8}$$

The spacing between adjacent lines in the line spectrum of $f_T(t)$ is (see Fig. 3.1c)

$$\Delta\omega = 2\pi/T. \tag{3.9}$$

Using this relation for T in Eq. (3.7), we get the alternate form

$$f_T(t) = \sum_{n=-\infty}^{\infty} F(\omega_n) e^{j\omega_n t} \frac{\Delta\omega}{2\pi}. \tag{3.10}$$

Now as T becomes very large, $\Delta\omega$ becomes small and the spectrum becomes denser (i.e., the lines become very close and the frequency components within a given frequency range increase). In the limit, as $T \to \infty$, the discrete lines in the spectrum of $f_T(t)$ merge and the frequency spectrum becomes continuous. Mathematically, the infinite sum in Eq. (3.10) becomes the ordinary Riemann integral so that

$$\lim_{T\to\infty} f_T(t) = \lim_{T\to\infty} \frac{1}{2\pi} \sum_{n=-\infty}^{\infty} F(\omega_n) e^{j\omega_n t}\, \Delta\omega$$

becomes

$$f(t) = \frac{1}{2\pi} \int_{-\infty}^{\infty} F(\omega) e^{j\omega t}\, d\omega. \tag{3.11}$$

In a similar manner, Eq. (3.8) becomes

$$F(\omega) = \int_{-\infty}^{\infty} f(t) e^{-j\omega t}\, dt. \tag{3.12}$$

The results expressed in Eqs. (3.11) and (3.12) are conmonly referred to as the *Fourier transform pair*. Equation (3.12) is known as the direct or forward Fourier transform of $f(t)$ (more commonly, just the *Fourier transform*). Equation (3.11) is known as the *inverse Fourier transform*. Symbolically, this suggests the following operator notation:

$$F(\omega) = \mathscr{F}\{f(t)\} = \int_{-\infty}^{\infty} f(t) e^{-j\omega t}\, dt, \tag{3.13}$$

$$f(t) = \mathscr{F}^{-1}\{F(\omega)\} = \frac{1}{2\pi} \int_{-\infty}^{\infty} F(\omega) e^{j\omega t}\, d\omega. \tag{3.14}$$

It is also useful to note that the complex exponential Fourier series coefficients can be evaluated in terms of the Fourier transform by combining Eqs. (3.5) and (3.6) to give

$$F_n = \frac{1}{T} F(\omega) \Big|_{\omega = n\omega_0}, \tag{3.15}$$

provided that $F(\omega)$ is finite at $\omega = n\omega_0$† and that $f(\pm T/2) \approx 0$.

Drill Problem 3.1.1   Find the Fourier transform of $f(t) - \exp(-at)\, u(t)$, $a > 0$.‡
Answer. $1/(a + j\omega)$.

---

† A sufficient condition for this to occur is that the energy in $f(t)$ is finite; see Section 3.3.

‡ For those acquainted with the Laplace transform, the Fourier transform can be found from the Laplace transform by setting $s = j\omega$ if *both* of the following conditions are true: (1) $f(t) = 0$ for $t < 0$; (2) $f(t)$ has finite energy, that is, $\int_{-\infty}^{\infty} |f(t)|^2\, dt < \infty$. [See, for example, Reference 1 at the end of this chapter.]

The Fourier transform can be visualized as the representation of a given signal in terms of an infinite sum of complex exponentials each weighted by $F(\omega)$ *df* [cf. Eq. (3.11)]. The concept of a phasor representation of the complex exponential and the addition of phasors using the rules for vector addition, as described in Section 2.8, aids in the visualization. To illustrate this, consider a signal $f(t)$ for which $F(\omega)$ is a constant for all ω. At $t = 0$, all of the complex exponential terms formed by $[F(\omega)\ df]e^{j\omega t}$ have identical magnitudes and zero phase angle. Because there is an infinite number of these, the real part of the resultant is infinitely large. At values of time for which $t \neq 0$, however, the complex exponential terms have identical magnitudes but phases that are proportional to frequency. The net result is that they add up to zero. Therefore we conclude that this signal $f(t)$ can be described by a function that is infinite at $t = 0$ and is zero for $t \neq 0$, suggesting description by an *impulse* function. Reference to Fourier transform pair 8 in Table 3.1 (and illustrated in Fig. 3.3) will show this conclusion is correct.

Reconstruction of $f(t)$ from a set of complex exponentials is useful for visualization purposes. However, these results are qualitative and must be visualized at various values of time for us to draw general conclusions. A convenient manner to present the same information more precisely and independent of time—but without the intuitive aspects—is to make a graph of $F(\omega)$ versus ω. This is investigated in the next section.

3.2   THE SPECTRAL DENSITY FUNCTION

Equation (3.14) represents $f(t)$ as a continuous sum of exponential functions with frequencies lying in the interval $(-\infty,\infty)$. The relative amplitude of the components at any frequency ω is proportional to $F(\omega)$. If the signal $f(t)$ represents a voltage, $F(\omega)$ has the dimensions of voltage multiplied by time. Because frequency has the dimensions of inverse time, we can consider $F(\omega)$ as a voltage-density spectrum or, more generally speaking, it is known as the *spectral-density function* of $f(t)$.

The *area* under the spectral-density function $F(\omega)$ has the dimensions of voltage (current). Each point on the $F(\omega)$ curve contributes nothing to the representation of $f(t)$; it is the area that contributes. On the other hand, each point does indicate the *relative weighting* of each frequency component. The contribution of a given frequency band to the representation of $f(t)$ may be found by integrating to find the desired area.

In contrast to this, a periodic waveform has all its amplitude components at discrete frequencies. At each of these discrete frequencies there is some definite contribution; to either side there is none. It follows, then, that to portray the amplitude components of a periodic waveform on a spectral-density graph requires a representation with area equal to the respective amplitude components yet occupying zero frequency width. We recognize that this can be done formally by representing each amplitude component of the periodic function by an impulse function. The area (weight) of the impulse is equal to the amplitude component and the position of the impulse is determined by the particular discrete frequency.

Summarizing, a signal of finite energy can be described by a continuous spectral-density function. This spectral-density function is found by taking the Fourier transform of the signal. A periodic signal of finite average power can be described either by a set of lines on a spectral graph or by a set of impulse functions on a spectral-density graph. Each impulse on the latter graph has an area corresponding to the height of each line, respectively, on the former graph.

For convenience in drawing spectral-density graphs of periodic functions, the height of each impulse is often made proportional to its area. Though technically not correct, this convention is very convenient because it makes the spectral-density plots resemble the spectral plots.

The above representation works out fine for spectral functions expressed in terms of inverse time (frequency). We choose to use radian frequency, ω. By the scaling property of the impulse function [see Eq. (2.99)], we can write

$$\delta(f) = 2\pi \, \delta(\omega),$$

and therefore we multiply each impulse function by 2π if it is expressed in radian frequency.

Let us find the Fourier transform (spectral density) of the unit gate function which is shown in Fig. 3.2(a) and is defined by[†]

$$\text{rect}(t) \triangleq \begin{cases} 1 & |t| < \tfrac{1}{2} \\ 0 & |t| > \tfrac{1}{2} \end{cases} \tag{3.16}$$

Using Eq. (3.12), we have

$$F(\omega) = \int_{-\infty}^{\infty} \text{rect}(t) e^{-j\omega t} \, dt = \int_{-1/2}^{1/2} e^{-j\omega t} \, dt$$

$$= (e^{j\omega/2} - e^{-j\omega/2})/j\omega = \frac{\sin(\omega/2)}{(\omega/2)}$$

or, using Eq. (2.82),

$$\mathscr{F}\{\text{rect}(t)\} = \text{Sa}(\omega/2). \tag{3.17}$$

---

† Some authors use the symbols $G(t)$ or $\pi(t)$ for the unit gate function.

This result is shown in Fig. 3.2(b).

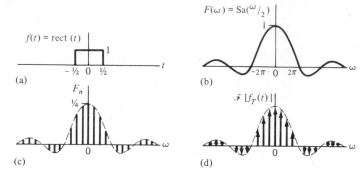

Fig. 3.2 (a) The unit gate function and (b) its Fourier transform.
(c) The line spectrum and (d) spectral density of the periodic
gate function.

Example 3.2.1   Determine tne coefficients of the Fourier exponential series (the
F_n) for the unit gate function if it is repeated every 4 seconds; plot on both a
spectral and a spectral-density graph.

Solution.   Using Eq. (3.15), we have

$$F_n = \frac{1}{T} F(\omega) \bigg|_{\omega = n\omega_0} = \frac{1}{4} \text{Sa}(n\pi/4).$$

The line spectrum and the spectral density are shown in Fig. 3.2(c) and (d).

Drill Problem 3.2.1   Find the Fourier transform of the two-sided exponential
$f(t) = \exp(-a|t|)$, $a > 0$.

Answer. $2a/(a^2 + \omega^2)$.

★ 3.3   EXISTENCE OF THE FOURIER TRANSFORM

Sufficient conditions for the existence of the Fourier series, known as the Dirichlet conditions, were given in Section 2.20. These conditions, appropriately
modified, also are sufficient for the existence of the Fourier transform:

1. $f(t)$ has only a finite number of maxima and minima in any finite time interval;
2. $f(t)$ has only a finite number of finite discontinuities in any finite time interval;
3. $f(t)$ is absolutely integrable; that is, $\int_{-\infty}^{\infty} |f(t)| \, dt < \infty$.

A somewhat weaker condition is $\int_{-\infty}^{\infty} |f(t)|^2 \, dt < \infty$. Because this condition
corresponds to our definition of an energy signal, we can state that the Fourier
transform can be used to uniquely represent any energy signal.

3.4　PARSEVAL'S THEOREM FOR ENERGY SIGNALS

The energy delivered to a one-ohm resistor is [Eq. (2.5)]:

$$E = \int_{-\infty}^{\infty} |f(t)|^2 \, dt = \int_{-\infty}^{\infty} f(t)f^*(t) \, dt. \tag{3.18}$$

We would like to express the energy in terms of the frequency components of $f(t)$. Using Eq. (3.11) in Eq. (3.18), we get

$$E = \int_{-\infty}^{\infty} f(t) \left[\frac{1}{2\pi} \int_{-\infty}^{\infty} F^*(\omega)e^{-j\omega t} \, d\omega \right] dt.$$

Interchanging the order of integration on t and ω,†

$$E = \frac{1}{2\pi} \int_{-\infty}^{\infty} F^*(\omega) \left[\int_{-\infty}^{\infty} f(t)e^{-j\omega t} \, dt \right] d\omega. \tag{3.19}$$

But from Eq. (3.12), this is

$$E = \frac{1}{2\pi} \int_{-\infty}^{\infty} F^*(\omega)F(\omega) \, d\omega. \tag{3.20}$$

Combining Eqs. (3.18) and (3.20), we obtain what is known as *Parseval's theorem for energy signals*:‡

$$\int_{-\infty}^{\infty} |f(t)|^2 \, dt = \frac{1}{2\pi} \int_{-\infty}^{\infty} |F(\omega)|^2 \, d\omega. \tag{3.21}$$

From this, we can find the energy of a signal in either the time domain or the frequency domain.

In the MKS system of units, energy is measured in joules. From Eq. (3.21) we observe that the quantity $|F(\omega)|^2$ is an energy density having the units of joules per Hz (note that $\omega/2\pi$ is measured in Hz).

Drill Problem 3.4.1　Calculate the energy in the signal of Drill Problem 3.2.1 (across a one-ohm resistance) in both the time domain and the frequency domain.

Answer.　$1/a$ joules.

3.5　SOME FOURIER TRANSFORMS INVOLVING IMPULSE FUNCTIONS

The procedure for finding the Fourier transform of signals of finite energy using Eq. (3.13) is straightforward. In contrast, this is not always true for signals of infinite energy. Of particular interest to us are those cases which involve the

---

† This assumes that the integrals are absolutely convergent, i.e., that they are convergent when the integrands are replaced by their absolute values—(a sufficient condition).

‡ This relation is also known as Plancherel's theorem or Rayleigh's theorem.

unit impulse function. Our discussions will be brief; the interested reader is referred to the references for more detailed discussions.

3.5.1 The Impulse Function

Using the integral properties of the impulse function, the Fourier transform of a unit impulse, $\delta(t)$, is

$$\mathscr{F}\{\delta(t)\} = \int_{-\infty}^{\infty} \delta(t)e^{-j\omega t} \, dt = e^{j0} = 1. \tag{3.22}$$

If the impulse is time-shifted, we have

$$\mathscr{F}\{\delta(t - t_0)\} = \int_{-\infty}^{\infty} \delta(t - t_0)e^{-j\omega t} \, dt = e^{-j\omega t_0}. \tag{3.23}$$

It is evident from Eqs. (3.22) and (3.23) that an impulse function has a uniform magnitude spectrum over the entire frequency interval $(-\infty, \infty)$. This type of spectrum is called "white" in an analogy to white light. The phase spectrum of the time-shifted impulse is linear with a slope which is proportional to the time shift.

3.5.2 The Eternal Complex Exponential

We could expect that the spectral density of $e^{\pm j\omega_0 t}$ will be concentrated at $\pm \omega_0$. That this is the case is demonstrated below.

$$\mathscr{F}^{-1}\{\delta(\omega \mp \omega_0)\} = \frac{1}{2\pi} \int_{-\infty}^{\infty} \delta(\omega \mp \omega_0)e^{j\omega t} \, d\omega$$

$$= \frac{1}{2\pi} e^{\pm j\omega_0 t} \tag{3.24}$$

Taking the Fourier transform of both sides of Eq. (3.24), we have

$$\mathscr{F}\mathscr{F}^{-1}\{\delta(\omega \mp \omega_0)\} = \frac{1}{2\pi} \mathscr{F}\{e^{\pm j\omega_0 t}\}$$

or, by interchanging sides,

$$\mathscr{F}\{e^{\pm j\omega_0 t}\} = 2\pi \mathscr{F}\mathscr{F}^{-1}\{\delta(\omega \mp \omega_0)\},$$

$$\mathscr{F}\{e^{\pm j\omega_0 t}\} = 2\pi \delta(\omega \mp \omega_0). \tag{3.25}$$

This result should hardly surprise us because the complex exponential describes a phasor whose angular rate is ω_0. The spectral description of such a phasor is a line at this angular rate (which is greater than $\omega = 0$ if rotating in a positive direction, less than $\omega = 0$ if in a negative direction). The spectral-density description is an impulse at these angular rates and the factor of 2π arises because we are using radian frequency.†

In a more general sense, Eq. (3.25) may be used to define the unit impulse function. Note that for $\omega_0 = 0$, Eq. (3.25) simplifies to

$$\mathscr{F}\{1\} = 2\pi\delta(\omega). \tag{3.26}$$

It follows that any signal with a nonzero average value over the infinite interval $(-\infty, \infty)$ has an impulse in its spectral-density function at $\omega = 0$.

3.5.3 Eternal Sinusoidal Signals

The sinusoidal signals $\cos \omega_0 t$ and $\sin \omega_0 t$ can be written in terms of the complex exponentials using Euler's identities [see Eqs. (2.43) and (2.44)]. Their Fourier transforms can be found directly from Eq. (3.25):

$$\mathscr{F}\{\cos \omega_0 t\} = \mathscr{F}\{\tfrac{1}{2}e^{j\omega_0 t} + \tfrac{1}{2}e^{-j\omega_0 t}\}$$

$$= \pi\delta(\omega - \omega_0) + \pi\delta(\omega + \omega_0), \tag{3.27}$$

$$\mathscr{F}\{\sin \omega_0 t\} = \mathscr{F}\left\{\frac{1}{2j}e^{j\omega_0 t} - \frac{1}{2j}e^{-j\omega_0 t}\right\}$$

$$= [\pi\delta(\omega - \omega_0) - \pi\delta(\omega + \omega_0)]/j. \tag{3.28}$$

Graphs of the trigonometric functions $\cos \omega_0 t$ and $\sin \omega_0 t$ together with their spectral-density functions are shown in Fig. 3.3. Note that it is convenient to multiply Eq. (3.28) by the factor j so that the result is real-valued and requires only one graph.

3.5.4 The Signum Function and the Unit Step

The signum function, sgn (t), is that function which changes sign when its argument is zero:

$$\text{sgn}\,(t) = \frac{|t|}{t} = \begin{cases} 1 & t > 0 \\ 0 & t = 0. \\ -1 & t < 0 \end{cases} \tag{3.29}$$

The signum function has an average value of zero and is piece-wise continuous, but not absolutely integrable. In order to make it absolutely integrable we multiply sgn(t) by $e^{-a|t|}$ and then take the limit as $a \to 0$:

$$\mathscr{F}\{\text{sgn}\,(t)\} = \mathscr{F}\left\{\lim_{a\to 0}[e^{-a|t|}\,\text{sgn}\,(t)]\right\}.$$

---

† If written in terms of frequency in Hz, the factor of 2π disappears in Eqs. (3.25) and (3.26) because $\delta(\omega) = \delta(2\pi f) = [1/(2\pi)]\,\delta(f)$.

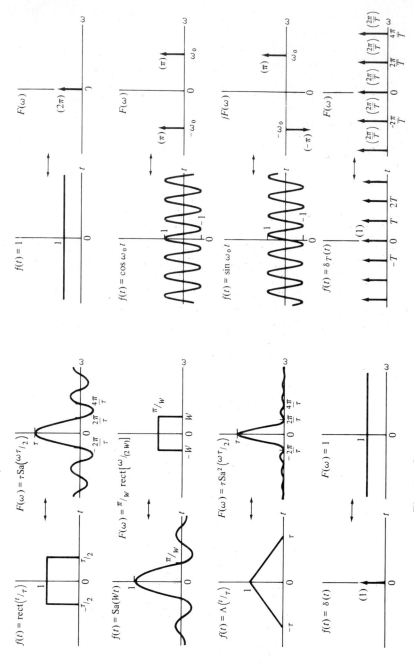

Fig. 3.3 Some functions of time and their spectral-density functions.

Interchanging the operations of taking the limit and integrating,

$$\mathcal{F}\{\text{sgn }(t)\} = \lim_{a \to 0} \left\{ \int_{-\infty}^{\infty} e^{-a|t|} \text{ sgn }(t) e^{-j\omega t} \, dt \right\}$$

$$= \lim_{a \to 0} \left\{ -\int_{-\infty}^{0} e^{(a - j\omega)t} \, dt + \int_{0}^{\infty} e^{-(a + j\omega)t} \, dt \right\}. \qquad (3.30)$$

Proceeding, we get

$$\mathcal{F}\{\text{sgn }(t)\} = \lim_{a \to 0} \left\{ \frac{-2j\omega}{a^2 + \omega^2} \right\}$$

$$= \frac{2}{j\omega}. \qquad (3.31)$$

The unit step function, expressed in terms of its average value and the signum function, is

$$u(t) = \tfrac{1}{2} + \tfrac{1}{2}\text{sgn }(t). \qquad (3.32)$$

The Fourier transform is

$$\mathcal{F}\{u(t)\} = \tfrac{1}{2}\mathcal{F}\{1\} + \tfrac{1}{2}\mathcal{F}\{\text{sgn }(t)\},$$

which, using Eqs. (3.26) and (3.31), becomes

$$\mathcal{F}\{u(t)\} = \pi \, \delta(\omega) + 1/j\omega. \qquad (3.33)$$

Therefore the spectral-density function of the unit step contains an impulse at $\omega = 0$ corresponding to the average value of $\tfrac{1}{2}$ in the step function. It also has all the high frequency components of the signum function reduced by one-half.

Note that Eq. (3.30) is the integral of a complex function taken symmetrically about the origin. Taking the integral of a complex function in a nonsymmetrical manner may lead to errors. For example, if one attempted to evaluate the Fourier transform of the unit step directly,

$$\lim_{a \to 0} \int_{0}^{\infty} e^{-at} e^{-j\omega t} \, dt = 1/j\omega,$$

which is not in agreement with the correct result in Eq. (3.33).

3.5.5   Periodic Functions

In Section 3.2 we discussed the characteristics of the spectral-density function of a periodic signal in a qualitative manner. Now we are prepared to formalize the relationship in a more quantitative way.

We can express a function $f(t)$ which is periodic with period T by its exponential Fourier series

$$f_T(t) = \sum_{n=-\infty}^{\infty} F_n e^{jn\omega_0 t} \qquad \text{where} \quad \omega_0 = 2\pi/T.$$

Taking the Fourier transform, we find

$$\mathscr{F}\{f_T(t)\} = \mathscr{F}\left\{ \sum_{n=-\infty}^{\infty} F_n e^{jn\omega_0 t} \right\}.$$

If we assume the operations of integration and summation can be interchanged,

$$\mathscr{F}\{f_T(t)\} = \sum_{n=-\infty}^{\infty} F_n \mathscr{F}\{e^{jn\omega_0 t}\}.$$

Using Eq. (3.25), we get

$$\mathscr{F}\{f_T(t)\} = 2\pi \sum_{n=-\infty}^{\infty} F_n \delta(\omega - n\omega_0). \tag{3.34}$$

Thus the Fourier transform (spectral density) of a periodic signal consists of a set of impulses located at the harmonic frequencies of the signal. The area (weight) of each impulse is 2π times the value of its corresponding coefficient in the exponential Fourier series. This result permits us to handle both periodic and nonperiodic functions in one unified treatment.

Example 3.5.1   Find the spectral-density function of an even periodic square wave whose average value is zero, whose period is two seconds, and whose peak-to-peak amplitude is A.

Solution.   We shall use known Fourier transform relations to obtain coefficients for the Fourier series and then find the Fourier transform of the series. For a rectangular pulse of unit width and height A, Eq. (3.17) gives

$$F(\omega) = A \, \text{Sa} \, (\omega/2).$$

Because $\omega_0 = \pi$ here, Eq. (3.15) becomes

$$F_n = \frac{1}{T} F(\omega)\big|_{\omega = n\omega_0} = (A/2) \, \text{Sa} \, (n\pi/2).$$

Writing the series and noting that the average value is zero, we get

$$f_T(t) = (A/2) \sum_{\substack{n=-\infty \\ n \neq 0}}^{\infty} \text{Sa} \, (n\pi/2) e^{jn\pi t}.$$

Using Eq. (3.34), the Fourier transform of this function is

$$\mathscr{F}\{f_T(t)\} = \pi A \sum_{\substack{n=-\infty \\ n \neq 0}}^{\infty} \text{Sa} \, (n\pi/2) \delta(\omega - \pi n).$$

The spectral-density function for this example is shown in Fig. 3.4. Note that the absence of an impulse at $\omega = 0$ is an indicator that the average value of the signal is zero.

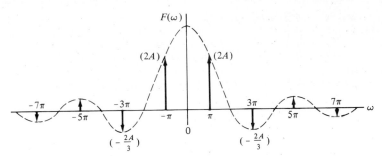

Fig. 3.4 The spectral-density function of a periodic square wave with zero average value.

A periodic function of considerable importance in later chapters is that of a periodic sequence of unit impulse functions (sometimes referred to as a "comb" function). For convenience, we write such a sequence with period T as

$$\delta_T(t) \triangleq \sum_{n=-\infty}^{\infty} \delta(t - nT). \tag{3.35}$$

Because this is a periodic function, we can express it in terms of a Fourier series by choosing $\omega_0 = 2\pi/T$,

$$\delta_T(t) = \sum_{n=-\infty}^{\infty} F_n e^{jn\omega_0 t},$$

where

$$F_n = \frac{1}{T} \int_{-T/2}^{T/2} \delta(t) e^{-jn\omega_0 t} \, dt = \frac{1}{T},$$

so that

$$\delta_T(t) = \frac{1}{T} \sum_{n=-\infty}^{\infty} e^{jn\omega_0 t}.$$

Thus the impulse train (comb) function contains components of the harmonic frequencies $\omega = 0, \pm\omega_0, \pm 2\omega_0, \ldots$ with equal weighting. Using Eq. (3.34), the Fourier transform of the unit impulse train function is

$$\mathcal{F}\{\delta_T(t)\} = \frac{2\pi}{T} \sum_{n=-\infty}^{\infty} \delta(\omega - n\omega_0)$$

$$= \omega_0 \sum_{n=-\infty}^{\infty} \delta(\omega - n\omega_0),$$

$$\mathcal{F}\{\delta_T(t)\} = \omega_0 \, \delta_{\omega_0}(\omega). \tag{3.36}$$

The Fourier transform of a periodic impulse train in the time domain gives an impulse train which is periodic in the frequency domain. The impulse train and its Fourier transform are shown in Fig. 3.3. A listing of some selected Fourier transform pairs appears in Table 3.1. (Not all entries in the latter table have been covered here, however.)

Table 3.1   Some Selected Fourier Transform Pairs

	$f(t)$	$F(\omega) = \mathcal{F}\{f(t)\}$
1.	$e^{-at}u(t)$	$1/(a + j\omega)$
2.	$te^{-at}u(t)$	$1/(a + j\omega)^2$
3.	$e^{-a\lvert t\rvert}$	$2a/(a^2 + \omega^2)$
4.	$e^{-t^2/2\sigma^2}$	$\sigma\sqrt{2\pi}\,e^{-\sigma^2\omega^2/2}$
5.	$\text{sgn}\,(t)$	$2/(j\omega)$
6.	$j/(\pi t)$	$\text{sgn}\,(\omega)$
7.	$u(t)$	$\pi\,\delta(\omega) + 1/j\omega$
8.	$\delta(t)$	1
9.	1	$2\pi\,\delta(\omega)$
10.	$e^{\pm j\omega_0 t}$	$2\pi\,\delta(\omega \mp \omega_0)$
11.	$\cos\omega_0 t$	$\pi\,[\delta(\omega - \omega_0) + \delta(\omega + \omega_0)]$
12.	$\sin\omega_0 t$	$-j\pi\,[\delta(\omega - \omega_0) - \delta(\omega + \omega_0)]$
13.	$\text{rect}\,(t)$	$\text{Sa}\,(\omega/2)$
14.	$\text{rect}\,(t/\tau)$	$\tau\,\text{Sa}\,(\omega\tau/2)$
15.	$\dfrac{1}{2\pi}\,\text{Sa}\,(t/2)$	$\text{rect}\,(\omega)$
16.	$\dfrac{W}{2\pi}\,\text{Sa}\,(Wt/2)$	$\text{rect}\,(\omega/W)$
17.	$\dfrac{W}{\pi}\,\text{Sa}\,(Wt)$	$\text{rect}\,[\omega/(2W)]$
18.	$\Lambda(t)$	$[\text{Sa}\,(\omega/2)]^2$
19.	$\Lambda(t/\tau)$	$\tau[\text{Sa}\,(\omega\tau/2)]^2$
20.	$\delta_T(t)$	$\omega_0\,\delta_{\omega_0}(\omega),$ where $\omega_0 = 2\pi/T$

(handwritten annotation: $\text{Sa}(x) = \dfrac{\sin x}{x}$)

3.6 PROPERTIES OF THE FOURIER TRANSFORM

The Fourier transform is a method of expressing a given function of time (or any other appropriate coordinate, for that matter) in terms of a continuous set of exponential components of frequency. The resulting spectral-density function gives the relative weighting of each frequency component.

We shall use the convention that a time function (signal) and its Fourier transform are indicated by lowercase and uppercase letters respectively.† A convenient shorthand notation to show the correspondence between the two

---

† A departure from this convention is allowed when we wish to form special symbols using more than one letter; e.g., rect (t), rect (ω), Sa (t), Sa (ω), etc.

domains is to use a double arrow: $f(t) \leftrightarrow F(\omega)$. The Fourier transform relationship between time and frequency is assumed to be unique; that is, $\mathcal{F}^{-1}\mathcal{F}[f(t)] = f(t)$ for all t.

What happens in one domain when an elementary operation is performed on the function in the other domain? One way to find out is to go through the transform relation, perform the desired operation, and then go through the inverse transform relation. However, this procedure is clumsy and often quite difficult. The purpose of this section is to introduce some properties of the Fourier transform in a general form.

3.6.1 Linearity (Superposition)

The Fourier transform is a linear operation based on the properties of integration and therefore superposition applies. Thus for any arbitrary constants a_1 and a_2,

$$\mathcal{F}\{a_1 f_1(t) + a_2 f_2(t)\} = a_1 F_1(\omega) + a_2 F_2(\omega), \tag{3.37}$$

or

$$a_1 f_1(t) + a_2 f_2(t) \leftrightarrow a_1 F_1(\omega) + a_2 F_2(\omega).$$

This follows from the integral definition of the Fourier transform. Although the proof is trivial, the consequences of this property to the study of linear systems are of major importance.

Some caution must be exercised in adding spectral densities. Recall that in general $F(\omega)$ is complex-valued and therefore it is important that this addition be complex. A common error is that of adding magnitudes only without regard for phase.

3.6.2 Complex Conjugate

For any complex-valued signal, we have

$$\mathcal{F}\{f^*(t)\} = F^*(-\omega). \tag{3.38}$$

Proof

$$\mathcal{F}\{f^*(t)\} = \int_{-\infty}^{\infty} f^*(t) e^{-j\omega t} \, dt$$

$$= \left[\int_{-\infty}^{\infty} f(t) e^{j\omega t} \, dt \right]^*$$

$$= F^*(-\omega)$$

An important consequence of this property is that if the signal $f(t)$ is real-valued, then $f^*(t) = f(t)$ and $F^*(-\omega) = F(\omega)$.

3.6.3 Symmetry

Any signal can be expressed as a sum of an even function, $f_e(t)$, and an odd function, $f_o(t)$ [cf. Eqs. (2.62) and (2.63)]. This gives rise to the following Fourier transform properties:

$$\mathscr{F}\{f_e(t)\} = F_e(\omega) \quad \text{(and real)}, \tag{3.39}$$

$$\mathscr{F}\{f_o(t)\} = F_o(\omega) \quad \text{(and imaginary)}. \tag{3.40}$$

Proof. For the first part, we have

$$\mathscr{F}\{f_e(t)\} = \int_{-\infty}^{\infty} f_e(t) e^{-j\omega t}\, dt$$

$$= \int_{-\infty}^{\infty} f_e(t) \cos \omega t\, dt - j\int_{-\infty}^{\infty} f_e(t) \sin \omega t\, dt$$

$$= 2\int_{0}^{\infty} f_e(t) \cos \omega t\, dt.$$

Because $\cos \omega t = \cos [(-\omega)t]$, this expression is even in ω. Proof of the second part follows in a similar manner.

For a causal signal (i.e., one which is zero for $t < 0$) we must have very specific combinations of even and odd functions. It follows that the real and imaginary parts of the spectrum of a causal signal must be related. In other words, both the real and imaginary parts of the spectrum of a causal signal cannot be specified arbitrarily.

3.6.4 Duality

A duality exists between the time domain and the frequency domain. This is exhibited in the Fourier transform relations [cf. Eqs. (3.13) and (3.14)] and can be stated explicitly in the following manner.

If

$$\mathscr{F}\{f(t)\} = F(\omega),$$

then

$$\mathscr{F}\{F(t)\} = 2\pi f(-\omega). \tag{3.41}$$

A proof of Eq. (3.41) involves an interchange in t and ω in the Fourier transform integrals. Its use is shown in the following example.

Example 3.6.1 If it is given that $\mathscr{F}\{\text{rect}\,(t)\} = \text{Sa}\,(\omega/2)$, determine $\mathscr{F}\{\text{Sa}\,(t/2)\}$.
Solution. Let $F(\omega) = \text{Sa}\,(\omega/2)$; then $F(t) = \text{Sa}\,(t/2)$. Using Eq. (3.41), we can write

$$\mathscr{F}\{F(t)\} = 2\pi \,\text{rect}\,(-\omega)$$

$$= 2\pi \,\text{rect}\,(\omega).$$

An illustration of this symmetry is shown in Fig. 3.5.

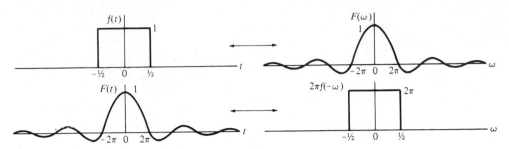

Fig. 3.5 Duality of the Fourier transformation.

3.6.5 Coordinate Scaling (Reciprocal Spreading)

The expansion or compression of a time waveform affects the spectral density of the waveform. For a real-valued scaling constant α and any pulse signal $f(t)$,

$$\mathscr{F}\{f(\alpha t)\} = \frac{1}{|\alpha|} F\left(\frac{\omega}{\alpha}\right). \tag{3.42}$$

Proof

$$\mathscr{F}\{f(\alpha t)\} = \int_{-\infty}^{\infty} f(\alpha t)e^{-j\omega t}\, dt$$

We consider positive and negative values of α separately. For α positive, and changing the variable of integration to $x = \alpha t$, we have

$$\mathscr{F}\{f(\alpha t)\} = \int_{-\infty}^{\infty} f(x)e^{-j\omega x/\alpha}\, dx/\alpha$$

$$= \frac{1}{\alpha} F\left(\frac{\omega}{\alpha}\right) \qquad \text{for } \alpha > 0.$$

When α is negative, the limits on the integral are reversed when the variable of integration is changed so that

$$\mathscr{F}\{f(\alpha t)\} = -\frac{1}{\alpha} F\left(\frac{\omega}{\alpha}\right) \qquad \text{for } \alpha < 0.$$

These two cases can be combined into the more compact form

$$\mathscr{F}\{f(\alpha t)\} = \frac{1}{|\alpha|} F\left(\frac{\omega}{\alpha}\right).$$

If α is positive and greater than unity, $f(\alpha t)$ is a compressed version of $f(t)$ and its spectral density is expanded in frequency by $1/\alpha$. The magnitude of the spectral density also changes—an effect necessary to maintain an energy balance between the two domains. If α is positive but less than unity, $f(\alpha t)$ is an expanded version of $f(t)$ and its spectral density is compressed. When α is negative, $f(\alpha t)$ is reversed in time compared to $f(t)$ and is expanded or compressed depending

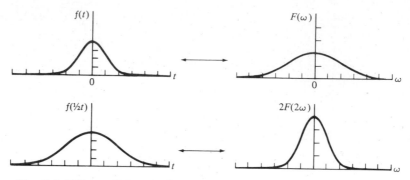

Fig. 3.6 Effects of time scale changes on the signal spectral density.

whether $|\alpha|$ is less than or greater than unity. (It may be helpful to note that the factor $1/|\alpha|$ is called the "magnification" in optics.) The effect of a time scale change on a pulse signal and its spectral density is shown in Fig. 3.6.

Drill Problem 3.6.1 A given signal is recorded on magnetic tape at $7\frac{1}{2}$ inches per second (ips) and played back at $3\frac{3}{4}$ ips. If we assume a flat frequency response, what is the magnitude spectral density of the signal on playback if the original spectral density was of the form (a) $\exp(-|\omega|)$; (b) $1/\sqrt{\omega^2 + 1}$; (c) $\delta(\omega - \omega_0) + \delta(\omega + \omega_0)$?

Answer. (a) $2 \exp(-2|\omega|)$; (b) $1/\sqrt{\omega^2 + 1/4}$; (c) $\delta(\omega - \omega_0/2) + \delta(\omega + \omega_0/2)$.

3.6.6 Time Shifting (Delay)

Another geometric operation is a translation of the time origin, causing the signal to be delayed (or advanced) in time by some time t_0. The corresponding effect on the signal spectral density is

$$\mathscr{F}\{f(t - t_0)\} = F(\omega)e^{-j\omega t_0}. \tag{3.43}$$

Proof

$$\mathscr{F}\{f(t - t_0)\} = \int_{-\infty}^{\infty} f(t - t_0)e^{-j\omega t}\, dt$$

Changing the variable of integration, let $x = t - t_0$,

$$\mathscr{F}\{f(t - t_0)\} = \int_{-\infty}^{\infty} f(x)e^{-j\omega(x + t_0)}\, dx$$

$$= e^{-j\omega t_0}\int_{-\infty}^{\infty} f(x)e^{-j\omega x}\, dx$$

$$= e^{-j\omega t_0}F(\omega).$$

Thus if a signal $f(t)$ is delayed in time by t_0, its magnitude spectral density remains unchanged and a negative phase $(-\omega t_0)$ is added to each frequency component. If t_0 is negative, the time function is advanced in time and the phase spectral density added has a positive slope.

Example 3.6.2   Determine the Fourier transform of the two rectangular pulses shown in Fig. 3.7.

Solution. $f(t) = \text{rect}[(t + \tau/2)/\tau] - \text{rect}[(t - \tau/2)/\tau]$

Using the coordinate scaling property, we get

$$\mathscr{F}\{\text{rect}\,(t/\tau)\} = \tau\,\text{Sa}\,(\omega\tau/2).$$

Using the delay property gives us

$$\mathscr{F}\{f(t)\} = \tau\,\text{Sa}\,(\omega\tau/2)[e^{j\omega\tau/2} - e^{-j\omega\tau/2}] = j(4/\omega)\sin^2(\omega\tau/2).$$

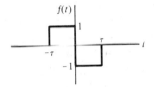

Fig. 3.7 Rectangular pulses
in Example 3.6.2.

Drill Problem 3.6.2   The spectral density of a given signal $f(t)$ is $F(\omega)$. The signal undergoes a distortion so that the new spectral density $F_1(\omega)$ is $F_1(\omega) = F(\omega)$ $[1 + 2\alpha \cos \omega T]$, where α and T are parameters. Determine $f_1(t)$ in terms of $f(t)$.

Answer. $f_1(t) = f(t) + \alpha f(t - T) + \alpha f(t + T).$

3.6.7   Frequency Shifting (Modulation)

The dual of the delay property is the *frequency-translation* property,

$$\mathscr{F}\{f(t)e^{j\omega_0 t}\} = F(\omega - \omega_0). \tag{3.44}$$

Proof

$$\mathscr{F}\{f(t)e^{j\omega_0 t}\} = \int_{-\infty}^{\infty} f(t)e^{j\omega_0 t}e^{-j\omega t}\,dt$$

$$= \int_{-\infty}^{\infty} f(t)e^{-j(\omega - \omega_0)t}\,dt$$

$$= F(\omega - \omega_0)$$

Therefore multiplying a time function by $e^{j\omega_0 t}$ causes its spectral density to be translated in frequency by ω_0 rad/sec.

Drill Problem 3.6.3   Determine the complex-valued constant k required to make the following Fourier transform pair relation hold:

$$\mathcal{F}\{\exp(j\pi t)f(t - \tfrac{1}{2})\} = k \exp(-j\omega/2)F(\omega - \pi).$$

Answer. $k = j$.

For real-valued $f(t)$, it is now a relatively simple matter to find the Fourier transform of $\mathcal{R}e\{f(t)e^{j\omega_0 t}\}$:

$$f(t)\cos\omega_0 t = \tfrac{1}{2}f(t)[e^{j\omega_0 t} + e^{-j\omega_0 t}].$$

Using the frequency translation property, we get

$$\mathcal{F}\{f(t)\cos\omega_0 t\} = \tfrac{1}{2}[F(\omega + \omega_0) + F(\omega - \omega_0)]. \tag{3.45}$$

This process of multiplying a signal by a harmonic function (i.e., a sine or cosine) to translate the spectral density is known as *amplitude modulation*. Equation (3.45) is known as the *modulation property* of the Fourier transform. Note that in the process of modulation one-half of the spectral density is moved up in frequency and one-half is moved down in frequency.

Example 3.6.3   Find the spectral density of the pulse waveform,

$$A \text{ rect }(t/\tau)\cos\omega_0 t.$$

Solution.   From Table 3.1,

$$\mathcal{F}\{A \text{ rect }(t/\tau)\} = A\tau \text{ Sa }(\omega\tau/2).$$

Use of the modulation property gives

$$\mathcal{F}\{A \text{ rect }(t/\tau)\cos\omega_0 t\} = \tfrac{1}{2}A\tau\{\text{Sa }[(\omega + \omega_0)\tau/2] + \text{Sa }[(\omega - \omega_0)\tau/2]\}.$$

The pulse and its spectral density are shown in Fig. 3.8. Note that the modulation property gives us a convenient way to handle sinusoids of finite length.

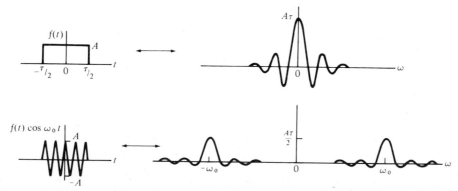

Fig. 3.8 Effects of modulation on the signal spectral density.

Drill Problem 3.6.4   Based on the knowledge that

$$\mathscr{F}\{\exp(-at)u(t)\} = 1/(a + j\omega),$$

determine the Fourier transform of $\exp(-at) \cos \omega_0 t \; u(t)$ using the modulation property.

Answer. $(a + j\omega)/[(a^2 + \omega_0^2 - \omega^2) + j2a\omega]$.

3.6.8   Differentiation and Integration

If $\dfrac{d}{dt} f(t)$ is absolutely integrable, then

$$\frac{d}{dt} f(t) \leftrightarrow j\omega F(\omega). \qquad (3.46)$$

Proof.   Using the inverse Fourier transform, we have

$$f(t) = \frac{1}{2\pi} \int_{-\infty}^{\infty} F(\omega)e^{j\omega t} \, d\omega,$$

$$\frac{d}{dt} f(t) = \frac{1}{2\pi} \frac{d}{dt} \int_{-\infty}^{\infty} F(\omega)e^{j\omega t} \, d\omega$$

$$= \frac{1}{2\pi} \int_{-\infty}^{\infty} \frac{\partial}{\partial t} [F(\omega)e^{j\omega t}] \, d\omega$$

$$= \frac{1}{2\pi} \int_{-\infty}^{\infty} j\omega F(\omega)e^{j\omega t} \, d\omega.$$

Taking the Fourier transform of both sides,

$$\mathscr{F}\left\{\frac{d}{dt} f(t)\right\} = j\omega F(\omega).$$

Therefore time differentiation enhances the high-frequency components of a signal.

The corresponding integration property is

$$\mathscr{F}\left\{\int_{-\infty}^{t} f(\tau) \, d\tau\right\} = \frac{1}{j\omega} F(\omega) + \pi F(0) \, \delta(\omega), \qquad (3.47)$$

where

$$F(0) = \int_{-\infty}^{\infty} f(t) \, dt. \qquad (3.48)$$

Integration in time suppresses the high-frequency components of a signal. This conclusion agrees with the time-domain viewpoint that integration smoothes out the time fluctuations in a signal.

If $F(0) \neq 0$, the proof of Eq. (3.47) must rely on a generalized function approach because the integrals may not converge. What is happening, of course, is that an ideal integrator senses any net positive or negative area in a pulse and holds it for an infinite time interval, thus creating a nonzero average value, as given in Eq. (3.48).

To illustrate the integration property, let us first define the *triangular function*:

$$\Lambda(t/\tau) = \begin{cases} 1 - |t|/\tau & |t| < \tau \\ 0 & |t| > \tau \end{cases}. \tag{3.49}$$

Equation (3.49) may be obtained by integrating the signal in Example 3.6.2 and dividing by the constant τ. Using Eq. (3.47) and the result of Example 3.6.2, we have

$$\mathcal{F}\{\Lambda(t/\tau)\} = \frac{1}{j\omega} \frac{1}{\tau} [j(4/\omega) \sin^2 (\omega\tau/2)]$$

$$= \tau \, \mathrm{Sa}^2 (\omega\tau/2). \tag{3.50}$$

This useful transform pair is also illustrated in Fig. 3.3.

Differentiation in the time domain is equivalent to multiplication by the algebraic factor $(j\omega)$ in the frequency domain. On the other hand, integration in the time domain is equivalent to division by $(j\omega)$ in the frequency domain. A combination of these two properties often proves convenient in determining the Fourier transform of piecewise continuous functions. The procedure is illustrated below. [Note that Eq. (3.48) is always zero following a derivative operation because a derivative deletes any nonzero average value.]

Example 3.6.4   Determine the Fourier transform of the trapezoidal pulse shown in Fig. 3.9(a).

Solution.   Taking two successive derivatives, as shown in Figs. 3.9(b) and (c), we can write by inspection

$$(j\omega)^2 F(\omega) = \frac{A}{\tau} (e^{j2\omega\tau} - e^{j\omega\tau} - e^{-j\omega\tau} + e^{-j2\omega\tau}).$$

Simplifying, we get

$$(j\omega)^2 F(\omega) = \frac{A}{\tau} (e^{j\omega\tau/2} - e^{-j\omega\tau/2})^2 (e^{j\omega\tau} + 1 + e^{-j\omega\tau}),$$

or

$$F(\omega) = A\tau \, \mathrm{Sa}^2 (\omega\tau/2)[1 + 2 \cos \omega\tau].$$

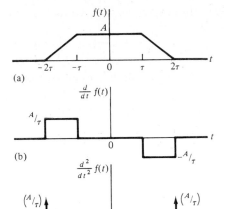

Fig. 3.9 Trapezoidal pulse and its derivatives.

Drill Problem 3.6.5   Determine the Fourier transform of the triangular pulse

$$f(t) = 2t[u(t) - u(t - 1)] - (t - 3)[u(t - 1) - u(t - 3)].$$

Answer. $F(\omega) = [2 + \exp(-j3\omega) - 3\exp(-j\omega)]/(j\omega)^2.$

3.6.9   Time Convolution

As discussed in Chapter 2, one method of characterizing a system is by its frequency transfer function; a second method is by its impulse response. We now wish to relate these methods using the principle of convolution.

For the test signal $f(t) = \delta(t - \tau)$, the system impulse response is defined as [cf. Eq. (2.11)]

$$\mathcal{T}\{\delta(t - \tau)\} = h(t, \tau), \qquad\qquad (3.51)$$

where τ is the delay or "age" variable. If the system is *time-invariant*, $h(t, \tau)$ takes the special form

$$\mathcal{T}\{\delta(t - \tau)\} = h(t - \tau). \qquad\qquad (3.52)$$

The input signal $f(t)$ may be expressed in terms of impulse functions by [cf. Eqs. (2.95) and (2.98)]

$$f(t) = \int_{-\infty}^{\infty} f(\tau)\, \delta(\tau - t)\, d\tau = \int_{-\infty}^{\infty} f(\tau)\, \delta(t - \tau)\, d\tau.$$

If the system is linear, the impulse responses corresponding to each value of the age variable τ may be added up to form the output response,

$$g(t) = \mathcal{T}\left\{ \int_{-\infty}^{\infty} f(\tau)\, \delta(t - \tau)\, d\tau \right\}.$$

From integration theory, we can rewrite this as

$$g(t) = \mathcal{T}\left\{ \lim_{\Delta\tau \to 0} \sum_{n=-\infty}^{\infty} f(\tau_n)\, \delta(t - \tau_n)\, \Delta\tau \right\}.$$

Using the principle of superposition, we move the system operator inside the summation. Also, the $f(\tau_n)$ are the weights (areas) of the impulse functions and are constants for each impulse. Therefore we have

$$g(t) = \lim_{\Delta\tau \to 0} \sum_{n=-\infty}^{\infty} f(\tau_n)\, \mathcal{T}\{\delta(t - \tau_n)\}\, \Delta\tau.$$

Returning to the integral form and using Eq. (3.51), we have

$$g(t) = \int_{-\infty}^{\infty} f(\tau)h(t, \tau)\, d\tau. \tag{3.53}$$

This is a key result in signal analysis for it links the input to the output by means of an integral operation and holds for any linear system. In the particular case in which the system is also time-invariant, Eq. (3.53) reduces to the important form,

$$g(t) = \int_{-\infty}^{\infty} f(\tau)h(t - \tau)\, d\tau. \tag{3.54}$$

This result is known as the *convolution integral*. We find it convenient to use the following shorthand notation for the convolution integral:

$$\int_{-\infty}^{\infty} f(\tau)h(t - \tau)\, d\tau \triangleq f(t) \circledast h(t).$$

An important property of the Fourier transform is that it reduces the convolution integral operation to an algebraic product. This is stated mathematically in the following way. If

$$f(t) \circledast h(t) = \int_{-\infty}^{\infty} f(\tau)h(t - \tau)\, d\tau$$

and

$$\mathcal{F}\{f(t)\} = F(\omega), \qquad \mathcal{F}\{h(t)\} = H(\omega);$$

then

$$\mathcal{F}\{f(t) \circledast h(t)\} = F(\omega)H(\omega). \tag{3.55}$$

Proof

$$\mathcal{F}\{f(t) \circledast h(t)\} = \int_{-\infty}^{\infty} \left[\int_{-\infty}^{\infty} f(\tau)h(t - \tau)\, d\tau \right] e^{-j\omega t}\, dt$$

Changing the order of integration and integrating with respect to t first yields

$$\mathcal{F}\{f(t) \circledast h(t)\} = \int_{-\infty}^{\infty} f(\tau) \left[\int_{-\infty}^{\infty} h(t - \tau)e^{-j\omega t} \, dt \right] d\tau.$$

Using the time-shift (delay) property, we have

$$\mathcal{F}\{h(t - \tau)\} = e^{-j\tau\omega}H(\omega).$$

Then proceeding with the remaining integration over τ, we have

$$\mathcal{F}\{f(t) \circledast h(t)\} = \int_{-\infty}^{\infty} f(\tau)H(\omega)e^{-j\omega\tau} \, d\tau$$

$$= H(\omega) \int_{-\infty}^{\infty} f(\tau)e^{-j\omega\tau} \, d\tau$$

$$= H(\omega)F(\omega).$$

Thus convolution in the time domain corresponds to multiplication in the frequency domain.

Drill Problem 3.6.6 Determine the Fourier transform of the signal $g(t) = \exp(-t)u(t) \circledast \exp(-2t)u(t)$ by (a) performing the convolution first and then taking the Fourier transform of the result; and (b) taking the Fourier transform of each term and then multiplying.

Answer. $1/(2 - \omega^2 + j3\omega)$.

3.6.10 Frequency Convolution

A dual to the preceding property is the following:

If

$$\mathcal{F}\{f_1(t)\} = F_1(\omega), \qquad \mathcal{F}\{f_2(t)\} = F_2(\omega);$$

then†

$$\mathcal{F}\{f_1(t)f_2(t)\} = \frac{1}{2\pi} [F_1(\omega) \circledast F_2(\omega)], \tag{3.56}$$

where

$$F_1(\omega) \circledast F_2(\omega) = \int_{-\infty}^{\infty} F_1(u)F_2(\omega - u) \, du.$$

---

† The factor of $(1/2\pi)$ arises here because we are using radian frequency, ω. It is helpful to remember that frequency, f, and time, t, are reciprocals dimensionally. Radian frequency is 2π times reciprocal time, thus suggesting a division by 2π to keep the correct units.

This property can be proved in the same way as that for the time convolution. Thus *the multiplication of two functions in the time domain is equivalent to the convolution of their spectra in the frequency domain.*

Drill Problem 3.6.7   Evaluate: rect $(\omega) \exp(-j\omega) \circledast \exp(-j\omega)$ at $\omega = \pi$.

Answer. -1.

3.6.11   Remarks on Properties of the Fourier Transform

The Fourier transform pair was obtained by a limiting operation on the complex exponential Fourier series. Assuming that an interchange of operations is valid, it follows that all of the properties discussed above for the Fourier transform also hold for the exponential Fourier series. The proofs are almost identical, the major change being the substitution of the discrete frequency variable $n\omega_0$ for the continuous variable ω.

A table of Fourier transform properties is given in Table 3.2. Note the symmetry between the time domain and the frequency domain.

Table 3.2   Some Fourier Transforms Corresponding to Given Mathematical Operations

Operation	$f(t)$ $\leftrightarrow$	$F(\omega)$
Linearity (superposition)	$a_1 f_1(t) + a_2 f_2(t)$	$a_1 F_1(\omega) + a_2 F_2(\omega)$
Complex conjugate	$f^*(t)$	$F^*(-\omega)$
Scaling	$f(\alpha t)$	$\dfrac{1}{|\alpha|} F\left(\dfrac{\omega}{\alpha}\right)$
Delay	$f(t - t_0)$	$e^{-j\omega t_0} F(\omega)$
Frequency translation	$e^{j\omega_0 t} f(t)$	$F(\omega - \omega_0)$
Amplitude modulation	$f(t) \cos \omega_0 t$	$\frac{1}{2} F(\omega + \omega_0) + \frac{1}{2} F(\omega - \omega_0)$
Time convolution	$\displaystyle \int_{-\infty}^{\infty} f_1(\tau) f_2(t - \tau)\, d\tau$	$F_1(\omega) F_2(\omega)$
Frequency convolution	$f_1(t) f_2(t)$	$\dfrac{1}{2\pi} \displaystyle \int_{-\infty}^{\infty} F_1(u) F_2(\omega - u)\, du$
Duality: time-frequency	$F(t)$	$2\pi f(-\omega)$
Symmetry: even-odd	$f_e(t)$	$F_e(\omega)$ [real]
	$f_o(t)$	$F_o(\omega)$ [imaginary]
Time differentiation	$\dfrac{d}{dt} f(t)$	$j\omega F(\omega)$
Time integration	$\displaystyle \int_{-\infty}^{t} f(\tau)\, d\tau$	$\dfrac{1}{j\omega} F(\omega) + \pi F(0)\, \delta(\omega),$
		where $F(0) = \displaystyle \int_{-\infty}^{\infty} f(t)\, dt$

Throughout this discussion we have been dealing with the Fourier transform pair, $f(t) \leftrightarrow F(\omega)$, where t represents time and ω represents radian frequency. The use of Fourier analysis, however, is not restricted to time-frequency variable pairs. In general, any pair of variables x, y may be used to form a Fourier transform pair $f(x) \leftrightarrow F(y)$ as long as their product (x, y) is dimensionless. The Fourier transform can also be extended to handle more than one dimension by choosing appropriate variable pairs in each dimension.

3.7 SOME CONVOLUTION RELATIONSHIPS

The convolution integral, as expressed in Eq. (3.54), holds as long as the system is linear and time-invariant. In addition, if the system is causal (i.e., physically realizable), then $h(t) = 0$ for all $t < 0$ and there is no contribution to the integration in Eq. (3.54) for $(t - \tau) < 0$:

$$g(t) = \int_{-\infty}^{t} f(\tau)h(t - \tau) \, d\tau; \qquad h(t) \text{ causal.} \tag{3.57}$$

Often it turns out that the input, $f(t)$, also satisfies the condition $f(t) = 0$ for $t < 0$, and Eq. (3.57) further simplifies to

$$g(t) = \int_{0}^{t} f(\tau)h(t - \tau) \, d\tau; \qquad f(t), h(t) \text{ causal.} \tag{3.58}$$

3.7.1 Properties of Convolution

Some useful properties of convolution are listed below.

Commutative Law

$$f_1(t) \circledast f_2(t) = f_2(t) \circledast f_1(t) \tag{3.59}$$

Distributive Law

$$f_1(t) \circledast [f_2(t) + f_3(t)] = f_1(t) \circledast f_2(t) + f_1(t) \circledast f_3(t) \tag{3.60}$$

Associative Law

$$f_1(t) \circledast [f_2(t) \circledast f_3(t)] = [f_1(t) \circledast f_2(t)] \circledast f_3(t) \tag{3.61}$$

Proofs of these properties follow readily from the integral definitions and possible changes in the order of integration.

3.7.2 Convolution Involving Singularity Functions

The unit step response is the indefinite integral of the unit impulse response. This can be shown as follows:

$$u(t) \circledast h(t) = \int_{-\infty}^{\infty} u(\tau)h(t - \tau) \, d\tau = \int_{0}^{\infty} h(t - \tau) \, d\tau.$$

Changing the variable of integration, let $x = t - \tau$; then

$$u(t) \circledast h(t) = \int_{-\infty}^{t} h(x) \, dx. \tag{3.62}$$

This result gives a method for determining the impulse response of a system in the laboratory. Although technically the unit step function exists forever, most systems have an impulse response which lasts only for a relatively short duration of time. If we use a low-frequency square-wave generator whose repetition rate is much longer than the duration of the impulse response of the system, the system for all practical purposes sees a step. After recording the step response of the system, one can take the derivative of the output graph as a function of time. This will be the impulse response of the system.

Convolution with the unit impulse function follows using the integral properties of the impulse function:

$$f(t) \circledast \delta(t - t_0) = \int_{-\infty}^{\infty} f(\tau) \, \delta(t - t_0 - \tau) \, d\tau$$

$$= f(t - t_0). \tag{3.63}$$

Therefore the convolution of a function with the unit impulse reproduces that function exactly except that it is delayed (or advanced) by the delay (or advance) of the impulse.

Example 3.7.1   Find $f(t) \circledast h(t)$ for the $f(t)$, $h(t)$ shown in Fig. 3.10(a), (b).

Solution.   From Fig. 3.10, $f(t) = A \sin \pi t \, u(t)$, $h(t) = \delta(t) - \delta(t - 2)$,

$$g(t) = f(t) \circledast h(t) = \int_{-\infty}^{\infty} [A \sin \pi \tau \, u(\tau)][\delta(t - \tau) - \delta(t - 2 - \tau)] \, d\tau$$

$$= [A \sin \pi t] u(t) - [A \sin \pi(t - 2)] u(t - 2)$$

$$g(t) = \begin{cases} 0 & t < 0 \\ A \sin \pi t & 0 < t < 2. \\ 0 & t > 2 \end{cases}$$

A sketch of $g(t)$ is shown in Fig. 3.10(c). This type of system is sometimes used in radar and navigation to determine the time-of-arrival of partially known wave-

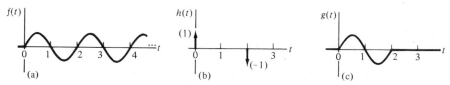

Fig. 3.10 The delay-differencing circuit discussed in Example 3.7.1.

forms. Note that the required $h(t)$ is that one used in Example 2.19.1 without the integrator.

An extension to Eq. (3.63) is that the convolution of an impulse function with an impulse function results in another impulse function. Stated mathematically, we have

$$A \, \delta(t - t_0) \circledast B \, \delta(t - t_1) = AB \int_{-\infty}^{\infty} \delta(\tau - t_0) \, \delta(t - t_1 - \tau) \, d\tau$$

$$= AB \, \delta(t - t_0 - t_1). \tag{3.64}$$

A proof of this result follows from an application of the method of generalized functions (cf. Section 2.18); we shall not pursue this topic here.

3.8 GRAPHICAL INTERPRETATION OF CONVOLUTION

The graphical interpretation of convolution permits one to grasp visually the results of the more abstract mathematical operations. Suppose we wish to find the convolution of two given functions $f_1(t)$ and $f_2(t)$. The operations to be performed are based on the convolution integral:

$$f_1(t) \circledast f_2(t) = \int_{-\infty}^{\infty} f_1(\tau) f_2(t - \tau) \, d\tau.$$

First of all we list the required operations, step by step:

1. Replace t by τ in $f_1(t)$, giving $f_1(\tau)$.
2. Replace t by $(-\tau)$ in $f_2(t)$. This folds the function $f_2(\tau)$ about the vertical axis passing through the origin of the τ axis.
3. Translate the entire frame of reference of $f_2(-\tau)$ by an amount t.† (As far as the integration is concerned, t is merely a parameter.) Thus the amount of translation, t, is the difference between the moving frame of reference and the fixed frame of reference. The origin in the moving frame is at $\tau = t$; the origin in the fixed frame is at $\tau = 0$. The function in the moving frame represents $f_2(t - \tau)$; the function in the fixed frame represents $f_1(\tau)$.
4. At any given relative shift between the frames of reference, for example, t_0, we must find the area under the product of the two functions, that is,

$$\int_{-\infty}^{\infty} f_1(\tau) f_2(t_0 - \tau) \, d\tau = [f_1(t) \circledast f_2(t)]_{t = t_0}.$$

---

† This translation is analogous to the relative motion between rigid frames of reference in relativistic mechanics. In the theory of relativity it is the velocity between coordinate systems which is important, whereas in convolution the important thing is the distance and direction between coordinate systems.

5. This procedure is to be repeated for different values of $t = t_0$ by successively progressing the movable frame and finding the values of the convolution integral at those values of t. For continuous functions, this can be accomplished by a straightforward integration. For piecewise continuous functions, the product will be piecewise continuous and we must integrate over each continuous section.

6. If the amount of shift of the movable frame is along the negative τ axis (i.e., to the left), t is negative. If the shift is along the positive τ axis (i.e., to the right), t is positive.

Example 3.8.1   Find the convolution of the rectangular pulse $f_1(t)$ and the triangular pulse $f_2(t)$ shown in Fig. 3.11.

Solution.   The various steps described above are shown in Fig. 3.11.

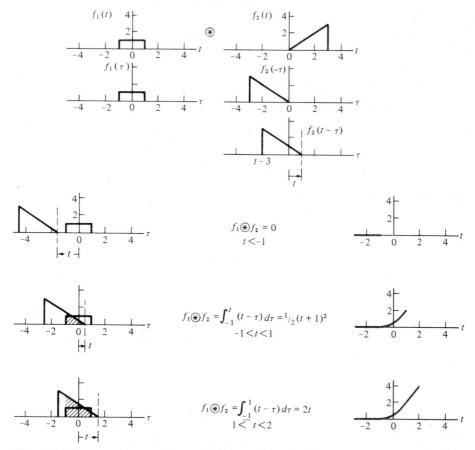

Fig. 3.11 The convolution of a rectangular and a triangular pulse (cont. on p. 109).

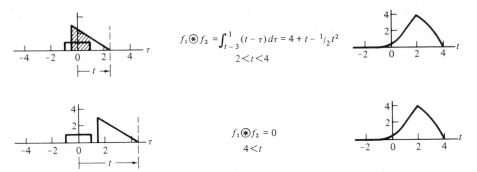

$$f_1 \circledast f_2 = \int_{t-3}^{1} (t - \tau)\, d\tau = 4 + t - \tfrac{1}{2} t^2$$
$$2 < t < 4$$

$$f_1 \circledast f_2 = 0$$
$$4 < t$$

Fig. 3.11 The convolution of a rectangular and a triangular pulse (cont. from p. 108).

Drill Problem 3.8.1   Find the convolution rect (t/α) ⊛ rect (t/α).

Answer. $\alpha\Lambda(t/\alpha)$.

Drill Problem 3.8.2   A group of five narrow rectangular pulses, each of amplitude A and spaced by an interval T, is applied to the system of Drill Problem 2.19.1. Using convolution, determine the peak signal output. (This type of system is useful in the detection of periodic signals if only the period is known.)

Answer. $A(1 + K + K^2 + K^3 + K^4)$.

The computation procedure for the convolution integral is easily programmed on a digital computer. Thus if both $f(t)$ and $g(t)$ are zero for $t < 0$, we can write

$$f(t) \circledast g(t) = \int_0^t f(\tau)g(t - \tau)\, d\tau$$

$$\cong \alpha \sum_{m=1}^{n} f(m\alpha)g(n\alpha - m\alpha),$$

where m, n are integers. Other cases follow in a similar manner.

3.9   FILTER CHARACTERISTICS OF LINEAR SYSTEMS

The application of an input signal $f(t)$ to a linear time-invariant system results in a corresponding output $g(t)$, as shown in Fig. 3.12. The spectral densities of

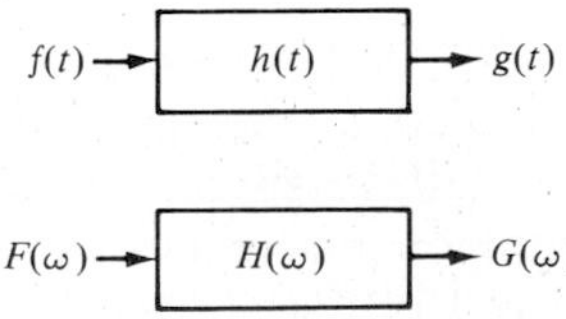

Fig. 3.12 Representation of a system in time and frequency.

$f(t) \longrightarrow \boxed{\quad h(t) \quad} \longrightarrow g(t)$

$F(\omega) \longrightarrow \boxed{\quad H(\omega) \quad} \longrightarrow G(\omega)$

the input and output are given by $F(\omega)$ and $G(\omega)$, respectively. The impulse response and the system frequency transfer function are designated by $h(t)$ and $H(\omega)$, respectively. The following relations hold:

$$g(t) = f(t) \circledast h(t), \tag{3.65a}$$

$$G(\omega) = F(\omega)H(\omega). \tag{3.65b}$$

Therefore the frequency transfer function is the Fourier transform of the impulse response of a linear time-invariant system. In the time domain, the system modifies the shape of the input signal. In frequency, the system modifies the spectral density of the input signal. These effects are related by the Fourier transformation.

A linear time-invariant system acts as a filter on the various frequency components applied to the system. Some frequency components may be amplified, some attenuated, and some remain unaffected. Each frequency component may be shifted in phase as it passes through the system. Rewriting Eq. (3.65b) to separate these two effects,

$$|G(\omega)| \, e^{j\theta_g(\omega)} = |F(\omega)| \, e^{j\theta_f(\omega)} \, |H(\omega)| \, e^{j\theta_h(\omega)},$$

$$|G(\omega)| = |F(\omega)| \, |H(\omega)|, \tag{3.66}$$

$$\theta_g(\omega) = \theta_f(\omega) + \theta_h(\omega). \tag{3.67}$$

In other words, the magnitude response is given by the product of the magnitude of the signal spectral density and the magnitude of the system transfer function. The phase response is given by the sum of the individual phase responses. This can be extended to the case of several systems connected in cascade as long as each system does not alter the transfer function of any other system.

Example 3.9.1   Determine the magnitude response of a RC low-pass filter to a gate function of unit amplitude and width τ; assume that $\tau = 4RC$.

Solution.

$$|F(\omega)| = \tau \, \text{Sa} \, (\omega\tau/2),$$

$$|H(\omega)| = \left| \frac{1}{1 + j\omega RC} \right| = \frac{1}{\sqrt{1 + (\omega\tau/4)^2}},$$

$$|G(\omega)| = |F(\omega)| \, |H(\omega)| = \frac{\tau}{\sqrt{1 + (\omega\tau/4)^2}} |\text{Sa} \, (\omega\tau/2)|$$

The magnitude plots are shown in Fig. 3.13. This system attenuates the high frequencies contained in the input spectral density and allows the lower frequencies to pass with relatively little attenuation. The unequal transmission of all frequency components results in a somewhat distorted replica of the input signal, as portrayed in Fig. 3.13(e).

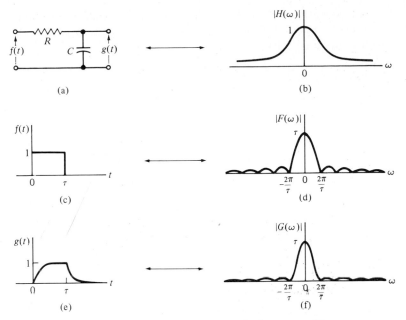

Fig. 3.13 Magnitude response of a low-pass filter.

Drill Problem 3.9.1   A given system delays the input signal by t_0 units of time and then subtracts the delayed version from the original signal. Calculate the frequency transfer function of this system.

Answer. $2 \sin (\omega t_0/2) \tan^{-1} [\sin \omega t_0/(1 - \cos \omega t_0)]$.

3.10 TRANSVERSAL FILTERS

The synthesis of a linear time-invariant system involves the design of a system to attain a given impulse response and frequency transfer function. One way in which this can be accomplished for the general case is by noting that the impulse response of a linear time-invariant system can be approximated by a delay line that is tapped at various points and weighted by a set of gain factors. To see this, we write the convolution integral for a causal system whose impulse response is $h(t)$ and whose signal input $f(t)$ is also causal:

$$g(t) = \int_0^t f(\tau)h(t - \tau)\, d\tau = \int_0^t f(t - \tau)h(\tau)\, d\tau. \qquad (3.68)$$

The integration in Eq. (3.68) can be approximated by

$$g(t) \approx \sum_{k=0}^{t/\Delta\tau} f(t - k\,\Delta\tau)h(k\,\Delta\tau)\,\Delta\tau. \qquad (3.69)$$

This result can be realized by using a delay line with taps at the delays $k \Delta\tau$; the output of each tap is multiplied by the preset weight $h(k \Delta\tau) \Delta\tau$. Such a system is shown in Fig. 3.14.

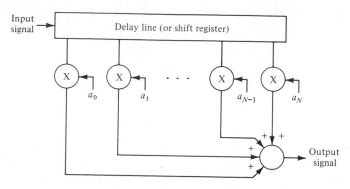

Fig. 3.14 A tapped-delay-line (transversal) filter.

A filter constructed using a tapped delay line, tap weights, and adder in the configuration shown in Fig. 3.14 is called a *transversal* filter. The theoretical constraints are more flexible on transversal filters than on lumped-RLC filters, and they are of increasing importance as new and more effective ways of implementing them are developed.

Both digital and analog methods can be used to implement transversal filters. In the digital implementation, delay is provided by a shift register, and the basic increment of delay is equal to the clock period. A computer can be used to implement the filter (in which case it is usually referred to as a digital filter). Another method used is the charge-coupled device (CCD), which eliminates the need for digitizing the signal.

The basic circuit realization of the CCD is a row of field-effect transistors (FET's) with drains and sources connected in series, and the drains capacitively coupled to the gates. This is illustrated in Fig. 3.15. Two clock lines are used to furnish alternating gate-control signals. The CCD samples the input signal and stores the sample as a charge on the first capacitor. This charge is then passed from stage to stage under the control of the clock pulses that turn on alternate FET's. This action transfers the charge to the next capacitor until eventually the charge reaches the end of the row of FET's. Taps at various delays may be

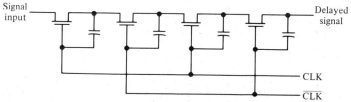

Fig. 3.15 A charge-coupled-device (CCD) delay line.

weighted and combined to synthesize a transversal filter. CCD's operate at clock frequencies up to 10 MHz and can provide large delays (up to 1 sec). The CCD is sometimes referred to descriptively as a "bucket brigade" device.

A popular way to synthesize transversal filters using analog methods is that of the surface-acoustic-wave (SAW) filter. The essential components of a transversal SAW filter are two transducers attached to opposite ends of a polished piezoelectric substrate, as shown in Fig. 3.16. When an electric signal is applied to the input transducer, it physically distorts the piezoelectric surface, creating acoustic traveling waves. As these waves reach the output transducer, their mechanical energy is transformed back into an electrical signal. The substrate used is either quartz or lithium niobate, and the velocity of propagation on the substrate is about 3000 m/sec. Size of substrate, convenience of fabrication, and acoustic attenuation limit operating frequencies between 2 MHz and 2 GHz. In general, frequencies for standard production devices go from about 20 MHz to 500 MHz.

Each SAW transducer is a set of interleaved fingers of thin metallic film, as shown in Fig. 3.16. The purpose of the transducer is not only to convert signals from electric to acoustic energy but also to filter them. The impulse response of the transducer is directly linked to its geometry; relative finger positions determine the phase, and the overlap between adjacent fingers determines the amplitude weighting.† Thus the tapped delay line is formed by the finger spacings and the tap weights are determined by the overlap between adjacent fingers. The overall frequency transfer function of the SAW filter is given by the product of the Fourier transform of the impulse response of each transducer.

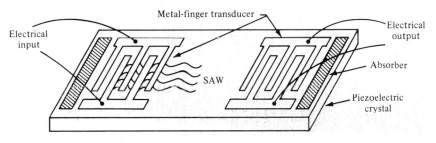

Fig. 3.16 A surface-acoustic-wave (SAW) device.

The SAW filters are used widely for high-quality custom filter applications. Once a SAW filter has been designed and constructed, of course, its characteristics cannot be adjusted. Research and development continues on other methods to build transversal filters.

---

† J. D. Maines and E. G. S. Paige, "Surface-Acoustic-Wave Devices for Signal Processing Applications," *Proceedings of the IEEE* [Special Issue on Surface Acoustic Wave Devices & Applications], 64 (May 1976): 639–671.

3.11   BANDWIDTH OF A SYSTEM

The constancy of the magnitude $|H(\omega)|$ in a system is usually specified by a parameter called its *bandwidth*. The bandwidth W of a system is defined as the interval of *positive* frequencies over which the magnitude $|H(\omega)|$ remains within a given numerical factor.† Although the exact criterion may vary, a popular numerical factor is -3 dB (that is, $1/\sqrt{2}$ in voltage or $\frac{1}{2}$ in power). Using this criterion, we refer to the bandwidth as the "-3 dB bandwidth" or the "half-power bandwidth" of the system.

According to this definition, the bandwidth of the system whose $|H(\omega)|$ is graphed in Fig. 3.17(a) is $W = \omega_1$ radians per second. The bandwidth of the system whose $|H(\omega)|$ plot is shown in Fig. 3.17(b) is $W = (\omega_2 - \omega_1)$ radians per second.

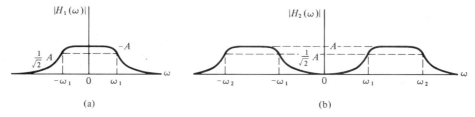

(a)   (b)

Fig. 3.17 The bandwidth of a system as measured to its -3 dB points.

3.12   REQUIREMENTS FOR DISTORTIONLESS TRANSMISSION

What general requirements must be met for a linear time-invariant system to behave like an ideal transmission system? To answer this question, suppose that a given signal $f(t)$ is to be transmitted through such a system and we insist that the resulting output is to look just like the input. We allow this replica to have a different magnitude and some time delay as long as the signal shape is unaltered. Thus if $f(t)$ is the input signal, the required output is

$$g(t) = Kf(t - t_0). \tag{3.70}$$

Taking the Fourier transform of both sides of Eq. (3.70) and using the time-shift property, we get

$$G(\omega) = Ke^{-j\omega t_0}F(\omega). \tag{3.71}$$

Comparing Eqs. (3.65b) and (3.71), we see that the required system transfer function for distortionless transmission is

$$H(\omega) = Ke^{-j\omega t_0}. \tag{3.72}$$

---

† The units of frequency can be expressed either in terms of radian frequency (radians per second) or in terms of cycles per second (Hz). As much as possible, we use W for radian frequency and B for frequency in Hz (hence $W = 2\pi B$).

Therefore, to achieve distortionless transmission, the overall system response must have a constant magnitude response and its phase shift must be linear with frequency. It is not enough that the system attenuate (or amplify) all frequency components equally. All these frequency components must also arrive with identical time delay in order to add up correctly. This demands a phase shift which is proportional to frequency.† Note that a factor of $e^{j2\pi n}$, for any integer n, may be included in Eq. (3.71) without any net effect.

In practice, a signal may be distorted in passing through some parts of a system. Phase or amplitude correction (equalization) networks may be introduced elsewhere in the system to correct for this distortion. It is the overall input-output characteristic of the system which determines its behavior.

3.13   TIME RESPONSE OF FILTERS

As a result of physical limitations, one cannot build a system with an infinite bandwidth. Despite this difficulty, the ideas developed for distortionless transmission are quite helpful in an understanding of filters. As an approximation, we choose a truncated distortionless transmission system and call this an ideal filter. More specifically, an ideal filter passes without distortion all frequency components between its lower cutoff frequency, which we shall designate as ω_ℓ, and its upper cutoff frequency, ω_u. Outside this range, the ideal filter is assumed to have zero magnitude response. The frequency range defined by $\omega_\ell < |\omega| < \omega_u$ is called the "passband" of the filter. The effective width of the passband is specified by the bandwidth $W = (\omega_u - \omega_\ell)$ radians per second (the bandwidth in Hz is designated by B, where $W = 2\pi B$). The ideal filter is allowed a gain factor K and a time delay t_0 within its passband.

The ideal filter described above is called an ideal bandpass filter (BPF). An ideal low-pass filter (LPF) has $\omega_\ell = 0$ and $\omega_u = W$. The ideal high-pass filter (HPF) has $\omega_u \to \infty$. Sketches of the ideal low-pass and bandpass filters are shown in Fig. 3.18.

We now consider the ideal low-pass filter in some detail. Letting the attenuation constant K be unity for convenience, the transfer function of such a filter is

$$H(\omega) = |H(\omega)|e^{j\theta(\omega)} = \text{rect}\left(\frac{\omega}{2W}\right) e^{-j\omega t_0}. \qquad (3.73)$$

The impulse response $h(t)$ of the ideal low-pass filter is (using transform pair number 17 in Table 3.1 and the time-shifting property)

$$h(t) = \mathcal{F}^{-1}\{H(\omega)\} = \frac{W}{\pi} \text{Sa}\,[W(t - t_0)]. \qquad (3.74)$$

---

† For a fixed frequency, phase shift is proportional to time delay. Therefore the phase shift must vary linearly with frequency to yield a fixed time delay.

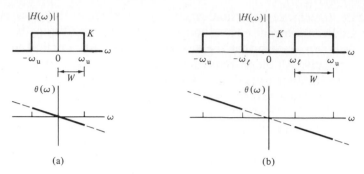

Fig. 3.18 The ideal filter: (a) low-pass; (b) bandpass.

This impulse response is plotted in Fig. 3.19. From this plot, we observe that the width of the impulse response and the bandwidth are inversely related (as a consequence of the scaling property of the Fourier transform). Also note that the response of the LPF appears for $t < 0$, indicating that the ideal low-pass filter is not causal (physically realizable). Similar reasoning shows that the ideal bandpass filter is also noncausal.

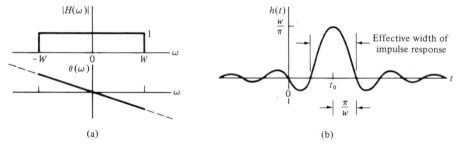

Fig. 3.19 The ideal low-pass filter: (a) its system transfer function and (b) its impulse response.

Why are the ideal filters noncausal? They have finite bandwidths and linear phase characteristics. The source of the problem with the ideal filter is actually more the infinite loss outside the pass band rather than the infinitely steep cutoff, although the latter is also impossible practically.† Of course, the very steep

---

† It can be shown that, for all $H(\omega)$ satisfying $\int_{-\infty}^{\infty} |H(\omega)|^2 \, d\omega < \infty$, a necessary and sufficient condition on the magnitude spectrum for a filter to be physically realizable is that

$$\int_{-\infty}^{\infty} \left| \frac{\ln |H(\omega)|}{1 + \omega^2} \right| \, d\omega < \infty.$$

This is called the Paley–Wiener criterion. From this result we can conclude that for a realizable filter the magnitude of $H(\omega)$ may not fall off toward zero faster than a function of simple exponential order (for example, $e^{-k|\omega|}$, but not as $e^{-k\omega^2}$) and the attenuation may not be infinite over any band of frequencies of nonzero width.

cutoff and the infinite loss outside the pass band are reasons that the ideal filter is so desirable. Therefore we often attempt to build realizable filters to approximate the ideal filter characteristic as closely as possible. One measure—called the "shape factor"—of how well this can be done is the ratio of the filter bandwidths at -60 dB and -6 dB. A sharp-cutoff bandpass filter can be made with a shape factor as low as about 2; in contrast, the shape factor of a simple *RC* low-pass filter is almost 600!

The concepts of impulse function and impulse response are valuable analytically, but a more practical system test consists in applying a unit step to the input of a system and observing its response. Using the results obtained above, the step response follows using the convolution property:

$$g(t) = h(t) \circledast u(t)$$

$$= \int_{-\infty}^{\infty} h(\tau)u(t - \tau)\, d\tau$$

$$= \int_{-\infty}^{t} h(\tau)\, d\tau. \tag{3.75}$$

Therefore the step response of a linear time-invariant system is the indefinite integral of its impulse response. Combining Eqs. (3.74) and (3.75), we have

$$g(t) = \frac{W}{\pi} \int_{-\infty}^{t} \mathrm{Sa}\,[W(\tau - t_0)]\, d\tau.$$

Changing variable, let $x = W(\tau - t_0)$ so that

$$g(t) = \frac{1}{\pi} \int_{-\infty}^{W(t - t_0)} \mathrm{Sa}\,(x)\, dx.$$

Using the fact that $\int_{-\infty}^{0} \mathrm{Sa}\,(x)\, dx = \pi/2$, we have

$$g(t) = \frac{1}{2} + \frac{1}{\pi} \int_{0}^{W(t - t_0)} \mathrm{Sa}\,(x)\, dx. \tag{3.76}$$

The integral in Eq. (3.76) appears frequently in analytical problems and is tabulated in standard numerical tables.† It is called the "sine integral," denoted by Si (x), and defined by

$$\mathrm{Si}\,(x) \triangleq \int_{0}^{x} \mathrm{Sa}\,(u)\, du. \tag{3.77}$$

The sine integral has odd symmetry about $x = 0$, is linear for small x, and approaches $\pi/2$ for large x.

---

† See, for example, E. Jahnke and F. Emde, *Tables of Functions*, New York: Dover, 1945.

Using the sine integral notation, the step response of the ideal low-pass filter is

$$g(t) = \frac{1}{2} + \frac{1}{\pi} \, \text{Si} \, [W(t - t_0)].$$ (3.78)

The impulse response of the LPF and its step response are shown in Fig. 3.20.

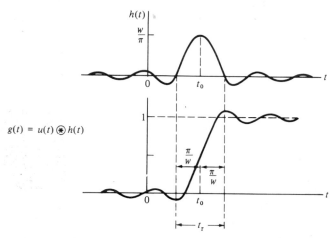

Fig. 3.20 The unit step response of an ideal low-pass filter.

Several important observations can be made from Fig. 3.20. Note that as the bandwidth W of the LPF is decreased, the filter output $g(t)$ rises to its peak value more slowly. A measure of the time it takes for this rise is called the "rise time" of the filter. Therefore we conclude that the rise time, t_r, of a filter is inversely proportional to its bandwidth. An exact measure of the rise time, of course, will depend on which particular definition we choose. For pulse circuit applications, a convenient measure of t_r is the time required for the output waveform to rise from 10% to 90% of its final value. For the LPF discussed above, this yields $t_r \approx 0.44/B$, where $B = W/2\pi$. A convenient choice for the LPF is the time required for the output waveform to rise from its minimum to its maximum. From Fig. 3.20, this criterion leads to $t_r = 1/B$. Other criteria could be chosen, but the conclusion is always the same—rise time is inversely proportional to filter bandwidth.

A useful approximation for practical low-pass filters of order greater than second-order is that:

$$t_r \approx 1/(2B),$$ (3.79)

where t_r is the 10% to 90% rise time in seconds and B is the -3 dB bandwidth in Hz. Because this is the type often used for filtering applications in communication systems, we shall use this approximation in later chapters.

Another important conclusion to be learned from Fig. 3.20 is that the rise time is a measure of an effective width of the system's impulse response. Recall that the impulse response is the derivative of the step response. The inverse relation between rise time and bandwidth is an example of the reciprocal spreading property of the Fourier transform. It is, therefore, a general property and not necessarily constrained to the example above.

Before leaving this discussion, we make one more rather general statement. Any strictly band-limited signal cannot be strictly time-limited. Conversely, a time-limited signal cannot be strictly band-limited. We can see that this is true for the rectangular-type signals and systems we have been discussing. That this is true in general is beyond the level of this book.

The concept of effective width of an impulse response is tied directly to that of the response of a linear time-invariant system in the above discussion. However, the basic underlying ideas are those of Fourier transform theory and are not constrained to systems and system responses. We wish to extend these concepts also to an effective width for signals. We call this effective width the *time duration*, T_d. If we were to use the same definition for T_d that was used for rise time, it would be that time over which a given signal has the central 80% of its area under a graph of signal waveform versus time. Signal bandwidth B can be defined in the same manner as the -3 dB bandwidth defined earlier for systems. Other definitions could be used, however, and more general analytical definitions are presented in the following section.

An important dimensionless parameter in signal analysis is the *time (duration)-bandwidth product*, $T_d B$. One reason for this importance is that the time-bandwidth product has a lower bound, but it is not constrained by an upper bound. Therefore the time-bandwidth product may be increased intentionally for a given signal, and systems that are intended for this purpose are called ''spread-spectrum'' systems. Because bandwidth is not dependent on phase, a method to accomplish this spectral spreading is to make alterations to the signal phase to increase the time-bandwidth product. Another reason for this importance is that the time-bandwidth product is a useful parameter in specifying some signal processing systems, particularly those using delay devices and those intended for spread-spectrum applications.

Drill Problem 3.13.1   Using the 10% to 90% rise time criterion, compare the rise time and the -3 dB bandwidth of the RC low-pass filter (cf. Example 2.12.1).
Answer. $t_r = 0.35/B$.

★ 3.14   MINIMUM TIME-BANDWIDTH PRODUCT

Investigation of a minimum time-bandwidth product requires some new definitions of time duration and bandwidth that are easier to handle analytically. We redefine these quantities in terms of a "radius of gyration" definition, so-called because it is analogous to the definition of the moment of inertia about a center of mass used in mechanics. Using this approach, we define a time duration T_d of a given signal $f(t)$ as

$$T_d^2 = \frac{\int_{-\infty}^{\infty} (t - t_0)^2 |f(t)|^2 \, dt}{\int_{-\infty}^{\infty} |f(t)|^2 \, dt}, \tag{3.80}$$

where t_0 is the delay time of $f(t)$. In a similar manner, the "radius-of-gyration" bandwidth W can be found from

$$W^2 = \frac{\int_{-\infty}^{\infty} (\omega - \omega_0)^2 |F(\omega)|^2 \, d\omega}{\int_{-\infty}^{\infty} |F(\omega)|^2 \, d\omega,} \tag{3.81}$$

where ω_0 is the center frequency. Explicit formulas for t_0 and ω_0 can be obtained by minimizing Eqs. (3.80), (3.81). However, because $|F(\omega)|^2$ is independent of time delay and $|f(t)|^2$ is independent of center frequency, we have deleted both t_0 and ω_0 in the following discussion.

Using Eqs. (3.80) and (3.81), and setting t_0 and ω_0 equal to zero, the time-bandwidth product (squared) is

$$(T_d W)^2 = \frac{\int_{-\infty}^{\infty} t^2 |f(t)|^2 \, dt \int_{-\infty}^{\infty} \omega^2 |F(\omega)|^2 \, d\omega}{\int_{-\infty}^{\infty} |f(t)|^2 \, dt \int_{-\infty}^{\infty} |F(\omega)|^2 \, d\omega}. \tag{3.82}$$

Using Parseval's theorem [cf. Eq. (3.21)], and using a dot above the function to indicate a derivative, we have

$$\int_{-\infty}^{\infty} |\dot{f}(t)|^2 \, dt = \frac{1}{2\pi} \int_{-\infty}^{\infty} \omega^2 |F(\omega)|^2 \, d\omega,$$

and Eq. (3.82) becomes

$$(T_d W)^2 = \frac{\int_{-\infty}^{\infty} t^2 |f(t)|^2 \, dt \int_{-\infty}^{\infty} |\dot{f}(t)|^2 \, dt}{\left[\int_{-\infty}^{\infty} |f(t)|^2 \, dt \right]^2}. \tag{3.83}$$

Fixing the signal energy, the term in the denominator of Eq. (3.83) is fixed.

Using the Schwarz inequality (cf. Ex. 2.5.2) in the numerator, we have

$$(T_d W)^2 \geq \frac{\left| \int_{-\infty}^{\infty} t f(t) \dot{f}(t)^* \, dt \right|^2}{\left[\int_{-\infty}^{\infty} |f(t)|^2 \, dt \right]^2}. \tag{3.84}$$

Using integration by parts and insisting that $f(t) \to 0$ as $t \to \pm\infty$, the numerator in Eq. (3.84) is

$$\left[\tfrac{1}{2} \int_{-\infty}^{\infty} |f(t)|^2 \, dt \right]^2,$$

and Eq. (3.84) becomes

$$(T_d W)^2 \geq \tfrac{1}{4}.$$

Thus the minimum time-bandwidth product for the above definitions of T_d, W is given by

$$T_d W \geq \tfrac{1}{2}. \tag{3.85}$$

Example 3.14.1   Find the time-bandwidth product of the gaussian-pulse signal: $f(t) = \exp[-t^2/(2\sigma^2)]$.

Solution.   Using Eqs. (3.80), (3.81), Appendix A, and setting t_0, ω_0 equal to zero we have

$$T_d^2 = \frac{\int_0^{\infty} t^2 e^{-t^2/\sigma^2} \, dt}{\int_0^{\infty} e^{-t^2/\sigma^2} \, dt} = \frac{\sigma^2}{2},$$

and

$$W^2 = \frac{\int_0^{\infty} \omega^2 e^{-\sigma^2 \omega^2} \, d\omega}{\int_0^{\infty} e^{-\sigma^2 \omega^2} \, d\omega} = \frac{1}{2\sigma^2}.$$

Thus for this example we have

$$T_d W = \tfrac{1}{2},$$

and we conclude that the gaussian-pulse signal is one that exhibits a minimum time-bandwidth product.

3.15   THE SAMPLING THEOREM

Signals bearing information may be available to us either in analog form or in digital or discrete form. We would like to determine the necessary conditions which will allow us to change an analog signal to a discrete one, or vice versa, without loss of information. As a criterion of how well this can be done, we shall insist that we be able to reconstruct the original signal completely by filtering.

   The link between the analog signal and the corresponding discrete signal is provided by what is known as the *sampling theorem*. The sampling theorem can be stated simply as follows:

   A real-valued band-limited signal having no spectral components above a frequency of B Hz is determined uniquely by its values at uniform intervals spaced no greater than $1/(2B)$ seconds apart.†

This statement is a sufficient condition such that an analog signal can be reconstructed completely from a set of uniformly spaced discrete samples in time.

   The validity of the sampling theorem can be demonstrated using either the modulation property or the frequency convolution property of the Fourier transform. Consider a band-limited signal $f(t)$ having no spectral components above B Hz. We shall sample this signal using the periodic gate function. Each rectangular sampling pulse is of unit amplitude, τ seconds in width, and occurs at intervals of T seconds.

   Denoting the sampled signal by $f_s(t)$ and the periodic gate function by $p_T(t)$, we can write

$$f_s(t) = f(t)p_T(t). \tag{3.86}$$

However, $p_T(t)$ is periodic and can be represented by a Fourier series

$$p_T(t) = \sum_{n=-\infty}^{\infty} P_n e^{jn\omega_0 t}, \tag{3.87}$$

where $\omega_0 = 2\pi/T$. Combining Eqs. (3.86) and (3.87), we have

$$f_s(t) = f(t) \sum_{n=-\infty}^{\infty} P_n e^{jn\omega_0 t}. \tag{3.88}$$

Taking the Fourier transform of both sides of Eq. (3.88) and interchanging the

---

† This statement is known as the "uniform sampling theorem." Requirements on the uniform spacing can be relaxed somewhat, yet the overall conclusions remain the same. See, for example, Chapter 4 of Reference 4.

order, we have

$$\mathcal{F}\{f_s(t)\} = \mathcal{F}\left\{f(t) \sum_{n=-\infty}^{\infty} P_n e^{jn\omega_0 t}\right\}$$

$$= \sum_{n=-\infty}^{\infty} P_n \mathcal{F}\{f(t)e^{jn\omega_0 t}\}. \tag{3.89}$$

Using the frequency translation property of the Fourier transform in Eq. (3.89), we now write the spectral density of $f_s(t)$, designated by $F_s(\omega)$, as

$$F_s(\omega) = \sum_{n=-\infty}^{\infty} P_n F(\omega - n\omega_0)$$

$$= P_0 F(\omega) + \sum_{\substack{n=-\infty \\ n \neq 0}}^{\infty} P_n F(\omega - n\omega_0). \tag{3.90}$$

Therefore we conclude that the spectral density (Fourier transform) of the sampled signal $f_s(t)$ is, within a constant factor, exactly the same as that of $f(t)$ within the original bandwidth. In addition, it repeats itself periodically in frequency every ω_0 radians per second. These replicas of the original spectral density are weighted by the amplitudes of the Fourier series coefficients of the sampling waveform, as shown in Fig. 3.21. Note that the spectral density of the original function $f(t)$ can be retrieved simply by using a LPF on $F_s(\omega)$.

Now let us investigate a change in the sampling rate. For an increase in the sampling rate, ω_0 increases, T decreases, and all replicas of $F(\omega)$ move farther apart. On the other hand, as we decrease the sampling rate, ω_0 decreases, T increases, and all replicas move closer. Soon a point is reached beyond which a reduction in sampling rate will result in overlap between spectral densities. From Fig. 3.21(f) it can be seen that this point is attained when

$$\frac{2\pi}{T} = 2W,$$

or, since $W = 2\pi B$,

$$T = \frac{1}{2B}. \tag{3.91}$$

At the point given by Eq. (3.91), all replicas of the original spectral density are just tangent to each other and an ideal low-pass filter can be used to retrieve the original from the sampled signal. However, if the sampling interval T becomes slightly larger than that given in Eq. (3.91), there is an overlap of spectral densities and the original signal cannot be retrieved by filtering the sampled signal. To avoid this condition, we must insist that

$$T < \frac{1}{2B}. \tag{3.92}$$

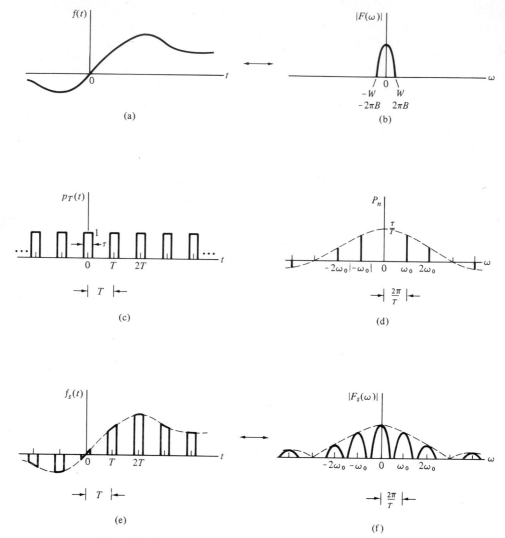

Fig. 3.21 Steps in the sampling of a band-limited signal.

Equation (3.92) is a concise mathematical statement of the sampling theorem. The maximum time interval T of sampling, given by Eq. (3.91), is called the *Nyquist interval*; its reciprocal is called the *Nyquist sampling frequency (rate)*.

In practice, the full potential of the sampling theorem usually cannot be realized and the statements above serve as upper bounds on actual performance. The first compromise we are forced to face is that we cannot build ideal low-pass filters. If the filter characteristic has a finite slope at the band edges, frequency components from the spectral replicas may be transmitted through the filter even though Eq. (3.92) is satisfied. These spurious frequencies can be suppressed by designing filters with as fast an attenuation rate as possible and then increasing the sampling frequency to allow some frequency space before the next replica appears. For this reason, practical systems oversample.

A second reason that the full potential of the sampling theorem cannot be realized arises from the fact that a time-limited signal is never strictly band-limited. When such a signal is sampled, there will be some unavoidable overlap of spectral components. In reconstruction of the signal, frequency components originally located above one-half the sampling frequency will appear below this point and will be transmitted by the low-pass filter. This is known as *aliasing* and results in distortion of the signal. The effects of aliasing can be combated by doing as good a job of filtering (i.e., band-limiting) as possible *before* sampling and by sampling at rates greater than the Nyquist rate. This is discussed in the next section.

Up to this point in our discussion of sampling we have assumed that all the signals under consideration were low-pass signals. Many signals of importance, however, are bandpass signals. Does the sampling theorem tell us that we must sample a bandpass signal at twice the highest frequency?

The answer to the question is "no" because the minimum sampling rate is dependent primarily on the bandwidth of a band-limited signal rather than on its highest frequency. For the low-pass case, of course, these two conditions coincide. When sampling a bandpass signal, we can position as many spectral replicas as possible between the original spectral densities, thus making more efficient use of spectral space. For uniform time sampling this results in a minimum sampling rate requirement between two and four times the bandwidth. For a given bandwidth, this minimum rate requirement for a bandpass signal approaches the limit of twice the bandwidth as the center frequency of the signal increases.† This is illustrated in Fig. 3.22. We shall assume that, unless specified otherwise, the minimum sampling rate for a bandpass signal is twice the bandwidth of the signal.

Fig. 3.22 Minimum periodic sampling rate for a signal of bandwidth of B Hz.

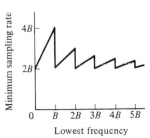

---

† See, for example, W. W. Harman, *Principles of the Statistical Theory of Communication*, New York, McGraw-Hill, 1963, Chapter 2.

A reconsideration of Fig. 3.21 shows that for the case of a complex-valued signal with a *one-sided* spectral density it is necessary to sample at a minimum rate equal to the highest frequency, not twice the highest frequency as for real-valued signals [and as given in Eq. (3.92)]. The basic principles illustrated in Fig. 3.21 remain the same, but one cannot jump to a quick conclusion based on Eq. (3.92). In reality, of course, the total number of samples taken remains the same for both cases because each sample of a complex-valued signal is an ordered pair (that is, a real part and an imaginary part). Sampling for the case of a one-sided spectral density is illustrated in the following example.

Example 3.15.1   It is possible to reconstruct a function from one of the spectral replicas after sampling. A common sampling device making use of this principle is the *stroboscope*. A stroboscope is a light source which is pulsed periodically. In a study of rotary motion, the repetition rate of the light is adjusted so that the rotary motion appears motionless. The eye, acting as a low-pass filter, reconstructs a low-frequency "alias" from the bandpass sampled signal.

For example, suppose you are watching a horse-drawn wagon in a western movie. The wagon is being chased by the bad guys *but the wheels do not appear to be rotating*! If each wheel is four feet in diameter and has twelve identical spokes, what is the minimum forward velocity of the wagon? The picture frame rate of the movie is 24 frames per second.

Solution.   The minimum required sampling frequency of the spokes is $\frac{24}{12} = 2$ Hz.

$$v = r\omega = (2)(4\pi) = 8\pi \text{ ft/sec} \approx 17 \text{ mph}$$

Note that velocities slightly lower than this will cause the wheels to appear to rotate *backwards*. The aliases here have a one-to-one correspondence with the phasor rotations discussed earlier in Chapter 2. (Sketch a spectrum of the sampled signal to convince yourself.)

Drill Problem 3.15.1   Beginning with Eq. (3.86), derive the sampling theorem using the frequency convolution property of the Fourier transform.

★ **3.16   ALIASING EFFECTS IN SAMPLING**

We have seen how the sampling of data introduces spectral replicas about multiples of the sampling frequency, and that the frequency band for spectral measurements should be limited to one-half the sampling rate. Now we return to the subject of aliasing which was discussed briefly in Chapter 2.

When a continuous function $f(t)$ is represented by M uniformly spaced samples, replicas of the spectral density of $f(t)$ are present, each spaced at a multiple of M units in frequency. The comparative magnitude of each spectral replica is dependent on the sampling pulse shape and the value of the sampling rate. If the sampling pulse shapes are impulse functions, all spectral replicas have equal area or weight (cf. Problem 2.18.1). We shall assume impulse sampling here both for convenience and because it is appropriate to the application of digital techniques in computation. Following the sampling theorem, we restrict the measurement or computation of spectral content to the interval $0 - M/2$. However, the existence of frequency components in the spectral replicas may fall within the measurement interval, causing errors.

We investigate these effects by considering the spectral density of a certain function $f(t)$, as shown in Fig. 3.23(a). If the waveform $f(t)$ is sampled using M point or impulse samples, the resulting spectral density is periodic in frequency, repeating at intervals of M, as shown in Fig. 3.23(b). Our measurement range is restricted to $M/2$, but it is seen that the measured frequency components include portions from the spectral replicas. This effect, known as aliasing, is shown by the shaded area in Fig. 3.23(b). (This effect is also known by the descriptive term "fold-over.") Note that for a given spectral density the aliasing effects decrease as the spectral replicas move apart and as the spectral components decrease more rapidly for frequencies higher than $M/2$. The former can be controlled by the choice of sampling rate. The latter can be controlled through use of a filter to attenuate those frequency components outside the band of interest. Filters used for this purpose are called "pre-alias" filters.

The above concepts can be extended easily to the Fourier series. The exponential Fourier series coefficients are related directly to the Fourier transform [cf. Eq. (3.15)] and the continuous spectral density becomes a line spectrum, as shown in Fig. 3.23(c).

Now let us estimate the accuracy of the measured coefficients in the exponential Fourier series. Using Eqs. (2.87), (2.88), and (2.93) we can write

$$F_n = \frac{1}{T} \int_0^T f(t) e^{-jn\omega_0 t} \, dt,$$

$$F_n \cong \frac{1}{M} \sum_{m=1}^M f(mT/M)[\cos(2\pi mn) - j \sin(2\pi mn)]$$

$$= \frac{a_n}{2} - j\frac{b_n}{2}. \tag{3.93}$$

Designating the true values of the exponential Fourier series coefficients by the F_n' and referring to Fig. 3.23(c), we can write an expression for the measured Fourier coefficients within the frequency range $-M/2 \le n \le M/2$ as

$$F_n = F_n' + F_{n-M}' + F_{n-2M}' + F_{n-3M}' + \cdots + F_{n+M}'$$
$$+ F_{n+2M}' + F_{n+3M}' \cdots . \tag{3.94}$$

However, for real-valued $f(t)$ we can also write [cf. Eq. (2.46)] $F_{-n} = F_n^*$ so that Eq. (3.94) becomes

$$F_n = F_n' + \sum_{k=1}^{\infty} [F_{kM+n}' + (F_{kM-n}')^*]. \tag{3.95}$$

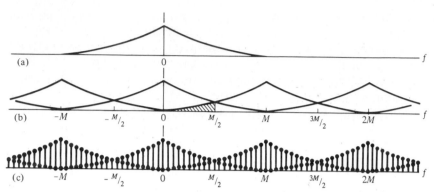

Fig. 3.23 (a) A given spectral density; (b) aliasing effects for continuous spectra; and (c) line spectra.

Note that the alias terms have been grouped into the second term of Eq. (3.95).

Equation (3.95) gives the measured (e.g., computed) value of a given Fourier series coefficient in terms of the true value and the alias terms. By use of Eq. (3.93) in Eq. (3.95), a similar set of relations can be obtained for the aliasing effects for the trigonometric series [cf. Eqs. (2.91) and (2.92)].

The effects of aliasing on the reconstructed waveform representing $f(t)$ depend on the phase relationships as well as the magnitude and are difficult to predict exactly. To generalize the approach, we shall use an energy criterion which disregards the phase. To obtain a simplified measure of aliasing on an energy basis we assume (1) that only the first spectral replica gives rise to aliasing components, and (2) that the aliasing effects are proportional to the spectral components weighted by their frequency span from the origin (this emphasizes those effects near $M/2$). Stating this in equation form, let F_n be the exponential Fourier series coefficients at the input of a pre-alias filter whose magnitude transfer function is $|H(\omega)|$. Therefore $|G_n| = |F_n||H(n\omega_0)|$ and we define the percent aliasing as

$$\% \text{ aliasing} \triangleq \frac{\sum_{n=1}^{M/2}(n - M/2)^2|G_{n+M/2}|^2}{\sum_{n=1}^{M/2} n^2|G_n|^2} \times 100\%. \qquad (3.96)$$

While the results of this calculation disregard the phase relationships, they turn out to be quite reliable on a comparative basis and are easily computed. An application of this result is demonstrated in the following example.

Example 3.16.1   A square wave is applied to a pre-alias filter whose magnitude transfer function is $|H(\omega)| = 1/\sqrt{1 + (\omega/\omega_c)^{2m}}$ and then sampled at a rate of 20 samples per period. (a) Compute the percent aliasing if the pre-alias filter is not used. (b) Compute the percent aliasing for $m = 1, 2, \ldots, 8$ and for ω_c, normalized to the half-sampling frequency, of 0.5, 0.6, 0.7, 0.8, 0.9, 1.0.

Solution.   a) Using Eq. (3.96), we can write

$$\% \text{ aliasing} = \frac{\sum_{n=1}^{10} (n - 10)^2 H^2(n + 10) \text{ Sa}^2 [(n + 10)\pi/2]}{\sum_{n=1}^{10} n^2 H^2(n) \text{ Sa}^2 (n\pi/2)} = 22.0877\%.$$

b) Here a pre-alias filter (known as an mth order Butterworth filter) with a variable -3 dB frequency (ω_c) and a variable attenuation rate ($\approx -20\ m$ dB/ decade) is used. The results are presented in tabular format below.

Percent Aliasing

m	$\dfrac{\omega_c}{\omega_s/2} =$	0.5	0.6	0.7	0.8	0.9	1.0
1		5.9830	7.2209	8.4080	9.5288	10.5723	11.5357
2		1.3154	2.2456	3.4790	4.9851	6.6896	8.4894
3		0.2676	0.6606	1.4020	2.6271	4.4042	6.6545
4		0.0538	0.1915	0.5584	1.3803	2.9390	5.3708
5		0.0108	0.0556	0.2221	0.7236	1.9719	4.4120
6		0.0022	0.0162	0.0885	0.3792	1.3251	3.6634
7		0.0004	0.0048	0.0354	0.1988	0.8902	3.0604
8		0.0001	0.0014	0.0142	0.1044	0.5976	2.5642

Note that the percent aliasing (or aliasing error) decreases with increasing attenuation rate in the pre-alias filter and as the -3 dB frequency is lowered. However, this also results in some undesirable attenuation within the desired band of frequencies. A solution is to sample at a higher rate, allowing some extra frequency band for the realizable filter characteristic.

---

c **Drill Problem 3.16.1**   A triangular waveform is sampled at a rate of 20 samples per period. (a) Compute the percent aliasing. (b) Repeat if a RC low-pass filter is used as a pre-alias filter with the -3 dB frequency set at the half-sampling frequency.

Answer.   (a) 0.6637%; (b) 0.2903%.

In summary, the accurate measurement of frequency spectral components from sampled data requires a sampling frequency high enough that a pre-alias filter can be used to attenuate the higher frequency components and yet not attenuate those frequency components appreciably that are within the band of interest. All practical measurements using sampled data involve careful compromises between these factors.

★ 3.17   THE DISCRETE FOURIER TRANSFORM

The increasing use of digital methods for computational aids and for signal processing applications has resulted in an increased emphasis on a discrete version of the Fourier transform. We examine this topic briefly here with a particular interest in how the discrete Fourier transform is related to the continuous Fourier transform.

The notation conventionally used for the discrete Fourier transform differs somewhat from that used for the continuous Fourier transform. We follow conventional usage as much as possible and the reader should be aware that the notation used in this section differs slightly from that used in the other sections of this chapter.

Let a sequence of N equally spaced samples over the interval $(0, NT)$ be represented by

$$f(kT) = f(0), f(T), f(2T), \ldots, f[(N - 1)T].$$

The *Discrete Fourier Transform* (DFT) is defined as that sequence of N complex-valued samples in the frequency domain given by†

$$F_D(n\Omega) = \sum_{k=0}^{N-1} f(kT)e^{-j\Omega Tnk}, \quad n = 0, 1, \ldots, N - 1, \tag{3.97}$$

where $\Omega = 2\pi/(NT)$. Note that $\Omega T = 2\pi/N$ and Ω and T do not appear explicitly in the DFT. These parameters enter only as scale factors for interpreting the results and are not needed in the computation steps.

In using a numerical approximation to the Fourier transform, it is necessary to restrict the observation interval to a finite length. Therefore let us define the truncated function $\tilde{f}(t)$ in terms of $f(t)$ by:

$$\tilde{f}(t) = \begin{cases} f(t) & 0 \le t < NT \\ 0 & \text{elsewhere} \end{cases}. \tag{3.98}$$

The Fourier transform, $\tilde{F}(\omega)$, of this truncated function is

$$\tilde{F}(\omega) = \int_0^{NT} f(t)e^{-j\omega t}\, dt. \tag{3.99}$$

Making the variable changes $\omega \to n\Omega$, $t \to kT$, $dt \to T$, we can approximate Eq. (3.99) by

$$\tilde{F}(n\Omega) \cong \sum_{k=0}^{N-1} f(kT)e^{-jn\Omega kT}T. \tag{3.100}$$

Equations (3.97) and (3.100) show that

$$\tilde{F}(\omega)|_{\omega=n\Omega} \cong TF_D(n\Omega). \tag{3.101}$$

A comparison with the continuous Fourier transform shows that the two are analogous if (1) the signal $f(t)$ is truncated to the interval $(0, NT)$; (2) within this

---

† Some authors use definitions which vary slightly from this, and the summation can go from $k = 1$ to $k = N$. Note that T here is used as the interval between samples.

interval the signal $f(t)$ is available as a sequence of N equally spaced values; and (3) the interval is extended periodically yielding the discrete harmonic frequencies $n\Omega = 2\pi n/(NT)$. Note that the second condition implies that the computed frequency spectra are periodic with a period of $N\Omega$. To demonstrate this, consider the calculation of $F_D(n\Omega)$ for $n > rN$, r any integer, so that $n = rN + n_1$, $n_1 < N$. Using this in Eq. (3.97), we have

$$F_D[(rN + n_1)\Omega] = \sum_{k=0}^{N-1} f(kT)e^{-j\Omega T(rN + n_1)k}.$$

But we can write $e^{-j\Omega TNr} = e^{-j2\pi r} = 1$ so that

$$F_D[(rN + n_1)\Omega] = \sum_{k=0}^{N-1} f(kT)e^{-j\Omega Tn_1 k} = F_D(n_1\Omega). \tag{3.102}$$

Therefore any value of $F_D(n\Omega)$ for $n > rN$ can be expressed in terms of a smaller argument $n_1\Omega$, where $n_1 = n$ modulo N.† In other words, $F_D(n\Omega)$ is periodic with a period of $N\Omega$.

As a result of the periodicity of $F_D(n\Omega)$, the accuracy of the computation of the continuous Fourier transform using the DFT is affected by aliasing. These effects are analogous to those encountered in our earlier discussion of the Fourier series (cf. Sections 2.17 and 3.16) and can be minimized by choosing a high sampling rate (i.e., small T).

The coefficients of the exponential Fourier series may be computed using the DFT and then multiplying by $1/N$ [cf. Eqs. (2.85), (2.88), and (2.94)]. The highest frequency component which can be determined corresponds to $n = N/2$, or $(N/2)\Omega = (2T)^{-1}$ Hz. This agrees with the sampling theorem.

Drill Problem 3.17.1 Derive Eq. (3.100) by multiplying $f(t)$ by $\sum_{k=-\infty}^{\infty} \delta(t/T - k)$ in Eq. (3.99) and letting $\omega \to n\Omega$. [Define the limits in Eq. (3.99) as 0_-, NT_-.]

In an analogy to the continuous case, the *Inverse Discrete Fourier Transform* (IDFT) is

$$f(kT) = \frac{1}{N} \sum_{n=0}^{N-1} F_D(n\Omega)e^{j\Omega Tkn}. \tag{3.103}$$

The DFT and the IDFT form an exact transform pair; it is only in the comparisons with the continuous transform that differences may arise.

Properties of the DFT are analogous to those of the continuous Fourier transform within the restrictions given above. As an example, the IDFT of the product of the DFT's of two sequences is the convolution of the sequences.

---

† A number is written as modulo N by expressing the remainder after all integral multiples of N have been subtracted; thus 14 modulo 4 is 2, 14 modulo 5 is 4, etc.

However, the resulting convolution is periodic.† In fact, because the IDFT is basically of the same form as the DFT, all functions having a DFT are automatically extended periodically with a period of NT. For a given truncated function, a convenient way in which the effects of this periodicity can be minimized is to add zeros as extra sample points to the sequence. These added zeros are called *augmenting zeros* and are placed at the end of the sequence. Augmenting zeros decrease the harmonic frequency spacings and the effects of aliasing for a given waveform at the expense of added computation time. Their use is illustrated in the following example.

Example 3.17.1   Compute the Fourier transform of $f(t) = u(t) - u(t - 1)$ using 4 samples over the interval $(0, 1)$.

Solution.   The sample sequence is $\{1, 1, 1, 1\}$ and the DFT is:

$$F_D(n\Omega) = \sum_{k=0}^{3} f(kT)e^{-j2\pi nk/4} = \begin{cases} 4 & n = 0 \\ 0 & n \neq 0 \end{cases}.$$

Using Eq. (3.101), we have

$$F(\omega) \cong \begin{cases} 1 & n\Omega = 0 \\ 0 & n\Omega \neq 0 \end{cases} \qquad n = 0, 1, 2, 3.$$

However, expressing this function as $f(t) = \text{rect}\,(t - \frac{1}{2})$, the continuous Fourier transform is (cf. Tables 3.1 and 3.2)

$$F(\omega) = e^{-j\omega/2}\,\text{Sa}\,(\omega/2).$$

These two results certainly do not look alike. Let us convert the latter result using Eq. (3.101),

$$F_D(n\Omega) = 4e^{-jn\pi/2}\,\text{Sa}\,(n\pi) = \begin{cases} 4 & n = 0 \\ 0 & n \neq 0 \end{cases} \text{ over } (0, 4),$$

which confirms our earlier result.

The apparent inconsistency here lies in an incorrect interpretation of the truncation of $f(t)$. A periodic extension of $f(t)$ does not yield the desired waveform, but rather gives a constant level of one unit. This constant has only one frequency component located at $\omega = 0$—and that is exactly what we obtained! Figure 3.24(a) illustrates what is happening by showing both $F(\omega)$ and $F_D(n\Omega)$. Note that the discrete frequency components are located at the nulls in the continuous spectrum for $\omega \neq 0$.

This situation can be improved by placing some augmenting zeros after the sequence of four samples. For example, if 4 zeros are used, the new sequence becomes $\{1, 1, 1, 1, 0, 0, 0, 0\}$. The DFT for this augmented sequence is shown

---

† Periodic convolution is often called circular convolution; see, for example, Chapter 5 in Reference 1.

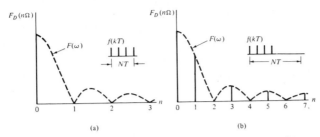

Fig. 3.24 Sampled spectrum of a unit pulse: (a) without augmenting zeros; (b) with augmenting zeros.

in Fig. 3.24(b). The addition of these augmenting zeros has increased the number of frequency components. Note that a periodic extension of this augmented sequence yields a square wave. (Is this verified by the resulting discrete spectrum?)

Drill Problem 3.17.2   Find the DFT of the following sequences: (a) $\{1, 0, 1, 0\}$; (b) $\{1, 1, 0, 0\}$.

Answer.   (a) $\{2, 0, 2, 0\}$; (b) $\{2, 1 - j1, 0, 1 + j1\}$.

We have assumed that the data is sampled over the integers $k = 0, 1, 2,$ . . . , $(N - 1)$ [or, equivalently, $k = 1, 2, 3, \ldots, N$†]. When a continuous transform is approximated by the DFT, samples for negative values of time may be assigned to the interval $(N/2, N)$. The resulting symmetry relations are shown in Fig. 3.25. If augmenting zeros are used, they may be added symmetrically about $k = N/2$. Analogous symmetry relations hold in the discrete frequency domain with the added stipulation that the negative and positive frequency components are complex conjugates for real-valued $f(t)$.

Often we are interested in making as good an estimate of the frequency components of $F(\omega)$ as possible based on data in the interval $(0, NT)$. For functions of finite duration, it is possible to choose the interval for computation to correspond to (or be greater than) the duration of the function $f(t)$. This was the case in Example 3.17.1. Another case of interest is when the interval chosen for computation is only a portion of the duration of $f(t)$. To investigate some of the effects of this truncation of $f(t)$, we rewrite Eq. (3.99) as

$$\tilde{F}(\omega) = \int_{-\infty}^{\infty} f(t) \, \text{rect} \, [(t - NT/2)/(NT)]e^{-j\omega t} \, dt. \qquad (3.104)$$

Written in this way, the truncation can be regarded as a window which permits us to observe only a finite interval of $f(t)$. Hence these functions, such as the second function in the integral in Eq. (3.104), are called *window functions*.

---

† As a result of periodicity by extension, the sample points $k = 0$ and $k = N$ are identical.

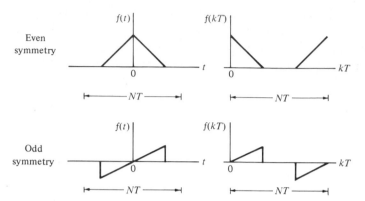

Fig. 3.25 Symmetry relations for discrete Fourier transform.

Equation (3.104) can also be expressed as a convolution in frequency:

$$\tilde{F}(\omega) = \frac{1}{2\pi} [F(\omega)] \circledast [NT \, \text{Sa} \, (\omega NT/2) e^{-j\omega NT/2}]. \tag{3.105}$$

Ideally, the second bracketed term in Eq. (3.105) should be an impulse function to give a correct measure of $F(\omega)$. However, this requires that $(NT) \to \infty$ and is not a practical alternative.

For a finite record length NT, a quantity of interest is the minimum measurable frequency separation between frequency components. This minimum separation is called the *frequency resolution* in the estimation of $F(\omega)$ from $\tilde{F}(\omega)$. If two adjacent frequency components of interest have equal amplitudes, the frequency resolution is simply set by the record length NT:

$$\Delta\Omega = 2\pi/(NT). \tag{3.106}$$

If the two components have differing amplitudes, we also require that the Fourier transform of the window function must decrease rapidly for $\omega \neq 0$. The $\sin x/x$ pattern corresponding to the rectangular window is relatively poor in this respect because the main lobe (that is, near $\omega = 0$) has unit magnitude and the first sidelobe has a peak magnitude of 0.217 (that is, -13 dB). Therefore adjacent frequency components whose magnitudes differ by more than about 5 may be indistinguishable even though Eq. (3.106) is satisfied.

A remedy is to choose a window function which truncates $f(t)$ and whose Fourier transform has low sidelobes. This has been a topic of much research and no ideal solution exists. One rather simple and yet effective window function is the Hanning window,

$$\begin{cases} \frac{1}{2}\left(1 + \cos\frac{2\pi t}{NT}\right) & |t| < NT/2 \\ 0 & |t| > NT/2 \end{cases}, \tag{3.107}$$

and is shown in Fig. 3.26. The Fourier transform of the Hanning window has lower sidelobes than that of the rectangular window at the expense of a wider main lobe and some attenuation. These compromises are typical, to differing extents, of the various window functions commonly employed in practice.

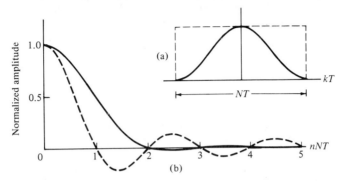

Fig. 3.26 (a) The rectangular and Hanning window functions and (b) their Fourier transforms.

Example 3.17.2 Compute the magnitude of the Fourier transform of $f(t) = \sin\sqrt{2}t + 0.1 \sin 7\pi t$ over the interval $(0, 2)$ using the DFT with 32 sample points and (a) a rectangular window; and (b) a Hanning window.

Solution. $\Omega = 2\pi/2 = \pi$, $T = \frac{2}{32} = \frac{1}{16}$;

$$F_D(n\Omega) = \tfrac{1}{2} \sum_{k=0}^{31} f(kT)\, \{1 + \cos\,[\pi(k - 16)/16]\}\, e^{-j\pi nk/16},$$

where $f(kT) = \sin(\sqrt{2}k/16) + 0.1 \sin (7\pi k/16)$ and $F(\omega)$ is given by Eq. (3.101). Results of this computation are listed below. Plotted to a logarithmic scale, they are shown in Fig. 3.27. Effects of the lower sidelobes, as well as the wider main lobe, of the Hanning window are clearly evident.

Computed Magnitude of the Fourier Transform

n	Rectangular	Hanning	n	Rectangular	Hanning
0	2.42383	.86525	8	.01787	.02490
1	.38146	.50423	9	.01622	.00009
2	.09842	.06028	10	.01499	.00006
3	.05336	.00735	11	.01408	.00005
4	.03679	.00211	12	.01340	.00003
5	.02842	.00086	13	.01291	.00002
6	.02343	.02476	14	.01258	.00002
7	.08971	.05015	15	.01239	.00001
			16	.01232	.00001

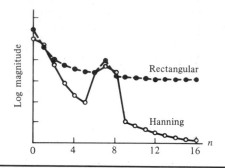

Fig. 3.27 Results of the rectangular and Hanning window functions in Example 3.17.2.

Drill Problem 3.17.3   Compute the Fourier transform of $\Lambda(t/2)$ using the DFT and eight samples, each taken at one second intervals.

Answer. $\{2, 1 + 1/\sqrt{2}, 1, 1 - 1/\sqrt{2}, 0, 1 - 1/\sqrt{2}, 1, 1 + 1/\sqrt{2}\}$.

★ 3.18   THE FAST FOURIER TRANSFORM

Computation of the DFT requires N^2 multiplications (that is, $0 \le k < N$, $0 \le n < N$) and the resulting computation time becomes excessive when N becomes large. The key to more efficient computational methods is to make use of as much symmetry of the complex exponential as possible before the multiplications are performed.

Recent advances in this area have resulted in a class of efficient algorithms, known as the *Fast Fourier Transform* (FFT), which offer significant reductions in computation time. The FFT is an algorithm (i.e., a systematic method of performing a series of computations in sequence) which enables the user to compute the DFT with a minimum of computation time. Aside from the algorithm itself, the interpretation of the FFT is the same as that for the DFT.

The particular algorithm introduced here is the commonly used Cooley-Tukey formulation.† This algorithm computes N discrete frequency components from N discrete time samples where $N = 2^r$, r any integer. This restriction to a power of two is not serious in practice as long as 2^r is greater than the number of data points—we can always fill the remainder with augmenting zeros.

As noted in the previous section, the parameters Ω and T are not really involved in the computation of the DFT. Therefore we adopt a shorthand notation by deleting them and Eq. (3.97) can be rewritten as

$$F_D(n) = \sum_{k=0}^{N-1} f(k) W^{nk},$$ (3.108)

where

$$W^\ell = e^{-j2\pi\ell/N}, \quad \ell = 0, 1, 2, \ldots.$$ (3.109)

---

† J. W. Cooley and J. W. Tukey, "An Algorithm for the Machine Calculation of Complex Fourier Series," *Math. Comput.* 19: 297–301 (April 1965).

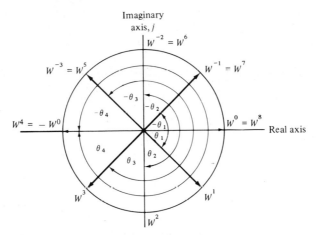

Fig. 3.28 Powers of the exponential function W for $N = 8$.

Equation (3.109) describes a phasor of unit magnitude and a phase angle of $\theta_\ell = -2\pi\ell/N$. As an example, let $N = 2^3 = 8$; the corresponding values of Eq. (3.109) are plotted in Fig. 3.28. From this figure, it can be seen that W^ℓ and $W^{-\ell}$ are symmetrically located about the real axis and the following symmetry properties hold:

$$W^N = W^0 = 1,$$

$$W^{N/2} = -1 = -W^0, \qquad (3.110)$$

$$W^{N-\ell} = [W^\ell]^*.$$

In order to utilize the full advantage of using $N = 2^r$, we express n, k as binary numbers. Suppose that $r = 2$ (that is, $N = 4$) and

$$k = (k_1, k_0) = \{00, 01, 10, 11\}, \qquad (3.111)$$

$$n = (n_1, n_0) = \{00, 01, 10, 11\},$$

where n_0, n_1, k_0, k_1 can take on values of 0 and 1 only. A compact method of writing the numerical value of k and n is

$$k = 2k_1 + k_0, \qquad (3.112)$$

$$n = 2n_1 + n_0.$$

Using Eqs. (3.111) and (3.112) in Eq. (3.108), we have

$$F_D(n_1, n_0) = \sum_{k_0=0}^{1} \sum_{k_1=0}^{1} f(k_1, k_0) W^{(2n_1+n_0)(2k_1+k_0)}, \qquad (3.113)$$

where the double summation is now needed to enumerate the complete binary representation of k.

Consider the exponential term in Eq. (3.113),

$$W^{(2n_1+n_0)(2k_1+k_0)} = W^{(2n_1+n_0)2k_1}W^{(2n_1+n_0)k_0}$$

$$= W^{2n_0k_1}W^{(2n_1+n_0)k_0}, \qquad (3.114)$$

because $W^{4n_1k_1} = 1$ for all integer n_1, k_1. This latter step is crucial to the efficiency of the FFT because now Eq. (3.113) can be rewritten as†

$$F_D(n_1, n_0) = \sum_{k_0=0}^{1}\left[\sum_{k_1=0}^{1} f(k_1, k_0)W^{2n_0k_1}\right]W^{(2n_1+n_0)k_0}. \qquad (3.115)$$

The algorithm can now be exhibited by rewriting the summation in the brackets of Eq. (3.115) as

$$f_1(n_0, k_0) = \sum_{k_1=0}^{1} f(k_1, k_0)W^{2n_0k_1}, \qquad (3.116)$$

and the outer summation as

$$f_2(n_0, n_1) = \sum_{k_0=0}^{1} f_1(n_0, k_0)W^{(2n_1+n_0)k_0}. \qquad (3.117)$$

From Eqs. (3.115) and (3.117) we have

$$F_D(n_1, n_0) = f_2(n_0, n_1). \qquad (3.118)$$

This last step is included because the ordering in the output is scrambled in this algorithm. A variant is to scramble the input ordering so that the output is in the correct ordering.

Equations (3.116)–(3.118) are the recursive relations which comprise the Cooley-Tukey algorithm for $N = 4$. A signal flow graph of these relations is shown in Fig. 3.29. Algorithms for higher powers of two follow in the same manner, with one summation in Eq. (3.113) for each power of two considered.

About N^2 complex multiply-and-add operations are required in a direct evaluation of the DFT while the FFT algorithm requires on the order of $N \log_2 N$ operations. The net savings become appreciable for large N. For example, the computation time required for a 1024-point DFT with direct evaluation is almost 100 times greater than that required using the FFT algorithm. However, the FFT algorithm requires considerable computer storage and this may limit its application in situations where only a limited amount of storage is available.

---

† For a matrix interpretation of this, see, for example, Reference 5, Chapters 10 and 11.

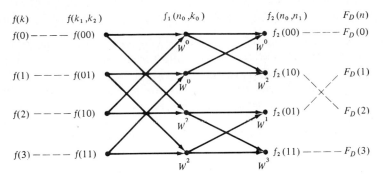

Fig. 3.29 A signal flow graph of the Cooley-Tukey algorithm for $N = 4$.

Finally, we list some points which prove helpful in processing continuous time functions with the FFT. Note that many of these arise from our earlier considerations of the DFT and the Fourier transform.

1. The number of samples N is chosen so that $N = 2^r$, r an integer. This number may include augmenting zeros [see (7) below].

2. For N time samples there are N discrete frequencies.

3. As a result of periodic extension, the sample points 0 and N are identical in both domains.

4. Positive frequency components are considered to be those over $(0, N/2)$; negative frequency components are those over $(N/2, N)$. This symmetry may also be used for time samples over negative and positive time.

5. For real-valued functions, the positive frequency components are the complex conjugates of the negative frequency components. The points $n = 0$, $N/2$ are common to both and are real-valued.

6. The highest frequency component (that is, $n = N/2$) is $(2T)^{-1}$ Hz; this may be increased by decreasing the spacing between time samples.

7. The spacing between frequency components is $(NT)^{-1}$ Hz; this may be decreased by adding augmenting zeros to the sample sequence.

8. The exact relationship of the FFT values depends on the particular multiplicative constants assigned in the algorithm; a fairly common procedure is to divide by N so that the computed values are $1/N$ times the DFT.

Drill Problem 3.18.1   (a) Repeat Example 3.17.2, using a FFT algorithm, and check your results with those given. (b) Add 224 augmenting zeros, rerun your program, and check results. (c) Determine the frequency component spacings and the maximum frequency computed for parts (a) and (b).

Answer.   (c) 0.5000 Hz, 8 Hz; 0.0625 Hz, 8 Hz.

3.19 SUMMARY

A nonperiodic signal $f(t)$ can be represented in terms of exponential functions over the entire real line $(-\infty, \infty)$. To be valid for all time, it must be represented by a continuous sum of exponential functions with frequencies lying in the interval $(-\infty, \infty)$. The Fourier transform is a method of expressing a given signal in terms of such a continuous set of exponential components of frequency. The resulting frequency function, $F(\omega)$, gives the relative weighting of each frequency component of the signal and is called the spectral-density function of the signal.

A signal of finite energy can always be described uniquely by a continuous spectral-density function. Periodic components (including dc levels) in the signal can be handled by introducing impulses in the spectral-density functions.

The absolute magnitude squared of the spectral-density function is an indicator of the relative energy per unit of frequency and is called the energy spectral-density function. The area under the energy spectral-density function is equal to the energy in the signal.

It is often convenient to use established properties of the Fourier transform pair without evaluating the transforms themselves. These properties are shown using the integral properties of the Fourier transform. One of the most useful properties is that of convolution. This property states that the Fourier transform reduces convolution operations to multiplications. A direct result of this property is that the system frequency transfer function is the Fourier transform of the impulse response of the system.

A linear time-invariant system acts as a filter on the various frequency components applied to the system. Transversal filters can be synthesized by using an addition of weighted tap outputs of a tapped delay line. Popular ways to construct transversal filters include charge-coupled devices (CCD) and surface-acoustic-wave (SAW) delay lines.

The bandwidth of a system is defined to be the interval of positive frequencies over which the magnitude of the system transfer function remains constant within a given numerical factor. For a distortionless transmission, the overall response of a system must have a constant (i.e., flat) magnitude response and a linear phase response with frequency.

A convenient experimental test of a linear system is to apply a step function to the input. The output is the indefinite integral of the impulse response of the system. The rise time of the system is that time required for the output to effectively rise to the level of the final value after the step function has been applied. The rise time is a measure of the width of the system impulse response and is inversely proportional to the system bandwidth.

Time duration is a measure of the width of a pulse signal. Time duration multiplied by bandwidth is called the signal's time-bandwidth product. Although there is a lower bound, the time-bandwidth product of a signal may be increased.

Any band-limited signal is determined by its values at uniform intervals spaced no greater than $1/(2B)$ seconds apart. This statement is called the sampling theorem. The minimum sampling rate, $(2B)$ seconds^{-1}, is called the Nyquist sampling rate.

The sampling of data introduces spectral replicas about multiples of the sampling frequency. Frequency components in these spectral replicas may fall within the bandwidth restrictions set by the sampling theorem, causing errors known as aliasing. These errors can be decreased by using higher sampling rates and by using filters before sampling to band-limit the signal.

The Discrete Fourier Transform (DFT) can be used to represent a sampled signal in terms of a finite set of complex exponentials. The DFT may be used to approximate the Fourier series and the Fourier transform. The accuracy of this approximation is limited by truncation and aliasing effects.

The Fast Fourier Transform (FFT) is an efficient algorithm for the numerical computation of the DFT. The most commonly used version of the FFT computes N frequency components from N time samples for $N = 2^r$, r any positive integer.

Selected References for Further Reading

1. C. D. McGillem and G. R. Cooper. *Continuous and Discrete Signal and System Analysis.* New York: Holt, Rinehart & Winston, 1974.
   A treatment paralleling much of this chapter with added emphasis on numerical methods of convolution.

2. R. A. Gabel and R. A. Roberts. *Signals and Linear Systems.* Second Ed. New York: John Wiley, 1980.
   A fairly thorough treatment of convolution and the discrete Fourier transform.

3. R. Bracewell. *The Fourier Transform and Its Applications.* Second Ed. New York: McGraw-Hill, 1978.
   Although more advanced, this book is very readable and stresses physical intuition.

4. H. S. Black. *Modulation Theory.* Princeton, N.J.: D. Van Nostrand, 1953.
   Presents an expanded discussion of sampling theory and applications; a classic in its field.

5. E. O. Brigham. *The Fast Fourier Transform.* Englewood Cliffs, N.J.: Prentice-Hall, 1974.
   A good and fairly complete discussion of the DFT and FFT as well as a concise treatment of the continuous Fourier transform.

6. A. Papoulis. *The Fourier Integral and Its Applications.* New York: McGraw-Hill, 1962.
   An advanced treatment of the Fourier transform and its application from a more mathematical perspective.

Problems

3.1.1 Determine the function $f(t)$ whose Fourier transform is shown in Fig. P–3.1.1.

3.1.2 Show that if $F(\omega) = \mathscr{F}\{f(t)\}$, then:

a) $F(0) = \displaystyle\int_{-\infty}^{\infty} f(t)\, dt,$

b) $|F(\omega)| \le \displaystyle\int_{-\infty}^{\infty} |f(t)|\, dt.$

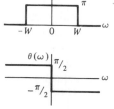

Fig. P–3.1.1.

3.2.1 A rectangular gate pulse of unit height, unit width, and centered at $t = 0$ has a Fourier transform $F(\omega) = \mathrm{Sa}(\omega/2)$.

a) Plot the magnitude spectral density; on the same graph plot the magnitude of the exponential Fourier series coefficients for a rectangular pulse train with the same shape of pulses and with a repetition period of two seconds.

b) Repeat for a period of three seconds and plot on the same graph.

c) What conclusions can you draw about finding a Fourier transform using a periodic sequence?

3.2.2 Find the spectral density of

$$f(t) = \begin{cases} \pm \exp(at) & t < 0 \\ \exp(-at) & t > 0 \end{cases}.$$

3.2.3 A pulse signal described by $f(t) = \exp(-10|t|) \ \mathrm{rect}\ (t)$ is repeated periodically every second.

a) Find the exponential Fourier series beginning with the Fourier transform of $\exp(-10|t|)$ and then converting to the series.

b) Find the exponential Fourier series of this periodic signal using a direct evaluation of the coefficients. Compare with your answer to part (a).

3.2.4 The time function $f(t) = (1/\sigma\sqrt{2\pi})e^{-t^2/2\sigma^2}$ ($\sigma = $ constant) is known as the gaussian function. This function has finite energy and thus is Fourier transformable. Find its Fourier transform. In carrying out your solution it will be helpful to combine exponents, complete the square in the exponent, and then use the definite integral $\int_{-\infty}^{\infty} e^{-u^2}\, du = \sqrt{\pi}$. Note that $f(t)$ and $F(\omega)$ have the same mathematical form; i.e., the gaussian function is its own Fourier transform.

★ **3.3.1** State if the Fourier transforms of the following functions, defined over $(-\infty, \infty)$, exist. Defend your answers.

a) $f(t) = |t|$

b) $f(t) = \exp(-t^2)$

c) $f(t) = \mathrm{rect}\ (1/t)$

d) $f(t) = \cos(1/t)$

e) $f(t) = \begin{cases} 1 & t = 0 \\ 0 & \text{elsewhere} \end{cases}$

3.4.1 The Fourier transform of a certain signal $f(t)$ is $F(\omega) = \pi \exp(-|\omega|)$. Calculate the percentage of the total energy in $f(t)$ contributed by frequency components up to one radian per second.

3.4.2 Determine the required numerical value of the positive real constant a if it is given that one-half the energy in $f(t) = \exp(-at)u(t)$ lies in the spectral range from zero to one Hz.

3.4.3 Evaluate the following definite integrals using Parseval's theorem:

a) $\displaystyle\int_{-\infty}^{\infty} [\sin x/x]^2 \, dx;$

b) $\displaystyle\int_{\infty}^{\infty} dx/(a^2 + x^2);$

c) $\displaystyle\int_{-\infty}^{\infty} dx/(a^2 + x^2)^2.$

3.5.1 Find the Fourier transform of

a) $f(t) = 2\delta(t + 2) - 2\delta(t - 2),$

b) $g(t) = \displaystyle\int_{-\infty}^{t} f(\tau) \, d\tau.$

3.5.2 The Fourier transform of a given signal $f(t)$ is

$$F(\omega) = (1 - j\omega)\frac{\omega^2 + 2}{\omega^4 + 3\omega^2 + 2} + \pi\delta(\omega).$$

Determine (a) the average (dc) value of the signal; and (b) the signal $f(t)$.

3.5.3 Suppose we define a new integral transform, which we shall call the *A*-transform, defined as follows:

$$F(a) = \lim_{T \to \infty} \int_{-T/2}^{T/2} f(t)(1 + at) \, dt \triangleq A\{f(t)\}.$$

Find the *A*-transform for the following functions:

a) $A\{\text{rect}(t/\tau)\}$
b) $A\{e^{-ct}u(t)\}$
c) $A\{e^{-c|t|}\}$
d) $A\{\delta(t - t_0)\}$

3.6.1 Two time signals, $f_1(t)$, $f_2(t)$, are shown in Fig. P–3.6.1 [$f_1(t) \leftrightarrow F_1(\omega)$; $f_2(t) \leftrightarrow F_2(\omega)$]. Sketch and compare $|F_1(\omega)| + |F_2(\omega)|$ to $|F_1(\omega) + F_2(\omega)|$. Should they be the same? Explain.

Fig. P–3.6.1.

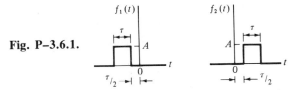

3.6.2 Use the modulation property to find the function $f(t)$ whose Fourier transform is shown in Fig. P–3.6.2.

Fig. P–3.6.2.

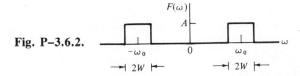

3.6.3 If $f(t) \leftrightarrow F(\omega)$, determine the Fourier transform of

a) $f(1 - t)$,

b) $f[(t/2) - 2]$,

c) $\dfrac{df}{dt} \cos t$,

d) $\dfrac{d}{dt} [f(-2t)]$.

3.6.4 Find the Fourier transform of the pulse waveform shown in Fig. P–3.6.4 using only the transform of an impulse together with appropriate transform properties.

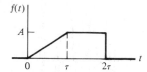

Fig. P–3.6.4.

3.6.5 A given pulse waveform is

$$f(t) = \begin{cases} \cos \pi t & |t| < \tfrac{1}{2} \\ 0 & \text{elsewhere} \end{cases}.$$

a) Sketch df/dt and d^2f/dt^2.

b) Using the result for d^2f/dt^2, write a differential equation in terms of $f(t)$ and then solve for $F(\omega)$.

3.6.6 The RC low-pass filter is sometimes used as an integrator. Using the integration property of the Fourier transform, determine the conditions when this is permissible.

3.6.7 A certain function of time, $f(t)$, has a Fourier transform given by

$$F(\omega) = [1/(\omega^2 + 1)] \exp [-2\omega^2/(\omega^2 + 1)].$$

Using the properties of the Fourier transform, write down the Fourier transform of

a) $f(2t)$,

b) $f(t - 2)e^{jt}$,

c) $4 \, df/dt$,

d) $\displaystyle\int_{-\infty}^{t} f(\tau) \, d\tau$.

3.6.8 Two functions of time, $f(t)$ and $g(t)$, are known to satisfy the following integral equation:

$$g(t) = \int_{-\infty}^{\infty} g(\tau)f(t - \tau) \, d\tau + \delta(t).$$

a) If $f(t) \leftrightarrow F(\omega)$, $g(t) \leftrightarrow G(\omega)$, find a relation between $F(\omega)$ and $G(\omega)$.

b) If $f(t) = e^{-2t}u(t)$, find $g(t)$.

3.6.9 The spectral density of the input to a given linear time-invariant system is $F(\omega) = [\exp (-j\pi\omega)]/(1 + j\omega)$. The corresponding output spectral density is $G(\omega) = [\exp (-j2\pi\omega)]/(1 - \omega^2 + 2j\omega)$.

a) Determine the transfer function of the system.

b) Determine the impulse response of the system.

3.6.10 Using the Fourier transform, show that the area under the convolution result of two given functions is equal to the product of the areas under the two functions.

3.7.1 Often an overall system is composed of groups of smaller systems, each designed to accomplish a specific objective. Using the properties of convolution, demonstrate the validity of the following rules as long as all operations remain linear and time-invariant:

a) The order of operation is immaterial.

b) The overall system impulse response is the convolution of impulse responses of all subsystems making up the system.

3.7.2 Show that

a) $f_1(t) \exp(j\omega_0 t) \circledast f_2(t) \exp(j\omega_0 t) = g(t) \exp(j\omega_0 t)$, if $g(t) = f_1(t) \circledast f_2(t)$.

b) the convolution of two delayed signals is delayed by the sum of the two individual delays.

3.8.1 Evaluate the following convolution integrals:

a) $u(t) \circledast u(t)$,

b) $u(t) \circledast tu(t)$,

c) $u(t) \circledast e^{-t}u(t)$,

d) $e^{-at}u(t) \circledast e^{-bt}u(t)$.

3.8.2 A running average or sliding integrator is defined by the integral operator

$$\frac{1}{T}\int_{t-T}^{t} f(\tau)\, d\tau,$$

where t indicates present time, T is the averaging time, and $f(t)$ is the function being averaged.

a) Express this running average as a convolution of $f(t)$ with a second function, $h(t)$. Plot and dimension $h(t)$.

b) Compare $h(t)$ to the impulse response of a low-pass RC filter (which is often used for a running average).

3.8.3 Consider two arbitrary signals $f_1(t)$ and $f_2(t)$ of finite widths t_1, t_2, respectively. In particular, let $f_1(t)$ be nonzero only over $(0, t_1)$, and let $f_2(t)$ be nonzero only over $(0, t_2)$. Show that the width of $f_1 \circledast f_2$ is the sum of the individual widths $t_1 + t_2$.

3.8.4 Evaluate $f_1 \circledast f_2$ for the functions shown in Fig. P–3.8.4.

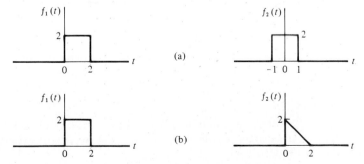

Fig. P–3.8.4 (Continued on p. 146.)

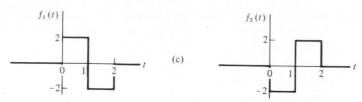

Fig. P–3.8.4 (Continued from p. 145.)

3.8.5 Sketch $f_1(t) \circledast h(t)$, $f_2(t) \circledast h(t)$, and $f_3(t) \circledast h(t)$ for the functions shown in Fig. P–3.8.5.

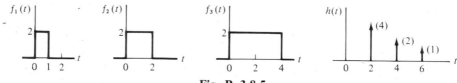

Fig. P–3.8.5.

3.8.6 Perform the convolution indicated in Fig. P–3.8.4(a), (c) by (1) differentiating each function once, (2) performing the convolution, and then (3) integrating the result twice. Why does this work out?

c **3.8.7** The signal shown in Fig. P–3.8.7 is applied to a filter whose impulse response is $h(t) = ae^{-at}u(t)$. Find the output signal in graphical form. Use numerical methods and space the sample points at 0.1 second intervals. Assume continuity to the left in the waveform. Compute and plot over (0, 5) for $a = 0.5, 1, 4$.

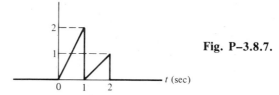

Fig. P–3.8.7.

3.8.8 A certain function of time, $f(t)$, has a Fourier transform shown in Fig. P–3.8.8.

a) Sketch the Fourier transform of $f(2t)$.
b) Sketch the Fourier transform of $[f(t)]^2$.

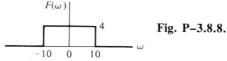

Fig. P–3.8.8.

3.8.9 Determine the Fourier transforms of the following finite-length sinusoids using the frequency-convolution property.

a) $f(t) = A \cos \omega_0 t \, u(t)$
b) $f(t) = A \, \text{rect} \, (t/\tau) \cos \omega_0 t$

3.9.1 The frequency transfer function of a given system is shown in Fig. P–3.9.1. If the input signal to this system is $f(t) = \text{Sa} \, (t - t_0)$, sketch the spectral density of the corresponding output signal.

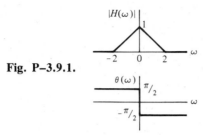

Fig. P–3.9.1.

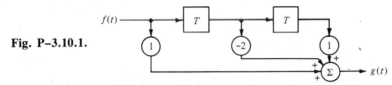

3.10.1 Determine the magnitude of the frequency transfer function of the transversal filter shown in Fig. P–3.10.1.

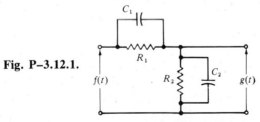

Fig. P–3.10.1.

3.10.2 A problem arising in high-frequency communication systems is that if the signal $f(t)$ is sent, the received signal may consist of $f(t)$ plus a delayed replica that arrives later via a longer transmission path. This is called *multipath*. Assuming that the received signal is

$$g(t) = f(t) + \alpha f(t - \tau),$$

design a transversal filter to compensate for this multipath if $\alpha \ll 1$.

3.11.1 Determine the -3 dB bandwidth, in Hz, of the filter whose frequency transfer function is $H(\omega) = 50/(\omega^2 + 100)$. Also calculate the ratio of the -60 dB to -6 dB bandwidths for this filter.

3.12.1 An attenuator network commonly used in measurement equipment is shown in Fig. P–3.12.1.

a) Show that the attenuation is distortionless if $R_1 C_1 = R_2 C_2$.
b) Derive a relation for the attenuation in terms of the constant k, where $R_2 = kR_1$ and $R_1 C_1 = R_2 C_2$.

Fig. P–3.12.1.

3.12.2 If we insist that the maximum deviation from the requirements for distortionless transmission is to be 1%, what is the maximum range of frequencies that may be handled by a RC low-pass filter? What is the time delay introduced by this filter?

3.13.1 Sketch the output of the filter whose frequency transfer function is shown in Fig. P–3.13.1 if the input signal is (a) $f(t) = 2u(t)$; (b) $f(t) = 10 \, \delta(t)$.

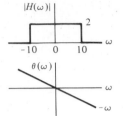

Fig. P–3.13.1.

3.13.2 Determine the 10%–90% rise time of the filter whose frequency transfer function is given in Problem 3.11.1 and compare with the reciprocal of the -3 dB bandwidth. [*Hint*: First find the impulse response.]

3.13.3 A common rule in the design and operation of scanning spectrum analyzers is that "the frequency sweep rate must be less than the square of the filter bandwidth." In fact, a warning light is used on some spectrum analyzers to warn the user when this inequality is not satisfied. Defend this rule.

★ **3.14.1** Using Eqs. (3.80) and (3.81), determine the time-bandwidth product of the pulse signal $f(t) = \text{rect }(t)$.

★ **3.14.2** Determine the time-bandwidth product of the pulse signal $f(t) = \exp(-a|t|)$.

3.15.1 A device commonly used to check the rotational rate of audio record turntables is a circular disc with equally spaced identical dark lines on a contrasting background and placed radially on the disc. When the disc is viewed under a strobe light, the pattern is designed to remain stationary if the rotational rate is correct. For example, suppose the turntable speed is set for $33\frac{1}{3}$ rpm. A strobe (e.g., a neon or even a fluorescent light) is connected to the 60-Hz ac line and flashes at 120 sec^{-1}. Calculate the minimum number of lines required on the disc. Repeat the calculation if the line frequency is 50 Hz.

3.15.2 A signal $f(t)$ whose spectral density is shown in Fig. P–3.15.2 is sampled using a periodic rectangular pulse waveform. The width of each sampling pulse is 20 msec and the period is equal to the Nyquist interval. Sketch the spectral density of the sampled signal from zero to 100 Hz, labeling important points.

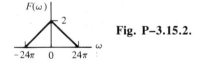

Fig. P–3.15.2.

3.15.3 A sampling system composed of an ideal multiplier and an ideal low-pass filter is shown in Fig. P–3.15.3. The signal $p(t)$ is a square wave whose Fourier cosine series is $p(t) = 2[\cos t - \frac{1}{3}\cos 3t + \frac{1}{5}\cos 5t - \cdots]$.

a) Sketch and label $Y(\omega)$ and $Z(\omega)$.
b) If $z(t)$ is applied as an input to a second system which is identical to the system described above, will the output of this second system resemble $f(t)$? Explain.

3.15.4 The spectral density of a given signal $f(t)$ is band-limited from 4 to 5 Hz, as shown in Fig. P–3.15.4. Determine the minimum sampling frequency for this signal.

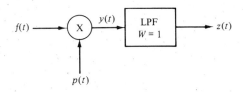

Fig. P–3.15.3.

Fig. P–3.15.4.

★ **3.16.1** The signal $f(t) = \cos 2\pi t + 0.1(\cos 5\pi t + \cos 7\pi t)$ is sampled (ideally) at a 4-Hz rate. The sampled waveform is then band-limited to 2 Hz using an ideal low-pass filter. Write an expression for the output.

c★ **3.16.2** The signal $f(t) = 50e^{-4t}$, $0 < t \le 2\pi$, is repeated periodically with $T = 2\pi$. Calculate the first five alias terms with the largest magnitudes if the sampling rate used is 20 samples per period (ideal sampling).

c★ **3.16.3** Compute the percent aliasing error in Problem 3.16.2; repeat using a RC low-pass pre-alias filter whose -3 dB frequency is 80% of the half-sampling frequency.

c★ **3.16.4** A RC low-pass filter is used as a pre-alias filter for an input signal whose magnitude spectral density is a constant. The -3 dB frequency of the RC filter is set at the highest frequency of interest in the input signal. Determine the minimum required sampling frequency in terms of the filter -3 dB frequency such that the percent aliasing error is less than 1%.

★ **3.17.1** Compute the DFT of the following sequences:
   a) $\{1, 0, 0, 0\}$,
   b) $\{0, 1, 0, -1\}$,
   c) $\exp(-k)$, $N = 4$.

c★ **3.17.2** a) Compute the magnitude of the Fourier transform of $f(t) = \exp(-t)u(t)$ over $(0, 2)$ using the DFT and 16 sample points. Assign a value of 0.5 to the point at $t = 0$.
   b) Compare your result with that of the magnitude of the continuous Fourier transform evaluated at the discrete frequency points.

c★ **3.17.3** Compute the magnitude of the Fourier transform of the following functions using the DFT and eight samples over the interval $(0, 1)$: (a) $\sin \pi t$; (b) $\sin 2\pi t$. Explain any significant differences in their spectral densities.

c★ **3.17.4** Using the DFT, compute and plot the energy spectral density of $f(t) = (1/t) - \frac{1}{4}$; $0 < t \le 5$. Use 50 equally spaced time samples and plot over the integer frequency increments: 0, 1, 2, . . . , 49.

c★ **3.17.5** Evaluate Problem 3.8.4(a) by taking the DFT of each, multiplying, and then taking the IDFT. Use sample spacings of 0.1, assume continuity to the left, and use as many augmenting zeros as samples in each sequence. Plot your results.

c★ **3.18.1** Using the FFT, compute and plot the Fourier transform of $f(t) = (20/\pi)$ Sa $(20t)$ rect (t/π); $-2\pi < t < 2\pi$. Let $N = 256$ and plot for $0 \le n \le 100$.

c★ **3.18.2** The Hamming window function is defined by

$$\begin{cases} 0.54 + 0.46 \cos[\pi t/(NT)] & |t| < NT \\ 0 & |t| > NT \end{cases}.$$

a) Using a FFT algorithm, compute and plot the spectral density of both the Hanning and the Hamming window functions. Choose 64 sample points for each, plus 192 augmenting zeros.

b) Compare the two spectral densities on the basis of (1) the heights of the largest sidelobes; and (2) the rate of decrease of the sidelobes.

c★ **3.18.3** Compute and plot the output signal in Problem 3.8.7 using the FFT with 100 sample points plus 156 augmenting zeros.

c★ **3.18.4** The discrete signal,

$$f(k) = \begin{cases} 1 & k = 0, 10 \\ 0 & \text{otherwise} \end{cases} \quad 0 \le k < 128,$$

is passed through a LPF such that only the first 5 harmonics are not attenuated to zero. Compute and plot the resulting output. [*Hint*: Be sure to observe the symmetry.]

CHAPTER 4

POWER SPECTRAL DENSITY

The Fourier methods of signal analysis which have been discussed in the preceding chapters give some valuable insights into the transmission of signals through systems. A topic of particular importance is that of the distribution of energy and power in both time and frequency. While our motivation for considering this topic is from a deterministic viewpoint, we find that the spectral representation of power can be applied also to a description of random signals and noise. In this chapter we develop some methods for handling the averaged effects of noise on a spectral basis. These methods will prove useful in the succeeding chapters.

4.1 ENERGY SPECTRAL DENSITY

Parseval's theorem gives a relation between a time signal $f(t)$ and its Fourier transform $F(\omega)$ [cf. Eq. (3.21)]:

$$\int_{-\infty}^{\infty} |f(t)|^2 \, dt = \frac{1}{2\pi} \int_{-\infty}^{\infty} |F(\omega)|^2 \, d\omega. \tag{4.1}$$

The integral on the left-hand side of Eq. (4.1) is the energy in $f(t)$ across a one-ohm resistance. Thus the quantity $|F(\omega)|^2$ appearing in Eq. (4.1) is the energy per unit of frequency normalized to a resistance of one ohm. For resistive loads not equal to one ohm the following reasoning is helpful.

The energy in a waveform can be found by integrating the product of the voltage and the current (i.e., the instantaneous power) over the interval under consideration. If we designate the voltage by $v(t)$ and the current by $i(t)$, then the energy is[†]

---

† We are using a generalized complex-valued notation of which the more familiar phasor notation, $v(t) = V \exp(j\omega_0 t)$, $i(t) = I \exp(j\omega_0 t)$, is a special case.

$$E = \int_{-\infty}^{\infty} v(t)i^*(t) \, dt. \qquad (4.2)$$

However, we are using the mathematical representation $f(t)$ for a signal whether it is a voltage or a current. If the resistive load is one ohm, then it doesn't make any difference and we can proceed. For resistive loads not equal to one ohm, the resistance scaling must be introduced using the familiar Ohm's law: $v(t) = Ri(t)$. Suppose, for example, that $f(t)$ represents a voltage. Using Parseval's theorem, we find the energy relations are

$$E_f = \frac{1}{R} \int_{-\infty}^{\infty} |f(t)|^2 \, dt \qquad (4.3a)$$

$$\left. \right\} \text{ for } f(t) \text{ a voltage.}$$

$$E_f = \frac{1}{2\pi R} \int_{-\infty}^{\infty} |F(\omega)|^2 \, d\omega \qquad (4.3b)$$

Analogous relations can be written for the case where $f(t)$ represents a current:

$$E_f = R \int_{-\infty}^{\infty} |f(t)|^2 \, dt \qquad (4.4a)$$

$$\left. \right\} \text{ for } f(t) \text{ a current.}$$

$$E_f = \frac{R}{2\pi} \int_{-\infty}^{\infty} |F(\omega)|^2 \, d\omega \qquad (4.4b)$$

Therefore the resistance is merely a scaling factor which can be taken into account later. The conversion factors are easy to supply for any given condition and we shall return to our assumption of a one-ohm resistive load. The dimension of energy in the MKS system is the joule.

For a one-ohm resistance, the quantity $|F(\omega)|^2$ appearing in the above expressions is in terms of energy per unit of frequency. For this reason the quantity $|F(\omega)|^2$ is called the *energy spectral density* of the signal $f(t)$, often shortened to the term energy density. Thus the energy spectral density is that function (1) that describes the relative amount of energy of a given signal versus frequency, and (2) whose total area under $|F(\omega)|^2$ is the energy of the signal. Note that the quantity $|F(\omega)|^2$ describes only the relative amount of energy at various frequencies. For continuous $|F(\omega)|^2$, the energy at any given frequency is zero—it is the *area* under $|F(\omega)|^2$ that contributes energy. To find energy, then, we must be given a range of frequencies over which to integrate and the amount of energy is dependent on the magnitude squared of $F(\omega)$ over that given range of frequencies.

The concept of energy spectral density is an important one for it permits us to account for relative spectral-energy attenuations through linear systems. To see this, let us apply a signal $f(t)$ to the input of a linear time-invariant system whose frequency transfer function is $H(\omega)$. Describing the (amplitude) spectral density of the output by $G(\omega)$, we have

$$G(\omega) = F(\omega)H(\omega).$$

Therefore the (normalized) energy density of $G(\omega)$ is[†]

$$|G(\omega)|^2 = |F(\omega)|^2 |H(\omega)|^2, \tag{4.5}$$

and the (normalized) energy in the output signal is

$$E_g = \frac{1}{2\pi} \int_{-\infty}^{\infty} |F(\omega)|^2 |H(\omega)|^2 \, d\omega. \tag{4.6}$$

In other words, the energy density of the system response is given by the energy density of the system input multiplied by the squared magnitude of the system transfer function. All phase information in both the input signal and the system transfer function is lost in the calculation of energy and energy density. Therefore only the magnitude of the system transfer function need be considered in any calculation of energy density.

An interesting and useful physical interpretation of energy density can be obtained through the use of Eq. (4.6). Consider a signal $f(t)$ applied to the input of a very narrow bandpass filter whose frequency transfer function $H(\omega)$ is shown in Fig. 4.1. Designating the output of the narrow-band filter as $g(t)$, we find the energy in $g(t)$ is

$$E_g = \frac{1}{2\pi} \int_{-\infty}^{\infty} |G(\omega)|^2 \, d\omega$$

$$= \frac{1}{2\pi} \int_{-\omega_0 - (\Delta\omega/2)}^{-\omega_0 + (\Delta\omega/2)} |F(\omega)|^2 |H(\omega)|^2 \, d\omega + \frac{1}{2\pi} \int_{\omega_0 - (\Delta\omega/2)}^{\omega_0 + (\Delta\omega/2)} |F(\omega)|^2 |H(\omega)|^2 \, d\omega$$

$$\cong \frac{1}{2\pi} |F(-\omega_0)|^2 \Delta\omega + \frac{1}{2\pi} |F(\omega_0)|^2 \Delta\omega, \tag{4.7}$$

where $\Delta\omega$ has been assumed to be small in Eq. (4.7).

If the signal $f(t)$ is real-valued, then $F^*(-\omega) = F(\omega)$ (see Section 3.6.2) and thus $|F(-\omega)| - |F(\omega)|$. An important consequence of this is that *for all real-valued signals the energy spectral density is an even function of ω*. Proceeding,

---

[†] The system frequency transfer function, $H(\omega)$, is not required to be dimensionless; i.e., it can convert a voltage to a current or a current to a voltage. The appropriate unit changes can be made along with the input resistance scale changes after all other calculations have been made. For our purposes, $H(\omega)$ is dimensionless and normalized to one ohm.

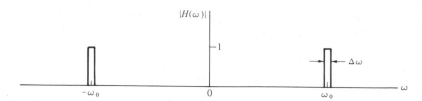

Fig. 4.1 An ideal narrow bandpass filter.

Eq. (4.7) simplifies to

$$E_g \cong \frac{1}{\pi} |F(\omega_0)|^2 \Delta\omega, \qquad (4.8)$$

if $f(t)$ is a real-valued signal. Note that half of the energy is contributed by the negative frequency components and half by the positive frequency components if the signal $f(t)$ is real-valued.

The practical significance of the above discussion can be realized by reversing the procedure. Given a pulse signal $f(t)$, how can we find its energy spectral density? One way is to extend the above reasoning and to construct a parallel bank of narrow-band filters, all filters positioned in frequency adjacent to each other. If we apply $f(t)$ to this parallel bank of filters, as shown in Fig. 4.2(a), we can obtain an approximation to the energy spectral distribution of $f(t)$. This is illustrated in Fig. 4.2(b). Assigning one-half the energy contributions to negative frequency components yields the final result shown in Fig. 4.2(c). A device available for performing this function is called a "multi-channel spectral analyzer."

In summary of our discussion, the energy spectral density of a signal represents its energy per unit of frequency and displays the relative energy contributions of the various frequency components. The area under the energy spectral density gives the energy within a given band of frequencies.

Example 4.1.1 A voltage signal described by $f(t) = e^{-5t}u(t)$ is applied to the input of an ideal low-pass filter. The low-frequency gain of the filter is unity, the bandwidth is 5 radians per second, and the resistance levels are 50 ohms. Calculate the energy of the input signal and of the output signal.

Solution. The energy in the input signal $f(t)$ is

$$E_f = \frac{1}{R} \int_{-\infty}^{\infty} |f(t)|^2 \, dt = \frac{1}{50} \int_{0}^{\infty} e^{-10t} \, dt$$

$$= \left(\frac{1}{50}\right)\left(\frac{1}{10}\right) = 0.002 \text{ joule.}$$

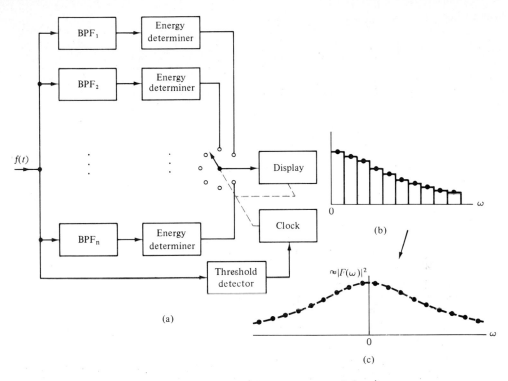

Fig. 4.2 Measurement of energy spectral density.

The energy in the output signal $g(t)$ is

$$E_g = \frac{1}{2\pi R} \int_{-\infty}^{\infty} |G(\omega)|^2 \, d\omega = \frac{1}{2\pi R} \int_{-\infty}^{\infty} |F(\omega)|^2 \, |H(\omega)|^2 \, d\omega$$

$$= \frac{1}{2\pi R} \int_{-5}^{5} \frac{d\omega}{\omega^2 + 25} = \frac{1}{\pi R} \int_{0}^{5} \frac{d\omega}{\omega^2 + 25}$$

$$= \frac{1}{5\pi R} \tan^{-1}(1) = \frac{1}{5\pi R}\left(\frac{\pi}{4}\right) = 0.001 \text{ joule.}$$

Example 4.1.2 A signal $f(t) = e^{-at}u(t)$ is applied to the input of a low-pass filter with a magnitude frequency transfer function $|H(\omega)| = b/\sqrt{\omega^2 + b^2}$. Determine the required relations between the constants a, b such that exactly 50% of the input signal energy, on a one-ohm basis, is transferred to the output.

Solution. The energy in the input signal $f(t)$ is (across one ohm)

$$E_f = \int_{-\infty}^{\infty} |f(t)|^2 \, dt = \int_{0}^{\infty} e^{-2at} \, dt = \frac{1}{2a}.$$

The energy in the output signal $g(t)$ is (across one ohm)

$$E_g = \frac{1}{2\pi} \int_{-\infty}^{\infty} |G(\omega)|^2 \, d\omega = \frac{1}{2\pi} \int_{-\infty}^{\infty} |F(\omega)|^2 \, |H(\omega)|^2 \, d\omega$$

$$= \frac{1}{2\pi} \int_{-\infty}^{\infty} \frac{b^2}{(\omega^2 + a^2)(\omega^2 + b^2)} \, d\omega$$

$$= \frac{b^2}{\pi} \int_{0}^{\infty} \frac{d\omega}{(\omega^2 + a^2)(\omega^2 + b^2)}.$$

Using a table of integrals (see, for example, Appendix A),

$$E_g = \frac{b^2}{\pi} \frac{\pi}{2ab(a + b)} = \frac{b}{2a(a + b)},$$

and we require $E_g = \frac{1}{2} E_f$ so that

$$\frac{b}{2a(a + b)} = \frac{1}{2} \frac{1}{2a}.$$

Solving, we find that the required relation is $a = b$.

4.2 POWER SPECTRAL DENSITY

Not all signals of interest have finite energy. Some signals have infinite energy but they may have a finite time-average of energy. This time-average of energy is called average power and such signals are called "power signals."

The time-averaged power of a signal (again, we assume a one-ohm resistance) is given by

$$P = \lim_{T \to \infty} \frac{1}{T} \int_{-T/2}^{T/2} |f(t)|^2 \, dt. \tag{4.9}$$

For a periodic signal, each period contains a replica of the function, and the limiting operation in Eq. (4.9) can be omitted as long as T is taken as the period. There are power signals which are not periodic, however, and we shall retain the indicated limiting operation for the general case.

The operation described by Eq. (4.9) is the average (or mean) of the squared signal $f(t)$. Thus this quantity is called the mean-square value of the signal $f(t)$, designated simply as $\overline{f^2(t)}$. Stated another way, the time-averaged power of a signal is its mean-square value, $\overline{f^2(t)}$, if the resistance is one ohm. The scaling for other resistance levels follows in the same manner as was discussed for energy and energy spectral density.

In an analogy with the energy signals discussed in the previous section, it would be very convenient and useful for us to define some new function in frequency which would give us some indication of the relative power contri-

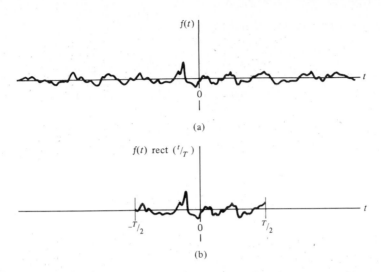

Fig. 4.3 (a) A power signal and (b) its finite-interval truncation.

butions at various frequencies. Let us define such a function as the *power spectral-density function*, $S_f(\omega)$. This function is in units of power (i.e., watts) per Hz and its integral yields the power in $f(t)$. Writing this mathematically, we have

$$P \triangleq \frac{1}{2\pi} \int_{-\infty}^{\infty} S_f(\omega)\, d\omega. \tag{4.10}$$

The power spectral-density function $S_f(\omega)$ describes the distribution of power versus frequency and hence it is an important measurement in practical systems. Although we are approaching it using deterministic signal theory, it is valid also for random signals and forms a basis for a meaningful measurement of many random signals.

We can use the following intuitive approach to relate the power spectral density $S_f(\omega)$ more closely to the signal $f(t)$. Suppose we are given a power signal $f(t)$ as shown in Fig. 4.3(a). Let us form a truncated version of this power signal by observing it only in the interval $(-T/2, T/2)$, as shown in Fig. 4.3(b). This truncated function can be written as $f(t)$ rect (t/T).

We shall assume that $f(t)$ is finite over the finite interval $(-T/2, T/2)$. Then the truncated function $f(t)$ rect (t/T) has finite energy and its Fourier transform $F_T(\omega)$ is†

$$F_T(\omega) = \mathscr{F}\{f(t) \text{ rect } (t/T)\}. \tag{4.11}$$

---

† We are departing from the notation used formerly where the subscript T was used to indicate a periodic function of period T. The alternative use of the subscript T here should not be confusing because it is used on functions of frequency. Both uses of the symbol T (i.e., to indicate truncation and periodicity) are widely used.

Parseval's theorem for this truncated function is

$$\int_{-T/2}^{T/2} |f(t)|^2 \, dt = \frac{1}{2\pi} \int_{-\infty}^{\infty} |F_T(\omega)|^2 \, d\omega.$$

Hence the average power P across a one-ohm resistor is given by

$$P = \lim_{T \to \infty} \frac{1}{T} \int_{-T/2}^{T/2} |f(t)|^2 \, dt = \lim_{T \to \infty} \frac{1}{T} \frac{1}{2\pi} \int_{-\infty}^{\infty} |F_T(\omega)|^2 \, d\omega. \qquad (4.12)$$

Combining Eqs. (4.10) and (4.12), we have

$$\frac{1}{2\pi} \int_{-\infty}^{\infty} S_f(\omega) \, d\omega = \lim_{T \to \infty} \frac{1}{T} \frac{1}{2\pi} \int_{-\infty}^{\infty} |F_T(\omega)|^2 \, d\omega. \qquad (4.13)$$

In addition, we insist that this relation should hold over each frequency increment so that Eq. (4.13) becomes

$$G_f(\omega) = \frac{1}{2\pi} \int_{-\infty}^{\omega} S_f(u) \, du = \lim_{T \to \infty} \frac{1}{T} \frac{1}{2\pi} \int_{-\infty}^{\omega} |F_T(u)|^2 \, du, \qquad (4.14)$$

where $G_f(\omega)$ represents the cumulative amount of power for all frequency components below a given frequency ω. For this reason, $G_f(\omega)$ is called the cumulative power spectrum or, equivalently, the integrated power spectrum of $f(t)$. The cumulative power spectrum $G_f(\omega)$ always exists for any power signal.

If an interchange in the order of the limiting operation and the integration is valid, Eq. (4.14) becomes

$$2\pi G_f(\omega) = \int_{-\infty}^{\omega} S_f(u) \, du = \int_{-\infty}^{\omega} \lim_{T \to \infty} \frac{|F_T(u)|^2}{T} \, du. \qquad (4.15)$$

Note that the average or mean power contained in any frequency interval (ω_1, ω_2) is $[G_f(\omega_2) - G_f(\omega_1)]$. In many cases of interest, $G_f(\omega)$ is differentiable and we have

$$2\pi \frac{dG_f(\omega)}{d\omega} = S_f(\omega). \qquad (4.16)$$

Under these conditions Eq. (4.15) gives

$$S_f(\omega) = \lim_{T \to \infty} \frac{|F_T(\omega)|^2}{T}. \qquad (4.17)$$

Equation (4.17) is our desired result for the power spectral density of $f(t)$.

An intuitive explanation of the result given in Eq. (4.17) can be given as follows. The energy of the truncated function increases (or at least does not decrease) with increasing T. Therefore the quantity $|F_T(\omega)|^2$ increases (or at least is not decreasing) with increasing T. As T becomes very large, the fluctuations due to end effects in the integration should become small and thus the quantity $|F_T(\omega)|^2/T$ may approach a limit. Assuming that such a limit exists, we find the power spectral density $S_f(\omega)$ is given by Eq. (4.17). Actually, it is only in a statistical-average sense that the limit in Eq. (4.17) is meaningful for signals with random components (see Chapter 8).

In practical usage the term "power spectral density" is often shortened to "power density spectrum" or just "power spectrum." Thus we speak of $S_f(\omega)$ as the power spectrum of the signal $f(t)$ even though this terminology is neither as descriptive nor as precise as power spectral density. Note that the power spectral density of a signal retains only the magnitude information and all phase information is omitted. It follows that all signals with the same magnitude spectral densities have identical power spectra, regardless of some possible differences in phase characteristics. Thus, in contrast to the amplitude spectral density $F(\omega)$, $S_f(\omega)$ does not uniquely describe $f(t)$. For a given signal there is a specified power spectral density, but there may be many signals having the same power spectral density.

Equation (4.17) suggests a method used to determine the power spectral density of a signal. A record length of T units is taken from the recording of a physical process—suppose, for example, that we are recording vertical wave motion in an ocean bay. Fixing a float to a recording mechanism, we record T seconds of data. Next, using computational techniques, the corresponding Fourier transform $F_T(\omega)$ is found. (If we use the sampling theorem, analog data can be converted to digital data so that digital computers can be employed.) Taking the magnitude squared, the ratio $|F_T(\omega)|^2/T$ is formed.

Obviously, the limit is not taken (to do so requires infinite record length) and errors are introduced in the measurement. One limitation is that the best resolution in frequency which can be obtained will be on the order of $1/T$ [cf., for example, Eq. (3.106)]. To get better frequency resolution, then, requires longer record lengths. Second, most physical processes have random components and therefore fluctuations arise in the power spectral density. If one is reasonably certain that the true power spectrum does not vary with time, then a possible remedy is to make several consecutive recordings, each of length T, and then average the power spectral densities after computation. These and other methods for obtaining power spectral densities and enhancing their frequency resolution form the subject of an area known as *spectral analysis*. A discussion of spectral analysis requires a knowledge of random signal processes and will have to be postponed.

The above discussion holds for any general power signal. We can proceed a little farther if we have a periodic power signal. Assume that $f(t)$ is periodic and that it is represented by the exponential Fourier series

$$f(t) = \sum_{n=-\infty}^{\infty} F_n e^{jn\omega_0 t}.$$

Using Parseval's theorem [cf. Eq. (2.69)], we have

$$\overline{f^2(t)} = \sum_{n=-\infty}^{\infty} |F_n|^2. \tag{4.18}$$

Equation (4.18) gives the power across a one-ohm resistor at each harmonic frequency for a given $f(t)$ and, when all terms are added, the total average power.

For a periodic signal, we can use Eq. (4.18) to plot a line power spectrum, as shown in Fig. 4.4(a). The corresponding cumulative power spectrum $G_f(\omega)$ is found simply by summing the terms in Eq. (4.18) over all harmonic numbers (n) up to the given frequency ω. Because the inclusion of each harmonic frequency adds a discrete amount of power, this $G_f(\omega)$ will be a series of step functions forming a staircase-type graph, as shown in Fig. 4.4(b). (By definition, the cumulative power spectrum is always a nondecreasing function of ω because power cannot be a negative quantity.)

Writing $G_f(\omega)$ in equation form, we have

$$G_f(\omega) = \sum_{n=-\infty}^{\omega/\omega_0} |F_n|^2 \, u(\omega - n\omega_0). \tag{4.19}$$

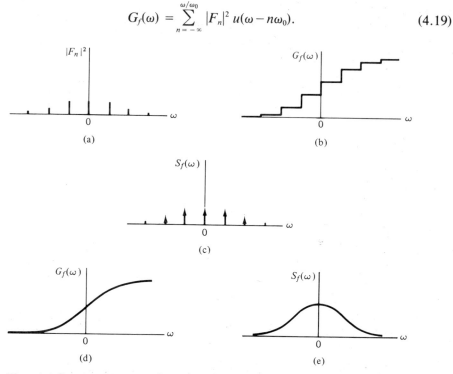

Fig. 4.4 Power spectra of periodic functions: (a) line power spectrum of a periodic function; (b) integrated power spectrum of a periodic function; (c) power spectral density of a periodic function; (d) integrated power spectrum of an aperiodic function; and (e) power spectral density of an aperiodic function.

According to our definition that the derivative of a step function is an impulse function, Eqs. (4.16) and (4.19) now yield the desired result

$$S_f(\omega) = 2\pi \sum_{n=-\infty}^{\infty} |F_n|^2 \delta(\omega - n\omega_0). \tag{4.20}$$

Therefore the power spectral density of a periodic function is a series of impulse functions with weights (areas) corresponding to the magnitude squared of the respective Fourier series coefficients. This is shown in Fig. 4.4(c).

Note that for aperiodic power signals, $G_f(\omega)$ will be a smoothly varying function of frequency and the power spectral density $S_f(\omega)$ follows directly from an application of Eq. (4.16). This case is illustrated in Fig. 4.4(d), (e). We conclude that if a given signal $f(t)$ contains a periodic component of frequency ω_c, then $G_f(\omega)$ will have a step discontinuity at ω_c and $S_f(\omega)$ will contain an impulse at ω_c. This assumes an infinite observation time, as noted earlier.

Now that we have obtained our result, we can formally convert any line power spectrum to a power spectral density simply by changing the lines to impulses. The weights (areas) of these impulses are equal to the squared magnitudes of the line heights and multiplied by 2π if in radian frequency. Integrating this power spectral density over all frequencies yields

$$P = \frac{1}{2\pi} \int_{-\infty}^{\infty} S_f(\omega) \, d\omega,$$

which, for a one-ohm resistor, gives

$$\overline{f^2(t)} = \frac{1}{2\pi} \int_{-\infty}^{\infty} 2\pi \sum_{n=-\infty}^{\infty} |F_n|^2 \delta(\omega - n\omega_0) \, d\omega = \sum_{n=-\infty}^{\infty} |F_n|^2.$$

This result, of course, checks with Parseval's theorem.

Example 4.2.1   Find the power spectral density of the periodic signal

$$f(t) = A \cos(\omega_0 t + \theta).$$

Solution

$$f(t) = \frac{A}{2} e^{j\theta} e^{j\omega_0 t} + \frac{A}{2} e^{-j\theta} e^{-j\omega_0 t}$$

Writing an exponential Fourier series for $f(t)$ (or using Euler's identities), we find

$$F_{-1} = \left(\frac{A}{2}\right) \exp(-j\theta); \qquad F_1 = \left(\frac{A}{2}\right) \exp(j\theta).$$

Using Eq. (4.20), we have

$$S_f(\omega) = \tfrac{1}{2}\pi A^2 \delta(\omega + \omega_0) + \tfrac{1}{2}\pi A^2 \delta(\omega - \omega_0).$$

Note that the mean power across a one-ohm load can be found from

$$\overline{f^2(t)} = \frac{1}{2\pi} \int_{-\infty}^{\infty} S_f(\omega) \, d\omega$$

$$= \frac{1}{2\pi} \int_{-\infty}^{\infty} \frac{1}{2} \pi A^2 \, \delta(\omega + \omega_0) \, d\omega + \frac{1}{2\pi} \int_{-\infty}^{\infty} \frac{1}{2} \pi A^2 \, \delta(\omega - \omega_0) \, d\omega$$

$$= \tfrac{1}{4} A^2 + \tfrac{1}{4} A^2 = \tfrac{1}{2} A^2.$$

This result can be checked easily in the time domain. Also note that the phase information is lost in the calculation.

Example 4.2.2 Develop an expression for the power spectral density of $f(t) = A \exp(j\omega_0 t) \, \text{rect}(t/T)$, both for finite T and as $T \to \infty$.

Solution. From Table 3.1, we have

$$\mathcal{F}\{A \, \text{rect}(t/T)\} = AT \, \text{Sa}(\omega T/2).$$

Using the frequency translation property of the Fourier transform, we have

$$\mathcal{F}\{A \exp(j\omega_0 t) \, \text{rect}(t/T)\} = AT \, \text{Sa}[(\omega - \omega_0)T/2].$$

Using Eq. (4.17) for a finite observation interval T, we have

$$S_f(\omega) \approx \frac{|F_T(\omega)|^2}{T} = |A|^2 T \, \text{Sa}^2[(\omega - \omega_0)T/2].$$

As the observation interval is made increasingly longer, the $[\sin x/x]^2$ pattern becomes more and more concentrated about $\omega = \omega_0$ and in the limit [cf. Eq. (2.110)],

$$\lim_{T \to \infty} \frac{|F_T(\omega)|^2}{T} = |A|^2 \lim_{T \to \infty} \{T \, \text{Sa}^2[(\omega - \omega_0)T/2]\}$$

$$= 2\pi |A|^2 \lim_{T \to \infty} \left\{ \frac{T}{2\pi} \, \text{Sa}^2[(\omega - \omega_0)T/2] \right\}$$

$$= 2\pi |A|^2 \, \delta(\omega - \omega_0).$$

This result agrees with that obtained by the Fourier series expression in Eq. (4.20). (If we were to derive Eq. (4.20) in this manner, however, we would have to use a sum of exponentials and then show that the cross-products vanish. Showing that the $[\sin x/x]^2$ functions become orthogonal as $T \to \infty$ is not easy.)

Example 4.2.3 The power spectral density of a certain signal is found using Eq. (4.17). Several recordings each of length T seconds are taken and the results averaged. The resulting graph is shown in Fig. 4.5. Determine what periodic

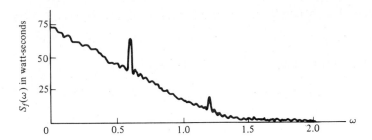

Fig. 4.5 Signal power spectral density of Example 4.2.3.

components, if any, are contained in the signal. Also estimate the length of record taken.

Solution. For an infinite observation time, the given signal will have a periodic component at any frequency for which its power spectral density has an impulse function. However, because the observation time here is finite, the periodic component will show up in the power spectral density in the form

$$T|F_n|^2 \, \text{Sa}^2 \, [(\omega - n\omega_0)T/2]$$

(see previous example). Therefore we can recognize periodic components at the frequencies

$$\omega_1 = 0.6 \, \text{rad/sec} \, (f_1 = 0.0955 \, \text{Hz}),$$

$$\omega_2 = 1.2 \, \text{rad/sec} \, (f_2 = 0.191 \, \text{Hz}).$$

Other small periodic components may be present but are masked in the fluctuations or noise in the measurement. From Fig. 4.5, the widths of these peaks are about 0.05 rad/sec so that the observation time is

$$T = \frac{4\pi}{0.05} \cong 250 \, \text{sec}.$$

The amplitudes of the periodic components are (across a one-ohm load)

$$2T|F_1|^2 = 25 \, \text{watts-sec}, \quad \text{or} \quad |F_1| = 0.22 \, \text{volt},$$

$$2T|F_2|^2 = 10 \, \text{watts-sec}, \quad \text{or} \quad |F_2| = 0.14 \, \text{volt}.$$

[The factor of two was introduced here because $S_f(\omega)$ was displayed as a one-sided spectrum.]

The transmission of power spectra through linear systems follows the same type of reasoning that was used for energy density. Suppose we apply our truncated function to a linear time-invariant filter whose frequency transfer function is $H(\omega)$. The truncated response function, $G_T(\omega)$, is

$$G_T(\omega) \approx F_T(\omega)H(\omega).$$

The power spectral density of the output signal is then

$$S_g(\omega) = \lim_{T \to \infty} \frac{|F_T(\omega)H(\omega)|^2}{T}$$

$$= \lim_{T \to \infty} \frac{|F_T(\omega)|^2}{T} |H(\omega)|^2,$$

$$S_g(\omega) = S_f(\omega) |H(\omega)|^2. \tag{4.21}$$

Therefore the power spectral density of the output signal is the power spectral density of the input signal modified by the squared magnitude of the system transfer function. Again, note that all phase information in both the input signal and the system transfer function is lost. The mean-square output signal is given by

$$\overline{g^2(t)} = \frac{1}{2\pi} \int_{-\infty}^{\infty} S_f(\omega) |H(\omega)|^2 \, d\omega. \tag{4.22}$$

Equations (4.21) and (4.22) give us some insight into why the magnitude transfer function is such a popular way to rate power amplifiers. Hi-fidelity audio amplifiers, for example, are rated on the basis of power-response curves—graphs of $\log |H(\omega)|^2$ vs. $\log \omega$.

Power spectral density has a physical interpretation very similar to that of energy spectral density. This resemblance can be based on a comparison of Eqs. (4.22) and (4.6). The MKS units of power spectral density are watts per Hz.

Drill Problem 4.2.1 Derive a relationship between the power spectral density of the input, $S_f(\omega)$, and the output, $S_g(\omega)$, of an ideal differentiator.

Answer. $S_g(\omega) = \omega^2 S_f(\omega)$.

4.3   TIME-AVERAGED NOISE REPRESENTATIONS

The concept of power spectra allows us to handle some of the averaged effects of the random fluctuations which are present in physical processes. These fluctuations in voltage or current tend to obscure and mask the desired signals and are commonly called noise. In a general sense, noise consists of any unwanted signals, random or deterministic, which interfere with the faithful reproduction of a desired signal in a system. These unwanted signals arise from a variety of sources and can be classified as man-made or naturally occurring.

Man-made types of interference (noise) include such things as electromagnetic pick-up of other radiating signals, inadequate power supply filtering, alias terms arising from poor sampling choices, mechanical vibrations resulting in electrical disturbances, etc. Man-made sources of noise all have the common property that their effects can be eliminated or at least minimized by careful engineering design and practice. We shall not concern ourselves here with these.

Interference caused by naturally occurring disturbances are not controllable in such a direct way and their characteristics can be best described statistically. For some, the fluctuations are very erratic (e.g., those arising from nearby electrical storms) and a valid analytical description is difficult. In others, the time-averaged power may remain quite constant and a treatment on an average power basis is possible and quite meaningful. The concept of a power spectral density is useful in the latter and enables us to treat the effects of noise on an average power basis.

In forming averages of any signal (random or nonrandom), we find parameters which tell us something about the signal. Much of the detailed information about the signal, of course, is lost in the process. In the case of random noise, however, "something is better than nothing" and we will be content with averaged quantities.

Suppose $n(t)$ is a noise voltage or current (again, assume a one-ohm resistive load). A typical waveform is illustrated in Fig. 4.6(a). We now define the following averages of $n(t)$.

1. *Mean value,* $\overline{n(t)}$:

$$\overline{n(t)} = \lim_{T\to\infty} \frac{1}{T} \int_{-T/2}^{T/2} n(t)\, dt. \qquad (4.23)$$

The parameter $\overline{n(t)}$ is often referred to as the dc, or average, value of $n(t)$. Typically the time interval T in Eq. (4.23) can be finite for a good estimate of $\overline{n(t)}$ if it is large enough to smooth the fluctuations of $n(t)$ adequately.

2. *Mean-square value,* $\overline{n^2(t)}$:

$$\overline{n^2(t)} = \lim_{T\to\infty} \frac{1}{T} \int_{-T/2}^{T/2} |n(t)|^2\, dt. \qquad (4.24)$$

The square root of $\overline{n^2(t)}$ is called the root-mean-square (rms) value of $n(t)$. The advantage of the rms notation is that the units of $\sqrt{\overline{n^2(t)}}$ are the same

Fig. 4.6 (a) A random noise waveform and (b) its ac component.

as those of $n(t)$. Aside from a resistance scaling factor, Eq. (4.24) gives the time-averaged power of $n(t)$. From the preceding section, we found it is also related to the integral of the power spectral density, $S_n(\omega)$.

3. *AC component, $\sigma(t)$:*

$$\sigma(t) \triangleq n(t) - \overline{n(t)}. \tag{4.25}$$

The ac, or fluctuation, component of $n(t)$ is that component which remains after the mean value, $\overline{n(t)}$, has been taken out; a typical waveform is shown in Fig. 4.6(b).

Using Eq. (4.25) in Eq. (4.24), we get

$$\overline{n^2(t)} = \lim_{T \to \infty} \frac{1}{T} \int_{-T/2}^{T/2} |\overline{n(t)} + \sigma(t)|^2 \, dt,$$

$$\overline{n^2(t)} = \lim_{T \to \infty} \frac{1}{T} \int_{-T/2}^{T/2} |\overline{n(t)}|^2 \, dt + \lim_{T \to \infty} \frac{1}{T} \int_{-T/2}^{T/2} |\sigma(t)|^2 \, dt. \tag{4.26}$$

In Eq. (4.26) we have used the fact that $\overline{n(t)}$ is a constant and the mean of $\sigma(t)$ is zero by definition. The term on the left-hand side of Eq. (4.26) is the time-averaged power in $n(t)$ across a one-ohm resistor. Likewise the first term on the right-hand side of Eq. (4.26) is the dc power and the second term, the ac power in $n(t)$. An inspection of Eq. (4.26) reveals that the rms value of $n(t)$ is equal to the rms value of $\sigma(t)$ only if the mean value $\overline{n(t)}$ is zero.

The foregoing definitions can be applied to any signal whether random or nonrandom. For periodic signals, T is taken as the period and the limiting operation is not necessary.

Example 4.3.1 Calculate the (a) average value, (b) ac power, and (c) rms value of the periodic waveform $v(t) = 1 + \cos \omega_0 t$.

Solution. Because $v(t)$ is periodic, we can integrate over one period rather than taking the limit.

a) $\overline{v(t)} = \dfrac{1}{T} \displaystyle\int_{-T/2}^{T/2} (1 + \cos \omega_0 t) \, dt = 1,$

b) $\overline{\sigma^2(t)} = \dfrac{1}{T} \displaystyle\int_{-T/2}^{T/2} (\cos \omega_0 t)^2 \, dt = \tfrac{1}{2},$

c) $\overline{v^2(t)} = \dfrac{1}{T} \displaystyle\int_{-T/2}^{T/2} (1 + \cos \omega_0 t)^2 \, dt$

$\qquad = \dfrac{1}{T} \displaystyle\int_{-T/2}^{T/2} (1 + 2 \cos \omega_0 t + \cos^2 \omega_0 t) \, dt = \tfrac{3}{2},$

$v_{\text{rms}} = \sqrt{\overline{v^2(t)}} = \sqrt{3/2}.$

An important measure of the performance of systems, particularly those involving the amplification of low-level signals, is how little noise is introduced in the system. Because the noise varies in an unpredictable manner from one point in time to the next, however, we prefer to analyze this on the basis of average noise power. It is convenient, in fact, to speak of a dimensionless ratio of signal power to noise power. This ratio, called the *signal-to-noise ratio*, can be formed by taking the ratio of the mean-square signal to the mean-square noise because the resistance factor drops. Designating the signal-to-noise ratio by S/N, we have

$$S/N = \overline{s^2(t)}/\overline{n^2(t)}. \tag{4.27}$$

It is quite common to express the signal-to-noise ratio in decibels†

$$[S/N]_{dB} = 10 \log_{10}\left[\overline{s^2(t)}/\overline{n^2(t)}\right]. \tag{4.28}$$

Both $\overline{s^2(t)}$ and $\overline{n^2(t)}$ are assumed to be measured at the same point. Because a dc level can be added arbitrarily, both of these mean-square values assume zero mean value unless stated otherwise. Thus the power spectral density of the signal and/or the noise may contain an impulse function at $\omega = 0$ which is usually neglected in the calculation of S/N.

Drill Problem 4.3.1　A given low-level amplifier has a bandwidth 0.01–10 Hz and a noise power spectral density of $S_n(\omega) = 10^{-14}\,|\omega|^{-1}$ W/Hz referred to its input. (Assume one ohm.) Calculate the signal-to-noise ratio of the amplifier output for a 1-μV rms sinusoidal input at 1 Hz.

Answer.　16.6 dB.

4.4　CORRELATION FUNCTIONS

In a previous section we have seen how signals can be handled using the power spectral-density function, $S_f(\omega)$. The question now naturally arises: Is there some operation in the time domain which is equivalent to finding the power spectral density in frequency?

To answer this important question, let us begin by assuming that our definition of power spectral density is satisfied, that is [cf. Eq. (4.17)],

$$S_f(\omega) = \lim_{T \to \infty} \frac{1}{T}|F_T(\omega)|^2. \tag{4.29}$$

The corresponding operation in time will be the inverse Fourier transform of

---

† Appendix B has a brief discussion of the decibel.

Eq. (4.29). Writing this out gives us

$$\mathscr{F}^{-1}\{S_f(\omega)\} = \frac{1}{2\pi} \int_{-\infty}^{\infty} \lim_{T \to \infty} \frac{1}{T} |F_T(\omega)|^2 e^{j\omega\tau} \, d\omega. \tag{4.30}$$

We have purposely chosen a new time variable, τ, in Eq. (4.30) because the time variable t is already in use in the definition of $F_T(\omega)$. Interchanging the order of operations yields

$$\mathscr{F}^{-1}\{S_f(\omega)\} = \lim_{T \to \infty} \frac{1}{2\pi T} \int_{-\infty}^{\infty} F_T^*(\omega) F_T(\omega) e^{j\omega\tau} \, d\omega$$

$$= \lim_{T \to \infty} \frac{1}{2\pi T} \int_{-\infty}^{\infty} \int_{-T/2}^{T/2} f^*(t) e^{j\omega t} \, dt \int_{-T/2}^{T/2} f(t_1) e^{-j\omega t_1} \, dt_1 \, e^{j\omega\tau} \, d\omega$$

$$= \lim_{T \to \infty} \frac{1}{T} \int_{-T/2}^{T/2} f^*(t) \int_{-T/2}^{T/2} f(t_1) \left[\frac{1}{2\pi} \int_{-\infty}^{\infty} e^{j\omega(t - t_1 + \tau)} \, d\omega \right] dt_1 \, dt. \tag{4.31}$$

The integration over ω within the brackets in Eq. (4.31) is now recognized as [cf. Eq. (3.25)] $\delta(t - t_1 + \tau)$, so that

$$\mathscr{F}^{-1}\{S_f(\omega)\} = \lim_{T \to \infty} \frac{1}{T} \int_{-T/2}^{T/2} f^*(t) \int_{-T/2}^{T/2} f(t_1) \, \delta(t - t_1 + \tau) \, dt_1 \, dt$$

$$= \lim_{T \to \infty} \frac{1}{T} \int_{-T/2}^{T/2} f^*(t) f(t + \tau) \, dt. \tag{4.32}$$

Equation (4.32) describes the operations in the time domain that correspond to the determination of $S_f(\omega)$ in frequency. The inverse Fourier transform of $S_f(\omega)$ is called the *autocorrelation function* of $f(t)$, designated by $R_f(\tau)$. Then our desired result can be written as[†]

$$R_f(\tau) = \lim_{T \to \infty} \frac{1}{T} \int_{-T/2}^{T/2} f^*(t) f(t + \tau) \, dt. \tag{4.33}$$

Also, taking the Fourier transform of both sides of Eq. (4.32) and using Eq. (4.33), we can write

$$S_f(\omega) = \mathscr{F}\{R_f(\tau)\}. \tag{4.34}$$

We now have another method to find the power spectral-density function, i.e., first determine the autocorrelation function and then take a Fourier trans-

---

† With a change of variable, this can also be written as

$$R_f(\tau) = \lim_{T \to \infty} \frac{1}{T} \int_{-T/2}^{T/2} f^*(t - \tau) f(t) \, dt.$$

form. This method is applicable to both random and deterministic signals. We have assumed that the power and the power spectral density do not vary with time. If this assumption is not valid, one can still find an autocorrelation function but it will depend on both τ and t.

For real-valued signals, a computation of the autocorrelation function $R_f(\tau)$ turns out to be similar to that of taking the convolution of $f(-t)$ with $f(t)$. Because these operations have been discussed in Chapter 3, we only illustrate with examples.

Example 4.4.1 Determine and sketch the autocorrelation function of a periodic square wave with peak-to-peak amplitude A, period T, and mean value $A/2$.

Solution. Because $f(t)$ is periodic, the limiting operation in the determination of $R_f(\tau)$ can be replaced by a computation over one period. Using Eq. (4.33) with this single change, we have

For $-T/2 < \tau < 0$:

$$R_f(\tau) = \frac{1}{T} \int_{-T/4}^{(T/4)+\tau} A^2 \, dt = A^2 \left(\frac{1}{2} + \frac{\tau}{T} \right),$$

For $0 < \tau < T/2$:

$$R_f(\tau) = \frac{1}{T} \int_{\tau-T/4}^{T/4} A^2 \, dt = A^2 \left(\frac{1}{2} - \frac{\tau}{T} \right).$$

A graph of $f(t)$ and $R_f(\tau)$ is shown in Fig. 4.7. Since $f(t + T) = f(t)$, all calculations repeat over every period. It follows that the autocorrelation function of a periodic waveform is periodic. Similarly, the autocorrelation function of an aperiodic waveform is aperiodic.

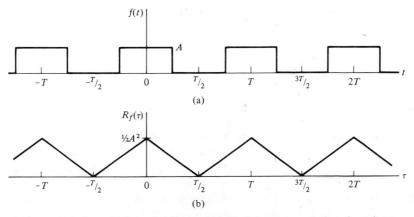

Fig. 4.7 (a) A periodic waveform and (b) its autocorrelation function.

Example 4.4.2 Find the autocorrelation function of $\sqrt{2} \cos (\omega_0 t + \theta)$.

Solution

$$R_f(\tau) = \frac{1}{T} \int_{-T/2}^{T/2} 2 \cos (\omega_0 t + \theta) \cos (\omega_0 t + \omega_0 \tau + \theta) \, dt,$$

$$R_f(\tau) = \frac{1}{T} \int_{-T/2}^{T/2} \cos \omega_0 \tau \, dt + \frac{1}{T} \int_{-T/2}^{T/2} \cos (2\omega_0 t + \omega_0 \tau + 2\theta) \, dt,$$

$$R_f(\tau) = \cos \omega_0 \tau.$$

Note that the autocorrelation function is independent of the phase θ.

The autocorrelation function is widely used in signal analysis. It is especially useful for the detection or recognition of signals which are masked by additive noise. For example, consider a periodic square wave such as that shown in Fig. 4.8(a). Its autocorrelation function, as developed in Example 4.4.1, is shown in Fig. 4.8(b). A band-limited white random-noise waveform is shown in Fig. 4.8(c) and its autocorrelation function is shown in Fig. 4.8(d). The random-noise waveform is added to the periodic square wave and the resulting waveform is shown in Fig. 4.8(e) and its autocorrelation function is shown in Fig. 4.8(f). Note that, even though the square wave is immersed in a considerable amount of noise, the autocorrelation function of the square wave is clearly recognizable in the final result.

One of the drawbacks in using autocorrelation for this type of application is that the autocorrelation of the noise is present in the output as well as the autocorrelation of the signal (e.g., compare Figs. 4.8d and f). This may make detection of aperiodic signals difficult. Also, the relative time shift (phase) between signals is lost (cf. Example 4.4.2). These drawbacks can be avoided by use of a closely related operation known as *crosscorrelation*.

Suppose that we have two waveforms $f(t)$ and $g(t)$. The crosscorrelation function $R_{fg}(\tau)$ is defined as

$$R_{fg}(\tau) = \lim_{T \to \infty} \frac{1}{T} \int_{-T/2}^{T/2} f^*(t) g(t + \tau) \, dt. \qquad (4.35)$$

As an example of an application of crosscorrelation, we choose a random waveform $f(t)$ as shown in Fig. 4.9(a). (Its autocorrelation function, $R_f(\tau)$, will be similar to that shown in Fig. 4.8d.) For the second function, $g(t)$, we choose a delayed replica of $f(t)$ plus a second random waveform $n(t)$ so that $g(t) = f(t - t_0) + n(t)$. The composite waveform $g(t)$ is shown in Fig. 4.9(b). It is presumed that the receiver has a replica of the waveform $f(t)$ available (e.g., in memory). On the basis of this knowledge of $f(t)$, we wish to have the receiver make a measurement of the time delay t_0. To do this, we take the crosscorrelation function $R_{fg}(\tau)$, as defined in Eq. (4.35). The result is shown in Fig. 4.9(c). The

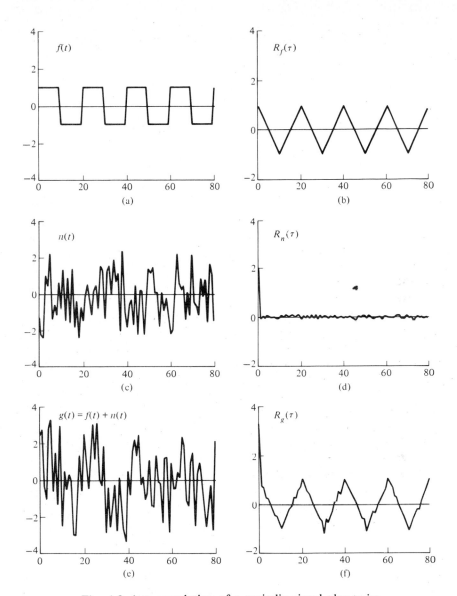

Fig. 4.8 Autocorrelation of a periodic signal plus noise.

value of the time delay t_0 is clearly evident by measuring the time delay from the origin to the large peak in the result.

Correlation functions furnish measures of the similarity of a signal $f(t)$ either with itself (in the case of autocorrelation) or with another signal (in the case of crosscorrelation) versus a relative shift by an amount τ. For dissimilar signals,

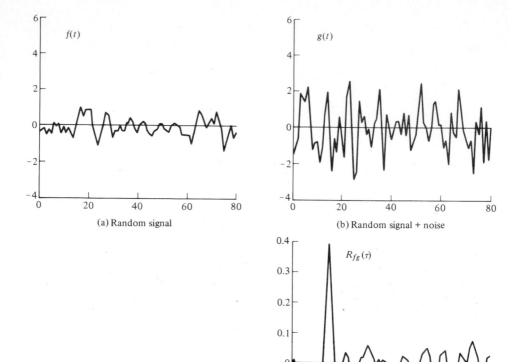

(a) Random signal

(b) Random signal + noise

(c) Crosscorrelation

Fig. 4.9 Crosscorrelation of a random signal plus noise.

the peak of the correlation function is an indicator of how good this match is between signals.

Both autocorrelation and crosscorrelation are powerful tools in signal analysis in analytical and practical work. We shall have occasion to use them in future chapters.

4.5   SOME PROPERTIES OF CORRELATION FUNCTIONS

We have already found, in the preceding section, that the Fourier transform of the autocorrelation function gives the power spectral density. We shall examine briefly several additional properties of correlation functions.

4.5.1 Symmetry

Examining the autocorrelation function for negative arguments, we have

$$R_f(-\tau) = \lim_{T\to\infty} \frac{1}{T} \int_{-T/2}^{T/2} f^*(t)f(t-\tau)\, dt,$$

$$R_f(-\tau) = \lim_{T\to\infty} \frac{1}{T} \int_{-T/2}^{T/2} f^*(\zeta + \tau)f(\zeta)\, d\zeta,$$

$$R_f(-\tau) = R_f^*(\tau). \tag{4.36}$$

Therefore the real part of $R_f(\tau)$ is an even function; and if $f(t)$ is real-valued, then $S_f(-\omega) = S_f^*(\omega)$ [cf. Eqs. (3.38) and (4.34)].

4.5.2 Mean-Square Value

The autocorrelation function $R_f(\tau)$ evaluated at $\tau = 0$ is equal to the mean-square value of the signal $f(t)$ [cf. Eq. (4.24)],

$$R_f(0) = \lim_{T\to\infty} \frac{1}{T} \int_{-T/2}^{T/2} f^*(t)f(t)\, dt,$$

$$R_f(0) = \overline{f^2(t)}. \tag{4.37}$$

It is left for the reader to verify that Eq. (4.37) agrees with Eqs. (4.10) and (4.34) when referenced to one ohm.

4.5.3 Periodicity

If $f(t + T) = f(t)$ for all t, then

$$R_f(\tau + T) = R_f(\tau) \qquad \text{for all } \tau. \tag{4.38}$$

Proof follows easily from writing out the integrals and using the definition of periodicity.

★ 4.5.4 Average Value

Let $f(t)$ be represented by a function $x(t)$ with zero average value, plus an average value designated by m_1. Similarly, we represent $g(t)$ as a function $y(t)$ with zero average value, plus an average value designated m_2. In equation form, this can be written as

$$f(t) = x(t) + m_1,$$

$$g(t) = y(t) + m_2.$$

The crosscorrelation of $f(t)$, $g(t)$ is

$$R_{fg}(\tau) = \lim_{T \to \infty} \frac{1}{T} \int_{-T/2}^{T/2} [x^*(t) + m_1][y(t + \tau) + m_2] \, dt.$$

Noting that the average values of $x(t)$ and $y(t)$ are zero by definition, we have

$$R_{fg}(\tau) = \lim_{T \to \infty} \frac{1}{T} \int_{-T/2}^{T/2} x^*(t)y(t + \tau) \, dt + m_1 m_2.$$

The average value of the crosscorrelation function is

$$\overline{R_{fg}(\tau)} = \lim_{T \to \infty} \frac{1}{T} \int_{-T/2}^{T/2} \lim_{T \to \infty} \frac{1}{T} \int_{-T/2}^{T/2} x^*(t)y(t + \tau) \, dt \, d\tau + m_1 m_2.$$

Interchanging the order of integration, we have

$$\overline{R_{fg}(\tau)} = \lim_{T \to \infty} \frac{1}{T} \int_{-T/2}^{T/2} x^*(t) \lim_{T \to \infty} \frac{1}{T} \int_{-T/2}^{T/2} y(t + \tau) \, d\tau \, dt + m_1 m_2.$$

Because $\overline{y(t + \tau)}$ is zero, we obtain the result

$$\overline{R_{fg}(\tau)} = m_1 m_2. \tag{4.39}$$

Therefore the average value of the crosscorrelation of two functions $f(t)$ and $g(t)$ is equal to the product of their average values. If the average value of either function is zero, then the average value of their crosscorrelation is zero. The case for autocorrelation follows readily from this result.

Drill Problem 4.5.1   Making use of Fig. 4.7, sketch the autocorrelation function of a symmetrical square wave (with zero average value).

Answer.   See Fig. 6.23(b).

★ 4.5.5   Maximum Value

We can show that $R_x(0) \geq R_x(\tau)$ for any τ [see, e.g., Fig. 4.7] by taking the magnitude squared of the autocorrelation function and using the Schwarz inequality (cf. Example 2.5.2). Thus we have

$$\left| \lim_{T \to \infty} \frac{1}{T} \int_{-T/2}^{T/2} f^*(t)f(t + \tau) \, dt \right|^2 \leq \lim_{T \to \infty} \frac{1}{T} \int_{-T/2}^{T/2} |f(t)|^2 \, dt \lim_{T \to \infty} \frac{1}{T} \int_{-T/2}^{T/2} |f(t + \tau)|^2 \, dt$$

$$\leq R_f(0) \, R_f(0).$$

Taking the square root of both sides of this result, we have

$$|R_f(\tau)| \leq R_f(0). \tag{4.40}$$

Therefore the autocorrelation function $R_f(\tau)$ is bounded by the mean-square value of the signal $f(t)$. For a periodic signal, the equality in Eq. (4.40) is valid at multiples of the period from the origin (see Fig. 4.7). For nonperiodic $f(t)$, $R_f(\tau)$ is strictly less than $R_f(0)$ for all $\tau \neq 0$.

4.5.6 Additivity

If two signals are added, the autocorrelation function of the sum may or may not be the sum of their respective autocorrelation functions. To investigate this, we write $z(t) = x(t) + y(t)$. The autocorrelation function for the sum of the two signals $x(t)$ and $y(t)$ is

$$R_z(\tau) = \lim_{T \to \infty} \frac{1}{T} \int_{-T/2}^{T/2} [x^*(t) + y^*(t)] [x(t + \tau) + y(t + \tau)] \, dt,$$

$$R_z(\tau) = R_x(\tau) + R_y(\tau) + R_{xy}(\tau) + R_{yx}(\tau). \tag{4.41}$$

We conclude that only if the crosscorrelation functions are zero [i.e., if $R_{xy}(\tau) = R_{yx}(\tau) = 0$] can we write

$$R_z(\tau) = R_x(\tau) + R_y(\tau). \tag{4.42}$$

For the condition $R_{xy}(\tau) = 0$ for all τ, we say that $x(t)$ and $y(t)$ are *uncorrelated*. Moreover, we can show that $R_{yx}(\tau) = R_{xy}^*(-\tau)$, so that if

$$R_{xy}(\tau) = 0, \text{ then } R_{yx}(\tau) = 0.$$

Note that if $x(t)$ and $y(t)$ are orthogonal, then they are uncorrelated. In addition, as will be discussed in Chapter 8, if $x(t)$ and $y(t)$ are statistically independent they are also uncorrelated.

Because power spectral density is the Fourier transform of the autocorrelation function, our conclusion here is that if two signals $x(t)$ and $y(t)$ are uncorrelated then their power spectral densities are additive. In other words, the average power of the sum of two signals is the sum of the average powers of the two signals only if the signals are uncorrelated. In cases in which the crosscorrelation functions are not zero, the signals must be added first and then the average power may be determined, or, equivalently, the crosscorrelations in Eq. (4.41) must be included.

Drill Problem 4.5.2 A sinusoidal waveform, $3\sqrt{2} \cos \omega_1 t$, is added to a second, $4\sqrt{2} \cos \omega_2 t$. Determine the rms value of the sum if (a) $\omega_1 = \omega_2$; (b) $\omega_1 \neq \omega_2$.

Answer. (a) 7; (b) 5.

4.6 CORRELATION FUNCTIONS FOR FINITE-ENERGY SIGNALS

The concept of correlation can be extended to signals of finite energy. Specifically, we define the autocorrelation function $r_f(\tau)$ for a signal $f(t)$ of finite energy as

$$r_f(\tau) = \int_{-\infty}^{\infty} f^*(t)f(t + \tau)\, dt. \tag{4.43}$$

Similarly, for signals $f(t)$ and $g(t)$, both of finite energy, we define the cross-correlation function $r_{fg}(\tau)$ as

$$r_{fg}(\tau) = \int_{-\infty}^{\infty} f^*(t)g(t + \tau)\, dt. \tag{4.44}$$

Note that for real-valued functions these operations are the same as for convolution, except that the second function is not reversed.

The Fourier transform of Eq. (4.43) gives

$$\mathscr{F}\{r_f(\tau)\} = \int_{-\infty}^{\infty} \int_{-\infty}^{\infty} f^*(t)f(t + \tau)\, dt\, e^{-j\omega\tau}\, d\tau. \tag{4.45}$$

Interchanging the order of integration in Eq. (4.45), we have

$$\mathscr{F}\{r_f(\tau)\} = \int_{-\infty}^{\infty} f^*(t) \int_{-\infty}^{\infty} f(t + \tau)e^{-j\omega\tau}\, d\tau\, dt. \tag{4.46}$$

Use of the time-delay property of the Fourier transform in the right-hand side of Eq. (4.46) gives

$$\int_{-\infty}^{\infty} f^*(t)F(\omega)e^{-j\omega t}\, dt = |F(\omega)|^2. \tag{4.47}$$

Combining Eqs. (4.46) and (4.47), we have

$$\mathscr{F}\{r_f(\tau)\} = |F(\omega)|^2. \tag{4.48}$$

Identifying the right-hand side of Eq. (4.48) as the energy spectral density of $f(t)$, we conclude that the energy spectral density is the Fourier transform of the autocorrelation function for finite-energy signals.

Drill Problem 4.6.1 Determine the autocorrelation function of the rectangular pulse waveform shown in Fig. 4.10(a).

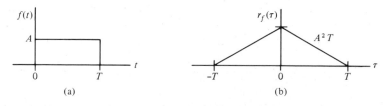

(a)   (b)

Fig. 4.10 (a) A pulse signal and (b) its autocorrelation function.

Answer. See Fig. 4.10(b).

4.7 BAND-LIMITED WHITE NOISE

The power spectral-density function plays a central role in our time-averaged description of random noise. The particular type of power spectral density in which we are interested is that one which tends to be a constant for all frequencies. Such a flat power spectrum contains all frequency components with equal power weighting and is designated as white, in an analogy to white light.

If we have a constant power spectral density of η watts per Hz (measured over positive frequencies), and if $n(t)$ has zero mean value, then the power spectral density of white noise is

$$S_n(\omega) = \eta/2 \qquad \text{for all } \omega. \tag{4.49}$$

The factor of one-half in Eq. (4.49) is necessary to have a two-sided power spectral density. Note that we are defining $S_n(\omega)$ on a power basis. In other words, for a resistor of R ohms, one must multiply Eq. (4.49) by R to convert to mean-square voltage and divide by R to convert to mean-square current.

Strictly speaking, Eq. (4.49) cannot be used to describe any physical process because it implies an infinite amount of power; that is,

$$\overline{n^2(t)} = \frac{1}{2\pi} \int_{-\infty}^{\infty} (\eta/2) \, d\omega \rightarrow \infty.$$

However, it turns out to be a good model for many cases in which the bandwidth of the measuring device is narrower than the bandwidth limitations of the physical process being observed. Because our measurements are restricted to finite bandwidths, what we are really interested in, then, is *band-limited white noise*. In other words, if a noise waveform has a flat power spectral density extending beyond the bandwidth of a given system, the noise appears to the system as if it were indeed white.

For band-limited white noise, the noise power is independent of the choice of operating frequency. For example, suppose that $n(t)$ is a zero-mean white noise whose power spectral density is $\eta/2$ watts per Hz. Across a bandwidth B, the noise power P_n is

$$P_n = \frac{1}{2\pi} \int_{-2\pi B}^{2\pi B} (\eta/2) \, d\omega = \eta B \text{ watts.} \tag{4.50}$$

Assuming that this is developed across a resistor R, we find the mean-square noise voltage is†

$$\overline{n^2(t)} = RP_n = \eta RB \text{ volts}^2. \tag{4.51}$$

---

† Note that here the average power is given (from the spectral density) and we wish to find the mean-square voltage.

If $n(t)$ is a current, then

$$\overline{n^2(t)} = P_n/R = \eta GB \text{ amperes}^2. \tag{4.52}$$

The transmission of white noise through linear time-invariant systems follows in the manner described for power spectral density. Suppose that we wish to find the rms voltage at the output of a filter whose transfer function $H(\omega)$ is known. Designating the input by $n_i(t)$ and the output by $n_o(t)$, we can write

$$S_{n_o}(\omega) = S_{n_i}(\omega) |H(\omega)|^2, \tag{4.53}$$

$$\overline{n_o^2(t)} = \frac{1}{2\pi} \int_{-\infty}^{\infty} S_{n_o}(\omega) \, d\omega$$

$$= \frac{1}{2\pi} \int_{-\infty}^{\infty} S_{n_i}(\omega) |H(\omega)|^2 \, \omega. \tag{4.54}$$

If the spectral density of the noise is white (one-ohm resistor assumed), Eq. (4.54) becomes

$$\overline{n_o^2(t)} = \frac{1}{2\pi} \int_{-\infty}^{\infty} \frac{\eta}{2} |H(\omega)|^2 \, d\omega$$

$$= \frac{\eta}{2\pi} \int_{0}^{\infty} |H(\omega)|^2 \, d\omega. \tag{4.55}$$

An important physical noise source giving rise to band-limited white noise over a very broad frequency range is discussed in the following section.

4.7.1   Thermal Noise

Thermal noise is produced as a result of the thermally excited random motion of free electrons in a conducting medium, such as a resistor. The path of each electron in motion is randomly oriented due to collisons. The net effect of the motion of all electrons is an electric current in the resistor which is random with a mean value of zero. From thermodynamic and quantum mechanical considerations, the power spectral density of thermal noise is given by

$$S_n(\omega) = \frac{h|\omega|}{\pi[\exp{(h|\omega|/2\pi kT)} - 1]}, \tag{4.56}$$

$$S_n(\omega) \cong 2kT \text{ watts per Hz} \qquad \text{for} \qquad |\omega| \ll 2\pi kT/h, \tag{4.57}$$

where

T = temperature of the conducting medium in Kelvin,
k = Boltzmann's constant = 1.38×10^{-23} joule/K,
h = Planck's constant = 6.625×10^{-34} joule-sec.

For frequencies above kT/h, thermal noise is no longer white. However, these frequencies are so high for electrical signals that we can safely assume that thermal noise is white for our purposes here (for example, $kT/h \approx 6000$ GHz for $T = 290$ K).

In practice, resistors may actually produce slightly more thermal noise than that indicated by the spectral density above. This increase is a function of materials and geometry and we shall neglect these factors here. Note that an ideal capacitor has no thermal noise source because there are no free electrons present in an ideal dielectric. On the other hand, an ideal inductor has no thermal noise source because an ideal conductor cannot have a lattice structure to impede the flow of electrons.

Using Eq. (4.57) and Eq. (4.51), we find the mean-square (open-circuit) voltage generated by a resistor R in a bandwith B is

$$\overline{v^2(t)} = RP_n = \frac{1}{2\pi} \int_{-2\pi B}^{2\pi B} 2kTR \, d\omega,$$

$$\overline{v^2(t)} = 4kTRB \text{ volts}^2. \tag{4.58}$$

The mean-square (short-circuit) current generated, using Eq. (4.57) and Eq. (4.52), is

$$\overline{i^2(t)} = GP_n = \frac{1}{2\pi} \int_{-2\pi B}^{2\pi B} 2kTG \, d\omega,$$

$$\overline{i^2(t)} = 4kTGB \text{ amperes}^2. \tag{4.59}$$

The voltage and current equivalent circuit models for band-limited thermal noise are shown in Fig. 4.11. The resistance R and the conductance G are assumed noise-free and the bandwidth B is that of the measurement device.

Noise circuit problems involving only resistive components can be solved

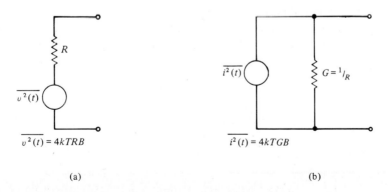

Fig. 4.11 Thermal-noise equivalent circuit models: (a) voltage model; (b) current model.

using these circuit models. The general procedure is to replace each noisy resistor in a series configuration with the equivalent voltage model and each noisy resistor in a parallel configuration with its equivalent current model. Note that voltage and current sources are added on a mean-square basis. The procedure is illustrated in the following example.

Example 4.7.1 Calculate the rms noise voltage arising from thermal noise in two resistors, 100 Ω and 150 Ω respectively, at $T = 300$ K within a bandwidth of 1 MHz if (a) the resistors are connected in series; and (b) the resistors are connected in parallel.

Solution
a) Placing the two equivalent circuit models in series and using Eq. (4.58), we get

$$\overline{v^2(t)} = 4kTB(R_1 + R_2)$$
$$= 4(1.38 \times 10^{-23})(300)(10^6)(250)$$
$$= 4.14 \times 10^{-12} \text{ volt}^2,$$
$$\sqrt{\overline{v^2(t)}} = 2.03 \ \mu\text{V}.$$

The equivalent source resistance is 250 Ω.

b) Placing the two in parallel and using Eq. (4.59) gives us

$$\overline{i^2(t)} = 4kTB(G_1 + G_2)$$
$$= 4(1.38 \times 10^{-23})(300)(10^6)(0.0167)$$
$$= 2.76 \times 10^{-16} \text{ ampere}^2,$$
$$\sqrt{\overline{i^2(t)}} = 0.0166 \ \mu\text{A}.$$

The equivalent parallel resistance is

$$R' = \frac{1}{G_1 + G_2} = 60 \ \Omega,$$

so that

$$\sqrt{\overline{v^2(t)}} = \sqrt{\overline{i^2(t)}}R'$$
$$= (0.0166 \times 10^{-6})(60)$$
$$= 0.997 \ \mu\text{V}.$$

Drill Problem 4.7.1 Repeat Example 4.7.1 with the sole exception that the 100-Ω resistor is at a temperature of 600 K.

Answer. (a) 2.41 μV; (b) 1.26 μV.

4.7.2   Transmission of Thermal Noise through Linear Systems

We have been assuming that there is no relation between the random motion of free electrons in different resistors. This means that noise contributions arising from this random motion add on a power basis for any given frequency band. Under this condition we can add the power spectral densities of different noise sources or, if the integration remains the same, we can add the mean-square values of the current or voltage directly. This was done above.

If any energy storage components (i.e., capacitors or inductors) are present, the bandwidth of the system may have more effect on the noise measurement than the measuring device itself. In this case we must be careful to "go back to the fundamentals" and treat the noise source(s) in terms of power spectral density, then integrate. The integration may change drastically as a function of the system.

Suppose, for example, that we connect a resistor to the input terminals of a linear system containing only noiseless components, as shown in Fig. 4.12(a). The procedure is the following. First we replace the input resistance with a noise voltage source and a noise-free resistor R, as shown in Fig. 4.12(b). The noise-free resistor is now included as part of the transfer function of the system. On a voltage basis,

$$S_{v_i}(\omega) = 2kTR \tag{4.60}$$

and

$$S_{v_o}(\omega) = S_{v_i}(\omega) |H(\omega)|^2. \tag{4.61}$$

The mean-square output voltage is

$$\overline{v_o^2(t)} = \frac{1}{2\pi} \int_{-\infty}^{\infty} S_{v_i}(\omega) |H(\omega)|^2 \, d\omega. \tag{4.62}$$

The analogous relations for current follow readily.

But what if the system itself contains noisy resistive components? If the system, in addition to being linear, is passive and bilateral, then the effective noise resistance referred to the input is[†]

$$R_{eq}(\omega) = \mathscr{R}e\{Z(\omega)\}, \tag{4.63}$$

where $Z(\omega)$ is the complex-valued input impedance of the system. All other calculations follow in a similar way to that discussed above. For example, the noise voltage spectral density would be [cf. Eq. (4.60)]

$$S_v(\omega) = 2kTR_{eq}(\omega). \tag{4.64}$$

Note that in general $R_{eq}(\omega)$ is a function of frequency.

[†] For more complicated networks involving active and unilateral devices it is common practice to state the equivalent noise voltage referenced to the input as a part of the measured data.

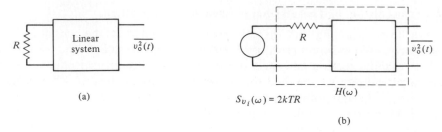

(a)

$S_{v_i}(\omega) = 2kTR$

$H(\omega)$

(b)

Fig. 4.12 Transmission of thermal noise through a linear system: (a) system model; (b) equivalent noise model.

Example 4.7.2   Calculate the rms noise voltage developed across a capacitor C when it is connected in parallel with a noisy resistor R.

Solution.   The circuit and its equivalent noise model are shown in Fig. 4.13. The resistor R is replaced by a noise-free resistor in series with a noise voltage source with power spectral density,

$$S_{v_i}(\omega) = 2kTR \text{ volts}^2/\text{Hz}.$$

The system transfer function is that of a RC low-pass filter:

$$\cdot H(\omega) = \frac{1/RC}{j\omega + 1/RC}.$$

The mean-square value of the output is given by Eq. (4.62):

$$\overline{v_o^2(t)} = \frac{1}{2\pi} \int_{-\infty}^{\infty} 2kTR \frac{(1/RC)^2}{\omega^2 + (1/RC)^2} \, d\omega$$

$$= \frac{1}{\pi} \int_{0}^{\infty} 2kTR \frac{(1/RC)^2}{\omega^2 + (1/RC)^2} \, d\omega$$

$$= \frac{2kT}{\pi C} \tan^{-1}(\omega RC)\big|_0^\infty$$

$$= kT/C.$$

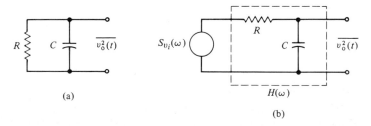

(a)

$S_{v_i}(\omega)$

$H(\omega)$

(b)

Fig. 4.13 (a) The RC network and (b) its equivalent noise model as discussed in Example 4.7.2.

Then the rms output noise voltage is

$$\sqrt{\overline{v_o^2(t)}} = \sqrt{\frac{kT}{C}}.$$

At first glance, it seems strange that this result should be independent of R because the resistance is the source of the noise! A closer look, however, reveals that even though the mean-square noise voltage is proportional to R, the bandwidth of the low-pass filter formed is inversely proportional to R. These two effects cancel, yielding the above result.

Drill Problem 4.7.2 Calculate the rms noise current developed through an ideal inductor L when it is connected to a noisy resistor R.

Answer. $\sqrt{kT/L}$.

4.7.3 Equivalent Noise Bandwidth

Expressions such as Eqs. (4.58) and (4.59) assume an ideal filter of bandwidth B for the noise measurement. In practice, it is convenient to combine the bandwidth-limiting characteristics of a system by defining an equivalent noise bandwidth. The equivalent noise bandwidth, B_N, is that ideal filter bandwidth which gives the same noise power as the actual system. The equivalent noise bandwidth for white noise can be determined easily as follows.

Within the white noise assumption, the input power spectral density is a constant, $\eta/2$. The mean-square voltage output of a linear system is given by Eq. (4.62). Then the mean-square output voltage, $\overline{v_o^2(t)}$, across a one-ohm resistor is [cf. Eq. (4.55)]

$$\overline{v_o^2(t)} = \frac{1}{2\pi} \int_{-\infty}^{\infty} \frac{\eta}{2} |H(\omega)|^2 \, d\omega$$

$$= \frac{\eta}{2\pi} \int_0^{\infty} |H(\omega)|^2 \, d\omega. \tag{4.65}$$

The definite integral in Eq. (4.65) is a constant for a given system frequency transfer function $H(\omega)$. (We are using the voltage transfer function here because it is the one most commonly used.)

We would like to use a simplified approach (such as was used for resistors) in which the power spectral density is white within some effective bandwidth of the system. We can do this by defining an *equivalent noise bandwidth* B_N such that (1) the power spectral density at the filter output is white within the bandwidth B_N and zero elsewhere, forming an equivalent rectangular spectral density; and (2) the area under this rectangular spectral density is equal to the area under the spectral density at the filter output. This is illustrated in Fig. 4.14.

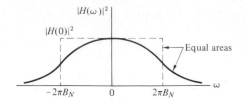

Fig. 4.14 A graphical definition of noise equivalent bandwidth.

Designating the midband frequency of the system by ω_0 ($\omega_0 = 0$ for a low-pass system), the midband system voltage gain is $|H(\omega_0)|^2$ and we can write

$$\overline{v_o^2(t)} = \frac{1}{2\pi} \int_{-2\pi B_N}^{2\pi B_N} (\eta/2) |H(\omega_0)|^2 \, d\omega,$$

$$\overline{v_o^2(t)} = \eta |H(\omega_0)|^2 B_N. \tag{4.66}$$

Equating the right-hand sides of Eqs. (4.65) and (4.66), we have

$$B_N = \frac{1}{2\pi} \frac{\int_0^{\infty} |H(\omega)|^2 \, d\omega}{|H(\omega_0)|^2}. \tag{4.67}$$

This definition of the equivalent noise bandwidth B_N allows us to discuss practical linear systems by using their idealized equivalents. For example, the equivalent noise bandwidth of the RC low-pass filter in Example 4.7.2 is $B_N = 1/(4RC)$. Therefore the mean-square voltage at the filter output is $\overline{v_o^2(t)} = 4kTRB_N = kT/C$.

Example 4.7.3   Compute the -3-dB bandwidth and the equivalent noise bandwidth of a filter with the following magnitude transfer characteristic:

$$|H(\omega)| = \frac{1}{\sqrt{1 + \omega^4}}.$$

Solution.   Because $H(0) = 1$, the -3 dB bandwidth is found by solving

$$|H(\omega_1)| = \frac{1}{\sqrt{2}}, \quad \text{or} \quad \omega_1 = 1; \quad f_1 = \frac{1}{2\pi} = 0.159 \text{ Hz.}$$

Using Eq. (4.67), we find the equivalent bandwidth is (cf. Appendix A)

$$B_N = \frac{1}{2\pi} \int_0^{\infty} \frac{1}{1 + \omega^4} \, d\omega = \sqrt{2}/8$$

$$= 0.177 \text{ Hz.}$$

Thus for this particular filter (known as a second-order Butterworth filter) the equivalent noise bandwidth is about 11% greater than the -3-dB bandwidth.

Drill Problem 4.7.3 Compute the equivalent noise bandwidth of the (RC-type) filter whose magnitude transfer characteristic is $|H(\omega)| = 1/\sqrt{1 + \omega^2}$. Compare your result to the -3-dB bandwidth.

Answer. 0.250 Hz; 57% greater.

[*Note*: As the order of the filter increases, the -3-dB bandwidth and the noise equivalent bandwidth agree more closely.]

4.7.4 Available Power and Noise Temperature

From Eqs. (4.49), (4.50), and (4.57), the thermal noise power generated in a resistor R is

$$P_n = 4kTB. \tag{4.68}$$

How much of this noise power can be extracted? Using a matched resistive load R (noise-free) for maximum power transfer, we find that the voltage transferred is exactly one-half of the open circuit voltage. The maximum available power, P_a, is then one-fourth that given in Eq. (4.68), or

$$P_a = kTB. \tag{4.69}$$

This available noise power, P_a, is the maximum thermal noise power that can be extracted from a noisy resistor.

Examining Eq. (4.69), we see that k is a constant and B is the equivalent noise bandwidth—constant for a given system. The temperature T is then directly related to the available noise power. A convenient way to describe the input noise power is to specify it as a noise temperature. Thus the *noise temperature* specifies the thermal noise power into a matched resistance.

The noise-free resistor used for matching purposes above is, of course, fictitious. In practice, we usually wish to connect an amplifier (receiver) which has an input resistance R for maximum power transfer. A simplified model of this amplifier is an input resistance R at an equivalent noise temperature T_e followed by a power gain G_p. In other words, the noise temperature T_e is the effective temperature of a white thermal noise source at the system input that would be required to produce the same noise power at the output of an equivalent noiseless system. Some very low-noise amplifiers, for example, have equivalent noise temperatures as low as 10 K to 30 K while standard broadcast receivers may have noise temperatures on the order of 1,000 K.

Note that the equivalent noise temperature is not necessarily the ambient temperature of the amplifier. Equivalent noise temperatures below ambient temperature are possible by use of amplifiers with a low resistive component in the gain characteristic (e.g., parametric amplifiers). Sometimes cryogenic cooling is also employed to lower the noise temperature.

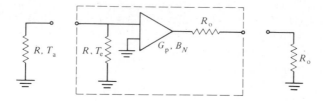

Fig. 4.15 Receiver equivalent circuit for thermal noise.

An equivalent circuit for a receiver operating in the presence of thermal noise is shown in Fig. 4.15. Here all of the noise (white) generated by the amplifier is included in the equivalent noise temperature T_e. Thermal-noise effects of the output resistance R_o can also be included in T_e. However, if the power gain G_p is appreciable, the effect of R_o on the overall noise output is negligible, as shown in the following example.

The noise temperature of the matched resistance connected to the receiver input is usually set at 290 K for test purposes. In operating systems, T_a represents the effective noise temperature of the sky and the surrounding noise environment as seen through the antenna, as well as of any transmission-line losses from the antenna to the receiver input terminals.

Example 4.7.4   We wish to design a high-gain cascaded-stage amplifier. The first stage power gain is fixed at 20 dB. Succeeding stages have provisions for gain control but the maximum gain per stage is 20 dB. The maximum net power gain of the amplifier is to be such that the thermal noise power level from internally generated noise is 20 milliwatts at the output. Determine the minimum number of stages needed if $T_e = 600$ K, $B_N = 10$ MHz.†

Solution.   The internally generated noise power in the first stage, referred to the input, is

$$P_a = kT_e B$$
$$= (1.38 \times 10^{-23})(600)(10^7) = 8.28 \times 10^{-14} \text{ watt.}$$

After 20 dB of gain, this becomes 8.28×10^{-12} watt of noise power at the input to the second stage. The available input noise power to the second stage is then 8.28×10^{-12} watt coming from the first stage, plus 8.28×10^{-14} watt originating from thermal noise in the second stage (and referred to its input). The contribution from the second stage is almost negligible ($= 1\%$) compared to the noise power coming from the first stage. Our conclusion, then, is that *the*

---

† The equivalent noise bandwidth does vary somewhat with the number of amplifier stages but we shall ignore this change here.

noise performance of cascaded amplifiers is primarily dependent only on the noise performance of the first stage if that stage has appreciable gain.

The required maximum power gain, G_p [$G_p = |H(\omega_0)|^2$] is

$$G_p = P_o/P_a$$
$$= 20 \times 10^{-3}/(8.28 \times 10^{-14}) = 2.42 \times 10^{11}$$
$$\cong 114 \text{ dB}.$$

Thus the amplifier will require a minimum of six (6) stages to meet the given specifications.

★ 4.7.5   Noise Figure

It is convenient to develop a concise way to state the equivalent noise temperature of an amplifier relative to a fixed standard. This leads to the definition of a noise figure which can then be used as a figure-of-merit in comparisons between amplifiers for low-level signals.

Let the input and output signal voltages (or currents) in a given system be $s_i(t)$, $s_o(t)$, respectively, and let the input and output noise voltages (or currents) be $n_i(t)$, $n_o(t)$. The input signal-to-noise ratio, $(S/N)_i$, is†

$$(S/N)_i = \overline{s_i^2(t)} \, / \, \overline{n_i^2(t)}, \tag{4.70}$$

and the output signal-to-noise ratio is

$$(S/N)_o = \overline{s_o^2(t)} \, / \, \overline{n_o^2(t)}. \tag{4.71}$$

The system always adds some noise so that the input signal-to-noise ratio is higher than the output signal-to-noise ratio. To measure the amount of degradation, we define a noise figure, F, to be the ratio of the input signal-to-noise ratio divided by the output signal-to-noise ratio:

$$F \triangleq \frac{(S/N)_i}{(S/N)_o}. \tag{4.72}$$

By definition, the input noise power in Eq. (4.72) is equivalent to the thermal noise power provided by a resistor matched to the input and at a temperature of $T_0 = 290$ K. The noise figure of a perfect system is unity, and the introduction of additional noise causes the noise figure to be larger than one.

Some convenient simplifications can be made when only thermal noise is present in the system. Let us apply a signal $s_i(t)$ and thermal noise $n_i(t)$ at a temperature T_0 to the input of an amplifier with a power gain G_p and noise bandwidth B. The available input noise power is

$$N_i = kT_0 B. \tag{4.73}$$

---

† The resistance factor drops out so that this is also a power ratio.

The output signal power is

$$S_o = S_i G_p. \tag{4.74}$$

The amplifier adds some thermal noise. Representing this noise by an equivalent noise temperature T_e referred to the amplifier input, we have

$$N_o = kT_0 BG_p + kT_e BG_p. \tag{4.75}$$

When we substitute (4.73)–(4.75) into Eq. (4.72), the noise figure of the amplifier is

$$F = 1 + \frac{T_e}{T_0}. \tag{4.76}$$

This result is so simple and easy to apply that even if not all noise sources are thermal, the effects are often included in an equivalent noise temperature using test data.† It is quite common to express the noise figure in decibels:

$$F_{\mathrm{dB}} = 10 \log_{10}(F).$$

Note that the noise figure of an ideal system which generates no noise itself is $F = 1$, and that portion of the noise figure of any system arising from internally generated noise is $(F - 1)$.

Example 4.7.5   A given amplifier has a 4-dB noise figure, a noise bandwidth of 500 kHz, and an input resistance of 50 Ω. Calculate the rms signal input which yields an output signal-to-noise ratio of unity when the amplifier is connected to a 50-Ω input at 290 K.

Solution.   The available input noise power is

$$P_n = kT_0 B = 2.00 \times 10^{-15} \text{ watt,}$$

so that

$$\overline{n_i^2(t)} = P_n R = 1.00 \times 10^{-13} \text{ volt}^2,$$
$$F = 4 \text{ dB} = 2.51.$$

Using Eq. (4.72), we have

$$(S/N)_i = F(S/N)_o = F.$$

The required input signal is

$$\overline{s_i^2(t)} = (2.51)(1.00 \times 10^{-13}) \text{volt}^2$$
$$\sqrt{\overline{s_i^2(t)}} = 0.501 \ \mu\text{V}.$$

---

† The standard value for T_0 is 290 K.

Example 4.7.6   A resistive attenuator (e.g., a coaxial cable or waveguide) at a temperature T_0 has matched input and output resistances and an attenuation (in power) of α, where $\alpha > 1$. Determine the equivalent noise temperature and noise figure of the attenuator when the input source and the attenuator are at temperature T_0.

Solution.   Using Eq. (4.75) and noting that $G_p = 1/\alpha$, we have

$$N_o = kT_0B(1/\alpha) + kT_eB(1/\alpha).$$

Looking back into the output terminals, we see that the attenuator appears entirely resistive and at a temperature T_0 so that

$$N_o = kT_0B.$$

Equating these two expressions, we have

$$T_e = (\alpha - 1)T_0.$$

Use of Eq. (4.76) then yields

$$F = \alpha.$$

A similar procedure can be used if the attenuator is at some other temperature.

Example 4.7.7   A multistage amplifier has stage power gains G_1, G_2, G_3, . . . , and stage noise figures F_1, F_2, F_3, . . . , respectively. Show that the overall noise figure is

$$F = F_1 + [(F_2 - 1)/G_1] + [(F_3 - 1)/G_1G_2] + \cdots$$

and hence that the first stage is the most significant in the determination of the overall noise figure if $G_1 \gg 1$.

Solution.   The result follows easily when one considers that the noise introduced by each stage is amplified only by the gains of that stage and succeeding stages. Assume a 3-stage amplifier with equivalent noise temperatures T_{e_1}, T_{e_2}, T_{e_3}:

$$S_o = S_iG_1G_2G_3,$$

$$N_i = kT_0B,$$

$$N_o = kT_0BG_1G_2G_3 + kT_{e_1}BG_1G_2G_3 + kT_{e_2}BG_2G_3 + kT_{e_3}BG_3$$

$$= N_iG_1G_2G_3 + (F_1 - 1)N_iG_1G_2G_3 + (F_2 - 1)N_iG_2G_3 + (F_3 - 1)N_iG_3$$

$$= F_1N_iG_1G_2G_3 + (F_2 - 1)N_iG_2G_3 + (F_3 - 1)N_iG_3,$$

$$F = \frac{S_i/N_i}{S_o/N_o} = F_1 + \frac{F_2 - 1}{G_1} + \frac{F_3 - 1}{G_1G_2}.$$

Extensions of this result are straightforward.

This result demonstrates the advantages of using a first stage of amplification with not only a low noise figure (F_1) but also high gain (G_1). From Example 4.7.6 we see that the performance of a lossy transmission line fails on both counts. Therefore some amplification with low-noise stages is often used near or at the antenna in low-noise receiving systems before the signal is transferred (via a transmission line) to the main receiver.

Drill Problem 4.7.4   A receiver for geostationary satellite transmissions at 2 GHz consists of an antenna preamplifier with a noise temperature of 127 K and a gain of 20 dB. This is followed by an amplifier with a noise figure of 12 dB and a gain of 80 dB. Compute the overall noise figure and equivalent noise temperature of the receiver.

Answer.   2.0 dB; 170 K.

★ 4.7.6 Sky-Noise Temperature

When an antenna is connected to the input of a receiver, it is convenient to represent it by a resistor which is matched to the receiver input and whose temperature represents the effective noise of the sky and the surrounding noise environment as seen through the antenna. This antenna temperature will in most cases differ substantially from 290 K.

If the input resistor, now representing an antenna, is at some arbitrary temperature T_a, we can modify Eq. (4.75) to give

$$N_o = kT_aBG_p + kT_eBG_p,$$

$$N_o = kB(T_a + T_e)G_p. \tag{4.77}$$

The equivalent noise temperature of an antenna is therefore easy to interpret and can be compared directly with the receiver noise temperature. Note that when T_a and T_e contribute equally to the output noise, the largest possible improvement by going to a perfect receiver amounts to only a factor of two in signal-to-noise improvement. Therefore efforts to obtain a low equivalent noise temperature in a receiver really pay large dividends only if the antenna noise temperature is low.

Average antenna noise temperatures are mainly a function of frequency and the pointing of the antenna. A graph illustrating some of the main sources of antenna noise temperature is shown in Fig. 4.16. Although the concept of noise figure is useful for the testing and comparisons of receivers, the concept of equivalent noise temperature is more useful in computations of actual system performance.

The principal source of antenna noise below 30 MHz is atmospheric noise which results mainly from lightning discharges. Propagation effects which make long-distance communication possible at these lower frequencies also provide

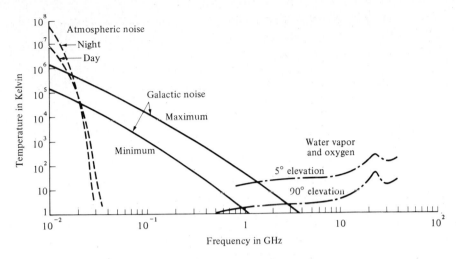

Fig. 4.16 Some average sky noise temperatures.

good propagation for electrical storm activities occurring throughout the world. Galactic or cosmic noise is a major contributor above 30 MHz and up to the GHz range. This noise arises from radiation from outer space; for narrow beamwidths, the intensity of galactic noise is a function of antenna pointing direction. Our sun is also an active source of radiation but its effects are more localized in angular direction.

    Water vapor and oxygen act as attenuators of rf energy, particularly around 23 GHz. These attenuation effects not only decrease signal strength but also act as thermal noise sources (see Example 4.7.6). The frequency range between about 2 to 8 GHz, bounded by the effects of galactic noise and noise due to oxygen and water vapor, is referred to as the "low-noise window." This is a preferred range for low-noise receivers for space telemetry and radio telescopes.

Example 4.7.8   The receiver in Drill Problem 4.7.4 has a bandwidth of 1 MHz. The receiving antenna gain is 40 dB and the antenna noise temperature is 59 K. If the satellite antenna gain is 6 dB and expected path losses are 190 dB, what is the minimum required satellite transmitter power to achieve a 14-dB S/N ratio at the output of the receiver?

Solution.   See Table 4.1.

★ 4.7.7   Other Sources of Band-limited White Noise

There are other types of noise which can be described in terms of white noise sources, although the approximation may not hold over as wide a range of frequencies as for thermal noise. One example is that of shot noise.

Table 4.1 *S/N* **Budget for the Ground-to-Geostationary Satellite Communications Problem of Example 4.7.8.**

Equivalent receiver noise temperature	170 K	
Antenna noise temperature (given)	59 K	
Total noise temperature	229 K	
Total noise temperature—dB above 1K	23.6	
Boltzmann's constant—dB/K/Hz	−228.6	
Bandwidth (given)—dB above 1 Hz	60.0	
Total receiver noise power	−145.0 dB	
Receiver *S/N* allowance (given)	14.0 dB	
Minimum receiver signal power	−131.0 dB	
Receiving antenna gain (given)	40 dB	
Satellite transmitting antenna gain (given)	6 dB	
Minimum detectable signal power		−177.0 dB
Path losses (given)†		190 dB
Minimum required satellite transmitter power		13.0 dB
		(20 W)

Shot noise, like thermal noise, is due to the discrete nature of physical matter. It arises in physical devices when a charged particle moves through a potential gradient without collisions and with a random starting time. Averaging over many such particles we have an average flow, but there will always be fluctuations about this average.

In vacuum tubes, shot noise arises as a result of the random emission of electrons from the cathode. In semiconductor devices, shot noise arises as a result of the random diffusion of minority carriers and the random generation and recombination of hole-electron pairs. For these cases, the power spectral density is approximately flat up to frequencies on the order of $1/\tau$, where τ is the transit time or lifetime of the charge carriers. In terms of mean-square current, the power spectral density is

$$S_i(\omega) = q\overline{i(t)} + 2\pi\overline{i(t)}^2 \delta(\omega) \tag{4.78}$$

where q is the charge of an electron = 1.6×10^{-19} coulomb. The first term in Eq. (4.78) corresponds to the ac or fluctuation part of the noise current and the second term corresponds to the nonzero mean value, $\overline{i(t)}$.

Example 4.7.9 If there are $\bar{n}$ electrons per second (time-averaged) crossing an evacuated region in a potential gradient, determine the spectral density of the fluctuation current due to shot noise. Assume a white spectral density.

Solution.   Each electron leaves at a random starting time and traverses the given region. Let the current induced in the circuit as a result of one electron be $i_e(t)$.

---

† Path loss formulas can be found in handbooks; see, for example, *Reference Data for Radio Engineers*, ITT, Sixth ed., Indianapolis: Howard W. Sams & Co., 1975, Chapter 28.

Because the charge on the electron is q coulombs,

$$\int_{-\infty}^{\infty} i_e(t)\, dt = q.$$

Let the Fourier transform of $i_e(t)$ be $I_e(\omega)$:

$$I_e(\omega) = \int_{-\infty}^{\infty} i_e(t) e^{-j\omega t}\, dt,$$

$$I_e(0) = \int_{-\infty}^{\infty} i_e(t)\, dt = q.$$

The energy spectral density of one electron is $|I_e(\omega)|^2$. Because the starting times are random, the induced current pulses are independent of each other and we can add their energy spectral densities† so that

$$S_i(\omega) = \frac{N}{T} |I_e(\omega)|^2 = \bar{n}|I_e(\omega)|^2.$$

Within the white noise assumption,

$$S_i(\omega) = \bar{n}\,|I_e(0)|^2 = \bar{n}q^2 = q(\bar{n}q) = q\overline{i(t)}.$$

Thus $S_i(\omega)$ varies in direct proportion to the average current $\overline{i(t)}$.

Another noise component that arises in the division of current in multielectrode devices is called partition noise. This partitioning, which is a random process, introduces noise because the charge carriers are discrete.

Both transistors and multielectrode vacuum tubes have all three sources of noise present—shot noise, partition noise, and thermal noise. All three can generally be treated as if their power spectral densities were flat across the bandwidths of interest. Sometimes the effects of shot noise and partitioning noise are accounted for by placing an equivalent thermal noise source in the equivalent circuit model.

4.8 SUMMARY

The quantity $|F(\omega)|^2$ describes the relative amount of energy of a given signal $f(t)$ versus frequency and is called the energy spectral density of $f(t)$. Likewise, the function describing the relative power contributions of a signal $f(t)$ versus frequency is called the power spectral density, $S_f(\omega)$. The power spectral density of an aperiodic signal is a continuous function of frequency. The power spectral density of a periodic signal approaches a series of impulse functions located at the harmonic frequencies of the signal as the observation time becomes long.

---

† This is an exception to the rule that spectral densities can be added without regard to phase only when their relative phases are 0° or 180°.

The mean-square value of a waveform can be found from the area under its power spectral-density function. The ratio of the mean-square signal to the mean-square noise is called the signal-to-noise ratio.

The inverse Fourier transform of the power spectral density is the autocorrelation function. For real-valued signals of finite duration, the autocorrelation function is, aside from a normalizing constant, given by the convolution of $f(-t)$ with $f(t)$. Similarly, the convolution of $f(-t)$ with a different signal is called the crosscorrelation function.

Two signals are uncorrelated if their crosscorrelation is zero. The power in the sum of two signals is equal to the sum of the powers in the two signals if they are uncorrelated. This rule can be extended to power spectral densities.

The inverse Fourier transform of the energy spectral density is the autocorrelation function for finite-energy signals.

Noise having a flat power spectral density is called *white*. Thermal noise, arising from the random motion of free electrons in a conducting medium, is white up to infrared frequencies with $S_n(\omega) = 2kT$ watts per Hz.

The available power from thermal noise is kTB watts, where B is the equivalent noise bandwidth of the system. A thermal-noise source may be specified as an equivalent temperature. Noise figure is a convenient figure-of-merit for systems when a reference temperature is set at $T_0 = 290$ K. Antenna input noise can be specified as an equivalent noise temperature, called the sky-noise temperature.

Shot noise and partition noise are two other types of noise which are white over fairly wide frequency ranges. All three types are present to varying degrees in thermionic and semiconductor devices. An equivalent thermal-noise source is sometimes used to account for these and other sources of band-limited white noise.

Selected References for Further Reading

1. S. J. Mason and H. J. Zimmerman. *Electronic Circuits, Signals, and Systems*. N.Y.: John Wiley & Sons, 1960.
   This book has a good discussion of autocorrelation and crosscorrelation for deterministic signals.

2. W. R. Bennett. *Electrical Noise*. N.Y.: McGraw-Hill, 1960.
   A general reference on the sources of electrical noise and the methods of handling the effects of noise in linear systems.

3. W. W. Mumford and E. H. Scheibe. *Noise Performance Factors in Communication Systems*. Dedham, Mass.: Horizon House-Microwave, 1968.
   A thorough treatment of noise and the use of noise figure in communication systems at the level of this chapter.

4. P. F. Panter. *Communication Systems Design: Line-of-Sight and Troposcatter Systems*. N.Y.: McGraw-Hill, 1972.
   Chapter 6 has a practical discussion of S/N calculations for communication systems design at a slightly more advanced level.

5. E. J. Baghdady (ed.). *Lectures on Communication Systems Theory*. N.Y.: McGraw-Hill, 1960.
   Chapters 15–17 discuss antenna noise and Chapters 21 and 22 discuss some system *S/N* considerations at a more advanced level.

6. A. Van der Ziel. *Noise*. Englewood Cliffs, N.J.: Prentice-Hall, 1954.
   A classic reference on noise and noise calculations.

Problems

4.1.1 The bandwidth of a given ideal low-pass filter is 4 rad/sec and the low-frequency gain is one. Calculate the output energy, on a one-ohm basis, if the input $f(t)$ is

a) $f(t) = \delta(t)$;
b) $f(t) = \exp(-4t)u(t)$.

4.1.2 The pulse signal $f(t)$ shown in Fig. P–4.1.2 is applied to the input of an ideal low-pass filter with a variable cut-off frequency, f_c. Find the approximate energy, on a one-ohm basis, in the output signal for

a) $f_c = 0.1$ Hz;
b) $f_c = 10$ Hz.

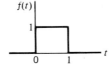

Fig. P–4.1.2.

4.1.3 The two-sided exponential voltage $f(t) = 10e^{-|t|}$ is developed across a 50-Ω resistor.

a) Calculate the total energy dissipated in the resistor.
b) What fraction of this energy is in the frequency range of zero to one radian per second?

4.1.4 In the circuit shown in Fig. P–4.1.4, $e_i(t) = \exp(-at)u(t)$. Determine the required value of the constant a such that

$$\frac{\int_0^\infty e_o^2(t)\, dt}{\int_0^\infty e_i^2(t)\, dt} = \tfrac{1}{27}.$$

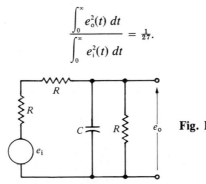

Fig. P–4.1.4.

4.2.1 A certain signal $f(t)$ has the following power spectral density (assume one-ohm load):

$$S_f(\omega) = \pi[e^{-|\omega|} + \delta(\omega - 2) + \delta(\omega + 2)].$$

a) What is the mean power in the bandwidth $\omega < 1$ rad/sec?
b) What is the mean power in $f(t)$ in the bandwidth 0.99 to 1.01 rad/sec?
c) What is the mean power in $f(t)$ in the bandwidth 1.99 to 2.01 rad/sec?
d) What is the total mean power in $f(t)$?

4.2.2 A given voltage signal is $f(t) = 4 \cos 20\pi t + 2 \cos 30\pi t$ across 1 Ω.

a) Determine the power spectral density of $f(t)$.
b) Sketch $S_f(\omega)$.
c) Calculate the mean (average) power, both in the time domain and in the frequency domain, that is dissipated by $f(t)$ across the one-ohm resistor.

4.2.3 A symmetric square wave (i.e., zero average value) with a peak amplitude of 1 V and period T is applied to the input of a filter whose transfer function is

$$H(\omega) = [(1 + \cos \omega)/2, \quad |\omega| < \pi; \quad 0 \text{ elsewhere}],$$

and whose input and output impedance levels are one ohm.

a) What is the average power in the input? Find the power spectral density of the output and calculate the ratio of the output power to the input power if T is
b) 4 sec;
c) 8 sec. [*Hint:* $F_n = \text{Sa } (n\pi/2)$ for $n \neq 0$ and $F_n = 0$ for $n = 0$.]

4.2.4 A source with a resistance of 3 kΩ feeds the parallel combination of a 50-μF capacitor and a 6-kΩ resistor. Compute the mean power across the 6-kΩ resistor if the power spectral density of the source (in watts per Hz) is (a) 10^{-2}; (b) $\delta(\omega + 10) + \delta(\omega - 10)$.

4.3.1 A symmetric triangular waveform has a peak amplitude of 1 V and a period T.

a) Find the power spectral density (assume one ohm).
b) Determine the mean-square value from the power spectral density.
c) Determine the mean-square value in the time domain.

4.3.2 White noise with a (two-sided) spectral density of 0.1 watt per Hz (one-ohm source) added to a signal described by $10 \cos 30\pi t$ is applied to a RC low-pass filter, where $RC = (30\pi)^{-1}$.

a) Calculate the rms noise voltage in the output of the filter.
b) Determine the ratio of the mean signal power to the mean noise power at the output of the filter.

4.4.1 Determine the power spectral density of $F_1 \exp (j\omega_0 t)$ by (a) first finding its autocorrelation function, then (b) taking the Fourier transform of the autocorrelation function to obtain the power spectral density.

4.4.2 Determine the autocorrelation function of the pulse waveform $e^{-at}[u(t) - u(t - T)]$ when it is repeated periodically every $2T$ seconds.

4.4.3 A periodic rectangular waveform is described by the constant amplitude A for a fraction α of the period T (where $0 < \alpha < \frac{1}{2}$) and the amplitude $(-A)$ for the remainder of the period. Determine the autocorrelation function of this waveform.

4.5.1 Determine the autocorrelation function of $f(t) = A(1 + m \cos \omega_m t) \cos \omega_c t$. From the result, now determine the mean-square value of $f(t)$. Assume that $\omega_c \gg \omega_m$.

4.5.2 A white noise voltage source is connected to the input of a RC low-pass filter. The resulting output has a mean-square value of one volt when referred to one ohm. Determine the autocorrelation functions of the input and the output of the filter in terms of the RC product.

4.5.3 A sinusoidal signal is transmitted; on reception, the signal is present together with additive noise. The autocorrelation function of the composite at the receiver is found to be $R(\tau) = a \cos \omega_0 \tau + b \exp(-c|\tau|)$. Determine (a) the mean-square signal-to-noise ratio; and (b) the spectral density of the additive noise present.

4.6.1 Determine the energy spectral density of the pulse signal

$$f(t) = [\exp(-at)]u(t)$$

by (a) finding its autocorrelation function first and then taking a Fourier transform; (b) finding the Fourier transform first and then taking the magnitude squared.

4.6.2 Find the autocorrelation function of the *derivative* of the pulse waveform shown in Fig. P–4.6.2. Will two successive integrations of your result give $r_f(\tau)$? Verify the results by finding $r_f(\tau)$ directly from $f(t)$.

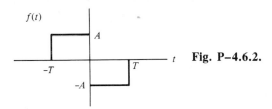

Fig. P–4.6.2.

4.6.3 Determine and sketch the autocorrelation function of the pulse signal $f(t) =$ Sa (Wt).

4.6.4 Determine the autocorrelation function of the gaussian pulse signal defined in Problem 3.2.4.

4.7.1 Two resistors R_1, R_2 are both at the temperature T. Determine the required relationship between R_1 and R_2 if the rms thermal noise voltage across the series combination is twice that across the parallel combination.

4.7.2 Two resistors connected in series and at differing temperatures are shown in Fig. P–4.7.2(a).

a) Derive relations for the noise-equivalent resistor and temperature, R_{eq}, T_{eq}.
b) Calculate R_{eq}, T_{eq} for the specific case where $R_1 = 1$ kΩ, $R_2 = 2$ kΩ, $T_1 = 300$ K, $T_2 = 390$ K.

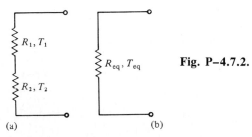

Fig. P–4.7.2.

4.7.3 A dentist wishes to have soothing "white noise" brought into his office to calm his patients. Suppose he decides to do this by amplifying the thermal noise of a resistor and then applying this to a speaker. (The sound of white noise resembles that of a waterfall.) Assuming an ideal voltage amplifier with a bandwidth of 100 kHz, calculate the voltage gain required if the resistor is 100 kΩ and the required output voltage level is 2 V rms. Assume the temperature of the dentist's office is 295 K.

4.7.4 The input of a voltage amplifier is connected to a 1-kΩ resistor and the output to a 100-kΩ resistor, as shown in Fig. P–4.7.4. The voltage gain A of the amplifier is 10, the input inpedance is 10 MΩ (assumed noise-free), and the output impedance is 100 Ω (noise-free). The bandwidth of both the amplifier and the measurement device is 1 MHz. The temperature of the entire system is 300 K.

a) Compute the rms thermal noise voltage at point 1.
b) Compute the rms thermal noise voltage at point 2.

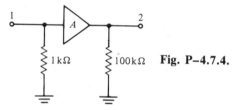

Fig. P–4.7.4.

4.7.5 A certain oscillator produces the open-circuit voltage waveform $A \cos \omega_0 t$ at its output terminals and has an internal resistance R. A capacitor C is placed across the output terminals in an effort to improve the ratio of the open-circuit signal voltage to rms noise voltage. What is the optimum value of C?

4.7.6 Develop an expression for the spectral density of the noise voltage e_n in Fig. P–4.7.6. Examine its behavior as $R_1 \to 0$; as $R_2 \to 0$. Also make a sketch of the spectral density.

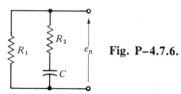

Fig. P–4.7.6.

4.7.7 Two identical resistors at a temperature T are connected to an ideal inductor, as shown in Fig. P–4.7.7.

a) Derive an expression for the voltage spectral density developed across one of the resistors.
b) What is the mean-square voltage across one of the resistors within the noise bandwidth B?

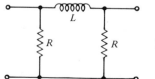

Fig. P–4.7.7.

4.7.8 Compute the noise-equivalent bandwidth of the third-order Butterworth filter whose magnitude frequency transfer function is $|H(\omega)| = 1/\sqrt{1 + \omega^6}$. Compare your answer to the -3-dB bandwidth. Also compare your answer to those found in Example 4.7.3 and Drill Problem 4.7.3.

4.7.9 Compute the noise-equivalent bandwidth of the gaussian filter whose magnitude frequency transfer function is $|H(\omega)| = \exp(-a\omega^2)$. Compare your answer to the -3-dB bandwidth. [*Hint:* See Problem 3.2.4.]

4.7.10 Starting with a noisy resistor R at a temperature T, show that the maximum power transferred is $P = kTB$ if it is matched by a noiseless resistor of the same value R.

4.7.11 The input resistance of a certain wideband oscilloscope is selectable at either 50 Ω or 1 MΩ. The noise bandwidth is 500 MHz and the equivalent noise temperature is 600 K. If the average input signal power must be at least ten times the average noise power for measurement accuracy, what is the minimum acceptable rms input signal at each resistance level under open-circuit conditions for no additional input noise? (Assume that the input resistance is a thermal noise source.)

4.7.12 A certain amplifier has input and output resistances of 50 Ω and a noise-equivalent bandwidth of 10 kHz. When connected to a matched source and a matched load, the net gain is 60 dB. If a 50-Ω resistor at 290 K is connected to the input, the output rms noise voltage across 50 Ω is 80 μV. Determine the equivalent noise temperature of the amplifier.

★ **4.7.13** An antenna is connected to a receiver having a noise figure of 3 dB. The available power gain of the receiver is 80 dB and the noise-equivalent bandwidth is 10 MHz. If the available output noise power is 10 μW, determine the equivalent antenna temperature.

★ **4.7.14** The input of an amplifier having an equivalent noise temperature of T_e is connected to a resistor R whose value is matched to the input resistance of the amplifier. When the temperature of the resistor R alone is raised to T_R, the output noise power of the amplifier is found to be twice that when the temperature of the resistor is at room temperature, T_0. Derive relationships for the equivalent noise temperature T_e and noise figure F of the amplifier on the basis of these measurements.

★ **4.7.15** The minimum-loss resistive attenuator shown can be used to match a 300-Ω antenna to a 75-Ω cable. Calculate the equivalent noise temperature and the noise figure of the attenuator. [*Hint:* Use the result of Example 4.7.6.]

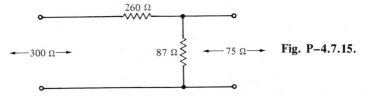

Fig. P–4.7.15.

★ **4.7.16** A proposed receiving station for a space experiment has a receiver (e.g., a cooled parametric amplifier) with a 0.2-dB noise figure, a bandwidth of 1 MHz, and a center frequency of 3 GHz. The receiving antenna has a gain of 48 dB and the spacecraft antenna gain is 6 dB. Path losses (in dB) are $\alpha = 37 + 20 \log_{10} f +$

20 $\log_{10} d$, where f is the center frequency in MHz and d is the distance in miles. If the transmitted power is 10 W and a minimum 15-dB S/N ratio is required at the receiver, estimate the maximum distance over which communication can be established (make use of Fig. 4.16).

★ **4.7.17** The average current through a thermionic diode is used to generate shot noise. The dc term is blocked by a capacitor and the output is used to simulate a thermal noise generator. Derive a relation between the average diode current and the effective thermal noise temperature.

★ **4.7.18** A traffic counter on a busy highway is designed to generate a rectangular pulse of height A V and of width τ sec every time a vehicle passes it. Thus the Fourier transform of a pulse occurring at $t = t_0$ is $F(\omega) = A\tau \exp(-j\omega t_0)$ Sa $(\omega\tau/2)$. Assume a one-ohm load at the output of the counter.

a) What is the energy spectral density of one pulse at the counter output?

b) The passing of a vehicle is a random event. We shall assume that the average number of vehicles passing the counter (at least for a given time of day) is a constant, $\bar{n}$. What is the power spectral density of the output of the counter?

c) A meter with an equivalent noise bandwidth of $B = 10^{-3}/\tau$ is connected to the above counter. This meter is calibrated in terms of power and its impedance is one ohm. Determine the reading of the meter in terms of A, τ, $\bar{n}$, and B. You may make reasonable approximations.

CHAPTER 5

AMPLITUDE MODULATION

Up to this point, we have worked with signals wherever they happened to be positioned in frequency. In particular, we have not been very concerned with altering signals in time and frequency to accomplish desired objectives. We begin an investigation of methods to alter continuous high-frequency sinusoidal signals in response to given low-frequency signals. This is often referred to as *continuous-wave* (CW) *modulation*. We shall continue this discussion of CW modulation in Chapter 6.

Generally speaking, modulation is that process by which a property or a parameter of a signal is varied in proportion to a second signal. The precise dependence is determined by the type of modulation employed. In amplitude modulation, the amplitude of a sinusoidal signal, whose frequency and phase are fixed, is varied in proportion to a given signal. This alters the given signal by translating its frequency components to higher frequencies.

The use of amplitude modulation may be advantageous whenever a shift in the frequency components of a given signal is desired. This need may arise, for example, in designing filters for very demanding requirements. Or consider the possibility of transmitting a voice communication through space via electromagnetic waves. If the maximum voice frequency is 3.3 kHz, the minimum wavelength is 100,000 meters. Because antennas with dimensions less than one-quarter wavelength are inefficient, it is a distinct advantage to be able to raise the frequency by at least several orders of magnitude before transmission is attempted.

By proper choices, a large number of signals can be transmitted at the same time without mutual interference. Commercial radio and television stations, for example, are assigned differing frequencies on which to operate and this allows the simultaneous operation of many stations. Because these signals are spaced in frequency, each receiver can easily separate out the desired signal. We shall discuss this more in a later section.

5.1 AMPLITUDE MODULATION: SUPPRESSED CARRIER

The equation of a general sinusoidal signal can be written as

$$\phi(t) = a(t) \cos \theta(t), \tag{5.1}$$

where $a(t)$ is the time-varying amplitude and $\theta(t)$ is the time-varying angle. It is convenient to write $\theta(t) = \omega_c t + \gamma(t)$ so that

$$\phi(t) = a(t) \cos [\omega_c t + \gamma(t)]. \tag{5.2}$$

We shall assume that $a(t)$ and $\gamma(t)$ are slowly varying compared to $(\omega_c t)$. The term $a(t)$ is called the envelope of the signal $\phi(t)$ and the term ω_c is called the carrier frequency; $\gamma(t)$ is the phase modulation of $\phi(t)$.

   In amplitude modulation, the phase term $\gamma(t)$ in Eq. (5.2) is zero (or a constant) and the envelope $a(t)$ is made proportional to the given signal $f(t)$. Letting the constant of proportionality be unity here, we have†

$$\phi(t) = f(t) \cos \omega_c t. \tag{5.3}$$

The term $\cos \omega_c t$ in Eq. (5.3) is called the carrier signal and $f(t)$ is called the modulating signal. The resultant signal, $\phi(t)$, is called the modulated signal.

   Applying the modulation property of the Fourier transform to Eq. (5.3), we find the spectral density of $\phi(t)$ is

$$\Phi(\omega) = \tfrac{1}{2}F(\omega + \omega_c) + \tfrac{1}{2}F(\omega - \omega_c). \tag{5.4}$$

Amplitude modulation therefore translates the frequency spectrum of a signal by $\pm\omega_c$ rad/sec but leaves the spectral shape unaltered, as shown in Fig. 5.1. This type of amplitude modulation is called *suppressed-carrier* because the spectral density of $\phi(t)$ has no identifiable carrier in it, although the spectrum is centered at the frequency ω_c.

   From Fig. 5.1(f), we see that both the positive and the negative frequency content of $f(t)$ are now displayed as positive frequencies. This implies that the bandwidth of $f(t)$ is doubled when this type of amplitude modulation is used, as we have indicated in Fig. 5.1(e), (f). In Fig. 5.1(f), the spectral content for positive frequencies above ω_c is called the *upper sideband* of $\phi(t)$ and the spectral content for positive frequencies below ω_c is called the *lower sideband* of $\phi(t)$. It is easily seen that, for positive frequencies, the upper sideband of $\phi(t)$ displays the positive frequency components of $f(t)$ and the lower sideband of $\phi(t)$ displays the negative frequency components of $f(t)$. Similar relations, though just reversed, hold for the negative frequencies in $\phi(t)$.

   From this discussion, we conclude that amplitude modulation with suppressed carrier provides us with a convenient means to observe the complete frequency spectrum of a given signal $f(t)$. All we must do is translate the signal

---

† Including a multiplier constant k_a, this can be written as $k_a f(t) \cos \omega_c t$.

using a carrier frequency ω_c which is higher than the spectral bounds (bandwidth) of the signal. This principle is widely used in spectral analysis. The fact that we have two sidebands—an upper and a lower—and no separate carrier present in $\phi(t)$ suggests the following convenient designation for this type of modulation: *double-sideband, suppressed-carrier*—abbreviated DSB-SC.

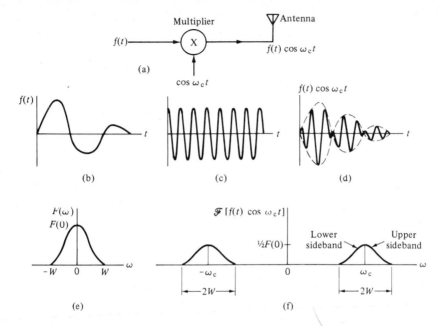

Fig. 5.1 An amplitude modulation suppressed-carrier transmission system.

Recovery of the original signal $f(t)$ from the DSB-SC signal $\phi(t)$ requires another translation in frequency to shift the spectrum to its original position. The process of retranslation of the spectrum to its original position in frequency is called demodulation or detection.

Because the modulation property of the Fourier transform proved useful in translating spectra for modulation, we try it again for demodulation. Assuming that $\phi(t) = f(t) \cos \omega_c t$ is the transmitted signal, we have

$$\phi(t) \cos \omega_c t = f(t) \cos^2 \omega_c t$$

$$= \tfrac{1}{2}f(t) + \tfrac{1}{2}f(t) \cos 2\omega_c t. \tag{5.5}$$

Taking the Fourier transform of both sides of Eq. (5.5) and using the modulation property, we get

$$\mathcal{F}\{\phi(t) \cos \omega_c t\} = \tfrac{1}{2}F(\omega) + \tfrac{1}{4}F(\omega + 2\omega_c) + \tfrac{1}{4}F(\omega - 2\omega_c). \tag{5.6}$$

This result can also be obtained by recalling that the operation of multiplication in the time domain is equivalent to that of convolution in the frequency domain. Thus the spectrum of $\phi(t) \cos \omega_c t$ can be obtained by convolving the spectrum of the received signal $\phi(t)$ with that of $\cos \omega_c t$ (i.e., with impulses at $\pm \omega_c$). Both approaches yield the same result and the spectrum is shown in Fig. 5.2(e). A low-pass filter is required to separate out the double-frequency terms from the original spectral components, as indicated in Fig. 5.2. Obviously, we require that $\omega_c > W$ for proper signal recovery, a requirement easily satisfied in practice.

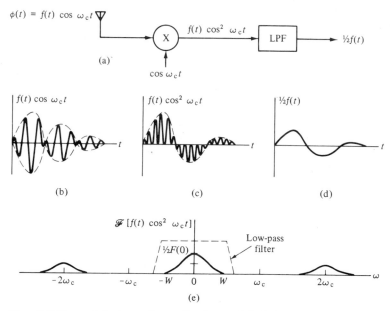

Fig. 5.2 Demodulation of amplitude modulation suppressed-carrier signals.

The principles developed above for DSB-SC modulation hold for any general modulating signal $f(t)$ as long as it is slowly varying with respect to the carrier frequency; that is, $W < \omega_c$. It is instructive to consider the special case of a single-frequency sinusoidal modulating signal to better understand the concepts. Several descriptive waveforms for this case are shown in Fig. 5.3.

It is evident from Fig. 5.3 that it is difficult for an observer to recognize $f(t)$ from an observation of the modulated signal $\phi(t)$ unless there is some knowledge of the phase relations involved. An observation of the amplitude alone is not sufficient. To be more precise, the relative phase of the modulated signal $\phi(t)$ conveys the sign of $f(t)$, as evidenced by the 180° phase reversals when $f(t)$ goes through zero. We conclude from this that both the correct phase and frequency must be known to correctly demodulate DSB-SC waveforms.

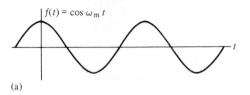

(a)

Fig. 5.3 The DSB-SC modulation and de-modulation of a sinusoidal signal.

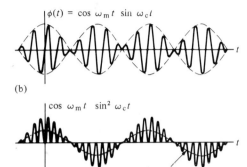

(b)

(c)

Term remaining
after LPF

To confirm this conclusion, let us introduce a small frequency error, $\Delta\omega$, and a phase error, θ_0, in the locally generated carrier signal at the receiver. The receiver then forms the product,

$$\phi(t) \cos\left[(\omega_c + \Delta\omega)t + \theta_0\right] = f(t) \cos\omega_c t \cos\left[(\omega_c + \Delta\omega)t + \theta_0\right]$$

$$= \tfrac{1}{2}f(t) \cos\left[(\Delta\omega)t + \theta_0\right]$$

$$+ \tfrac{1}{2}f(t) \cos\left[(2\omega_c + \Delta\omega)t + \theta_0\right]. \qquad (5.7)$$

The second term on the right-hand side of Eq. (5.7) is centered at $\pm(2\omega_c + \Delta\omega)$ and is filtered out by the low-pass filter. The output of this filter will then be given by the remaining term in Eq. (5.7). Thus instead of recovering the original signal $f(t)$, we obtain

$$e_o(t) = \tfrac{1}{2}f(t) \cos\left[(\Delta\omega)t + \theta_0\right]. \qquad (5.8)$$

It is evident from Eq. (5.8) that the output signal is not $\tfrac{1}{2}f(t)$, as we had before, unless both $\Delta\omega$ and θ_0 are zero (i.e., no frequency or phase error).

We shall consider two special cases of phase and frequency errors here. In the first, let $\Delta\omega = 0$ so that there is only a phase error. Equation (5.8) simplifies to

$$e_o(t) = \tfrac{1}{2}f(t) \cos\theta_0.$$

Thus a phase error in the local carrier causes an attenuation of the output signal proportional to the cosine of the phase error. For small fixed phase errors, this is quite tolerable. For phase errors approaching $\pm 90°$, however, the received signal is wiped out. In some cases, the phase error varies randomly, resulting in unacceptable performance.

In the second special case, let $\theta_0 = 0$ so that there is only a frequency error. Equation (5.8) simplifies to

$$e_o(t) = \tfrac{1}{2} f(t) \cos (\Delta \omega) t.$$

Instead of recovering the original signal $f(t)$, we obtain the signal $f(t)$ multiplied by a low-frequency sinusoid. This results in undesirable and unacceptable distortion.

The effects of both frequency errors and random phase errors render this demodulation of the signal unsatisfactory. It is necessary, therefore, to have synchronization in both frequency and phase between the transmitter and the receiver when DSB-SC modulation is used. The synchronization of the carrier signals presents no major problem when the transmitter and the receiver are in close proximity. Methods for overcoming this limitation when the receiver is remotely located will be discussed in a later section.

Recovering the original signal $f(t)$ from the modulated signal $\phi(t)$ using a synchronized oscillator is called *synchronous* detection, or *coherent* detection. If only one synchronized oscillator is used (as we have done), it is also called *homodyne* detection.[†]

Example 5.1.1   Using the orthogonality of sines and cosines makes it possible to transmit and receive two different signals simultaneously on the same carrier frequency. A scheme for doing this, known as *quadrature multiplexing*, is shown in Fig. 5.4. Show that each signal can be recovered by synchronous detection of the received signal using carriers of the same frequency but in phase quadrature.

Solution

$$\phi(t) = f_1(t) \cos \omega_c t + f_2(t) \sin \omega_c t,$$

$$\phi(t) \cos \omega_c t = f_1(t) \cos^2 \omega_c t + f_2(t) \sin \omega_c t \cos \omega_c t$$

$$= \tfrac{1}{2} f_1(t) + \tfrac{1}{2} f_1(t) \cos 2\omega_c t + \tfrac{1}{2} f_2(t) \sin 2\omega_c t,$$

$$\phi(t) \sin \omega_c t = f_1(t) \cos \omega_c t \sin \omega_c t + f_2(t) \sin^2 \omega_c t$$

$$= \tfrac{1}{2} f_1(t) \sin 2\omega_c t + \tfrac{1}{2} f_2(t) - \tfrac{1}{2} f_2(t) \cos 2\omega_c t.$$

---

† A discussion of homodyne detection, with specific applications to measurement systems, can be found in: R. J. King, *Microwave Homodyne Systems*. Stevenage, Herts (England): Peter Peregrinus Ltd., 1978.

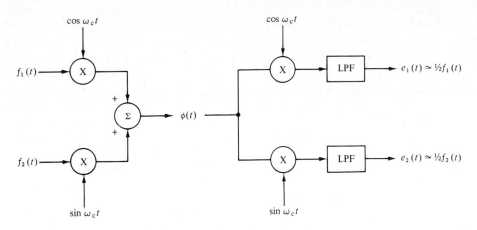

Fig. 5.4 The quadrature multiplexing system described in Example 5.1.1.

In the low-pass filter, all terms at $2\omega_c$ are attenuated, yielding

$$e_1(t) = \tfrac{1}{2}f_1(t),$$

$$e_2(t) = \tfrac{1}{2}f_2(t).$$

Quadrature multiplexing is an efficient method of transmitting two message signals within the same bandwidth. It requires precise phase synchronization of transmitter and receiver, as shown by the following drill problem. Quadrature multiplexing is used in the transmission of color (chroma) information signals in commercial television broadcasts. (See Appendix D for a description of commercial television signal transmissions.)

Drill Problem 5.1.1   Practical use of the quadrature multiplexing scheme of Example 5.1.1 is limited by the rigid demands on the phase synchronization required. For example, let the local carrier generated in the receiver in Fig. 5.4 have a phase error of θ_0 radians. Derive an expression for $e_1(t)$ if θ_0 is small; also estimate how small θ_0 must be to keep the interference caused by $f_2(t)$ at least -40 dB (1%) of $f_1(t)$ if $f_1(t)$, $f_2(t)$ are of equal magnitude.

Answer. $e_1(t) \approx \tfrac{1}{2}[f_1(t) \pm \theta_0 f_2(t)];$ $\pm 0.57°$.

Drill Problem 5.1.2   Show that $f_1(t)$ and $f_2(t)$ essentially control both the amplitude and the phase of $\phi(t)$ in the quadrature multiplexing system of Example 5.1.1. In other words, derive an expression for $\phi(t)$ which exhibits the envelope and phase angle in terms of $f_1(t)$ and $f_2(t)$. [*Hint*: Recall a Fourier series conversion.] Also find $\phi(t)$ for the special case where $f_1(t) = \cos\theta$, $f_2(t) = \sin\theta$.

Answer. $\phi(t) = \sqrt{f_1^2(t) + f_2^2(t)} \cos\{\omega_c t + \tan^{-1}[-f_2(t)/f_1(t)]\};$ $\cos(\omega_c t - \theta)$.

5.1.1   Generation of DSB-SC Signals

As a result of the modulation property of the Fourier transform, we have observed that the spectrum of any signal may be translated $\pm \omega_c$ rad/sec in frequency by multiplying it with a sinusoidal frequency of ω_c rad/sec. It is not necessary, however, that we multiply it with a waveform that is sinusoidal. In fact, we shall now demonstrate that a signal spectrum can be translated an amount $\pm \omega_c$ by multiplying the signal with *any* periodic waveform whose fundamental frequency is ω_c rad/sec. Our reasoning here will closely parallel that followed in the discussion of the sampling theorem in Chapter 3.

Any periodic signal $p_T(t)$ with finite average power can be represented by the Fourier series,

$$p_T(t) = \sum_{n=-\infty}^{\infty} P_n e^{jn\omega_0 t}, \tag{5.9}$$

where $\omega_0 = 2\pi/T$. Choosing $\omega_0 = \omega_c$ and multiplying Eq. (5.9) by $f(t)$ gives us

$$f(t)p_T(t) = \sum_{n=-\infty}^{\infty} P_n f(t) e^{jn\omega_c t}. \tag{5.10}$$

Applying the frequency translation property of the Fourier transform to Eq. (5.10), we have

$$\mathscr{F}\{f(t)p_T(t)\} = \sum_{n=-\infty}^{\infty} P_n F(\omega - n\omega_c). \tag{5.11}$$

From this, we see that the spectrum of $f(t)p_T(t)$ contains $F(\omega)$ and $F(\omega)$ translated by $\pm\omega_c$, $\pm 2\omega_c$, $\ldots$, $\pm n\omega_c$, $\ldots$. The amplitudes of these successive spectral replicas of $F(\omega)$ are scaled by the constants P_0, P_1, P_2, $\ldots$, P_n, $\ldots$. This is illustrated in Fig. 5.5 using a periodic square wave. These results can also be demonstrated using the frequency convolution property of the Fourier transform.

In amplitude modulation, we are interested only in the spectrum centered around $\pm\omega_c$. This can be obtained by using a bandpass filter following the multiplication. The bandpass filter allows those frequency components centered at $\pm\omega_c$ to pass and attenuates all other frequency components. If ω_c is chosen much larger than W, the bandwidth of $f(t)$, the undesired frequency components are widely separated from the desired frequency components and simple filters can be used.

The process of frequency translation is also called *frequency conversion*, *frequency mixing*, and *heterodyning*. Systems that perform this operation are called frequency converters or frequency mixers. In order to translate a spectrum, such systems must produce new frequencies which are different from those present in the input signal. However, the response of linear time-invariant systems cannot contain frequencies other than those present in the input signal (cf.

Section 2.12). Therefore we conclude that to produce modulation we must use systems which are either time-varying or nonlinear (or both).

This fact can be used to advantage in the testing of linear amplifiers. The sum of two sinusoids, say, at 400 Hz and 1 kHz, is applied to the input. Any output at frequencies other than 400 Hz or 1 kHz indicates the presence of nonlinearities in the amplifier. In particular, the output at 1400 Hz (or 600 Hz) is the amount of modulation present which in turn is a relative measure of the nonlinearities. This test is called an *intermodulation test* and is used in the testing of high-fidelity audio equipment.

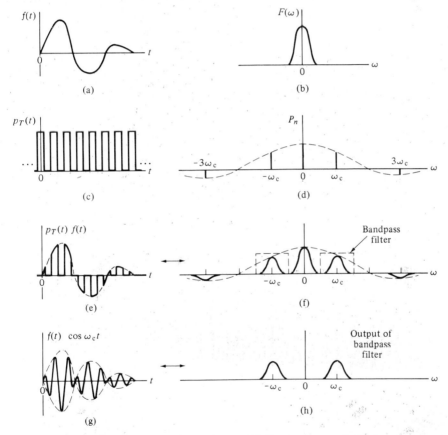

Fig. 5.5 Amplitude modulation using multiplication by periodic functions.

Example 5.1.2   A periodic signal consists of the exponentially decreasing waveform e^{-at}, $0 \le t < T$, repeated every T seconds. A given signal $f(t)$ is multiplied by this periodic signal. Determine an expression describing the spectrum and the time waveform of the resulting amplitude-modulated signal if all components except those centered at $\pm \omega_c$, $\omega_c = 2\pi/T$, are discarded.

Solution.   We can write a Fourier series,

$$p_T(t) = \sum_{n=-\infty}^{\infty} P_n e^{jn\omega_c t}$$

where

$$P_n = \frac{1}{T} \int_0^T e^{-at} e^{-jn\omega_c t}\, dt = \frac{1}{T} \frac{1 - e^{-aT}}{a + jn\omega_c}.$$

Using Eq. (5.11), the spectrum of the product $p_T(t)f(t)$ is

$$\frac{1}{T}(1 - e^{-aT})\left[\frac{1}{a + j\omega_c} F(\omega - \omega_c) + \frac{1}{a - j\omega_c} F(\omega + \omega_c) \right].$$

The corresponding terms in the Fourier series are

$$\frac{1}{T}(1 - e^{-aT})\left[\frac{1}{a + j\omega_c} f(t)e^{j\omega_c t} + \frac{1}{a - j\omega_c} f(t)e^{-j\omega_c t} \right].$$

Combining [or using Eqs. (2.57), (2.60), and (2.61)] yields the time waveform

$$\frac{2}{T} \frac{1 - e^{-aT}}{\sqrt{a^2 + \omega_c^2}} f(t) \cos(\omega_c t + \theta_0),$$

where

$$\theta_0 = \tan^{-1}(-\omega_c/a),$$
$$\omega_c = 2\pi/T.$$

Drill Problem 5.1.3   Repeat Example 5.1.2 if the periodic signal $p_T(t)$ is a sequence of unit impulse functions spaced T sec apart with one impulse function at $t = 0$. Check your result using frequency convolution.

Answer. $\dfrac{1}{T} F(\omega - \omega_c) + \dfrac{1}{T} F(\omega + \omega_c); \qquad \dfrac{2}{T} f(t) \cos \omega_c t.$

The Chopper Modulator

As an example of a modulator which operates as a linear time-varying system, consider a system whose gain (i.e., transfer function) is varied with time. The simplest example of such a system is a switch which is opened and closed at a given rate or frequency, ω_c. This type of system is often descriptively called a "chopper" and is illustrated in Fig. 5.6(a).

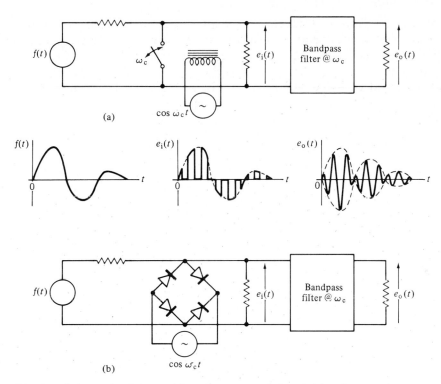

Fig. 5.6 Schematic diagram of a chopper modulator: (a) using an electrome-chanical switch and (b) using diodes as switches.

The operation of the chopper type of modulator shown in Fig. 5.6(a) can be explained briefly as follows. The switch alternates between the open and closed states, driven by a source at a frequency of ω_c rad/sec. We shall assume that the transition time between contacts is negligible. For one-half the period, the switch allows the relatively slowly varying signal $f(t)$ to be applied to the filter input. During the next half-period, the input to the filter is grounded. Therefore the signal waveform at the filter input has literally been chopped at a frequency of ω_c rad/sec. This chopping operation is essentially a multiplication of the signal $f(t)$ with a periodic square wave $p_T(t)$ whose two levels are 0 and 1. As discussed earlier, such a chopped waveform contains the spectrum of $f(t)$ translated in frequency by multiples of ω_c. The desired amplitude modulated signal $\phi(t)$ may be obtained by passing this chopped signal through a bandpass filter centered at $\pm\omega_c$. Note that the sampling theorem must be satisfied to allow the proper filtering of the waveform.

The chopper modulator is widely used in practical DSB-SC modulation systems. At low frequencies, the chopper may be a magnetically driven switch.

Photo diodes, FET's, and mechanically interrupted light beams are also used. At higher frequencies diodes are usually used as switches. A typical circuit using four diodes, sometimes known as a ring modulator, is shown in Fig. 5.6(b). It is easily seen that these diodes alternately switch the filter input between the input signal and ground at the frequency of ω_c rad/sec. This is the same action as that of the switch used above and the analysis follows in exactly the same manner. For best operation, the diodes should be identical (i.e., balanced) so that the bridge circuit is balanced when $f(t) = 0$ and there is no carrier output. Several configurations of the ring modulator are in common usage.

A popular configuration, known as the double-balanced (i.e., full-wave balanced) ring modulator, is shown in Fig. 5.7. Popularity of this modulator is attributable to the fact that it does not require ideal components, provided that their characteristics are matched, and the output is often usable without additional filtering.

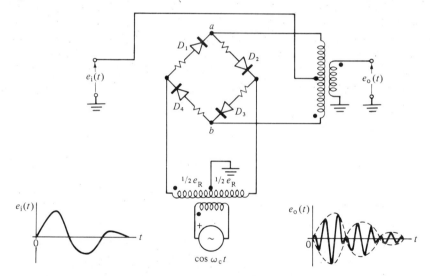

Fig. 5.7 The double-balanced ring modulator.

The operation of the double-balanced ring modulator is briefly described as follows. First we assume that $e_i(t) = 0$. On the positive half-cycles of $\cos \omega_c t$, diodes D_1 and D_2 will conduct and point a is connected to the input through half of the output-transformer secondary. If the secondary of the reference transformer is accurately center-tapped and if the impedances of D_1 and D_2 are identical, no current will flow through the output transformer and no voltage will be developed at the output. During the next half-cycle, diodes D_3 and D_4 conduct and point b is connected to the input through the opposite half of the output-transformer secondary. Again, no current will flow and no output results. Thus we see that the carrier is indeed suppressed; i.e., the modulator is balanced.

Now let us apply a positive-polarity input signal whose peak amplitude is much smaller than that of the reference e_R. On the positive half-cycles of cos $\omega_c t$, point a is essentially at ground potential and a current will flow upwards through half of the output-transformer secondary, inducing a positive output voltage. On the negative half-cycles of cos $\omega_c t$, point b is essentially at ground potential and a current will flow downward through the opposite half of the output-transformer secondary, inducing a negative output voltage. The peak positive and negative output voltages will be identical for a given fixed input-signal amplitude if the output transformer is also accurately center-tapped. Therefore we have developed a signal which alternates in sign at a rate determined by the carrier (reference) frequency and whose amplitude is proportional to the input signal amplitude. It can be seen that the phase of the output changes by 180° for negative-polarity input signals. The output signal, then, is an example of DSB-SC modulation.

The description above assumed a reference signal large enough to control the diode switching. Referring to Fig. 5.7, we see this requires that

$$[e_i(t)]_{max} < \tfrac{1}{2}[e_R(t)]_{max}.$$

In practice, an imbalance between diode characteristics and inaccuracies in trans-former center-taps results in nonideal performance and some carrier does appear in the output unless balancing controls are added. The transformer impedances are matched to the geometric mean of the forward and back resistances of the diodes for best performance. The use of hot-carrier diodes can result in good carrier suppression without a need for balancing controls.

Use of Nonlinear Devices

Amplitude modulation can also arise in nonlinear systems. A common nonlinearity in electrical devices is a nonlinearity between the voltage and the current. A semiconductor diode is a good example of such a device; tubes and transistors also exhibit similar characteristics under large-signal conditions.

A possible arrangement for the use of nonlinear devices to obtain DSB-SC modulation is shown in Fig. 5.8(a). Approximating the nonlinearity with a power series, we get†

$$i(t) = a_1 e(t) + a_2 e^2(t) + a_3 e^3(t) + \cdots.$$

Referring to Fig. 5.8(a), we find

$$e_1(t) = \cos \omega_c t + f(t),$$
$$e_2(t) = \cos \omega_c t - f(t).$$

---

† A discontinuity may exist if a piece-wise linear model is chosen. The discontinuity prevents the use of a power-series expansion. In this case, the nonlinear characteristic can be approximated by a polynomial series following the methods used in Problem 2.5.4.

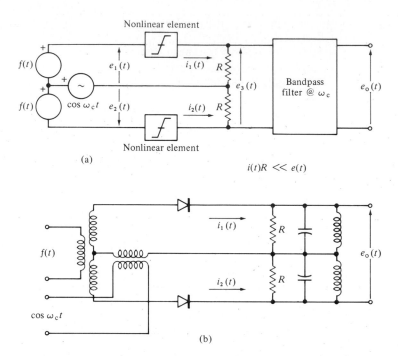

Fig. 5.8 The balanced modulator using nonlinear devices: (a) block diagram; (b) schematic diagram.

Retaining the first two terms in the power series for each nonlinearity and assuming that the nonlinearities are identical gives us

$$i_1(t) = a_1[\cos \omega_c t + f(t)] + a_2[\cos \omega_c t + f(t)]^2,$$

$$i_2(t) = a_1[\cos \omega_c t - f(t)] + a_2[\cos \omega_c t - f(t)]^2.$$

The voltage $e_3(t)$ is then

$$e_3(t) = [i_1(t) - i_2(t)]R$$

$$= 4a_2 R \left[f(t) \cos \omega_c t + \frac{a_1}{2a_2} f(t) \right]. \qquad (5.12)$$

The second term on the right-hand side of Eq. (5.12) can be filtered out by using a bandpass filter centered at $\pm \omega_c$, yielding the desired output.

Semiconductor diodes may be used for the nonlinear devices in such modulators, as shown in Fig. 5.8(b). The performance of this type of modulator, as for the ring modulator, is dependent on how close the characteristics of the diodes can be matched. All of the modulators discussed above generate DSB-SC signals and are known as *balanced modulators*.

Drill Problem 5.1.4 The model of a possible DSB-SC modulator is shown in Fig. 5.9. Determine the required value of the constant k if the bandpass filter has unity gain at $\pm\omega_c$.

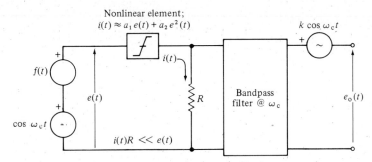

Fig. 5.9 The modulator described in Drill Problem 5.1.4.

Answer. $k = a_1 R$.

5.1.2 Demodulation (Detection) of DSB-SC Signals

To recover the original signal $f(t)$ from the modulated signal $\phi(t)$, we must translate the spectrum again. We have seen that this can be achieved by multiplying $\phi(t)$ by $\cos \omega_c t$ (i.e., synchronous detection). Therefore the same circuits as those used for modulation can be used for demodulation with only minor differences.

The first difference arises because the desired output spectrum of the modulator was centered about the carrier frequency. A bandpass filter centered at $\pm\omega_c$ was used to reject all other frequency components. In the demodulator, however, the desired output spectrum is centered about $\omega = 0$ and therefore a low-pass filter is needed at the output.

A second difference arises from the fact that the oscillator in the demodulator must be synchronized to the oscillator in the modulator to achieve proper demodulation. This is usually accomplished by either a direct connection if the modulator and demodulator are in close proximity or by supplying a sinusoid displaced in frequency but related to the modulator-oscillator frequency. The sinusoid in the latter method is called a "pilot carrier." We shall have occasion to illustrate both methods.

The Chopper Amplifier

It is difficult to construct amplifiers with dc or very low frequency response yet with negligible drift and reasonable-size components. The principle of DSB-SC modulation can be used to advantage to shift the spectrum of the input signal

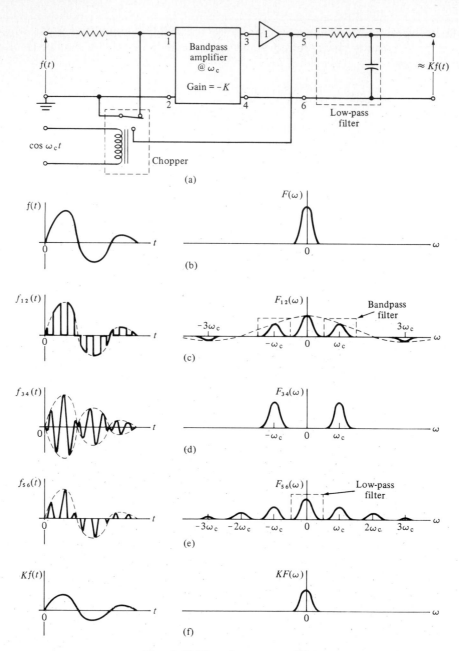

Fig. 5.10 The chopper amplifier.

to higher frequencies where it can be easily amplified. If a low-frequency output is desired, a synchronous demodulator can be used because the modulation oscillator signal is readily available. The chopper-type modulator is generally used for this application.

A typical chopper amplifier is shown in Fig. 5.10(a). The chopper grounds the input signal $f(t)$ for every half-cycle, producing the waveform and spectral density of Fig. 5.10(c). The bandpass amplifier amplifies those spectral components about $\pm\omega_c$, producing the output shown in Fig. 5.10(d). The chopper grounds the amplifier output every half cycle, synchronously demodulating the output signal. As shown, the grounding half-cycles at the input and output are complementary. If a 180° phase reversal in the amplifier is assumed, the output signal is an amplified replica of the input. Note that the sampling theorem must be obeyed to allow adequate filtering.

The primary advantage of the chopper amplifier is that it allows one to build very stable high-gain amplifiers. As an example, a simplified diagram of a low-level, chopper-stabilized voltmeter is shown in Fig. 5.11. Separate choppers for input and output can be used if they are synchronized.

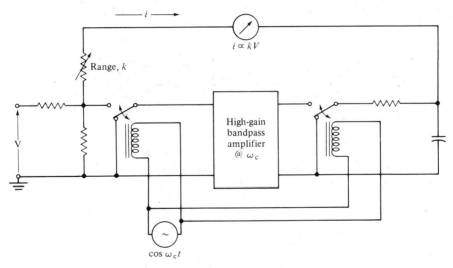

Fig. 5.11 A chopper-stabilizer voltmeter.

The use of DSB-SC modulation in the chopper amplifier is advantageous mainly because both the modulator and the demodulator are in close proximity and can operate from a common source. When the demodulator is remotely located, a problem arises in deriving the necessary phase and frequency synchronization. The use of a pilot carrier to facilitate this is discussed next.

Pilot Carrier Systems

A method commonly used in DSB-SC modulation to maintain synchronization between modulator and demodulator is to send a sinusoidal tone whose frequency and phase are related to the carrier frequency. This tone is generally sent outside the passband of the modulated signal so it will not alter the frequency response capability of the system. A tuned circuit in the receiver detects the tone, translates it to the proper frequency, and uses it to correctly demodulate the DSB-SC signal.

   The stereo multiplexing used in commercial frequency modulation (FM) stations is an example of a pilot tone system.† An FM station is permitted the use of 75 kHz before using frequency modulation to transmit the composite signal. Upon reception, two demodulators are used in tandem, one for the FM, the other to separate the audio signals. We shall consider frequency modulation later; here we are concerned only with the stereo multiplexing.

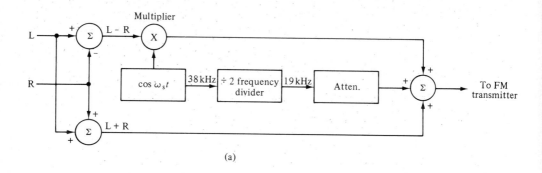

(a)

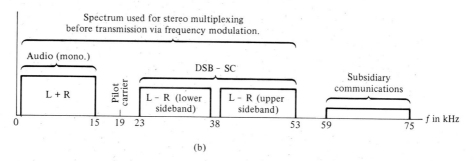

(b)

Fig. 5.12 (a) Block diagram and (b) spectrum of a stereo-multiplex system.

---

† Frequency modulation is discussed in Chapter 6.

In stereo broadcasting, it is necessary to transmit and receive both the left (L) and the right (R) audio channels while also furnishing the sum (L + R) to monophonic receivers. To serve both types of receivers, the (L + R) signal occupies the normal audio spectrum up to 15 kHz and the (L − R) signal, also in the same spectral range, is shifted up in frequency using DSB-SC modulation. A system designed to accomplish this is shown in Fig. 5.12(a). The carrier frequency used is 38 kHz and a sketch of the resulting composite spectrum is shown in Fig. 5.12(b). This leaves the frequency band 15–23 kHz unoccupied and a pilot carrier is placed at 19 kHz. In the receiver, a narrow bandpass filter centered at 19 kHz picks out the pilot carrier.† This frequency is doubled to 38 kHz and used to synchronize the demodulator. Finally, an addition and subtraction of the two signals (this is known as matrixing) yields the desired L and R audio signals.

Drill Problem 5.1.5   Let us designate the four channels of a quadruplexed audio system by RF (right front), RR, LF, and LR (left rear). A proposed method for transmission of quadruplexed sound is to use quadrature multiplexing (cf. Example 5.1.1) at 38 kHz for two signals and shift one signal using DSB-SC centered at 76 kHz.

a) If the quadrature signal at 38 kHz is composed of LF − LR − RF + RR, determine what the other signals must be such that each audio output can be found by algebraic addition (matrixing) of these signals; compatibility with existing monophonic and stereo systems must be maintained. [*Hint*: One signal must contain differences between the front and rear signals.]

b) Sketch a block diagram of your required receiver operations.

Answer.   (a)

$$0\text{–}15 \text{ kHz:} \quad \text{LF} + \text{LR} + \text{RF} + \text{RR}$$
$$23\text{–}53 \text{ kHz:} \quad \text{LR} + \text{LF} - \text{RF} - \text{RR}$$
$$61\text{–}91 \text{ kHz:} \quad \text{LF} - \text{LR} + \text{RF} - \text{RR}$$

5.1.3   Uses in Filter Design

Although the emphasis in our discussion of DSB-SC modulation thus far has been on signals, the same principles can be used to advantage in the design of specialized filters. An important class of filters using these principles is that of the electrically controlled filter.

---

† Detection of this 19-kHz pilot carrier is used to switch on the stereo indicator light on the receiver panel.

A problem arising in the design of some systems is that of a bandpass filter whose center frequency is movable and controllable. The design of such filters by the usual filter design methods often proves difficult. An attractive alternative is to use a fixed filter and to move the signal frequency components instead, using DSB-SC modulation. Let us examine two examples.

The Phase-Locked Loop

A question that arises in pilot tone systems is: What is the best way to synchronize one sinusoid to another? Narrow bandpass filters are a possibility but they tend to drift in tuning and are not optimum in rejecting noise. A well-known solution to the synchronization problem is the so-called *phase-locked loop* (PLL).

The basic phase-locked loop consists of a signal multiplier (i.e., a balanced modulator), a low-pass filter, and a voltage-controlled oscillator (VCO), as shown in Fig. 5.13. Operation of the phase-locked loop when the VCO frequency is near the incoming frequency is as follows. The incoming sinusoidal signal, $\cos \omega_c t$, is multiplied by the output of the voltage-controlled oscillator (VCO). The low-frequency component of the output of the multiplier is a voltage whose magnitude and sign are proportional to the phase difference, for small differences, between the incoming sinusoid and the VCO. This voltage is used to control the VCO and the loop attempts to keep the phase difference small (ideally, zero) between the VCO signal and the incoming signal. The objective of synchronization between the two signals has thus been accomplished.

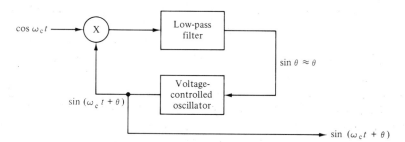

Fig. 5.13 The phase-locked loop.

The bandwidth of the phase-locked loop is controlled by the bandwidth of the low-pass filter and can be made very small to reject noise. In practice, some compromises have to be made, however, because the loop may never find the signal if the bandwidth is too small. Generally the loop will pull into lock within one bandwidth. The operation of the phase-locked loop when it is not locked is very nonlinear, however, and we shall not continue farther into the subject.

The phase-locked loop is widely used in many types of measurement and communication equipment. It will be discussed in more detail in Chapter 6. A simplified block diagram of a phase-locked loop for the demodulation of stereo multiplexing is shown in Fig. 5.14. Note that the use of a divide-by-n circuit following the VCO in the feedback signal path results in a VCO frequency exactly n times the input frequency. This technique is often used in frequency synthesizers.†

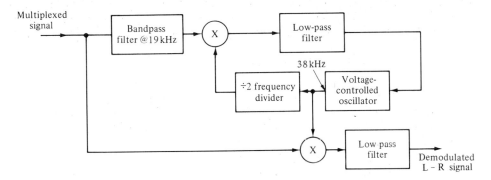

Fig. 5.14 A simplified phase-locked loop stereo demodulator.

The Scanning Spectrum Analyzer

As a second example, suppose we wish to measure the magnitude of a power spectral density experimentally. Because power is proportional to area under the power spectral density, a very narrow bandpass filter is needed to slowly scan all frequencies of interest. This scanning filter would be difficult to build. Instead, it is much easier to move the spectrum and keep the filter fixed. This is normally accomplished in the following manner. A ramp voltage is generated and applied to a voltage-controlled oscillator (VCO), as shown in Fig. 5.15. This produces a sinusoid whose amplitude is constant and whose frequency is proportional to the input voltage. This sinusoid is used to multiply $f(t)$ in a balanced modulator, shifting the spectrum of $f(t)$. As the ramp progresses, the net effect is to move the signal spectrum past the (fixed) bandpass filter (rather than vice versa). The ramp voltage also serves to sweep the beam of a cathode-ray display horizontally and the magnitude of the bandpass-filter output is used to deflect the beam vertically. The resulting display is an approximation to the spectrum of $f(t)$.

---

† A good treatment of the use of phase-locked loops for frequency synthesis can be found in U. Manassewitsch, *Frequency Synthesizers,* Second Edition, Somerset, NJ: Wiley-Interscience, 1980.

Accuracy of the approximation improves with decreasing filter widths and slower ramp-sweep rates. To permit adequate rise-time allowances, the rate of frequency change is usually no greater than the square of the filter bandwidth in Hz (see Problem 3.13.3).

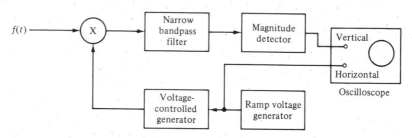

Fig. 5.15 The scanning spectrum analyzer.

5.2   AMPLITUDE MODULATION: LARGE CARRIER (AM)

We have seen that the use of suppressed-carrier signal waveforms requires fairly complicated circuitry at a remotely located receiver in order to acquire and maintain the necessary phase synchronization. If we wish to use very inexpensive receivers, some alternative must be found to this synchronization problem even at the expense of a less efficient transmitter.

To this end, suppose we incorporate the carrier information as a part of the waveform being transmitted and in the same spectral width as the desired signal. In fact, it is convenient to let the amplitude of this carrier term be larger than any other part of the signal spectral density. But if we add this carrier, it will ruin the low-frequency response of the system. For some signals, e.g., audio signals, however, we really do not need a frequency response down to zero. (We have seen that a dc, that is, $\omega = 0$, frequency response shows up in the DSB-SC signal as the spectral density at ω_c.) For systems not requiring a frequency response down to zero, then, we can place a large carrier term at ω_c. To differentiate this case from the previous case, we shall designate this as *double-sideband large-carrier* (DSB-LC). Because commercial broadcast stations use this method of transmission, it is commonly known as just *amplitude modulation* (AM).

The modulated waveform of a DSB-LC signal can be described mathematically simply by adding a carrier term, $A \cos \omega_c t$, to a DSB-SC signal:

$$\phi_{AM}(t) = f(t) \cos \omega_c t + A \cos \omega_c t. \tag{5.13}$$

The spectral density of $\phi_{AM}(t)$ is

$$\Phi_{AM}(\omega) = \tfrac{1}{2}F(\omega + \omega_c) + \tfrac{1}{2}F(\omega - \omega_c)$$
$$+ \pi A \, \delta(\omega + \omega_c) + \pi A \, \delta(\omega - \omega_c). \tag{5.14}$$

The spectral density of $\phi_{AM}(t)$ is the same as that of the DSB-SC signal $f(t) \cos \omega_c t$ with the addition of impulses at $\pm \omega_c$. This is illustrated in Fig. 5.16.

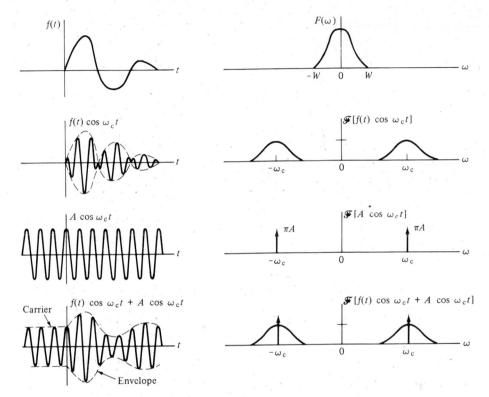

Fig. 5.16 Addition of carrier to produce DSB-LC waveform.

The amplitude-modulated signal $\phi_{AM}(t)$ described by Eq. (5.13) can be rewritten in the form

$$\phi_{AM}(t) = [A + f(t)] \cos \omega_c t. \tag{5.15}$$

Written in this form, $\phi_{AM}(t)$ may be regarded as a carrier signal $\cos \omega_c t$ whose amplitude is given by the quantity $[A + f(t)]$. If A is large enough, the envelope (magnitude) of the modulated waveform will be proportional to $f(t)$. Demodulation in this case simply reduces to the detection of the envelope of a sinusoid, with no dependence on the exact phase or frequency of the sinusoid (i.e., of the carrier). If A is not large enough, however, the envelope of $\phi_{AM}(t)$ is not always proportional to $f(t)$, as shown in Fig. 5.17.

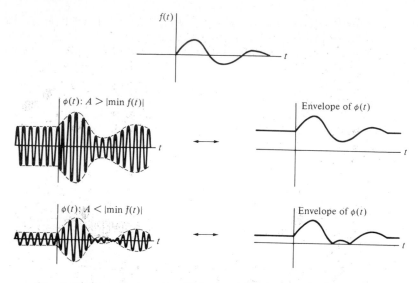

Fig. 5.17 Importance of sufficient carrier in DSB-LC waveform.

The carrier amplitude A must be made large enough so that $[A + f(t)] \geq 0$ at all times, or

$$A \geq |\min \{f(t)\}|. \tag{5.16}$$

If Eq. (5.16) is not satisfied, then $f(t)$ cannot be recovered by the relatively simple process of envelope detection.† Synchronous detection, however, will still correctly demodulate such signals. Note that the dc response of the signal $f(t)$ has been lost in the demodulation as a result of the addition of the carrier.

To emphasize these concepts, we now return to the special case of a single-frequency sinusoidal tone, $\cos \omega_m t$, as the modulating signal. Because the relative magnitudes of the sideband and carrier portion of the signal are variable, we define a dimensionless scale factor, m, to control the ratio of the sidebands to the carrier:

$$m = \frac{\text{peak DSB-SC amplitude}}{\text{peak carrier amplitude}}, \tag{5.17}$$

so that‡

$$\begin{aligned}
\phi_{AM}(t) &= A \cos \omega_c t + mA \cos \omega_m t \cos \omega_c t \\
&= A(1 + m \cos \omega_m t) \cos \omega_c t. \tag{5.18}
\end{aligned}$$

---

† Methods of envelope detection will be discussed in Section 5.2.3.

‡ Including a multiplier constant k_a, if $f(t) = a \cos \omega_m t$, then $m = ak_a$.

A graph of each of the terms appearing in Eq. (5.18) and their addition for different values of the parameter m is shown in Fig. 5.18. Note that those portions of the DSB-SC waveform which are in phase [that is, $f(t) > 0$] with the carrier add while those portions which are 180° out of phase [that is, $f(t) < 0$] with the carrier subtract. The maxima in the composite waveform envelope are $(1 + m)A$ and the minima for $m \le 1$ are $(1 - m)A$. As long as $m \le 1$, the peak values of the modulated carrier trace out an envelope waveform which is proportional to the modulating tone, $\cos \omega_m t$. When $m = 0$, of course, the envelope reduces to the constant A, as it should.

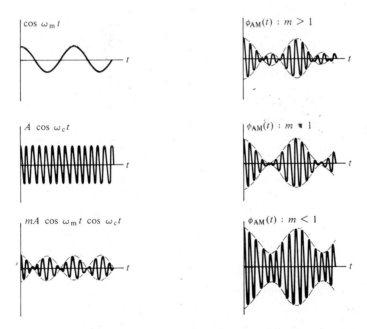

Fig. 5.18 Effects of varying modulation indices.

Often it is convenient to define a percentage of modulation for a DSB-LC signal with sinusoidal modulation as

$$\% \text{ mod.} = \frac{(\text{max. magnitude}) - (\text{min. magnitude})}{(\text{max. magnitude}) + (\text{min. magnitude})} \times 100\%$$

$$= \frac{(1 + m)A - (1 - m)A}{(1 + m)A + (1 - m)A} \times 100\% = m \times 100\%. \qquad (5.19)$$

The parameter m controlling the relative proportions of sideband to carrier is called the *modulation index* of the AM waveform. From Fig. 5.18 one can readily see that $m \leq 1$ is required for envelope detection to take place without severe distortion. If $m > 1$, the waveform is said to be overmodulated.

The student should beware not to confuse the waveform of the summation of two sinusoids with that of a sinusoidal amplitude-modulated waveform with a small modulation index. Both cases are depicted in Fig. 5.19. Note that their spectral densities are clearly very different.

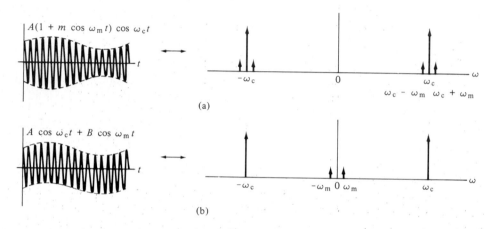

Fig. 5.19 (a) Amplitude modulation and (b) addition for the sinusoidal case.

5.2.1   Carrier and Sideband Power in AM

In AM signal waveforms, the carrier term does not contain any information about the modulating signal $f(t)$. Therefore the power expended in this carrier is wasted for any transfer of information. It is the price one must be willing to pay in order to make cheap receivers available.

A general AM signal waveform can be described by

$$\phi_{AM}(t) = A \cos \omega_c t + f(t) \cos \omega_c t. \qquad (5.20)$$

For a one-ohm load, the total average power is given by the mean-square value of Eq. (5.20),

$$\overline{\phi_{AM}^2(t)} = A^2 \overline{\cos^2 \omega_c t} + \overline{f^2(t) \cos^2 \omega_c t} + 2A\overline{f(t) \cos^2 \omega_c t}, \qquad (5.21)$$

where the bar indicates time averaging. We shall assume that $f(t)$ varies slowly with respect to $\cos \omega_c t$. If we also assume that the average value of $f(t)$ is zero (the usual case), then the last term in Eq. (5.21) is zero so that†

$$\overline{\phi^2_{AM}(t)} = A^2 \overline{\cos^2 \omega_c t} + \overline{f^2(t) \cos^2 \omega_c t}$$

$$= A^2/2 + \overline{f^2(t)}/2. \tag{5.22}$$

Thus the total power P_t can be expressed as the sum of a carrier power P_c and a sideband power P_s:

$$P_t = \tfrac{1}{2}A^2 + \tfrac{1}{2}\overline{f^2(t)} = P_c + P_s. \tag{5.23}$$

That fraction of the total power contained in the sidebands, μ, is given by

$$\mu = \frac{P_s}{P_t} = \frac{\overline{f^2(t)}}{A^2 + \overline{f^2(t)}}. \tag{5.24}$$

Returning to the case in which $f(t)$ is a single sinusoid [cf. Eq. (5.18)],

$$\phi_{AM}(t) = A(1 + m \cos \omega_m t) \cos \omega_c t$$
$$= A \cos \omega_c t + mA \cos \omega_m t \cos \omega_c t,$$

we have

$$\overline{\phi^2_{AM}(t)} = \tfrac{1}{2}A^2 + (\tfrac{1}{2})(\tfrac{1}{2})m^2 A^2, \tag{5.25}$$

and

$$\mu = \frac{m^2}{2 + m^2}. \tag{5.26}$$

Because $m \leq 1$, we see from Eq. (5.26) that the transmission efficiency of an AM (DSB-LC) system is at best 33%. Under the best condition, i.e., $m = 1$, 67% of the total power is expended in the carrier and represents wasted power as far as the transfer of information is concerned. For lower modulation indices, the efficiency is less than 33%. For example, a modulation index of 0.50 yields an efficiency of 11.1%. In contrast, the transmission efficiency of a DSB-SC system is 100%. (If a pilot carrier is transmitted, the efficiency is slightly less than 100%.)

Example 5.2.1   A given AM (DSB-LC) broadcast station transmits an average carrier power output of 40 kilowatts and uses a modulation index of 0.707 for sine-wave modulation. Calculate (a) the total average power output; (b) the

---

† If $\overline{f(t)}$ is not zero, then the first and last terms on the right-hand side of Eq. (5.21) can be combined in an effective carrier power. In other words, a nonzero average value of $f(t)$ shows up in the AM waveform as a carrier term.

transmission efficiency; and (c) the peak amplitude of the output if the antenna is represented by a 50-Ω resistive load.

Solution

a) Using Eq. (5.25), we get

$$P_t = P_c(1 + m^2/2).$$

For $m = 0.707$,

$$P_t = (1 + \tfrac{1}{4})P_c = \tfrac{5}{4}P_c,$$
$$P_t = 50 \text{ kW}.$$

b) Using Eq. (5.26) gives us

$$\mu = \frac{(0.707)^2}{2 + (0.707)^2} = \frac{0.500}{2.500} = 20\%.$$

c) Then

$$P_c = \frac{A^2}{2R},$$

$$A^2 = 2RP_c = 4 \times 10^6,$$

$$(1 + m)A = 3414 \text{ V}.$$

Drill Problem 5.2.1 The amplitude of the modulating tone is decreased in the above example until the total power is 45 kW. Assuming that the carrier power remained constant, compute the new modulation index and transmission efficiency.

Answer.   (a) 0.500; (b) 11.1%.

5.2.2   Generation of DSB-LC Signals

Conceptually, the easiest way to generate a DSB-LC signal is to first generate a DSB-SC signal and then add some carrier. This is shown in block diagram form in Fig. 5.20. It turns out, however, that DSB-LC signals are generally easier to generate directly so that the system shown in Fig. 5.20 has more analytical than practical value.

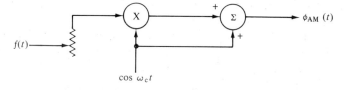

Fig. 5.20 Block diagram of DSB-LC generation.

As in the case of DSB-SC signals, the generation of DSB-LC signals can be divided into the chopper (switch) type modulators and those modulators using nonlinear characteristics of devices.

The Chopper (Rectifier) Modulator

A straightforward extension of the chopper modulator for generating AM is to add a dc level to $f(t)$ before chopping. If this dc level A is large enough so that the condition $[A + f(t)] > 0$ holds, then it is easy to see that the output signal will be a DSB-LC signal.

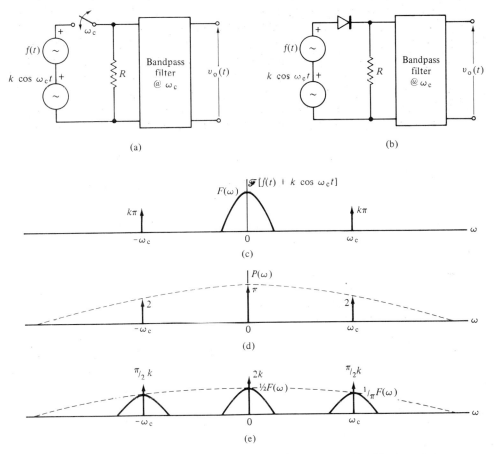

Fig. 5.21 Generation of an AM signal using the chopper (rectifier) modulator.

Another possibility, which really amounts to the same thing, is to add some carrier to $f(t)$ before chopping. This is shown in Fig. 5.21(a). The chopper action may be viewed as a multiplication of the input signal with a periodic square wave $p_T(t)$ whose fundamental frequency is ω_c rad/sec. The spectral density of the resulting chopped signal can be found by using the modulation property of the Fourier transform or by taking the frequency convolution of the spectral density of $[f(t) + k \cos \omega_c t]$ with the spectral density of $p_T(t)$, as shown in Fig. 5.21. The result yields a term: $[\frac{1}{2}k \cos \omega_c t + (2/\pi)f(t) \cos \omega_c t]$ centered about $\pm\omega_c$. Unwanted frequency components at $\omega = 0$, $\pm3\omega_c$, $\pm5\omega_c$, etc., can be filtered out by proper choice of a bandpass filter.

It is not necessary that the chopper be driven by a separate source if the carrier amplitude is made much greater than $f(t)$. An ideal diode will act as a good switch and, if the carrier is large, will turn on and off at the carrier rate. A diagram of a modulator using a diode as a switch is shown in Fig. 5.21(b). The spectral characteristics are the same as for the chopper modulator described above. Because the diode eliminates the negative part of the composite signal $[f(t) + k \cos \omega_c t]$, it essentially performs a half-wave rectification of the input signal. For this reason, this type of modulator is also known as a rectifier-type modulator.

Amplitude modulation with the rectifier-type modulator is relatively easy to accomplish at high-power levels by using amplifiers operated in class C conditions. A parallel LC filter on the output performs the required bandpass filtering operation. If overmodulation is attempted with this type of modulator, no output will result instead of reversals of the carrier phase. This results in abrupt discontinuities in the modulated waveform and introduces much harmonic content in the spectrum. To prevent this, high-power AM broadcast transmitters operated in this mode are required to maintain a modulation index less than 100% on negative peaks in programming material.

Modulator Using Nonlinearities

The system shown in Fig. 5.21(b) can be used to generate AM even if the diode is not operated as an ideal switch. In this case, the nonlinearities in the diode characteristic may be approximated with a power series of the form

$$i(t) = a_1 e(t) + a_2 e^2(t) + \cdots, \qquad i(t)R \ll e(t). \tag{5.27}$$

Retaining only the first two terms, we find that the voltage at the input to the bandpass filter (neglecting effects of any finite input impedance) is

$$i(t)R = a_1 R[f(t) + k \cos \omega_c t] + a_2 R[f(t) + k \cos \omega_c t]^2 + \cdots. \tag{5.28}$$

Expanding and collecting terms at the carrier frequency, we have

$$v_0(t) = a_1 Rk \cos \omega_c t + 2a_2 Rkf(t) \cos \omega_c t. \tag{5.29}$$

Equation (5.29) is the desired result for a DSB-LC signal. A semiconductor diode operates as a combination of the rectifier modulator and the modulator using a nonlinear characteristic.

Drill Problem 5.2.2   Let $f(t)$ be equal to $\cos \omega_m t$. If $a_1 = 0.010$ and $a_2 = 0.001$ in Eq. (5.28), determine the modulation index of the resulting DSB-LC signal when all terms except those near the carrier frequency are filtered out.

Answer. $m = 0.20$.

5.2.3   Demodulation (Detection) of DSB-LC (AM) Signals

In DSB-LC (AM) signals, the desired signal waveform $f(t)$ is available in the envelope of the modulated signal. The use of synchronous detection will, of course, yield the desired waveform but it is possible to demodulate AM signals by much simpler techniques. The simplest and most popular method is that one which detects the envelope of the modulated waveform directly and is called the envelope detector.

The Envelope Detector

Any circuit whose output follows the envelope of the input signal waveform will serve as an envelope detector. The simplest form of an envelope detector is a nonlinear charging circuit with a fast charge time and a slow discharge time. It can easily be constructed using a diode in series with a capacitor, as shown in Fig. 5.22(b). A resistor placed across the capacitor controls the discharge time

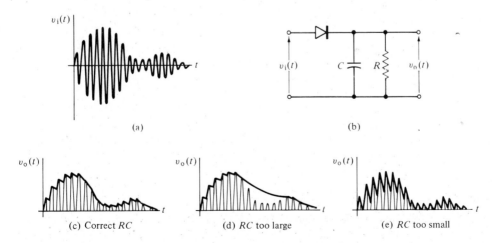

Fig. 5.22 The envelope detector.

constant. Effects of various discharging time constants are shown in Fig. 5.22(c), (d), (e).

The operation of the envelope detector is as follows. On the positive half-cycles of the input signal, the capacitor C charges to the peak value of the input signal waveform. As the input signal falls below this value, the diode is turned off. The capacitor slowly discharges through the resistor until the next positive half-cycle when the input signal becomes greater than the capacitor voltage and the diode again conducts. The capacitor charges to the new peak value, etc.

For best operation, the discharge time constant, RC, should be adjusted so that the maximum negative rate of the envelope will never exceed the exponential discharge rate. If this time constant is too large, the envelope detector may miss some positive half-cycles of the carrier and thus will not reproduce the envelope faithfully. If the time constant is much too small, the envelope detector generates a very ragged waveform, losing some of its efficiency. The resultant detected signal is usually passed through a low-pass filter to eliminate the unwanted harmonic content. A coupling capacitor can be used to remove the dc level introduced by the carrier. As a result of the presence of this dc level, DSB-LC methods are not suitable if a frequency response for $f(t)$ down to $\omega = 0$ is desired.

The envelope detector is simple, efficient, and cheap to build. It is almost universally used for the purpose of detecting DSB-LC signals.

The ease and simplicity in using the envelope detector suggests that perhaps these techniques could also be applied to the detection of DSB-SC signals. Because there is no carrier term in such signals, enough carrier must be added in the receiver to make the envelope detection possible. This is sometimes done to simplify receiver design; this type of receiver is called an *injected-carrier* type receiver. A major problem, of course, is to keep the injected-carrier at the correct frequency and thus such systems also have synchronization problems.

Other Methods of Detection

If a ground return is provided in the envelope detector circuit described above, the diode can be made to operate essentially like a switch, turning on and off at the carrier rate. Because the modulated waveform has been rectified by this action, this detector is called a rectifier detector. A low-pass filter removes the high-frequency terms after rectification, as shown in Fig. 5.23. The rectifier detector is not as efficient as the envelope detector and is rarely used in commercial receivers. It finds application in the amplitude detection of pulsed signals where the slow exponential discharge time of the envelope detector is undesirable.

The synchronous detection methods used for DSB-SC signal waveforms can, of course, also be used for DSB-LC signals. For example, a phase-locked loop can be used to synchronize an oscillator to the carrier frequency and this can be used to synchronously demodulate the signal. Such elaborate receiving techniques are rarely used, however, as a result of the required additional complexity in receiver design and operation.

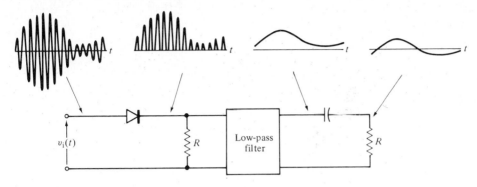

Fig. 5.23 The rectifier detector.

5.3 FREQUENCY-DIVISION MULTIPLEXING (FDM)

From our discussion of amplitude modulation thus far, we see that it is possible to send several signals simultaneously by choosing a different carrier frequency for each signal. These carrier frequencies are chosen so that the signal spectra are not overlapping. This mode of transmission is called *frequency-division multiplexing* (FDM). Frequency-division multiplexing, then, is the positioning of signal spectra in frequency such that each signal spectrum can be separated out from all the others by filtering. We shall emphasize the use of amplitude modulation here but frequency-division multiplexing does not preclude the use of other modulation methods.

To illustrate the principles of frequency-division multiplexing (FDM), suppose we wish to transmit several signals simultaneously, using DSB-LC or DSB-SC modulation. Let us assume that there are N different signals and, for convenience, that each signal is bandlimited to ω_m rad/sec. In order to separate these N signals in frequency, each is modulated with a carrier frequency ω_1, ω_2, . . . , ω_N. Thus the spectral density of every modulated signal has a bandwidth of $2\omega_m$ and each is centered at the various carrier frequencies ω_1, ω_2, . . . , ω_N. These carrier frequencies are chosen far enough apart such that each signal spectral density is separate from all the others. This requires that each carrier frequency be separated from an adjacent carrier frequency by at least $2\omega_m$. Figure 5.24(a), (b), (c) illustrates the case in which three signals are translated in frequency and transmitted simultaneously.

At the receiver, we consider two possibilities. In the first, the receiver processes the various signal spectra simultaneously, separating them in frequency with the appropriate bandpass filters and then demodulating. This is shown in Fig. 5.24(d). In practice, the composite signal formed by spacing several signals in frequency may, in turn, be modulated using another carrier frequency. To distinguish the choice of the first carrier frequencies in this case, the ω_1, ω_2, . . . , ω_N are called *subcarriers*.

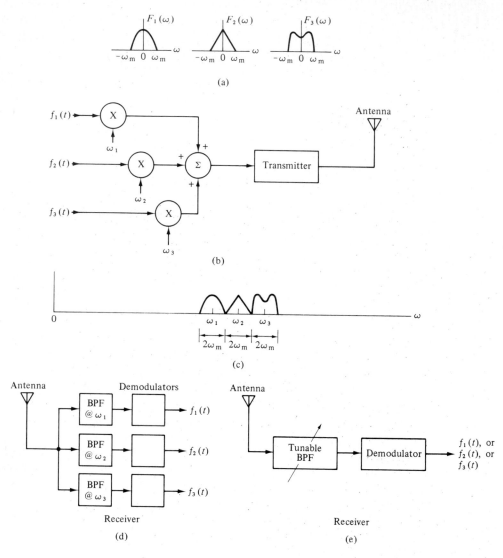

Fig. 5.24 Frequency-division multiplexing (FDM).

The second possibility is to have each receiver select only one of all possible signals. This is accomplished by tuning a bandpass filter to the center frequency of the desired signal and then demodulating. A receiver diagram for this possibility is shown in Fig. 5.24(e).

Each of these possibilities has applications and the choice largely determines whether large-carrier or suppressed-carrier amplitude modulation is employed. In the first, several signals are processed simultaneously so that at most only one large carrier, or several pilot carriers, need be sent to the receiver. This allows the use of the more efficient DSB-SC modulation. The stereo multiplexing discussed earlier is a familiar example of this. On the other hand, in the second method there is only one signal being received at a time and, to keep the receiver simple, DSB-LC is usually used for this type of operation. Commercial radio and television receivers provide familiar examples of this type of FDM. It is worthwhile considering the standard radio receiver to see how this is accomplished in practice.

Each commercial AM broadcast transmitter sends out a DSB-LC signal whose carrier frequency is separated from the carrier frequencies of other stations. Carrier frequencies are assigned at 10-kHz spacings from 540 kHz to 1600 kHz.† A receiver can pick up any one of these signals by tuning to the appropriate carrier frequency. The selected signal is then demodulated using an envelope or rectifier detector to produce the desired signal.

Early AM receivers performed these operations in exactly this manner. An antenna, an LC tuning arrangement for station selection, a diode for a detector, and a pair of headphones constituted the first receivers. Unless one lived near a broadcast station, the received signals at the antenna terminals were small and amplification and filtering were needed to improve the sensitivity (i.e., the ability to receive weaker signals) and the selectivity (i.e., the ability to separate the signals of different stations).

The next logical step in receiver improvement was the tuned-radio-frequency (TRF) receiver, shown in Fig. 5.25(a). All stages of amplification (usually about three) were tuned simultaneously to select stations. Varying three separate tuning controls, however, made changing stations a frustrating chore. Ganging all three together, on the other hand, created additional problems because all stages didn't always change at the same rate (i.e., track) unless the bandwidth of each stage was broadened, ruining the sensitivity and selectivity. What was needed was the amplification of a narrow band of frequencies which would remain fixed regardless of which station was selected. This gave rise to the heterodyne receiver that is widely used today.

Heterodyning means the translating or shifting in frequency. In the heterodyne receiver the incoming modulated signal is translated in frequency, thus

---

† Frequency allocations in the U.S. are described in Appendix C.

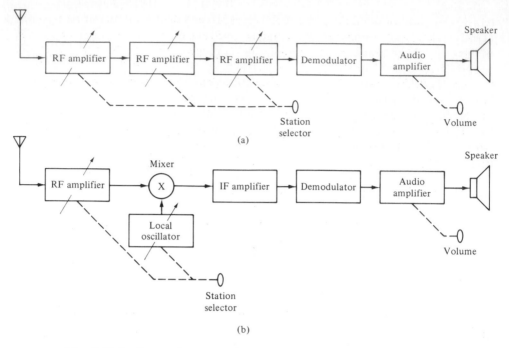

Fig. 5.25 Radio receivers: (a) the TRF and (b) the superheterodyne.

occupying an equal bandwidth centered about a new frequency. This new center frequency, known as an intermediate frequency (IF), is fixed and is not dependent on the received signal center frequency. The signal is amplified at the IF before demodulation. If this intermediate frequency is lower than the received carrier frequency but above the final output signal frequency, it is called a super-heterodyne receiver. In most commercial AM broadcast receivers of the super-heterodyne type, the intermediate frequency is 455 kHz. AM broadcast frequency assignments are from 540 kHz to 1600 kHz. The required frequency translation to the intermediate frequency is accomplished by mixing the incoming signal with a locally generated signal which differs from the incoming carrier by the intermediate frequency (in this case 455 kHz). The received signal, now translated to a fixed intermediate frequency, can easily be amplified, filtered, and demodulated.† The primary advantage of the superheterodyne receiver over the TRF

---

† For reception of higher frequencies, the translation to an intermediate frequency may be repeated a second time. Receivers using this method are called dual-conversion receivers.

receiver is that the amplification and filtering is performed at a fixed frequency regardless of station selection. A block diagram of the superheterodyne receiver is shown in Fig. 5.25(b).†

To translate the spectrum of the incoming signal to the fixed intermediate frequency, the locally generated carrier must be at a frequency either above or below the incoming carrier frequency by the intermediate frequency (455 kHz). In the AM superheterodyne receiver, the locally generated frequency is chosen to be 455 kHz higher than the incoming signal. The reason for this choice is simply because it is easier to build oscillators which are reasonably linear in tuning over a 1–2 MHz range than over a 0.1–1.1 MHz range.

The advantages of the superheterodyne receiver are not without disadvantages. Suppose, for example, that reception of an AM station at 600 kHz is desired. This means that the locally generated carrier, assuming it is 455 kHz above the station carrier, is at 1055 kHz. Now if there is another station at 1510 kHz, it also will be received (note that 1510 kHz − 1055 kHz = 455 kHz). This second frequency, 1510 kHz, is called the *image frequency* of the first, and after the heterodyning operation it is impossible to distinguish the two.

It is instructive to use the modulation property (or the frequency-convolution property) of the Fourier transform to see how the image-frequency problem arises. The spectral densities are shown in Fig. 5.26 for cases in which the station center frequency is either above or below the local oscillator frequency by the amount of the intermediate frequency. If *both* station transmissions are present, both will be received equally well. One is called the *image* of the other. Note that the image frequency is displaced by twice the intermediate frequency from the desired station frequency.

There are two ways to minimize the image-frequency problem. Noting that the image frequency is separated from the desired signal by exactly twice the intermediate frequency, we may choose the intermediate frequency as high as possible and practical. The other way is to attenuate the image frequency before heterodyning. This is the primary function of a selective radio-frequency (RF) amplifier placed before the mixer. It must, of course, be tunable over the frequency range of the incoming signals.

Another disadvantage of the superheterodyne receiver is that the high-gain IF stages are tuned outside the assigned frequency band. Thus one must be careful that the intermediate frequency chosen is free from other strong transmissions or otherwise the receiver will amplify these spurious signals as they leak into the high-gain IF stages. A combination of these considerations makes the choice of 455 kHz a popular one for standard AM broadcast receivers.

---

† In some receiver circuits, the mixer and the local oscillator are combined (e.g., in one transistor); this is called an autodyne circuit.

Although the details of modulation differ, commercial FM and television receivers also use the superheterodyne principle with intermediate frequencies usually at 10.7 MHz and about 44 MHz, respectively.

Because the heterodyne receiver generates an internal carrier, it is also a miniature transmitter of energy at the local oscillator frequency. This can be demonstrated by placing two AM superheterodyne receivers in close proximity and tuning one 455 kHz below the second. When close to this value, a difference-frequency tone (beat) between the two local oscillators can be heard in the speaker. This radiation of the heterodyne receiver has been used to advantage in electronic detection and countermeasures work. On the other hand, it can cause undesirable interference effects. These effects can be minimized by proper shielding in the offending receiver.

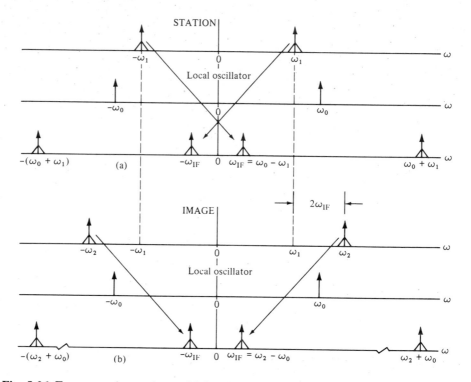

Fig. 5.26 Frequency heterodyne with local oscillator above station carrier showing (a) translation of the received station signal to the intermediate frequency and (b) translation of the image frequency to the intermediate frequency.

Example 5.3.1   A given radar receiver operating at a frequency of 2.80 GHz and using the superheterodyne principle has a local oscillator frequency of 2.86 GHz.

A second radar receiver operates at the image frequency of the first and interference results.

a) Determine the intermediate frequency of the first radar receiver.

b) What is the carrier frequency of the second receiver?

c) If you were to redesign the radar receiver, what is the minimum intermediate
   frequency you would choose to prevent image-frequency problems in the
   2.80–3.00 GHz radar band?

Solution

a) $f_{IF} = f_{LO} - f_c = 2.86 \text{ GHz} - 2.80 \text{ GHz} = 60 \text{ MHz}$
b) $f_{IMAGE} = f_c + 2f_{IF} = 2.80 \text{ GHz} + 0.12 \text{ GHz} = 2.92 \text{ GHz}$
c) $2f_{IF} \geq f_{MAX} - f_{MIN} = 3.00 \text{ GHz} - 2.80 \text{ GHz} = 0.20 \text{ GHz}; f_{IF} \geq 100 \text{ MHz}$

Drill Problem 5.3.1   A telemetry receiver is designed to receive satellite transmissions at 136 MHz. The receiver uses two heterodyne operations with intermediate frequencies of 30 MHz and 10 MHz. (This type of receiver, commonly
known as a dual conversion receiver, is shown in block diagram form in Fig.
5.27.) The first local oscillator is designed to operate below the incoming carrier
frequency; the second, above the first (30 MHz) intermediate frequency. Determine all possible image frequencies. Do not assume filters are ideal. (In other
words, what possible input frequencies could result in images for both the first
and second frequency mixers if the filters were not ideal?) Note that the image
frequencies of the second can be filtered more with the addition of the first
intermediate frequency.

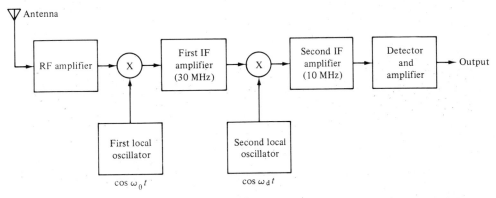

Fig. 5.27 Block diagram of a dual-conversion receiver (cf. Drill Problem 5.3.1).

Answer.   56 MHz, 76 MHz, 156 MHz.

5.4 SINGLE-SIDEBAND (SSB) MODULATION

We have seen in previous sections that DSB modulation results in a doubling of the bandwidth of a given signal. However, this doubling of the bandwidth is a disadvantage when a given band of frequencies is crowded. Thus we wish to investigate whether this doubling of bandwidth is really necessary.

It will be helpful to first recall the spectral densities of DSB signals (we assume suppressed-carrier for convenience). The spectral density of any real-valued signal must exhibit the symmetry condition (cf. Fig. 5.28a)

$$F(-\omega) = F^*(\omega). \tag{5.30}$$

After multiplication by a sinusoid at ω_c rad/sec, half of this spectral density is translated up in frequency and centered about ω_c and half is translated down to $(-\omega_c)$, as shown in Fig. 5.28(b).

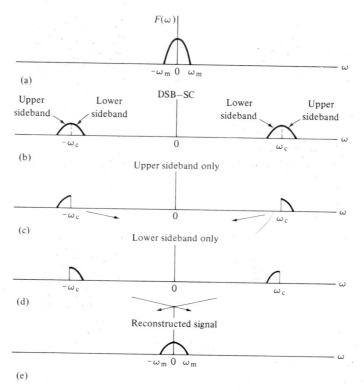

Fig. 5.28 Spectra of DSB and SSB signals.

From Fig. 5.28(a), (b), it can be seen that what was originally positive frequency content in $F(\omega)$ now becomes the upper sideband for $\omega > 0$ and the lower sideband for $\omega < 0$. Conversely, that which was originally negative frequency content now becomes the lower sideband for $\omega > 0$ and the upper sideband for $\omega < 0$. Equation (5.30) has been satisfied throughout this process. Each pair of sidebands (i.e., upper or lower) contains the complete information of the original signal.

Now we entertain an intriguing thought: Why not transmit only the upper or the lower pair of sidebands since each pair contains all the information about the signal? We note that Eq. (5.30) will still be satisfied and that the original signal can be recovered again from either the upper or lower pair of sidebands by an appropriate frequency translation, as suggested in Fig. 5.28(e). This type of modulation is called *single-sideband* (SSB) modulation. Single-sideband modulation is efficient because it requires no more bandwidth than that of the original signal and only half that of the corresponding DSB signal. We shall investigate SSB modulation and demodulation methods in more detail.

5.4.1   Generation of SSB Signals

The most straightforward way conceptually to generate an SSB signal is to first generate a DSB signal and then suppress one of the sidebands by filtering. This procedure is shown in Fig. 5.29.

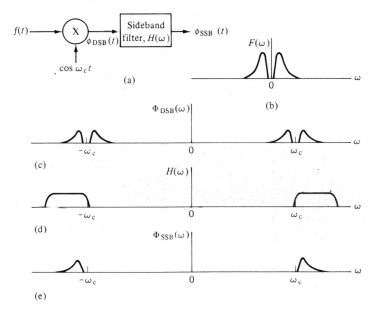

Fig. 5.29 SSB modulator using filtering for the upper sideband.

In practice, these operations are not as easy as they may appear and the primary source of difficulty is in meeting the filter requirements. The sideband filter required must have a very sharp cutoff characteristic at ω_c to reject all frequency components on one side of ω_c and to accept all frequency components on the other side of ω_c. Because it is impossible to obtain an ideal filter characteristic, we must settle for some compromises. First we note that if the modulating signal $f(t)$ does not contain significant low-frequency components there will be no significant frequency components in the vicinity of ω_c after modulation. Therefore a response which does not extend down to zero frequency will allow the use of a filter with a less steep slope. Second, it is usually easier to build a sideband filter at a frequency dictated by the filter components and not by the transmitted frequency. Heterodyning can be employed to translate the spectrum, as desired, after the sideband filtering. Even with these provisions, the design of sideband filters is not easy. In practice, electromechanical-resonator filters are used in the .05–.5 MHz range; crystal-lattice filters are used in the 1–10 MHz range.

It is also possible to generate SSB by a proper phasing of signals. Because it affords some insight into the nature of SSB signals, we shall investigate this method in some detail, first with sinusoidal signals and then in more generality.

Consider now a single-frequency complex exponential signal. This signal has a one-sided spectral density and it is only in taking the real part that the spectral density becomes two-sided. To be specific, let the modulating signal $f(t)$ be $f(t) = e^{j\omega_m t}$ and let the carrier signal be $e^{j\omega_c t}$. The spectral densities of these signals are shown in Fig. 5.30. Multiplying, we get $f(t)\, e^{j\omega_c t} = e^{j\omega_m t} e^{j\omega_c t}$, shifting the spectral density up in frequency according to the frequency-translation property of the Fourier transform. Now taking the real part (cf. Appendix A), we have

$$\mathscr{R}e\{e^{j\omega_m t} e^{j\omega_c t}\} = \mathscr{R}e\{e^{j\omega_m t}\}\mathscr{R}e\{e^{j\omega_c t}\} - \mathscr{I}m\{e^{j\omega_m t}\}\mathscr{I}m\{e^{j\omega_c t}\}$$
$$= \cos \omega_m t \cos \omega_c t - \sin \omega_m t \sin \omega_c t.$$

The procedure of taking the real part restores the two-sided characteristic to the spectral density so that this result can be realized using real-valued signals. Because this represents the upper sideband, we write

$$\phi_{\text{SSB}_+}(t) = \cos \omega_m t \cos \omega_c t - \sin \omega_m t \sin \omega_c t. \tag{5.31}$$

It is left to the reader to verify that the lower sideband can be obtained in a similar manner by using $f(t) = e^{-j\omega_m t}$, yielding

$$\phi_{\text{SSB}_-}(t) = \cos \omega_m t \cos \omega_c t + \sin \omega_m t \sin \omega_c t. \tag{5.32}$$

Although these results have been derived for the sinusoidal case, they hold true for the more general case and we can write

$$\phi_{\text{SSB}_\mp}(t) = f(t) \cos \omega_c t \pm \hat{f}(t) \sin \omega_c t, \tag{5.33}$$

where $\hat{f}(t)$ is that signal obtained by shifting the phase of $f(t)$ by 90° at each frequency. We shall investigate this in more detail in the following section.

The signals described by Eqs. (5.31) and (5.32) can be produced by balanced modulators provided that both the modulating signal and the carrier are shifted in phase by 90° to form the second terms in these equations. This phase-shift method of generating SSB signals is diagrammed in Fig. 5.31. It is used in low-frequency SSB generation and in the digital generation of SSB signals. A major

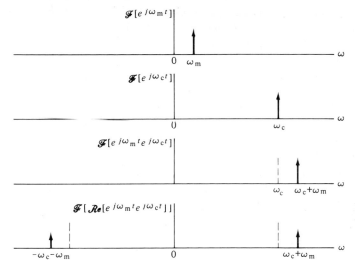

Fig. 5.30 Analytical generation of SSB — upper sideband.

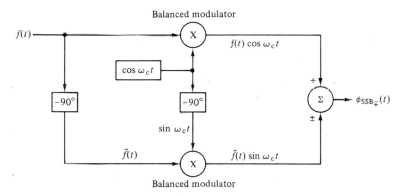

Fig. 5.31 Phase-shift method of generating SSB.

problem in the design of a phase-shift SSB system is the practical realization of the 90° phase-shift network for $f(t)$ because all frequency components must be shifted by exactly 90°. Thus such systems have very restricted bandwidths.

Drill Problem 5.4.1   A DSB-LC signal is generated using a 1-kHz carrier and the input $f(t) = \cos 200\pi t$. The modulation index is 80%. The lower sideband is attenuated (assume ideal filtering). Find an expression for the resulting SSB-LC signal if it develops 0.58 watt across a one-ohm resistive load.

Answer. $\cos(2000\pi t) + 0.4 \cos(2200\pi t)$.

★ 5.4.2   Analytic Signals and the Hilbert Transform

The previous discussion on SSB signals demonstrated that real-valued bandpass signals can be represented in terms of complex-valued signals with one-sided spectral densities. Such signals are called *analytic signals* and are useful in describing the general behavior of signals. In other words, an analytic signal is a complex-valued signal whose spectral density is one-sided and whose real part is the original real-valued signal. Not all complex-valued signals are analytic signals but all analytic signals must be complex-valued.

Consider a given real-valued signal $f(t)$; let its analytic signal representation be $z(t)$. Maintaining $f(t)$ as the real part of $z(t)$, we can write

$$z(t) = f(t) + j\hat{f}(t), \tag{5.34}$$

where $\hat{f}(t)$ is yet to be determined. The Fourier transform of Eq. (5.34) is

$$Z(\omega) = F(\omega) + j\hat{F}(\omega). \tag{5.35}$$

To be an analytic signal, $Z(\omega)$ must be one-sided. Let us require that $Z(\omega) = 0$ for $\omega < 0$; a moment's reflection then shows that [recall that the real part of Eqs. (5.34) and (5.35) cannot be changed arbitrarily]

$$\hat{F}(\omega) = jF(\omega), \qquad \omega < 0. \tag{5.36}$$

To maintain a phase characteristic which is an odd function of frequency then requires that

$$\hat{F}(\omega) = -jF(\omega), \qquad \omega > 0. \tag{5.37}$$

Combining Eqs. (5.36) and (5.37), we have

$$\hat{F}(\omega) = \begin{Bmatrix} -jF(\omega) & \omega > 0 \\ jF(\omega) & \omega < 0 \end{Bmatrix} = -jF(\omega)\,\text{sgn}\,(\omega). \tag{5.38}$$

Using Eq. (5.38) in Eq. (5.35) we can now verify that the spectral density of $z(t)$ is one-sided:

$$Z(\omega) = \begin{cases} 2F(\omega) & \omega > 0 \\ 0 & \omega < 0 \end{cases}. \tag{5.39}$$

It is left to the reader to verify that $\mathscr{F}\{z^*(t)\}$ is similar to Eq. (5.39) with the exception that the inequalities are reversed.

The function $\hat{f}(t)$ can be found by taking the inverse Fourier transform

$$\hat{f}(t) = \frac{1}{2\pi} \int_{-\infty}^{\infty} \hat{F}(\omega)e^{j\omega t} \, d\omega. \qquad (5.40)$$

The function $\hat{f}(t)$ is called the quadrature function of $f(t)$ because each frequency component of $\hat{f}(t)$ is in phase quadrature (i.e., 90°) with that of $f(t)$. We see now that this is a general relationship and not limited to the sinusoids used in the previous section.

The conditions described by Eq. (5.38) define what is known as the *Hilbert transform*; i.e., the function $\hat{f}(t)$ is the Hilbert transform of the function $f(t)$ if Eq. (5.38) holds true. In specific cases, Eqs. (5.38) and (5.40) can be used to find $\hat{f}(t)$ from $f(t)$. For example, let $f(t)$ be the rectangular pulse signal: $f(t) =$ rect (t/τ). Using (cf. No. 14, Table 3.1) $F(\omega) = \tau$Sa $(\omega\tau/2)$, Eqs. (5.38) and (5.40) yield

$$\hat{f}(t) = \frac{1}{\pi} \ln \left| \frac{t - \tau/2}{t + \tau/2} \right|,$$

which is shown in Fig. 5.32. Note that the Hilbert transform becomes infinite at points of finite discontinuity in the signal $f(t)$. Thus a very large instantaneous peak power is required for the transmission of pulse data via SSB.

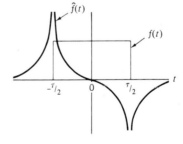

Fig. 5.32 The Hilbert transform of a rectangular pulse.

Drill Problem 5.4.2　Show that (a) $\hat{\hat{f}}(t) = -f(t)$; (b) $\overline{\hat{f}^2(t)} = \hat{f}^2(t)$.

We can also write relationships in the time domain by using (cf. transform pair No. 6, Table 3.1)

$$\text{sgn} \, (\omega) \leftrightarrow \frac{j}{\pi t}.$$

Application of the time convolution property to Eq. (5.38) then yields

$$\hat{f}(t) = f(t) \circledast \frac{1}{\pi t}$$

$$= \frac{1}{\pi} \int_{-\infty}^{\infty} \frac{f(\tau)}{t - \tau} \, d\tau. \tag{5.41}$$

Equation (5.41) is an alternate definition of the Hilbert transform, though not as useful as the phase-shift interpretation as a result of the improper integral form.

Now let us return to the SSB signal. Given a signal $f(t)$, we form the analytic signal $z(t)$ by adding $\hat{f}(t)$ in quadrature, that is, $z(t) = f(t) + j\hat{f}(t)$. The spectral density of $Z(\omega)$ is one-sided, as shown in Fig. 5.33. Multiplying $z(t)$ by $e^{j\omega_c t}$ translates $Z(\omega)$ to $Z(\omega - \omega_c)$. Finally, the real part of $Z(\omega - \omega_c)$ places half of the spectral density, with reversed phase, symmetrically about $\omega = 0$ as shown in Fig. 5.33. The time domain equivalent is

$$\mathscr{Re}\{z(t)e^{j\omega_c t}\} = \mathscr{Re}\{[f(t) + j\hat{f}(t)]e^{j\omega_c t}\}$$

$$= f(t) \cos \omega_c t - \hat{f}(t) \sin \omega_c t,$$

verifying Eq. (5.33). These relations are for the upper sideband case; it is left to the reader to verify that the use of $z^*(t)$ yields the lower sideband.

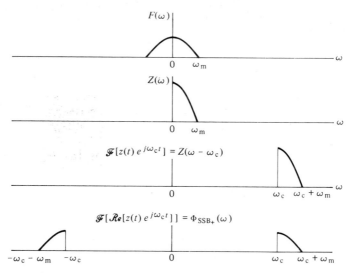

Fig. 5.33 Generation of SSB using analytic signals.

5.4.3   Demodulation of SSB Signals

For demodulation, the spectral density of the SSB signal must be translated back to $\omega = 0$. As observed before, this operation can be achieved using synchronous detection. Multiplication of the SSB signal by $\cos \omega_c t$ translates half of each spectral density up in frequency by ω_c rad/sec and half down by the same amount, as shown in Fig. 5.34. That portion shifted up to a frequency of $2\omega_c$ can be filtered out with a low-pass filter. We conclude that the synchronous detector will properly demodulate SSB-SC signals.†

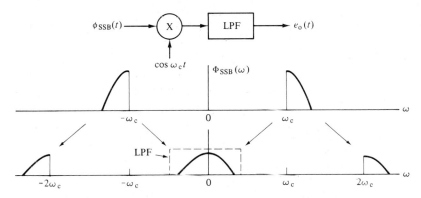

Fig. 5.34 Synchronous demodulation of an SSB signal.

    The above case, of course, assumed ideal conditions. Now let us reconsider when frequency and phase errors are present in the demodulation process. The incoming SSB-SC signal can be expressed as [cf. Eq. (5.33)]

$$\phi_{SSB_\mp}(t) = f(t) \cos \omega_c t \pm \hat{f}(t) \sin \omega_c t.$$

Let the locally generated carrier signal be $\phi_d(t) = \cos [(\omega_c + \Delta\omega)t + \theta]$, where $(\Delta\omega)$ is the frequency error and θ is the phase error. Multiplying, we have

$$\phi_{SSB_\mp}(t)\phi_d(t) = [f(t) \cos \omega_c t \pm \hat{f}(t) \sin \omega_c t] \cos [(\omega_c + \Delta\omega)t + \theta]$$

$$= \tfrac{1}{2}f(t)\{\cos [(\Delta\omega)t + \theta] + \cos [(2\omega_c + \Delta\omega)t + \theta]\}$$

$$\mp \tfrac{1}{2}\hat{f}(t)\{\sin [(\Delta\omega)t + \theta] - \sin [(2\omega_c + \Delta\omega)t + \theta]\}. \quad (5.42)$$

The double-carrier frequency terms can be eliminated using a low-pass filter.

---

† Although the upper-sideband case is illustrated, a similar result holds for the lower-sideband case.

Designating the output of the low-pass filter by $e_o(t)$, we have†

$$e_o(t) = \tfrac{1}{2}f(t) \cos{[(\Delta\omega)t + \theta]} \mp \tfrac{1}{2}\hat{f}(t) \sin{[(\Delta\omega)t + \theta]}. \qquad (5.43)$$

Checking, if the frequency error ($\Delta\omega$) and the phase error θ are both zero, the output is

$$e_o(t) = \tfrac{1}{2}f(t).$$

Therefore a synchronous detector composed of a local oscillator, multiplier, and low-pass filter will correctly demodulate SSB-SC signals if it is correctly synchronized.

To investigate the effects of a phase error, we set ($\Delta\omega$) $= 0$ and Eq. (5.43) becomes

$$e_o(t) = \tfrac{1}{2}[f(t) \cos\theta \mp \hat{f}(t) \sin\theta]. \qquad (5.44)$$

The first term in Eq. (5.44) is the same as occurred for a phase error in the DSB-SC case. However, now there is also a second term present which cannot be filtered out. To investigate, let us rewrite Eq. (5.44) as (cf. Appendix A)

$$e_o(t) = \tfrac{1}{2}\mathcal{R}e\{[f(t) \pm j\hat{f}(t)]e^{j\theta}\}. \qquad (5.45)$$

Because both $f(t)$ and $\hat{f}(t)$ are present now in the output, we conclude that a phase error in the locally generated carrier gives rise to phase distortion in the receiver output.

It turns out that the human ear is relatively insensitive to phase changes in signals and thus this phase distortion in SSB-SC demodulation is quite tolerable for voice communications. In fact, even slowly varying frequency errors are tolerable for voice, changing the quality of speech reproduction but maintaining intelligibility. (This distortion gives a Donald Duck voice effect.) However, for other data signals—particularly pulse data—this type of distortion limits the use of SSB-SC systems.

Proceeding in the same way to see the effects of frequency errors, we set $\theta = 0$ and Eq. (5.43) becomes

$$e_o(t) = \tfrac{1}{2}[f(t) \cos{(\Delta\omega)t} \mp \hat{f}(t) \sin{(\Delta\omega)t}], \qquad (5.46)$$

or

$$e_o(t) = \tfrac{1}{2}\mathcal{R}e\{[f(t) \pm j\hat{f}(t)]e^{j\Delta\omega t}\}. \qquad (5.47)$$

Thus frequency errors give rise to spectral shifts as well as to phase distortion

---

† Note that the upper set of signs refers to the upper-sideband case, the lower to the lower-sideband.

in the demodulated output. As long as these spectral shifts are small they can be tolerated in voice communications.

For values of frequency errors ($\Delta\omega$) approaching the bandwidth of the modulating signal $f(t)$, an interesting effect can occur, as demonstrated in Fig. 5.35. Here the frequency error is equal to the bandwidth of $f(t)$ and chosen in such a way that the spectra end up being exactly reversed! This system is a spectrum inverter, exchanging high-frequency spectral components for lows and vice versa. All the signal energy is present but it has been rearranged. This spectral inversion renders speech quite unintelligible and the principle is used in speech scramblers to ensure communication privacy. It is left to the reader to verify that a repetition of these operations will completely unscramble the signal again.

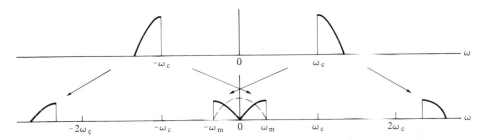

Fig. 5.35 Some effects of frequency errors in demodulation of SSB signals.

As with DSB signals, the possibility exists here also of adding large amounts of carrier to SSB signals, which we designate as SSB-LC. An expression for such an SSB-LC signal is

$$\phi(t) = A \cos \omega_c t + f(t) \cos \omega_c t \mp \hat{f}(t) \sin \omega_c t. \qquad (5.48)$$

Apart from a dc level introduced by the carrier term, the original signal $f(t)$ can always be recovered from $\phi(t)$ using synchronous detection. If the carrier is large, however, envelope detection can also be used. The envelope of Eq. (5.48) can be written as

$$r(t) = \sqrt{[A + f(t)]^2 + [\hat{f}(t)]^2}. \qquad (5.49)$$

Equation (5.49) may be rewritten as

$$r(t) = A\sqrt{1 + \frac{2f(t)}{A} + \frac{f^2(t)}{A^2} + \frac{\hat{f}^2(t)}{A^2}}. \qquad (5.50)$$

If the carrier is much larger than the SSB-SC envelope, the last two terms in

Eq. (5.50) can be dropped to give

$$r(t) \simeq A\sqrt{1 + \frac{2f(t)}{A}}.$$ (5.51)

When we use the binomial expansion† and discard higher-order terms, Eq. (5.51) becomes

$$r(t) \approx A\left[1 + \frac{f(t)}{A}\right] = A + f(t).$$ (5.52)

When we discard the dc term introduced by the carrier, the SSB-LC signal can then be demodulated correctly, using an envelope detector.

We conclude that SSB-LC signals can be demodulated by using an envelope detector. Such signals could be received on commercial AM receivers and would require one-half the bandwidth of present AM transmissions. However, the proportionate amount of carrier required is considerably more than that used for DSB-LC.

Now we pose an interesting question. Does a modulation method exist that affords the spectral economy of SSB and also provides a signal that can be detected using envelope detection without a large carrier term? The answer, surprisingly, is—within some restrictions—yes, providing that a combination of amplitude modulation and phase modulation is used. Although the theory is not easy, a modulation method that approximately achieves the above objectives has the form‡

$$\phi(t) = a(t)\cos[\omega_c t + k_p \widehat{\log a(t)}]$$ (5.53)

where

$$a(t) = \sqrt{f(t)},$$

k_p = phase modulation constant (cf. Section 6.1).

The spectral density of the signal described by Eq. (5.53) is not strictly band-limited to one sideband, and there is some spectral spillover. However, it can be demodulated by using an envelope detector, preferably one with a square-law dependence, and without a large carrier term. Note that the generation of such a signal requires the nonlinear processing of the input signal $f(t)$. Signals

---

† See Appendix A.
‡ See, e.g., K. H. Powers, "The Compatibility Problem in Single-Sideband Transmission," *Proceedings of the IRE*, vol. 48 (August 1960): 1431; or B. F. Logan and M. R. Schroeder, "A Solution to the Problem of Compatible Single-Sideband Transmission," *IRE Transactions on Information Theory*, vol. IT-8 (September 1962): S252.

that permit the use of envelope detection without using bandwidth in excess of that of the modulating signal arc sometimes called *compatible single-sideband* (CSSB) signals.

Drill Problem 5.4.3   A given DSB-SC signal is generated using a 20.000 MHz carrier and a 0–3.0 kHz input signal. One sideband is attenuated using an SSB filter (assume ideal). The resulting SSB-SC signal is to be demodulated using a superheterodyne receiver whose local oscillator operates at a frequency f_o (which is above the incoming signal). The IF amplifier passes all frequency components within the frequency range 10.000–10.003 MHz (assume ideal). Following the IF amplifier is a product detector operating at a frequency f_d followed by a low-pass filter. Determine the numerical values of the frequencies f_o, f_d such that this receiver will correctly demodulate the SSB signal if (a) it is an upper-sideband transmission; (b) it is a lower-sideband transmission.

[The small frequency shifts required are often obtained by heterodyning with an audio frequency oscillator, known as a *beat frequency oscillator* (BFO)—cf. Problem 5.1.9.]

Answer.   (a) 30.003; 10.003; (b) 30.000; 10.000 MHz.

Drill Problem 5.4.4   Determine and plot the envelope of each of the following large-carrier signals and compare (a) DSB-LC; (b) SSB-LC. Assume an input of the form $f(t) = \cos \omega_m t$ and normalize your answer such that the unmodulated carrier magnitude is unity and the peak envelope magnitude at $t = 0$ is 1.30.

Answer.   (a) $1.00 + 0.30 \cos \omega_m t$; (b) $\sqrt{1.09 + 0.60 \cos \omega_m t}$; the plots will differ slightly.

5.5   VESTIGIAL-SIDEBAND MODULATION

The generation of SSB signals may be quite difficult when the modulating signal bandwidth is wide or where one cannot disregard the low-frequency components. To conserve spectrum space, a compromise can be made between SSB and DSB in what is known as *vestigial-sideband* (VSB) modulation. In VSB modulation, only a portion of one sideband is transmitted in such a way that the demodulation process reproduces the original signal. The partial suppression of one sideband reduces the required bandwidth from that required for DSB but does not match the spectrum efficiency of SSB. If a large carrier is also transmitted, the desired signal can be recovered using an envelope detector. If no carrier is sent, the signal can be recovered using a synchronous detector or the injected carrier method discussed in the preceding section.

This filtering operation can be represented by a filter $H_V(\omega)$ which passes some of the lower (or upper) sideband and most of the upper (or lower) sideband.

The magnitude characteristic of such a filter is shown in Fig. 5.36. The spectral density of the resulting vestigial-sideband signal is

$$\Phi_{VSB}(\omega) = [\tfrac{1}{2}F(\omega - \omega_c) + \tfrac{1}{2}F(\omega + \omega_c)]H_V(\omega). \qquad (5.54)$$

We wish to determine the requirements on this filter such that a synchronous detector will yield $f(t)$ without distortion.

Fig. 5.36 VSB filtering.

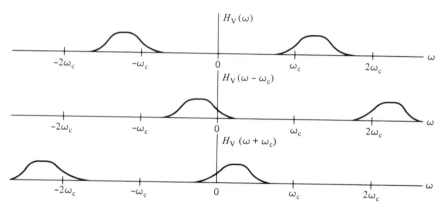

Fig. 5.37 Characteristics of VSB filtering.

The output of the synchronous detector is

$$e_o(t) = [\phi_{VSB}(t) \cos \omega_c t]_{LP},$$

or

$$E_o(\omega) = \tfrac{1}{4}F(\omega)H_V(\omega + \omega_c) + \tfrac{1}{4}F(\omega)H_V(\omega - \omega_c). \qquad (5.55)$$

For faithful reproduction of $f(t)$, we then require that

$$[H_V(\omega - \omega_c) + H_V(\omega + \omega_c)]_{LP} = \text{constant}, \qquad |\omega| < \omega_m. \qquad (5.56)$$

The frequency translations indicated in Eq. (5.56) are illustrated in Fig. 5.37. Note that, at least on a magnitude basis, Eq. (5.56) is satisfied if $|H_V(\omega)|$ is antisymmetric about the carrier frequency ω_c. Motivated by this observation, we let the constant in Eq. (5.56) be $2H_V(\omega_c)$. Under this condition Eq. (5.56) becomes

$$[H_V(\omega - \omega_c) - H_V(\omega_c)] = -[H_V(\omega + \omega_c) - H_V(\omega_c)], \qquad (5.57)$$

showing the required antisymmetry about the carrier frequency. Therefore the required filtering for a VSB signal is a complementary filter symmetry about the carrier frequency, as shown in Fig. 5.36. The required phase relationships are not as easy to satisfy and VSB is generally used in cases where the correct phase response is not of prime concern. Although synchronous detection is assumed here, these same principles hold also when a large carrier is present and an envelope detector is used.

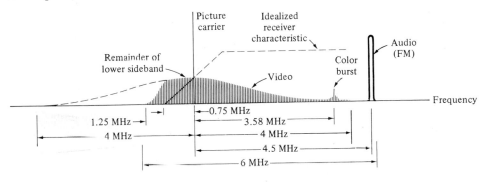

Fig. 5.38 Spectrum of a public television transmission.

Vestigial-sideband modulation is used to advantage in the video portion of public television systems. In a television transmission, 525 lines of video (picture) information are sent each $\frac{1}{30}$ of a second (that is, 15,750 lines per second—the horizontal trace frequency).† Allowing time for retrace and synchronization, this requires a minimum video bandwidth of 4 MHz to transmit an array of picture elements. (The spectral density of typical video, however, is not flat but heavily weighted toward the lower frequencies.) Use of DSB modulation for video transmission would require a frequency allocation of 8 MHz per channel. This is judged excessive and a version of VSB modulation is used to decrease the required video bandwidth to about 5 MHz. The spectrum of a public television channel is illustrated in Fig. 5.38.‡ The filter characteristic is positioned not

---

† Standards for U.S. and Japan; a standard of 625 lines sent each $\frac{1}{25}$ of a second is generally used elsewhere where 50-Hz power is used. Commercial television signal transmissions are described in Appendix D.

‡ Note that the video spectrum appears about the 15.75-kHz sampling lines. Here the (horizontal) sampling frequency is much lower than the signal bandwidth. The sampling theorem does *not* state that the signal cannot be recovered but merely that the signal cannot be recovered by filtering alone. Thus a television receiver must utilize time synchronization techniques to recover the original video information.

about the carrier in this case but 1 MHz below the carrier. The receiver filtering is used to complete the vestigial-sideband characteristic, as indicated in Fig. 5.38.

5.6 A TIME-REPRESENTATION OF BANDPASS NOISE

At this point, we return to a discussion of bandpass noise. We have not attempted to develop a description of noise in terms of time functions because the noise is random. Yet for bandpass systems, which limit the bandwidth of the noise, some of the noise fluctuations are restricted and it is possible to write some rather general time relations.

As the bandwidth of the noise becomes small compared to the center frequency, it becomes possible to approximate it with a phasor representation. Such a phasor representation is shown in Fig. 5.39. Here $n_c(t)$ is that portion of the noise which is the in-phase component and $n_s(t)$ is the quadrature component in the phasor representation. Both $n_c(t)$ and $n_s(t)$ are random and thus the phasor sum has a random amplitude a_n and phase θ_n.

Fig. 5.39 A phasor representation of narrowband random noise.

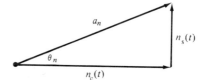

We can write out the complete phasor representation of the narrowband random noise using complex notation,

$$[n_c(t) + jn_s(t)]e^{j\omega_0 t}, \tag{5.58}$$

where ω_0 is the center frequency. Both $n_c(t)$ and $n_s(t)$ are slowly varying compared to $e^{j\omega_0 t}$ as a result of the narrow bandwidth restriction. Taking the real part of Eq. (5.58) to obtain an expression for $n(t)$, we have

$$n(t) = \mathcal{R}e \left\{ [n_c(t) + jn_s(t)]e^{j\omega_0 t} \right\}.$$

Using an identity from complex algebra (cf. Appendix A) gives us†

$$n(t) = n_c(t) \cos \omega_0 t - n_s(t) \sin \omega_0 t. \tag{5.59}$$

---

† Given two complex quantities z_1, z_2; then

$$\mathcal{R}e\{z_1 z_2\} = \mathcal{R}e\{z_1\}\mathcal{R}e\{z_2\} - \mathcal{I}m\{z_1\}\mathcal{I}m\{z_2\}.$$

Equation (5.59) is the desired representation of $n(t)$ Although derived for the narrowband case here, this representation can be extended to the more general bandpass case by properly defining $n_c(t)$ and $n_s(t)$. Therefore Eq. (5.59) is known as the bandpass representation of noise. The terms $n_c(t)$ and $n_s(t)$ are low-pass voltages whose fluctuations are limited by the bandwidth of the bandpass noise.

A comparison of Eqs. (5.58) and (5.59) suggests that $n_c(t)$ and $n_s(t)$ could be visualized as the bandpass noise $n(t)$ shifted down to zero frequency. Motivated by this observation and recalling the frequency translation properties of the Fourier transform, let us multiply both sides of Eq. (5.59) by $\cos \omega_0 t$:

$$n(t) \cos \omega_0 t = n_c(t) \cos^2 \omega_0 t - n_s(t) \sin \omega_0 t \cos \omega_0 t. \tag{5.60}$$

If we use the trigonometric identities $2 \cos^2 A = 1 + \cos 2A$, $2 \sin A \cos A = \sin 2A$, and if we retain only the low-pass (LP) terms, Eq. (5.60) becomes

$$[n(t) \cos \omega_0 t]_{\text{LP}} = (\tfrac{1}{2}) n_c(t). \tag{5.61}$$

The power spectral density of Eq. (5.61) is, using Eqs. (3.45) and (4.17),

$$S_{n_c}(\omega) = \left\{ \lim_{t \to \infty} \left[\frac{|N_T(\omega - \omega_0) + N_T(\omega + \omega_0)|^2}{T} \right] \right\}_{\text{LP}}. \tag{5.62}$$

For random noise, the average of the cross products in Eq. (5.62) goes to zero, giving the result

$$S_{n_c}(\omega) = [S_n(\omega - \omega_0) + S_n(\omega + \omega_0)]_{\text{LP}}. \tag{5.63}$$

The case for $n_s(t)$ follows in the same manner by first multiplying both sides of Eq. (5.59) by $\sin \omega_0 t$, yielding

$$S_{n_s}(\omega) = [S_n(\omega - \omega_0) + S_n(\omega + \omega_0)]_{\text{LP}}. \tag{5.64}$$

Combining Eqs. (5.63) and (5.64), we have

$$S_{n_c}(\omega) = S_{n_s}(\omega) = [S_n(\omega - \omega_0) + S_n(\omega + \omega_0)]_{\text{LP}}. \tag{5.65}$$

A sketch of a given bandpass noise spectral density and the resulting low-pass $S_{n_c}(\omega)$ and $S_{n_s}(\omega)$ is shown in Fig. 5.40.

It can be seen from Fig. 5.40 that the area under $S_n(\omega)$ is equal to the area under $S_{n_c}(\omega)$ or $S_{n_s}(\omega)$. Therefore it follows that their mean-square values are equal; that is,

$$\overline{n^2(t)} = \overline{n_c^2(t)} = \overline{n_s^2(t)}. \tag{5.66}$$

These results apply to bandpass *random* noise. It is the random nature of the noise which tends to distribute the noise components over both the cosine and sine terms. This is in direct contrast to a deterministic signal in which we can control the phase to give only cosine or sine terms.

The mean-square value of bandpass random noise, using Eq. (5.59), is

$$\overline{n^2(t)} = \tfrac{1}{2}\overline{n_c^2(t)} + \tfrac{1}{2}\overline{n_s^2(t)}. \tag{5.67}$$

Our conclusion, based on Eqs. (5.66) and (5.67), is that the mean noise power is divided equally between the cosine and sine terms.

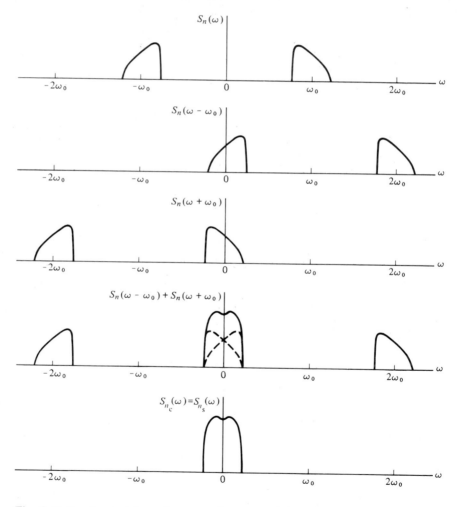

Fig. 5.40 In-phase and quadrature noise spectral densities for bandpass noise.

5.7   EFFECTS OF NOISE IN AM SYSTEMS

In order to discuss and compare various modulation systems, we need a standard basis for the measurement of performance. A convenient quantity is the ratio of the mean-squared signal to the mean-squared noise, where both quantities are measured at the same point in the system. This signal-to-noise (power) ratio is designated by the ratio of the capital letters S/N, with the appropriate subscripts.

We assume that the signals that are transmitted and received are bandpass signals centered about a carrier frequency, f_c. This allows us to make use of the in-phase and quadrature representations of bandpass noise that are described in the preceding section.

5.7.1   DSB-SC

For DSB-SC signals, the receiver consists of a synchronous detector — a sinusoidal signal generator at $f = f_c$, a multiplier, and a low-pass filter. The input signal to the detector is $f(t) \cos \omega_c t$, so that

$$S_i = \overline{[f(t) \cos \omega_c t]^2} = \tfrac{1}{2}\overline{f^2(t)}. \tag{5.68}$$

The useful output signal is $\tfrac{1}{2}f(t)$ [the $\tfrac{1}{2}$ arises from $\overline{\cos^2 \omega_c t}$] so that

$$S_o = \overline{[\tfrac{1}{2}f(t)]^2} = \tfrac{1}{4}\overline{f^2(t)} = \tfrac{1}{2}S_i. \tag{5.69}$$

The noise output of the detector is $n_i(t) \cos \omega_c t$, filtered by the low-pass filter. Letting $n_d(t)$ be the output of the detector and $n_o(t)$ be the output of the low-pass filter, and using the bandpass noise representation [cf. Eq. (5.59)] gives us

$$n_d(t) = \tfrac{1}{2}n_c(t) + \tfrac{1}{2}n_c(t) \cos 2\omega_c t - \tfrac{1}{2}n_s(t) \sin 2\omega_c t, \tag{5.70}$$
$$n_o(t) = \tfrac{1}{2}n_c(t). \tag{5.71}$$

Defining

$$\overline{n_i^2(t)} = N_i,$$

we have

$$N_o = \overline{n_o^2(t)} = \tfrac{1}{4}\overline{n_c^2(t)} = \tfrac{1}{4}\overline{n_i^2(t)} = \tfrac{1}{4}N_i, \tag{5.72}$$

where use has been made of Eq. (5.66). Combining Eqs. (5.69) and (5.72) yields

$$\frac{S_o}{N_o} = 2\frac{S_i}{N_i} \qquad \text{for DSB-SC.} \tag{5.73}$$

Thus the detector improves the signal-to-noise ratio in a DSB-SC system by a factor of two. This improvement results from the fact that the coherent (synchronous) detector rejects the out-of-phase (quadrature) noise component in the input noise, thereby halving the mean-square noise power.

The use of a synchronous detector for the large-carrier case can be derived easily from the above result by noting that $f(t)$ is replaced by $[A + f(t)]$, so that $S_i = \frac{1}{2}A^2 + \frac{1}{2}\overline{f^2(t)}$ [assuming that $\overline{f(t)}$ is zero] and

$$\frac{S_o}{N_o} = \frac{2\overline{f^2(t)}}{A^2 + \overline{f^2(t)}}\frac{S_i}{N_i}. \tag{5.74}$$

Because $|f(t)| < A$ for large-carrier systems, the signal-to-noise ratio is somewhat poorer than for a suppressed-carrier system.

5.7.2 SSB-SC

Using the general representation for an SSB-SC signal [cf. Eq. (5.33)],

$$\phi(t) = f(t)\cos\omega_c t \pm \hat{f}(t)\sin\omega_c t,$$

we have

$$S_i = \overline{\phi^2(t)} = \frac{1}{2}\overline{f^2(t)} + \frac{1}{2}\overline{\hat{f}^2(t)}. \tag{5.75}$$

However, we also have [because $\hat{F}(\omega)$ is only shifted in phase from $F(\omega)$]

$$|F(\omega)|^2 = |\hat{F}(\omega)|^2. \tag{5.76}$$

If we integrate and use Parseval's theorem, Eq. (5.76) yields

$$\overline{f^2(t)} = \overline{\hat{f}^2(t)}. \tag{5.77}$$

Equation (5.75) then becomes

$$S_i = \overline{f^2(t)}. \tag{5.78}$$

The useful output signal is $\frac{1}{2}f(t)$ so that

$$S_o = \overline{[\frac{1}{2}f(t)]^2} = \frac{1}{4}\overline{f^2(t)} = \frac{1}{4}S_i. \tag{5.79}$$

Combining Eqs. (5.72) and (5.79) we get

$$\frac{S_o}{N_o} = \frac{S_i}{N_i} \qquad \text{for SSB-SC.} \tag{5.80}$$

Because the process of SSB detection is a straightforward frequency translation, all of the signal and noise components are transferred, unmodified, to low frequencies. Therefore the relationship between the signal and the noise remains unchanged by SSB detection and the predetection and postdetection S/N ratios are equal.

Is the DSB-SC system then superior to the SSB system? Not where noise power is proportional to bandwidth because the DSB-SC system requires twice the bandwidth of the SSB system and therefore has twice the noise power. We conclude that *the performance of* SSB-SC *is identical to that of* DSB-SC *from the point of view of noise improvement in the presence of white noise.*

5.7.3 DSB-LC: The Envelope Detector

The synchronous detector used in the SSB-SC and DSB-SC systems above could also be used here, but to utilize the full advantages of AM the envelope detector is more commonly used. However, the envelope detector is nonlinear and at best we shall only be able to establish some bounds on its performance.

Using the bandpass noise representation, the input signal and noise can be written as

$$s_i(t) + n_i(t) = [A + f(t)] \cos \omega_c t + n_c(t) \cos \omega_c t - n_s(t) \sin \omega_c t. \quad (5.81)$$

The envelope of this signal is (see Fig. 5.41)

$$r(t) = \sqrt{\{[A + f(t)] + n_c(t)\}^2 + \{n_s(t)\}^2}. \quad (5.82)$$

For high input signal-to-noise, this can be approximated by use of the binomial expansion to give

$$r(t) \approx A + f(t) + n_c(t). \quad (5.83)$$

The useful signal here is $f(t)$, while the noise is $n_c(t)$. The detector output gives

$$S_o = \overline{f^2(t)}, \quad (5.84)$$

$$N_o = \overline{n_c^2(t)} = \overline{n_i^2(t)} = N_i; \quad (5.85)$$

also

$$S_i = \overline{\{[A + f(t)] \cos \omega_c t\}^2} = \tfrac{1}{2}A^2 + \tfrac{1}{2}\overline{f^2(t)}. \quad (5.86)$$

Combining Eqs. (5.84) and (5.86), we get

$$\frac{S_o}{N_o} = \frac{2\overline{f^2(t)}}{A^2 + \overline{f^2(t)}} \frac{S_i}{N_i}, \quad \text{for AM, large signal-to-noise.} \quad (5.87)$$

Note that Eq. (5.87) is identical to a previous result using synchronous detection, as expressed in Eq. (5.74). Thus for the large signal-to-noise case, envelope demodulation in the presence of noise has the same performance quality as synchronous demodulation.

In the particular case of sinusoidal modulation, $f(t) = mA \cos \omega_m t$ and Eq. (5.87) can be rewritten as

$$\frac{S_o}{N_o} = \frac{2m^2}{2 + m^2} \frac{S_i}{N_i}, \quad \text{for AM, large signal-to-noise, sinusoidal modulation.} \quad (5.88)$$

The maximum improvement in the signal-to-noise ratio is $\tfrac{2}{3}$ at 100% modulation.

Sometimes it is convenient to refer the output signal-to-noise ratio to the *carrier-to-noise* ratio (CNR) at the output of the IF amplifiers. Noting that the mean-square value of the carrier is $S_c = A^2/2$ and that the mean-square value of the noise at the output of the IF amplifiers is $N_c = N_i$, we find Eqs. (5.84) and (5.85) give

$$\frac{S_o}{N_o} = \frac{2\overline{f^2(t)}}{A^2} \frac{S_c}{N_c}. \tag{5.89}$$

In the particular case of sinusoidal modulation, $f(t) = mA \cos \omega_m t$ and Eq. (5.89) can be rewritten as

$$\frac{S_o}{N_o} = m^2 \frac{S_c}{N_c}. \tag{5.90}$$

In other words, the maximum output signal-to-noise ratio of an AM system is equal to the carrier-to-noise ratio (CNR) at the output of the IF amplifiers. This form proves convenient in comparing the performance of AM with other types of modulation.

For the large-noise case, we can no longer make the approximation of Eq. (5.83). Rewriting the expression for the signal envelope of Eq. (5.82), we have

$$r(t) = \left\{ A^2 \left[1 + \frac{2f(t)}{A} + \frac{2n_c(t)}{A} + \frac{2n_c(t)f(t)}{A^2} + \frac{n_c^2(t)}{A^2} + \frac{f^2(t)}{A^2} \right] + n_s^2(t) \right\}^{1/2}. \tag{5.91}$$

This result shows that not only are there additive noise terms present but the desired signal is multiplied by noise. Thus for large-noise conditions the desired signal is hopelessly mutilated by the envelope detector and, as far as reception is concerned, the received signal is nonexistent. This complete loss of message occurs at low signal-to-noise ratios and is called the *threshold effect*. This name arises because there is some value of input signal-to-noise above which this mutilation becomes negligible and below which the system performance deteriorates.

The threshold effect becomes quite evident when the modulated carrier power is on the same order of magnitude as the average noise power. This effect may be visualized with the aid of Fig. 5.41. The random noise $n(t)$ adds to the modulated carrier signal with a random phase and magnitude. The envelope $r(t)$ is controlled primarily by the modulated carrier for high S/N conditions. When the magnitude of the noise (on the average) is on the same order as that of the modulated carrier signal, however, this predominance in the control of the envelope vanishes. In fact, for $S/N < 1$, the noise takes over the control of the envelope.

We can combine some of the preceding results for the case of a white input noise power-spectral density of $\eta/2$ W/Hz and a modulating bandwidth of f_m Hz.

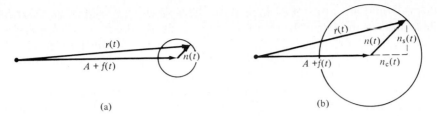

Fig. 5.41 Phasor diagram of DSB-LC signals for (a) large S/N and (b) small S/N ratios.

For this case, we have $N_i = 2\eta f_m$ for DSB systems and $N_i = \eta f_m$ for SSB systems. Thus Eqs. (5.73) and (5.80) can be written as

$$\frac{S_o}{N_o} = \frac{S_i}{\eta f_m} \tag{5.92}$$

for both systems using synchronous detection. Note that for synchronous detection, the S/N improvement is independent of the input S/N ratio and that the effects of filtering before detection (predetection filtering) and after detection (postdetection filtering) are identical. A graph of Eq. (5.92) is shown in Fig. 5.42. The output S/N ratio for envelope detection for DSB and SSB systems is

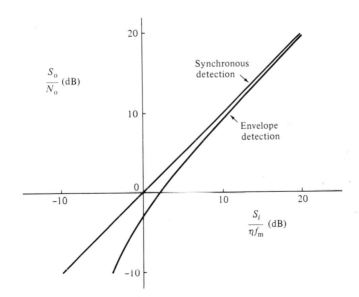

Fig. 5.42 Signal-to-noise performance of AM systems.

shown also in Fig. 5.42.† For high input S/N ratio, the output S/N ratio for envelope detection approaches that for synchronous detection. For lower input S/N ratios, however, the output S/N decreases more rapidly than for the synchronous detection case with an S/N ratio that is less than about 10–15 dB. However, this threshold effect is not very abrupt in AM systems.

For voice communications, intelligibility demands a high signal-to-noise ratio (e.g., 30 dB) and the threshold effect is usually not a limit on system performance. This is not necessarily true, however, for data communications.

The threshold effect is a property of envelope detectors; we observed no such effect for synchronous detectors. With synchronous detection, the output signal and noise always remain additive and the signal-to-noise improvement holds under all noise conditions.

We conclude that the performance of the envelope detector approaches that of the synchronous detector for large signal-to-noise conditions. But for large noise, the envelope detector suffers from a threshold effect and its performance is inferior to the synchronous detector, becoming completely unacceptable at low signal-to-noise conditions.

Finally, if we normalize all of the above AM detection methods to equal sideband input power and do not include the carrier power after detection, the output signal-to-noise ratios are all the same for large input signal-to-noise conditions. Synchronous detection methods continue to give this performance for low signal-to-noise conditions whereas the output of the envelope detector becomes unusable.

5.8 PROPAGATION EFFECTS

Communication systems frequently use electromagnetic propagation to convey information from point to point. Because the choice of modulation may be influenced by the propagation, we briefly digress to investigate some of these effects.

Electromagnetic radiation through space has its average power divided equally between an electric field and a magnetic field. In free space the directions of the electric field, the magnetic field, and the direction of propagation are all at right angles. The polarization of an electromagnetic wave is taken to be in the direction of the electric field. Hence a wave is said to be vertically polarized if its electric field is in the vertical direction. In the presence of a ground plane, vertical polarization is usually preferred when antennas are within a wavelength of the ground, while horizontal is preferred when antennas are several wavelengths above ground.

---

† The S/N analysis for envelope detection becomes quite involved; see, among others, W. D. Gregg, *Analog and Digital Communications*, New York: Wiley, 1977, Chapter 7.

As a result of electromagnetic theory, the minimum dimension of an antenna for efficient radiation of radio frequency energy is one-half the wavelength of the radio frequency. At low frequencies, this length becomes excessive and for vertical polarization the earth is used as one half of the antenna, as shown in Fig. 5.43. This type of antenna is used, for example, in the commercial AM broadcast band. At higher frequencies (i.e., shorter wavelengths), horizontally polarized half-wave antennas are commonly used, often in arrays to improve their performance. At very high frequencies, physical dimensions of the antenna become large compared to a wavelength, and optical approximations can be used.

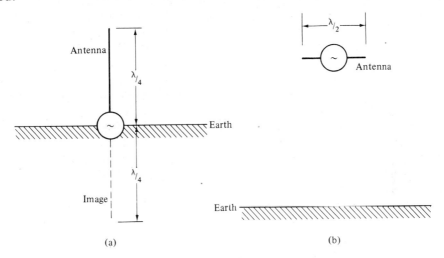

(a)   (b)

Fig. 5.43 The basic half-wave antenna: (a) vertical polarization, low frequency; (b) horizontal polarization, high frequency.

In general, an electromagnetic wave radiated from an antenna is composed of a ground wave which arises from currents induced in the earth and travels along the earth's surface and a sky wave which is propagated through space. At very low frequencies, the ground wave can be used to communicate reliably over long distances while sky-wave propagation is generally used at high frequencies.

The ground wave loses energy as a result of dispersion along the earth's surface and energy dissipated in the earth. These losses vary directly with frequency and the resistance of the earth. The attenuation above 10 MHz is so high that ground wave propagation is of little practical value at these frequencies. Because the tangential component of the electric field cannot exist along the ground interface, vertical polarization is used where propagation via ground wave is desired.

The sky wave leaves the antenna and travels essentially in a straight path. Some of the sky-wave energy leaves the antenna at an angle to the horizontal and travels upward until it enters an ionized layer, known as the ionosphere, about 30–70 miles above the earth's surface. In this region the path of the wave is altered by refractive effects which are dependent on the intensity of ionization and the wave frequency. Depending on the conditions, the path of the sky wave may be bent downward enough to cause the radiated energy to return to earth, as illustrated in Fig. 5.44. The losses in this mechanism are small and the signal strength of the sky wave may be very strong. As a result, long-distance communications can be accomplished by use of the sky wave.

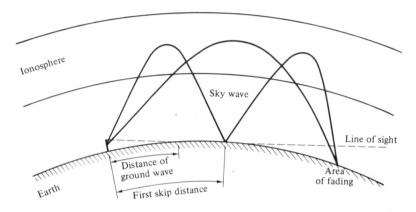

Fig. 5.44 Propagation of ground wave and sky wave from an antenna. (Not drawn to scale.)

The ionization within the ionosphere varies with solar activity so that these effects vary with the time of day and season of the year. These effects are more pronounced at night and it is not unusual to receive strong signals in the AM broadcast band over a thousand-mile distance at night while the ground wave signal strength is weak after a few hundred miles. To avoid undue interference from these effects, some AM stations are permitted to operate only during daylight hours.

When two or more radio waves arrive at a point of reception along different paths, their phase relations may not be the same. The results of this multipath propagation can cause wide variations in signal intensity at the point of reception. Cancellation of one wave by another results in a loss of signal strength and is called *fading*. If the fading is highly sensitive to the frequency of propagation, different parts of the spectrum of a signal may undergo different amounts of fading. This effect is called *selective fading*. The effects of fading can be minimized by proper choices of carrier frequency and modulation, the use of several different carrier frequencies (frequency diversity), and the use of several different antennas (space diversity) or highly selective antenna arrays.

The direction of the sky wave above a frequency of about 30 MHz is not altered enough to return to earth and propagation above this frequency is predominantly straight-line from transmitter to receiver. Multipath propagation is still present as a result of reflected waves from the earth and man-made structures but is not as frequency dependent (e.g., the "ghosts" in television video). The distance of communication is governed predominantly by the height of the transmitting and receiving antennas. Some refraction or bending of the path does occur for low elevation angles, and communications beyond the line of sight are still possible at very high frequencies, although not as efficient as at the lower frequencies. Multipath propagation and fading may vary as a function of meteorological conditions in this mode of propagation.

5.9   COMPARISON OF VARIOUS AM SYSTEMS

Various amplitude-modulated systems have been discussed and it is interesting to compare their performance characteristics. First we compare large-carrier and suppressed-carrier systems.

Large-carrier systems have the advantage of simpler detectors at the receiver and thus are a natural choice where the emphasis is on inexpensive receivers. It also turns out that high-power modulators are easier to build for large-carrier systems because the carrier terms arising in generation do not have to be balanced or filtered out.

In contrast, suppressed-carrier systems require much less power to transmit the same information. Receivers for suppressed-carrier transmissions are more complicated because they must generate a carrier of the correct frequency and phase. Where communication efficiency is a prime consideration, however, the added receiver complexity may be justified. Suppressed-carrier systems are less susceptible to the effects of selective fading because a fade of the carrier can result in severe distortion in an envelope detector.

In comparing double-sideband and single-sideband systems, we note that the required receiver operations are identical. Double-sideband systems have an advantage in ease of generation, not requiring sideband suppressions. Double-sideband systems can be used to transmit information frequencies down to zero frequency with good fidelity.

In contrast, single-sideband systems require half the bandwidth of double-sideband systems, thus affording good spectrum efficiency. Because selective fading may disturb the phase relationships between sidebands, single-sideband systems are affected less by selective fading. Some of the advantages of both single-sideband and double-sideband can be realized using vestigial sideband systems.

For SSB-SC and DSB-SC signals, all of the transmitted power is in the sidebands. For equal-input sideband power, the output S/N ratio is the same for both on an average power basis. In addition, of course, the SSB signal occupies only half the bandwidth. For sinusoidal modulation, DSB-LC results in at most only 1/3 the output S/N ratio of the corresponding DSB-SC signal [e.g., compare Eqs. (5.73) and (5.88)]. For typical voice modulation, the difference is greater because the peak-to-rms ratio is about 1/8. Taking the peak to be equal to one (for 100% modulation), the S/N ratio of DSB-LC for voice is more like about 1/65 that for a comparable DSB-SC signal on an average-power basis.

Another comparison can be made on a peak-power basis rather than on an average-power basis. The ratio of the peak amplitude in a DSB-LC signal compared to the corresponding DSB-SC signal, for 100% modulation on the peaks, is two [e.g., see Eq. (5.18)]. Thus for the same peak power limitation the DSB-SC signal has four times more average sideband power than the corresponding DSB-LC signal and therefore a 6-dB S/N advantage. Comparisons between SSB-SC and DSB-SC on a peak-power basis for general signals is more difficult. For sinusoidal modulation, the SSB-SC signal has an average power twice that of the DSB-SC signal for equal peak amplitudes. For sinusoidal modulation, then, the SSB-SC signal has a 3-dB S/N advantage over DSB-SC for the same peak power. However, this possible advantage varies with the type of modulating signal used.

In conclusion, we see that each type of amplitude modulation has its advantages and disadvantages. Each circumstance must be evaluated on its own merits and the choice of modulation tailored to that situation. We will continue this discussion when S/N ratios of angle-modulation systems are discussed in Chapter 6.

5.10 SUMMARY

Multiplication of a waveform with a sinusoid translates its spectral density in frequency. This effect, known as the modulation property, can be used to generate amplitude-modulated waveforms and to demodulate them again. Because such waveforms have no separate carrier frequency and are symmetric about a center frequency, they are called double-sideband suppressed-carrier (DSB-SC) waveforms. Proper demodulation of DSB-SC waveforms demands an exact knowledge of the frequency and phase used in modulation.

The addition of a carrier frequency is used to aid in waveform demodulation. If the added carrier amplitude is large, an envelope detector can correctly demodulate the waveform. This principle is used in commercial AM broadcasting.

Choice of different carrier frequencies allows several message spectra to be placed side by side in frequency and transmitted simultaneously. This is called frequency-division multiplexing (FDM).

To conserve bandwidth, one sideband of a DSB waveform can be suppressed;

this type of modulation is called single-sideband (SSB). This demands exacting filter design or phase balance. An economical way to gain some of the advantage of SSB is to partially suppress one sideband. This is known as vestigial-sideband modulation.

Bandpass random noise may be described in terms of random amplitudes and phases of sines and cosines at the center frequency. In random noise, the average noise power is divided equally between the two components.

For a given input sideband power and noise spectral density, the output S/N ratios for amplitude modulation are all the same as long as the S/N ratios are large. Synchronous detection may be used for low S/N ratios whereas envelope detection becomes unusable for low S/N.

While large-carrier systems allow the use of inexpensive envelope detectors, the performance of these systems is not optimum in terms of total power or the signal-to-noise ratios required for good communication. The use of synchronous demodulation is preferable in low signal-to-noise conditions.

The phases of signals arriving by various paths at the receiver cause the signal strength to vary. Cancellation of these signals results in loss of signal strength or fading. Fading which is sensitive to frequency is called selective fading. The effects of selective fading are least detrimental to amplitude-modulated systems when the bandwidth of the signal is narrow and the maintenance of a large carrier is not necessary, making SSB-SC an obvious choice.

Selected References for Further Reading

1. R. L. Shrader. *Electronic Communication*. Fourth Ed. New York: McGraw-Hill, 1980.
   A fairly detailed discussion of the circuits and operating characteristics of practical communication systems; easy reading. Chapter 20 has a brief discussion of antennas and propagation.

2. M. Mandl. *Principles of Electronic Communications*. Englewood Cliffs, N.J.: Prentice-Hall, 1973.
   A detailed discussion of the signals and circuits used in commercial television and stereo multiplexing; easy reading.

3. B. P. Lathi. *Communication Systems*. New York: John Wiley & Sons, 1968.
   Chapters 3 and 7 are a parallel treatment of amplitude modulation, stressing the use of general signal notation and convolution.

4. A. B. Carlson. *Communication Systems: An Introduction to Signals and Noise in Electrical Communications*. Second Ed. New York: McGraw-Hill, 1975.
   Chapter 5 uses phasor representations to analyze amplitude modulation, particularly envelope detection.

5. K. K. Clarke and D. T. Hess. *Communication Circuits: Analysis and Design*. Reading, Mass.: Addison-Wesley, 1971.
   Chapters 7–10 present an advanced treatment of the circuits and theory used for amplitude modulation and demodulation.

6. M. Schwartz, W. R. Bennett, and S. Stein. *Communication Systems and Techniques.*
New York: McGraw-Hill, 1966.
Chapter 9 includes an extensive treatment of fading.

7. W. D. Gregg. *Analog and Digital Communication.* New York: John Wiley & Sons,
1977.
Chapters 2, 4, and 7 present an advanced analysis of AM generation and detection
with examples of current practice.

Problems

5.1.1 A certain transformer is constructed to allow the secondary winding to be rotated
mechanically with respect to the primary. Sketch the output time waveform if the
primary is connected to a 60-Hz sinusoidal source and the rotational rate is 600
rpm.

5.1.2 The modulating signal $f(t) = 2 \cos 100\pi t + \cos 400\pi t$ is applied as the input to
a double-sideband suppressed-carrier modulator operating at a carrier frequency
of 1 kHz. Sketch the spectral density of $f(t)$ and the resulting DSB-SC waveform,
identifying the upper and lower sidebands.

5.1.3 The spectral density of the input $f(t)$ to the system shown in Fig. P–5.1.3 is band-
limited to 100 Hz.

a) Sketch the spectral density of the output for an assumed input spectral density.
b) Write an expression for the output spectral density in terms of the input spectral
density if the sequence of sinusoidal generators and mixers were extended
indefinitely.

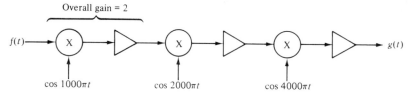

Fig. P–5.1.3.

5.1.4 The system shown in Fig. P–5.1.4 is an alternative to the one shown in Fig. 5.4
for sending two messages on one carrier.

a) If $f_1(t) = \cos \omega_1 t$, $f_2(t) = \cos \omega_2 t$, derive an expression for $\phi(t)$.
b) Devise block diagrams for a suitable demodulator for $\phi(t)$.

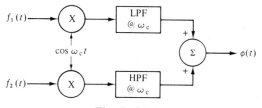

Fig. P–5.1.4.

5.1.5 A sinusoidal signal $f(t)$ is multiplied by a periodic sawtooth waveform $g(t)$ as shown in Fig. P–5.1.5.

a) Determine the minimum and maximum bandwidth of the low-pass filter if $e_o(t)$ is to be a DSB-SC waveform corresponding to $f(t)$.

b) Determine $e_o(t)$ if the low-pass filter has unity gain within the pass band and a bandwidth of 150 kHz.

Fig. P–5.1.5.

5.1.6 When the input to a given audio amplifier is $(4 \cos 800\pi t + \cos 2000\pi t)$ mV, the measured frequency component at 1 kHz in the output is 1 V and the frequency component at 600 Hz is 1 mV. Represent the amplifier output-input characteristic by $e_o = a_1 e_i + a_2 e_i^2$ and evaluate the numerical values of a_1, a_2 from the data given. (This type of test is called an intermodulation distortion test.)

5.1.7 In the balanced modulator shown in Fig. P–5.1.7, $\phi_1(t) = \cos(\omega_c t + \theta)$ and $\phi_2(t) = A \cos \omega_c t$. Show that the output voltage contains a term proportional to $\cos \theta$ (and hence that the balanced modulator can be used as a phase detector). Assume that the diode characteristics are piece-wise linear [that is, $i(t) = e(t)/r_d$ for $e(t) > 0$; $i(t) = 0$ for $e(t) < 0$] and that $A \gg 1$.

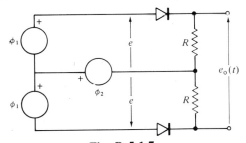

Fig. P–5.1.7.

5.1.8 The balanced modulator shown in Fig. 5.8 is used and a piece-wise linear diode characteristic is assumed so that $i(t) = e(t)/r_d$ for $e(t) > 0$; $i(t) = 0$ for $e(t) < 0$. Using the first three terms of a Legendre series (see Problem 2.5.4) to obtain a series representation for $i(t)$, determine the modulator output corresponding to Eq. (5.12) over the range $(-1, 1)$ of $e(t)$.

5.1.9 The device shown in Fig. P–5.1.9 (see p. 270) is called a *beat-frequency oscillator* (BFO).

a) Describe its operation.

b) Compare the percentage frequency change in $e_o(t)$ to the percentage change in ω_1 with ω_2 fixed.

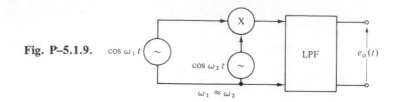

Fig. P–5.1.9. $\cos \omega_1 t$ X   LPF $e_o(t)$ $\cos \omega_2 t$ $\omega_1 \approx \omega_2$

5.1.10 Two measurement systems are shown in Fig. P–5.1.10 below. Find the output of each system if $e_1(t) = \cos \omega_1 t$. What type of modulation is present in each and how should it be detected?

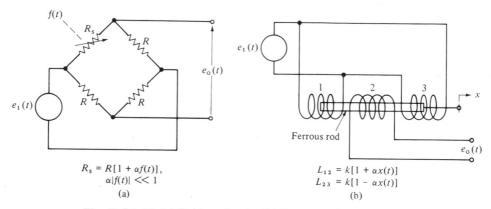

$R_s = R[1 + \alpha f(t)],$
$\alpha|f(t)| \ll 1$
(a)

$L_{12} = k[1 + \alpha x(t)]$
$L_{23} = k[1 - \alpha x(t)]$
(b)

Fig. P–5.1.10 (a) Bridge circuit; (b) linear motion transducer.

5.1.11 Show that the chopper system shown in Fig. P–5.1.11 will produce essentially the correct signals for FM stereo transmission.

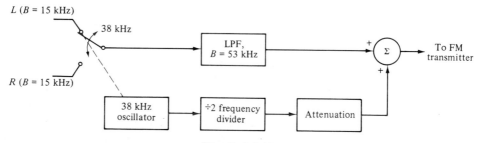

L ($B = 15$ kHz)   38 kHz   LPF, $B = 53$ kHz Σ To FM transmitter

R ($B = 15$ kHz)   38 kHz oscillator   ÷2 frequency divider   Attenuation

Fig. P–5.1.11.

5.1.12 A proposed frequency synthesizer to control the output frequency of a stable oscillator digitally is shown in Fig. P–5.1.12. The stable oscillator operates at $f_c = 100$ kHz and the integer values of N_1, N_2 are selectable between 1 and 10.

a) What is the output frequency when $N_1 = 4$, $N_2 = 2$?

b) Determine the frequency range and minimum frequency increment of the synthesizer.

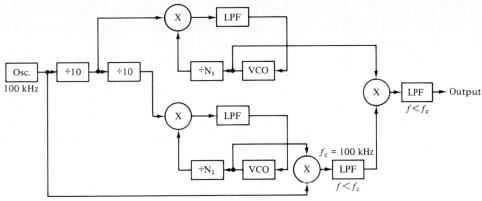

Fig. P–5.1.12.

c★ **5.1.13** Run a 256-pt inverse DFT of the discrete spectrum

$$F(n) = \begin{cases} 100 & n = 62, 66, \\ 0 & \text{elsewhere.} \end{cases}$$

Plot the real part of your result and explain.

5.2.1 The modulating signal shown in Fig. P–5.2.1 is modulated using DSB-SC modulation with a carrier frequency of 10 kHz.

a) Sketch the resulting line spectrum.
b) Sketch the spectrum when this signal is used for DSB-LC.
c) Can a dc level in the input be distinguished from the carrier in the modulated DSB-LC waveform?

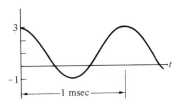

Fig. P–5.2.1.

5.2.2 Determine the modulation index of the output DSB-LC waveform in the system shown in Fig. P–5.2.2 (see p. 272). Assume that the time spent between switch contacts is negligible.

5.2.3 Consider the two amplitude-modulated signals, where $\omega_c \gg \omega_m$,

$$\phi_1(t) = (2 + E_1 \cos \omega_m t) \cos \omega_c t,$$

$$\phi_2(t) = E_2 \cos \omega_m t \cos \omega_c t.$$

a) Sketch the spectral density of each signal.
b) Determine the required numerical values of E_1, E_2 to produce 100% modulation in the large-carrier signal and the same average power in both signals.

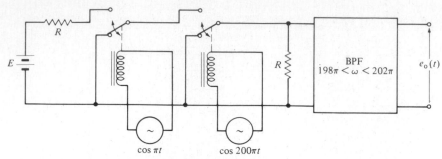

Fig. P–5.2.2.

c) Find the ratio of the respective outputs when these signals are applied to a synchronous detector.

5.2.4 For the sinusoidally modulated DSB-LC waveform shown in Fig. P–5.2.4,

a) Find the modulation index.
b) Sketch a line spectrum.
c) Calculate the ratio of the average power in the sidebands to that in the carrier.
d) Determine the amplitude of the additional carrier which must be added to attain a modulation index of 10%.

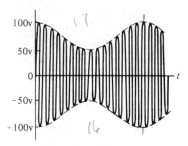

Fig. P–5.2.4.

5.2.5 One method of measuring the modulation index of a large-carrier AM waveform is to apply the modulated waveform to the vertical amplifier of an oscilloscope and the modulating waveform to the horizontal amplifier. A typical oscilloscope trace is shown in Fig. P–5.2.5.

a) Derive a formula for the modulation index in terms of A, B assuming that $m < 1$.
b) Is the trapezoidal pattern affected by the choice of modulating waveform?
c) Can the method be extended to $m > 1$? If so, sketch a pattern for $m = \frac{3}{2}$.

Fig. P–5.2.5.

5.2.6 A given AM (DSB-LC) transmitter develops an unmodulated power output of 100 W across a 50-Ω resistive load. When a sinusoidal test tone with a peak amplitude of 5.0 V is applied to the input of the modulator, it is found that the average power output increases by 50%; under these conditions, determine (a) the average power output in each sideband; (b) the modulation index; (c) the peak amplitude of the modulated waveform; (d) the peak amplitude of the upper sideband; and (e) the total average power in the output if the amplitude of the modulating sinusoid is reduced to 2.0 V.

5.2.7 An AM (DSB-LC) transmitter is tested using a dummy (resistive) load. With no audio input the average output power is 3 kW. With a 1-kHz sinusoidal input of 1.0-V peak amplitude, the output power is 4 kW. Assume that typical programming material has a peak-to-rms ratio of 1/8.

a) If the peaks are set to 100% modulation, what is the resulting average output power for programming?

b) Calculate the equivalent sinusoidal modulation index for part (a).

5.2.8 A sinusoidal carrier waveform with 80-V peak amplitude is added to a low-frequency sinusoid. This sum is applied to the input of a class-C amplifier whose output-input gain characteristic is

$$e_o(t) = \begin{cases} 25[e_i(t) - 20] & \text{for} \quad e_i(t) > 20 \text{ V}, \\ 0 & \text{for} \quad e_i(t) < 20 \text{ V}. \end{cases}$$

The output of the amplifier is applied to a narrowband filter tuned to the carrier frequency (cf. Fig. 5.21).

a) Determine the maximum peak amplitude of the low-frequency sinusoid which will not result in envelope distortion at the amplifier output.

b) Find the modulation index of the output waveform if the peak amplitude of the low-frequency sinusoid is 40 V.

c) Sketch the resulting modulated waveform if the peak amplitude of the low-frequency sinusoid is 80 V.

5.2.9 The balanced modulator in Fig. P–5.2.9 is to be investigated for the possible generation of a DSB-LC signal with $m \le 1$. Each diode has the characteristic $i(t) = a_1 e(t) + a_2 e^2(t)$.

a) Determine the maximum allowable value of A.

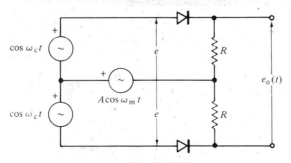

Fig. P–5.2.9.

b) Determine the minimum and maximum bandwidth of an ideal bandpass filter required on the output if one of the diodes is open-circuited.

5.2.10 Investigate the possible use of full-wave rectification of a DSB-LC waveform and compare to the performance of the half-wave rectifier detector shown in Fig. 5.23. [*Hint*: Full-wave rectification is equivalent to multiplication by a square wave of zero mean value.]

5.2.11 The waveform shown in Fig. P–5.2.11(a) is the input to the envelope detector shown in Fig. P–5.2.11(b).

a) Sketch the output waveform assuming an ideal diode characteristic.
b) Repeat using a diode back resistance of 10 kΩ.

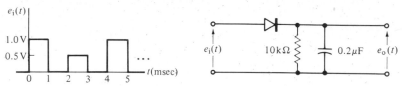

Fig. P–5.2.11.

5.2.12 The capacitor C of an envelope detector should be large enough to filter out the ripple at the carrier frequency present in the demodulated signal. However, if C is too large, the time constant RC of the discharge path becomes too large and the detector output may be unable to follow the envelope of the modulated signal. Determine the largest value of RC which will enable the detector to follow the envelope of the waveform shown in Fig. P–5.2.12. [*Hint*: Approximate the RC exponential decay by the first two terms of a Taylor series.]

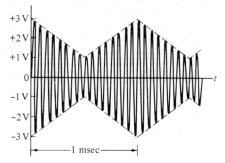

Fig. P–5.2.12.

c **5.2.13** In practice, a semiconductor diode is often approximated by a one-sided square-law behavior for low voltages and a linear behavior for large voltages. Suppose that the voltage $\phi(t) = A(1 + 0.50 \cos \omega_m t) \cos \omega_c t$ is applied to a diode whose current-voltage characteristic is described by (in mA)

$$i(t) = \begin{cases} 0 & e(t) < 0, \\ 10e^2(t) & 0 < e(t) < 0.50 \text{ V}, \\ 10[e(t) - 0.25] & 0.50 < e(t). \end{cases}$$

Using the digital computer, calculate and plot the ratio of the magnitudes of the second harmonic to the first harmonic of the modulating waveform present in $i(t)$

for $A = 0$–1.25, in .05 step increments. Can one tell from the output when operation is entirely within the linear range? (Assume ideal low-pass filtering after the diode.)

5.3.1 Sketch a typical spectrum of a frequency-division multiplexing system using DSB-LC for both the subcarrier and the main carrier modulation. What are the advantages and disadvantages of such a system?

5.3.2 Twelve voice signals, each band-limited to 3 kHz, are frequency-multiplexed using 1-kHz guard bands between channels and between the main carrier and the first channel. The modulation of the main carrier is DSB-LC. Calculate the bandwidth of the composite signal if the subcarrier modulation is (a) DSB-SC; (b) SSB-SC.

5.4.1 The signal $f(t) = \cos 2000\pi t$ is used to modulate a 5-kHz carrier. Sketch the time waveforms and line spectra if the modulation used is (a) DSB-SC; (b) DSB-LC, $m = 0.5$; (c) SSB-SC$_-$; (d) SSB-SC$_+$.

5.4.2 The VHF television allocation for Channel 6 calls for a video carrier at 83.25 MHz and a sound carrier at 87.75 MHz. A television receiver using the superheterodyne principle has the video IF carrier at 45.75 MHz and the sound carrier at 41.25 MHz.

a) What must be the TV local oscillator frequency when tuned to Channel 6?
b) Why is this the only choice possible?
c) What channel forms the image frequencies for Channel 6? (Use frequency listings in Appendix C.)

5.4.3 Assume that the first IF amplifier in the dual-conversion receiver shown in Fig. 5.27 passes all frequencies between ω_1 and $\omega_1 + W$. Determine the receiver oscillator frequencies, ω_0 and ω_a, required to properly demodulate the following transmissions:

a) a SSB$_+$ transmission occupying the frequency range from ω_c to $\omega_c + W$;
b) a SSB$_-$ transmission occupying the frequency range from $\omega_c - W$ to ω_c.
c) What happens if the receiver is set for receiving upper sideband and the actual received signal is lower sideband?

5.4.4 Assuming that $f(t) = a(t) \cos \omega_1 t$, show that the system illustrated in Fig. 5.31 serves as an image-free frequency mixer. What possible advantages are there for this type of mixer? List them. (Consider cases with and without noise.)

5.4.5 The system shown in Fig. P–5.4.5 is a simplified speech scrambler used to aid in communication privacy.

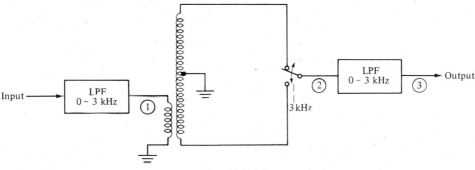

Fig. P–5.4.5.

a) Analyze the system by sketching spectra at the points: 1, 2, 3. Sketch to scale assuming that the low-pass filters have unity gain within their pass band.
b) Show that the same system can be used for unscrambling the signal.
c) Will the system work if a single-ended chopper is substituted for the symmetrical (balanced) chopper shown?

5.4.6 In the SSB-SC modulation system shown in Fig. 5.31, the signal $f(t)$ is shifted in phase ideally by exactly 90° at all frequencies. This is difficult to accomplish in practice. Assume here that $f(t) = \cos \omega_m t$ is shifted by $(-90 - \alpha)°$. Show that the ratio of the undesired output sideband magnitude to the desired output sideband magnitude is given by $\tan (\alpha/2)$.

★ **5.4.7** Determine the Hilbert transform of $f(t) = \cos (\omega_c t + \theta)$.

★ **5.4.8** a) Find the Hilbert transform of $f(t) = A \, Sa \, (Wt)$.
b) Determine the magnitude of $z(t) = f(t) + j\hat{f}(t)$.

5.5.1 A vestigial sideband signal is generated by transmitting a DSB-LC signal ($m = 1$) with a carrier frequency of 10 kHz through the vestigial sideband filter shown in Fig. P–5.5.1. Find an expression for the resulting vestigial sideband signal and sketch the spectral density if the modulating signal $f(t)$ is (a) $f(t) = \cos 1000\pi t$; (b) $f(t) = \cos 2000\pi t$.

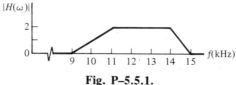

Fig. P–5.5.1.

5.5.2 Calculate the envelope of the resulting VSB signal for each case in Prob. 5.5.1.

5.6.1 A noise signal with a power spectral density

$$S_n(\omega) = \begin{cases} 10^{-7}\left(1 - \dfrac{|\omega|}{\alpha}\right) & |\omega| < 2\pi \times 10^7 \\ 0 & \text{elsewhere} \end{cases} \text{W/Hz},$$

where $\alpha = 2\pi \times 10^7/\text{sec}$, is passed through an ideal BPF with unity gain, unity resistance levels, and a bandwidth of 100 kHz centered at 5 MHz. Express the output signal in terms of a bandpass time representation and find the mean-square values of the in-phase and quadrature components.

5.6.2 Repeat Problem 5.6.1 for the case of maximum expected galactic noise at 100 MHz. (See Fig. 4.16.)

5.7.1 A DSB-SC and an SSB-SC transmission are each sent at 1 MHz in the presence of noise. The modulating signal in each case is band-limited to 3 kHz. The received signal power in each case is 1 mW and the received noise is assumed to be white with a (two-sided) power spectral density of $10^{-3} \, \mu\text{W/Hz}$. The receiver consists of a bandpass filter whose bandwidth matches the bandwidth of each transmission, followed by a synchronous detector.

a) Compare the signal-to-noise ratios at the detector input.

b) Compare the signal-to-noise ratios at the detector output.

c) Repeat (a) if the (two-sided) power spectral density were: $10^3(1/|f|)\mu W/Hz$. Would a "white noise" assumption be valid here?

5.7.2 The DSB-LC signal $\phi(t) = 3\cos 10,000\pi t + \cos (1000\pi t)\cos (10,000\pi t)$ V is present with additive band-limited white noise whose (two-sided) power spectral density is 1 $\mu W/Hz$ up to 10 kHz and zero at higher frequencies. This signal-plus-noise is passed through an ideal bandpass filter with a bandwidth of 100 Hz centered at 5500 Hz. Assume all resistance levels are one ohm.

a) Compute the average signal-to-noise ratio at the input of the bandpass filter.

b) If the desired output signal is that portion of the input signal with spectral components within the passband of the filter, what is the average signal-to-noise ratio at the output of the filter?

c) Assume that the desired output signal is the amplitude of the signal in item (b) above. Compute the output signal-to-noise ratio of a synchronous detector with an output bandwidth of 50 Hz that can make this measurement.

★ **5.7.3** A superheterodyne receiver with a 10-dB noise figure is used to receive a DSB-SC transmission at 10 MHz. A signal power of 10^{-10} watt is present at the receiver input. The IF filter, before the demodulator, has a rectangular transfer characteristic with a 10-kHz bandwidth. The IF filter is followed by a synchronous detector and a low-pass filter which attenuates all frequency components above 5 kHz. Compute the expected output signal-to-noise ratio in dB. (Assume daytime operation and make use of Fig. 4.16.)

5.8.1 When an electromagnetic wave arrives via two differing paths which result in approximately the same attenuation and 180° difference, the signals cancel resulting in a "deep fade" condition.

a) If one requires an automatic gain control to hold the receiver output constant over 179° out of 180° phase shift (and equal amplitudes), what must be the dynamic range (in dB) of the automatic gain control?

b) Assume that the distance from transmitter to receiver is 200 km and a strong reflector at midpath and at a height of 2 km (e.g., an airplane) is providing the second path for a deep fade condition. Calculate how far one needs to move the receiving antenna from the receiver (along a line joining the transmitter and receiver) to change the phase difference by 180° if the transmitter frequency is 100 MHz. Assume a flat earth and a direct line-of-sight propagation.

★ **5.8.2** A certain SSB-SC transmitter operates at 10 MHz with an effective output power of 100 W. Bandwidth of transmission is 3 kHz. Assume that transmission is primarily by sky wave and that the path loss formula given in Problem 4.7.16 is accurate for this purpose. (Neglect possible transmitting and receiving antenna gain factors, as well as losses resulting from ground wave propagation.) Find the maximum range, in miles, for acceptable day/night reception under the following conditions:

a) Receiver noise figure is 6 dB.

b) Antenna noise temperature is given by Fig. 4.16.

c) A minimum S/N of 10 dB is required.

5.9.1 A certain station uses DSB-SC with an average transmitter power of P watts. If SSB-SC were used instead, what must be the average transmitted power for (a) the same received signal strength; (b) the same received S/N ratio. Assume synchronous detection with the same local oscillator signal strengths for both cases.

5.9.2 Amplitude modulation is to be considered for each of the following objectives. State which type of amplitude modulation you would recommend and the reason(s) for your choice.

   a) A voice intercom system using the 60-Hz power lines to conduct the modulated signal.
   b) Remote control of landing flaps from the 400-Hz supply in an aircraft.
   c) Weather broadcasts to local marine users.
   d) Biweekly voice contact between two hospitals in the U.S. and Peru, South America.
   e) Transmission of weather photofax information (bandwidth of 3200 Hz centered at 2400 Hz) over a standard voice-grade telephone circuit (response of 300–3300 Hz).

CHAPTER 6

ANGLE
MODULATION

A continuous-wave (CW) sinusoidal signal can be varied by changing its amplitude and its phase angle. Recalling Eq. (5.2), we write

$$\phi(t) = a(t) \cos [\omega_c t + \gamma(t)].$$

In Chapter 5 we kept $\gamma(t)$ constant and varied $a(t)$ proportional to $f(t)$. This introduced the concept of amplitude modulation. Now we shall investigate the case in which $a(t) = A$ (a constant) and the phase angle $\gamma(t)$ is varied in proportion to $f(t)$. This introduces the concept of angle modulation.

6.1 FM AND PM

The angle of a sinusoidal signal is described in terms of a frequency and/or a phase angle. Before proceeding here, however, we must decide precisely what we mean by the frequency of a sinusoid. If a sinusoid has a constant angular rate ω_0, then we say that the frequency of the sinusoid is ω_0 radians per second. But what happens if the angular rate is not constant? It is helpful at this point to return to a phasor representation.

The phasor representation of a constant-amplitude sinusoid is shown in Fig. 6.1. This phasor has a magnitude A and a phase angle $\theta(t)$. If $\theta(t)$ increases linearly with time [that is, $\theta(t) = \omega_0 t$], we say that the phasor has an angular rate, or "frequency," of ω_0 radians per second. If the angular rate is not constant, we can still write a relation between the instantaneous angular rate $\omega_i(t)$ and $\theta(t)$:

$$\theta(t) = \int_0^t \omega_i(\tau) \, d\tau + \theta_0.$$ (6.1)

Taking the derivative of both sides of Eq. (6.1), we have

$$\omega_i(t) = \frac{d\theta}{dt}.$$ (6.2)

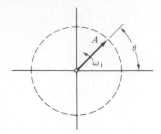

Fig. 6.1 A general phasor representation.

Therefore we conclude that the instantaneous frequency of a sinusoidal signal is given by the time derivative of its phase. Note that this definition agrees with our usual concept of frequency when the phase is linear with time.

Example 6.1.1 Determine the instantaneous frequency of the signal $\phi(t) = A \cos(10\pi t + \pi t^2)$.

Solution

$$\theta(t) = 10\pi t + \pi t^2$$

$$\omega_i(t) = \frac{d\theta}{dt} = 10\pi + 2\pi t = 2\pi(5 + t)$$

The frequency of $\phi(t)$ is 5 Hz at $t = 0$ and increases linearly at a rate of 1 Hz per second. Thus a quadratic phase shift gives a linear frequency dependence.

Drill Problem 6.1.1 Determine the instantaneous frequency of the following signal at $t = 0$: $\phi(t) = 5 \cos(10t + \sin 5t)$.

Answer.   15 rad/sec.

The concept of instantaneous frequency now permits us to describe two obvious possibilities for angle modulation (there are many more). If the phase angle $\theta(t)$ is varied linearly with the input signal $f(t)$, we can write

$$\theta(t) = \omega_c t + k_p f(t) + \theta_0 \tag{6.3}$$

where ω_c, k_p, θ_0 are constants. Because the phase is linearly related to $f(t)$, this type of angle modulation is called *phase modulation* (PM). The instantaneous frequency of this phase-modulated signal is

$$\omega_i = \frac{d\theta}{dt} = \omega_c + k_p \frac{df}{dt}. \tag{6.4}$$

Another possibility is to make the instantaneous *frequency* proportional to the input signal,

$$\omega_i = \omega_c + k_f f(t), \tag{6.5}$$

where ω_c, k_f are constants. Because the frequency is linearly related to $f(t)$, this type of angle modulation is called *frequency modulation* (FM). The phase angle of this frequency-modulated signal is

$$\theta(t) = \int_0^t \omega_i(\tau) \, d\tau = \omega_c t + \int_0^t k_f f(\tau) \, d\tau + \theta_0. \tag{6.6}$$

A comparison of Eqs. (6.3)–(6.6) shows that PM and FM are closely related. In PM the phase angle of the carrier signal is varied linearly with the modulating signal. In FM the phase angle of the carrier signal is varied linearly with the integral of the modulating signal. Therefore if we integrate the modulating signal $f(t)$ first and then use it to phase modulate a carrier, we will obtain a frequency-modulated signal. Figure 6.2 is an illustration of FM and PM waveforms for given $f(t)$.

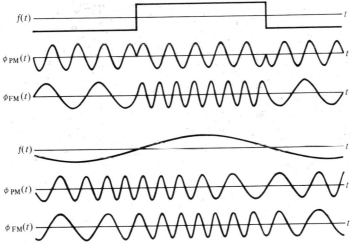

Fig. 6.2 Examples of frequency and phase modulation.

Because frequency and phase modulation are so closely related, any variation in phase will necessarily result in a variation in frequency and vice versa. The essential difference between FM and PM is the nature of the dependency on the modulating signal. Although we shall discuss FM in more detail, our discussion is also valid for PM with only minor differences and these are pointed out in a later section.

In the case of AM signals, there was always a one-to-one correspondence between the modulated signal and the modulating signal. When this condition holds, the modulation is said to be *linear*.† For PM and FM this is not always true, however, as can be seen from the following reasoning.

---

† More formally, if $f(t)$ is the modulating signal and $\phi(t)$ is the modulated signal, the modulation is linear if $d\phi/df$ is independent of $f(t)$.

A general PM (or FM, with the appropriate modifications) signal can be represented by (note our return to complex notation)†

$$\phi_{PM}(t) = Ae^{j\theta(t)} = Ae^{j(\omega_c t + \theta_0)} e^{jk_p f(t)}. \tag{6.7}$$

Using a series expansion for the exponential modulation factor in Eq. (6.7), we have

$$\phi_{PM}(t) = Ae^{j(\omega_c t + \theta_0)} \left[1 + jk_p f(t) - \frac{1}{2!} k_p^2 f^2(t) - j\frac{1}{3!} k_p^3 f^3(t) + \cdots \right]. \tag{6.8}$$

From this result we conclude that, unless $|k_p f(t)| \ll 1$, angle modulation—in this case PM—is not linear. Therefore we can expect that, in general, the side-bands arising in angle modulation will not obey the principle of superposition. An analysis of spectra, etc., will have to be carried out choosing a particular waveform. When confronted by this choice, we shall use the sinusoidal waveform unless otherwise specified.

6.2   NARROWBAND FM

The linear condition in Eq. (6.8) maintains a linear modulation for FM and this appears to be a good place to begin. To lay the groundwork for the nonlinear modulation case we shall use a sinusoidal modulating signal. To be specific, let

$$f(t) = a \cos \omega_m t. \tag{6.9}$$

Because we are dealing with FM [cf. Eq. (6.5)],

$$\omega_i = \omega_c + k_f f(t)$$
$$= \omega_c + ak_f \cos \omega_m t, \tag{6.10}$$

where k_f is the frequency modulation constant; typical units are in radians per second per volt. Defining a new constant called the *peak frequency deviation,*

$$\Delta\omega = ak_f, \tag{6.11}$$

we can rewrite Eq. (6.10) as

$$\omega_i = \omega_c + \Delta\omega \cos \omega_m t. \tag{6.12}$$

The phase of this FM signal is [cf. Eq. (6.6)] (let $\theta_0 = 0$ for convenience)

$$\theta(t) = \omega_c t + \frac{\Delta\omega}{\omega_m} \sin \omega_m t = \omega_c t + \beta \sin \omega_m t, \tag{6.13}$$

---

† Although these are written in terms of PM signals, the conclusions are applicable to FM as well by substituting

$$k_f \int_0^t f(\tau)\, d\tau \qquad \text{for} \qquad k_p f(t).$$

where

$$\beta = \Delta\omega/\omega_m \tag{6.14}$$

is a dimensionless ratio of the peak frequency devi_____ e modulating frequency.

The resulting FM signal is

$$\phi_{FM}(t) = Ae^{j(\omega_c t + \beta \sin \omega_m t)}, \tag{6.15a}$$

$$\mathscr{R}e\{\phi_{FM}(t)\} = A \cos(\omega_c t + \beta \sin \omega_m \qquad \tag{6.15b}$$

Expanding the exponential modulation terms of Eq. (6.1___) ___ ____ ___ies, we have

$$\phi_{FM}(t) = Ae^{j\omega_c t}\left(1 + j\beta \sin \omega_m t - \frac{1}{2!}\beta^2 \sin^2 \omega_m t - j\frac{1}{3!}\beta^3 \text{ s}\qquad \cdots\right). \tag{6.16}$$

It is fairly obvious that the bandwidth of $\phi_{FM}(t)$ as re_____ in Eq. (6.16) is dependent on the value of β. For small values of β, only the constant and first-order term are significant and the bandwidth will be $2\omega_m$. However, as the value of β increases, more terms become significant and the bandwidth will increase accordingly. The condition where β is small enough to make all terms after the first two in Eq. (6.16) negligible is the condition for *narrowband* FM (NBFM). Note that NBFM is an example of linear modulation; usually a value of $\beta < 0.2$ is taken to be sufficient to satisfy this condition, although at times values as high as 0.5 are used.

Proceeding with the narrowband condition, we see that Eq. (6.16) becomes

$$\phi_{NBFM}(t) = Ae^{j\omega_c t}(1 + j\beta \sin \omega_m t). \tag{6.17}$$

It is instructive to compare Eq. (6.17) with an equivalent expression for an AM signal:

$$\phi_{AM}(t) = Ae^{j\omega_c t}(1 + m \cos \omega_m t). \tag{6.18}$$

As suggested by a comparison of Eqs. (6.17) and (6.18), β is called the *modulation index* of the FM signal.

Although the narrowband FM signal and the AM signal have similarities, they are distinctly different methods of modulation. The similarities and differences can be portrayed by considering their phasor representations. Expanding Eqs. (6.17) and (6.18) in phasor form, we have

$$\phi_{NBFM}(t) = Ae^{j\omega_c t}(1 + \tfrac{1}{2}\beta e^{j\omega_m t} - \tfrac{1}{2}\beta e^{-j\omega_m t}),$$

$$\phi_{AM}(t) = Ae^{j\omega_c t}(1 + \tfrac{1}{2}me^{j\omega_m t} + \tfrac{1}{2}me^{-j\omega_m t}).$$

Taking the term $Ae^{j\omega_c t}$ as the reference (i.e., suppressing the continuous ω_c rotation), we show the phasor representation of each of these waveforms in Fig. 6.3. The resultant waveform can be found by rotating the entire phasor diagram

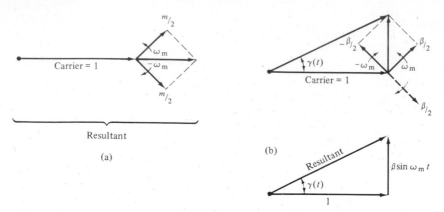

Fig. 6.3 Phasor representation of (a) AM and (b) NBFM.

at an angular rate of ω_c rad/sec and then taking the projection of the resultant on the real axis.

From Fig. 6.3, the differences between Eqs. (6.17) and (6.18) become quite evident. In the AM waveform, the modulation is added in phase with the carrier whereas in NBFM the modulation is added in quadrature with the carrier. The NBFM case gives rise to phase variations with very little amplitude change whereas the AM case gives amplitude variations with no phase deviation.

It is instructive to determine the limits on β from the results of the phasor diagrams in Fig. 6.3(b); the phase angle from the carrier is

$$\gamma(t) = \tan^{-1}(\beta \sin \omega_m t). \tag{6.19}$$

The instantaneous frequency deviation from the carrier frequency should be equal to $\Delta\omega \cos \omega_m t = \beta\omega_m \cos \omega_m t$ and is found by taking the derivative of this phase angle, or,

$$\frac{d\gamma}{dt} = \frac{\beta\omega_m \cos \omega_m t}{1 + \beta^2 \sin^2 \omega_m t} \approx \beta\omega_m \cos \omega_m t, \quad \text{if} \quad \beta^2 \sin^2 \omega_m t \ll 1. \tag{6.20}$$

The amplitude of the resultant phasor should be a constant (A); checking from the phasor diagram, we find

$$A\sqrt{1 + \beta^2 \sin^2 \omega_m t} \approx A \quad \text{if} \quad \beta^2 \sin^2 \omega_m t \ll 1. \tag{6.21}$$

Because $\sin^2 \omega_m t \leq 1$, these approximations are valid if $\beta^2 \ll 1$. Choosing $\beta^2 < 0.1$, we find that $\beta < 1/\sqrt{10} = 0.316$ is a reasonable bound for the narrowband approximation. Values as high as 0.50 can be used in practice if the resulting amplitude modulation is removed by amplitude-limiting the angle-modulated waveform.

The addition of the modulation in quadrature with the carrier in NBFM, in contrast to that in phase in AM, is emphasized when we take the real parts of

Eqs. (6.17) and (6.18):

$$\mathscr{R}e\,\{\phi_{NBFM}(t)\} = A(\cos \omega_c t - \beta \sin \omega_m t \sin \omega_c t), \tag{6.22}$$
$$\mathscr{R}e\,\{\phi_{AM}(t)\} = A(\cos \omega_c t + m \cos \omega_m t \cos \omega_c t). \tag{6.23}$$

Equations (6.22) and (6.23) suggest a method of generation for the NBFM or NBPM case using phase shifters and balanced modulators as shown in Fig. 6.4. This method is commonly used in the generation of NBFM and NBPM signals. Note that even though we have been discussing the FM case, the PM case follows in the same manner.

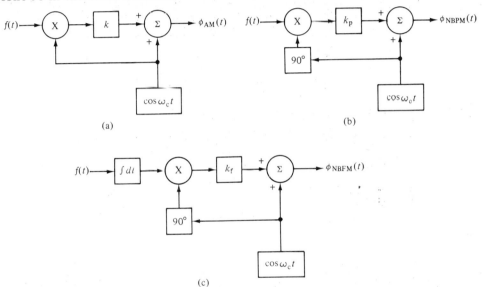

Fig. 6.4 Generation of signals using balanced modulators: (a) AM; (b) NBPM; (c) NBFM.

Summarizing, narrowband FM (and PM), like AM, is an example of linear modulation. A major difference is that the modulation is added in phase with the carrier in AM whereas it is added in quadrature with the carrier in NBFM. Both systems require a bandwidth of $W = 2\omega_m$ to transmit a signal of ω_m rad/sec in spectral width. The modulation index in FM is $\beta = \Delta\omega/\omega_m$ and a useful criterion for NBFM is $\beta < 0.2$.

Drill Problem 6.2.1   Calculate the maximum (peak)-percentage amplitude, phase, and frequency error incurred in using the phasor approximations to narrowband FM for the sinusoidal case when (a) $\beta = 0.20$; (b) $\beta = 0.50$.

Answer.   (a) 2.0%, 1.3%, 3.8%; (b) 11.8%, 7.3%, 20.0%.

Example 6.2.1   A comparison of the two phasor representations in Fig. 6.3 motivates us to investigate the simultaneous use of both amplitude modulation and narrowband frequency modulation for the possible elimination of one side-band. Investigate the possible use of this technique, for sinusoidal modulation, to generate SSB-LC signals.

Solution. For simultaneous amplitude and frequency modulation we write

$$\phi(t) = \mathcal{R}e\,\{A(1 + m\cos\omega_m t)\exp\,[j(\omega_c t + \beta\sin\omega_m t)]\,\},$$

where m, β are the amplitude- and frequency-modulation indices, respectively. Using the identity $\mathcal{R}e\{z_1 z_2\} = \mathcal{R}e\{z_1\}\mathcal{R}e\{z_2\} - \mathcal{I}m\{z_1\}\mathcal{I}m\{z_2\}$, we get

$$\phi(t) = A\,(1 + m\cos\omega_m t)[\cos\omega_c t\cos(\beta\sin\omega_m t) - \sin\omega_c t\sin(\beta\sin\omega_m t)].$$

For the NBFM condition, $\beta \ll 1$ so that $\cos(\beta\sin\omega_m t) \approx 1$ and $\sin(\beta\sin\omega_m t) \approx \beta\sin\omega_m t$; then

$$\phi(t) \approx A(1 + m\cos\omega_m t)(\cos\omega_c t - \beta\sin\omega_m t\sin\omega_c t),$$

$$\phi(t) \approx A(\cos\omega_c t + m\cos\omega_m t\cos\omega_c t - \beta\sin\omega_m t\sin\omega_c t$$
$$- m\beta\sin\omega_m t\cos\omega_m t\sin\omega_c t),$$

$$\phi(t) \approx A\cos\omega_c t + \tfrac{1}{2}A(m + \beta)\cos(\omega_c + \omega_m)t + \tfrac{1}{2}A(m - \beta)\cos(\omega_c - \omega_m)t$$
$$- m\beta A\sin\omega_m t\cos\omega_m t\sin\omega_c t.$$

Now if we set $\beta = m$ and then note that the last term is a second-order effect because $\beta \ll 1$, we obtain the approximate SSB-LC signal,

$$\phi(t) \approx A\cos\omega_c t + mA\cos(\omega_c + \omega_m)t.$$

Drill Problem 6.2.2 Sometimes when AM is the desired modulation, a combination of AM and FM can actually occur as a result of an imperfect modulator. Combined AM and NBFM is characterized on a spectrum-analyzer display by two sidebands of unequal amplitude. This arises because the AM sidebands are of the same phase but the NBFM sidebands are of opposite phase. Because the intended modulation is AM, the incidental FM introduced is assumed to be the smaller of the two effects.

As an example, suppose that a spectrum-analyzer measurement of the output of a modulator using sinusoidal modulation indicates a carrier line of unit magnitude, an upper sideband line of magnitude 0.45 and a lower sideband line of magnitude 0.35. Calculate the percent AM and the percent FM present using the result of Ex. 6.2.1.

Answer. 80%; 10%.

6.3 WIDEBAND FM

Up to this point we have relied heavily on the use of the Fourier transform of a general signal $f(t)$ to give us the spectral density $F(\omega)$. However, if the value of β is not small, the Fourier transform of a general angle-modulated waveform cannot be evaluated. For specific cases the integration can be performed numerically or in terms of tabulated values. Therefore we shall first try to establish some bounds on the spectral density before we are forced to restrict the analysis to a few given modulating signal waveforms.

A measure of the peak amplitude-to-frequency conversion is the peak frequency deviation, $\Delta\omega$. This represents the maximum amount that ω_i deviates from the "average" value of ω_c. This is demonstrated for two differing cases in Fig. 6.5.

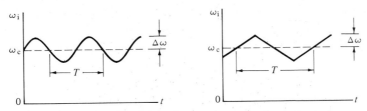

Fig. 6.5 Definition of maximum (peak) frequency deviation.

There are two identifiable mechanisms in the description of the spectrum of an FM waveform. The first is attributable to the rate of change of the modulating signal; i.e., the frequency content of the modulating signal. The second effect, peculiar to FM, is the proportionality between the amplitude of the modulating signal and the instantaneous frequency of the FM signal. The instantaneous frequency follows the amplitude of the modulating signal, but this does not imply necessarily that the spectral density follows the same pattern. The concept of instantaneous frequency and the frequency used in the Fourier transform are not identical.

In the NBFM approximation, it is seen that the second effect was neglected in favor of the first since $\Delta\omega \ll \omega_m$. In fact, we now see that for the sinusoidal case the modulation index $\beta = \Delta\omega/\omega_m$ gives us a relative measure as to the importance of these two effects in FM.

The idea of a modulation index can be extended to more general waveforms. For a general pulse waveform we can define a peak frequency deviation $\Delta\omega$ and a time duration T; if the waveform is periodic, then T is the period. The product $\beta_1 = (\Delta\omega/2\pi)T$ is a dimensionless number, called a *dispersion index,* which takes the place of the modulation index for more general modulation waveforms. It is easy to see that $\beta_1 \rightarrow \beta$ for sinusoidal modulation. For very low dispersion indices the spectral content of a modulating signal largely controls the magnitude of the FM spectral density. For very high dispersion indices, the amplitude-to-frequency conversion largely controls the magnitude spectral density. Phase effects are not as predictable because they depend on the relative phasing between signals. What happens for intermediate values must be examined on the basis of each given type of signal.

Returning to the sinusoidal case of FM, let $f(t) = a \cos \omega_m t$ and $\omega_i = \omega_c + \Delta\omega \cos \omega_m t$. The spectral content of the modulating signal is at ω_m rad/sec.

The peak amplitude-to-frequency conversion is $\Delta\omega$ rad/sec. Then for very low values of $\beta = \Delta\omega/\omega_m$ (that is, $\Delta\omega \ll \omega_m$) the spectrum will be band-limited to $2\omega_m$. On the other hand, for very high values of $\beta = \Delta\omega/\omega_m$ (that is, $\Delta\omega \gg \omega_m$) the amplitude-to-frequency conversion will predominate and we would expect the bandwidth to be on the order of $2\Delta\omega$. We therefore have some rather intuitive bounds on the bandwidth at both extremes.

★ 6.3.1   General Approximations

Another general intuitive comment can be made here before restricting ourselves to specific waveforms. If we let $\beta_1 \to \infty$ ($\beta \to \infty$ for the sinusoidal case), we would expect the amplitude-to-frequency conversion to completely predominate. From the concept of a spectral density we would then expect the spectral magnitudes to be in proportion to the fractional time spent at each frequency.† For example, let $f(t) = a\cos\omega_m t$ so that the frequency deviation about the carrier, $\omega_i' = \omega_i - \omega_c$, is

$$\omega_i' = \Delta\omega \cos\omega_m t, \tag{6.24}$$

or

$$t = \frac{1}{\omega_m}\cos^{-1}\left(\frac{\omega_i'}{\Delta\omega}\right) \qquad \text{for} \quad |\omega_i'| \le \Delta\omega. \tag{6.25}$$

The fractional amount of time per unit of frequency is:‡

$$\frac{1}{T}\left|\frac{dt}{d\omega_i'}\right| = \frac{1/2\pi}{\sqrt{1-(\omega_i'/\Delta\omega)^2}} \qquad \text{for} \quad |\omega_i'| \le \Delta\omega. \tag{6.26}$$

Therefore as $\beta_1 \to \infty$ (in this case $\beta \to \infty$) the magnitude weighting of the spectral density of the FM waveform will approach the shape shown in Fig. 6.6 over

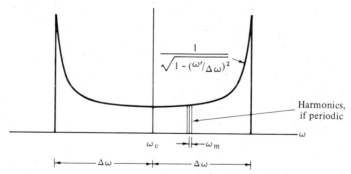

Fig. 6.6 Approximation to the magnitude FM spectral density as $\beta \to \infty$, sinusoidal case.

---

† This is sometimes referred to as Woodward's theorem.

‡ The reader with some knowledge of probability will recognize this as the probability density function of the modulating waveform for uniform phase.

band limits $2\Delta\omega$ in width. Note that this is based on a signal T units long; for the periodic case the spectral density will be composed of impulses with weights determined from this curve. Effects of phase may cause the individual components to vary somewhat from this approximation.

Example 6.3.1   A sinusoidal signal at a frequency of ω_c rad/sec is frequency-modulated by the sawtooth waveform shown in Fig. 6.7(a). The peak frequency deviation on each side of the carrier is $\Delta\omega$ rad/sec, as shown in Fig. 6.7(b).

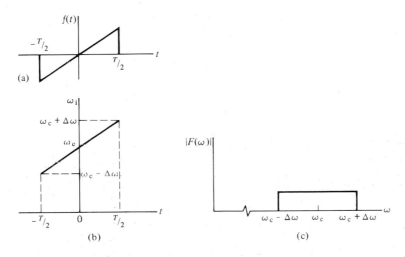

Fig. 6.7 Example of FM spectral density as $\beta_1 \rightarrow \infty$.

Describe the approximate magnitude spectral density as the dispersion index of the system becomes very large.†

Solution.   As $\beta_1 \rightarrow \infty$, the bandwidth approaches $2\Delta\omega$ and the magnitude spectrum is approximated by

$$\omega_i' = \frac{\Delta\omega}{T/2} t, \qquad -T/2 < t < T/2,$$

$$t = \frac{T}{2\Delta\omega} \omega_i',$$

$$\frac{1}{T}\left|\frac{dt}{d\omega_i'}\right| = \frac{1}{2\Delta\omega}, \qquad -\Delta\omega < \omega_i' < \Delta\omega.$$

This is shown in Fig. 6.7(c). If the modulating signal were repeated periodically,

---

† This is a simplified version of the type of modulation that a bat uses (at ultrasonic frequencies) for navigation and target location. It is also used for radar purposes.

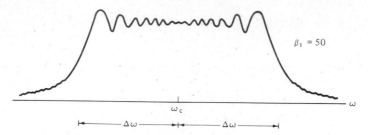

Fig. 6.8 Computed magnitude spectrum for the FM discussed in Example 6.3.1.

a series of impulses would be present spaced by ω_0 units. A numerical example of the computation of a magnitude spectrum for $\beta_1 = 50$ is shown in Fig. 6.8.

Drill Problem 6.3.1   A sinusoidal signal at a frequency of ω_c rad/sec is frequency-modulated by a symmetrical square wave. The peak frequency deviation is $\Delta\omega$. Describe the approximate magnitude spectral density as the dispersion index becomes very large.

Answer. $\frac{1}{2}\delta(\omega - \omega_c + \Delta\omega) + \frac{1}{2}\delta(\omega - \omega_c - \Delta\omega)$.

6.3.2   Sinusoidal Case

Having obtained some intuitive insights into the mechanisms of FM, we now seek to extend our knowledge by using the Fourier transform. As pointed out earlier, however, it is not possible to evaluate the Fourier transform of a general FM waveform and so we restrict the analysis here to pure sinusoids. Although pure sinusoids make for rather uninteresting communications, the results of the analysis hopefully will permit us to draw some more general conclusions. Because FM is a nonlinear modulation we cannot appeal to superposition here.

As before, we choose $f(t) = a\cos\omega_m t$; for FM,†

$$\omega_i(t) = \omega_c + ak_f\cos\omega_m t$$

$$= \omega_c + \Delta\omega\cos\omega_m t,$$

and

$$\theta(t) = \int_0^t \omega_i(\tau)\,d\tau$$

$$= \omega_c t + \frac{\Delta\omega}{\omega_m}\sin\omega_m t$$

$$= \omega_c t + \beta\sin\omega_m t.$$

---

† A constant term is introduced if the lower limit in the integral is not zero; this does not change the analysis and will be omitted for convenience.

Using complex notation,

$$\phi_{FM}(t) = \mathcal{R}e\ \{Ae^{j\theta(t)}\}$$

$$= \mathcal{R}e\ \{Ae^{j\omega_c t}e^{j\beta\sin\omega_m t}\}. \tag{6.27}$$

Note that Eq. (6.27) can be rewritten as

$$\phi_{FM}(t) = A\cos(\omega_c t + \beta\sin\omega_m t). \tag{6.28}$$

Alternatively, an identity for the real part of a product (cf. Appendix A) can be used to rewrite Eq. (6.27) as

$$\phi_{FM}(t) = A\cos\omega_c t\cos(\beta\sin\omega_m t) - A\sin\omega_c t\sin(\beta\sin\omega_m t). \tag{6.29}$$

Equations (6.28) and (6.29) are both valid forms but the complex forms are generally easier to use and we shall prefer Eq. (6.27).

The second exponential in Eq. (6.27) is a periodic function of time with a fundamental frequency of ω_m rad/sec. It can be expanded in a Fourier series,

$$e^{j\beta\sin\omega_m t} = \sum_{n=-\infty}^{\infty} F_n e^{jn\omega_m t}, \tag{6.30}$$

where

$$F_n = \frac{1}{T}\int_{-T/2}^{T/2} e^{j\beta\sin\omega_m t}e^{-jn\omega_m t}\ dt. \tag{6.31}$$

Making a change of variable $\xi = \omega_m t = (2\pi/T)t$, we get

$$F_n = \frac{1}{2\pi}\int_{-\pi}^{\pi} e^{j(\beta\sin\xi - n\xi)}\ d\xi. \tag{6.32}$$

This integral can be evaluated numerically in terms of the parameters n and β, and because it occurs in many physical problems it has been tabulated extensively.† It is a function of n and β, denoted by $J_n(\beta)$, and is called the Bessel function of the first kind (signified by the "J") of order n and argument β. Note that in our case n is an integer (negative and positive) and β is a continuous variable (positive values only). Some of these functions are plotted in Fig. 6.9. Though we do not wish to get involved in discussing the detailed characteristics of Bessel functions, we do have use for the following properties:

1. $J_n(\beta)$ are real valued,
2. $J_n(\beta) = J_{-n}(\beta)$, for n even,
3. $J_n(\beta) = -J_{-n}(\beta)$, for n odd, $\qquad\qquad\qquad\qquad\qquad$ (6.33)
4. $\displaystyle\sum_{n=-\infty}^{\infty} J_n^2(\beta) = 1.$

---

† A table of Bessel functions is given in Appendix I.

$\Sigma F_m = J_n(\beta)\cos(\omega_c + n\omega_m)t)$

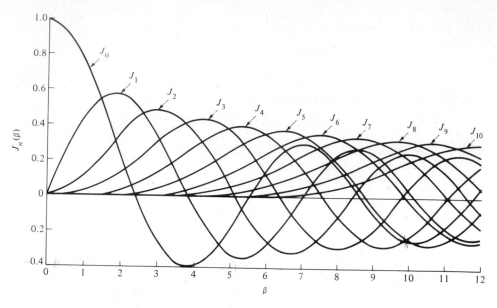

Fig. 6.9 Plot of Bessel function of the first kind, $J_n(\beta)$.

Using these results we can rewrite Eq. (6.30) as

$$e^{j\beta\sin\omega_m t} = \sum_{n=-\infty}^{\infty} J_n(\beta)e^{jn\omega_m t}, \qquad (6.34)$$

and Eq. (6.27) becomes

$$\phi_{FM}(t) = \mathcal{R}e\left\{ Ae^{j\omega_c t} \sum_{n=-\infty}^{\infty} J_n(\beta)e^{jn\omega_m t}\right\}, \qquad (6.35)$$

$$\phi_{FM}(t) = A\sum_{n=-\infty}^{\infty} J_n(\beta)\cos(\omega_c + n\omega_m)t. \qquad (6.36)$$

This can be expanded, if desired, using properties (2) and (3) of Eq. (6.33):

$$\begin{aligned}
\phi_{FM}(t) = A\{ & J_0(\beta)\cos\omega_c t \\
& + J_1(\beta)[\cos(\omega_c + \omega_m)t - \cos(\omega_c - \omega_m)t] \\
& + J_2(\beta)[\cos(\omega_c + 2\omega_m)t + \cos(\omega_c - 2\omega_m)t] \\
& + J_3(\beta)[\cos(\omega_c + 3\omega_m)t - \cos(\omega_c - 3\omega_m)t] \\
& + \dots
\end{aligned} \qquad (6.37)$$

From these results, it is evident that an FM waveform with sinusoidal modulation, in contrast to AM, has an infinite number of sidebands. However, the magnitudes of the spectral components of the higher-order sidebands become negligible and, for all practical purposes, the power is contained within a finite bandwidth. Plots of the sideband magnitudes for several different values of β are shown in Fig. 6.10. Note that β can be varied by varying $\Delta\omega$ or by varying ω_m, as demonstrated in Fig. 6.10.

How many sidebands are important to the FM transmission of a signal? This will depend on the intended application and the fidelity requirements. A rule commonly adopted is that a sideband is *significant* if its magnitude is equal to or exceeds 1% of the unmodulated carrier, i.e., if

$$|J_n(\beta)| \geq 0.01. \tag{6.38}$$

The actual number of significant sidebands for different values of β can be found from a plot or a table of Bessel functions. It can be seen from the plots of Fig. 6.10 that the $J_n(\beta)$ diminish rapidly for $n > \beta$, particularly as β becomes large. A graph of the ratio n/β for $|J_n(\beta)| \geq 0.01$ is shown in Fig. 6.11 and it is seen that the ratio approaches one as β becomes very large. The bandwidth for very large β can then be approximated by taking the last significant sideband at $n = \beta$ so that

$$W = 2n\omega_m \approx 2\beta\omega_m = 2\frac{\Delta\omega}{\omega_m}\omega_m,$$

or

$$W \approx 2\Delta\omega \qquad \text{for large} \quad \beta. \tag{6.39}$$

For very small values of β, the only Bessel functions of significant magnitude (see Fig. 6.9) are $J_0(\beta)$ and $J_1(\beta)$. Therefore the bandwidth for the narrowband case (verifying an earlier result) is

$$W \approx 2\omega_m \qquad \text{for small} \quad \beta. \tag{6.40}$$

We now have bounds on the limiting cases. It would be convenient to have a more general rule to take care of the intermediate cases and, if possible, also approach the limiting cases in a continuous manner. One such rule was proposed by J. R. Carson (one of the first to investigate FM in the 1920's):

$$W \approx 2(\Delta\omega + \omega_m), \tag{6.41a}$$

which can also be written as

$$W \approx 2\omega_m(1 + \beta). \tag{6.41b}$$

Carson's rule approaches the correct limits for both very large and very small β; it is widely used in practice because it gives a very convenient approximation that is reasonably accurate. It always gives less bandwidth than our definition of significant sidebands, with a maximum bandwidth error in the neigh-

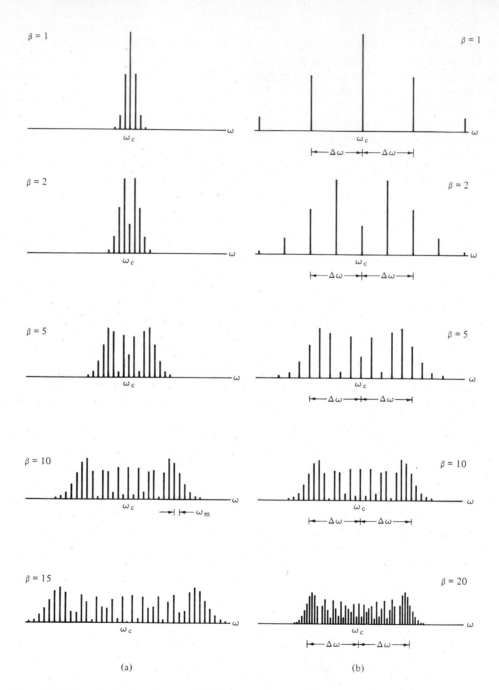

Fig. 6.10 Magnitude line spectra for FM waveforms with sinusoidal modulation: (a) for constant ω_m; (b) for constant $\Delta\omega$.

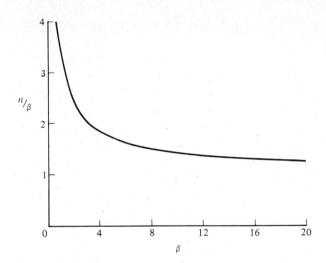

Fig. 6.11 Number of FM sidebands for which $|J_n(\beta)| \geq 0.01$.

borhood of $\beta = 1$. The average power in the sidebands neglected, however, is small and less than 1% of the total average power in the FM waveform. In fact, the approximation is good enough that we now release the restriction that the modulating signal be purely sinusoidal and make the wide generalization that Carson's rule holds for general modulating signals that are band-limited and have finite power. An intuitive justification for this is that the two terms in Carson's rule display the effects of the two mechanisms in the generation of FM, and that these effects on the bandwidth are additive. This is strictly intuitive reasoning and it is doubtful if one could argue with such conviction if one did not previously know the desired result. Perhaps the best justification is that it works!

Example 6.3.2   A 10 MHz carrier is frequency-modulated by a sinusoidal signal such that the peak frequency deviation is 50 kHz. Determine the approximate bandwidth of the FM signal if the frequency of the modulating sinusoid is (a) 500 kHz; (b) 500 Hz; (c) 10 kHz.

Solution

a) $\beta = \dfrac{\Delta f}{f_m} = \dfrac{50}{500} = 0.10$

 This is a narrowband FM signal; $B \approx 2f_m = 1$ MHz.

b) $\beta = 100$; this is the wideband case and

$$B \approx 2\Delta f = 100 \text{ kHz (Carson's rule gives 101 kHz)}.$$

c) $\beta = 5$; use of Carson's rule gives $B \approx 2(\Delta f + f_m) = 120$ kHz. A more accurate method is to use Fig. 6.9 or Fig. 6.11 to find the number n of significant sidebands:

$$B = 2nf_m = 2(8)(10 \text{ kHz}) = 160 \text{ kHz}.$$

A magnitude line spectrum for $\beta = 5$ is shown in Fig. 6.10; in this case the spacing between lines would be 10 kHz.

Drill Problem 6.3.2   Repeat Example 6.3.2 if the peak frequency deviation were decreased to 20 kHz.

Answer.   (a) 1 MHz; (b) 41 kHz; (c) 80 kHz (60 kHz if you use Carson's rule).

Drill Problem 6.3.3   A given FM signal is

$$\phi_{\text{FM}}(t) = 10 \cos [10^6 \pi t + 8 \sin (10^3 \pi t)].$$

Determine the following: (a) the carrier frequency, f_c; (b) the modulation index, β; (c) the peak frequency deviation, Δf.

Answer.   (a) 500 kHz; (b) 8; (c) 4 kHz.

★ **Example 6.3.3**   The analytical method used above for sinusoidal modulation can be used for a general periodic modulating signal with zero mean. Here we consider an FM system in which the instantaneous frequency may take on only two possible values—known as *frequency-shift keying* (FSK). We can analyze FSK using the above methods if the frequency shift is periodic. For convenience we choose the modulating signal to be a periodic square wave of unit amplitude, as shown in Fig. 6.12(a). We wish to find the resulting FM spectrum.

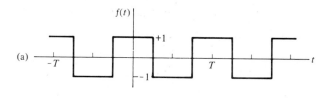

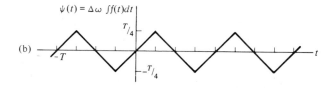

Fig. 6.12 The frequency and phase modulation used in Example 6.3.3.

Solution

$$\omega_i(t) = \omega_c + k_f f(t) = \omega_c + \Delta\omega f(t)$$

$$\theta(t) = \int_0^t \omega_i(\tau) \, d\tau = \omega_c t + \Delta\omega \int_0^t f(\tau) \, d\tau$$

$$= \omega_c t + \psi(t)$$

where

$$\psi(t) = \begin{cases} (\Delta\omega)t & -T/4 < t < T/4 \\ (\Delta\omega)\left(\dfrac{T}{2} - t\right) & T/4 < t < 3T/4 \end{cases}$$

$$\phi_{FM}(t) = \mathcal{R}e \, \{Ae^{j\theta(t)}\} = \mathcal{R}e \, \{Ae^{j\omega_c t} e^{j\psi(t)}\}.$$

The phase function $\psi(t)$ is periodic with period T and can be expressed in a Fourier series:

$$e^{j\psi(t)} = \sum_{n=-\infty}^{\infty} F_n e^{jn\omega_0 t}, \qquad \omega_0 = 2\pi/T,$$

$$F_n = \frac{1}{T} \int_{-T/4}^{3T/4} e^{j\psi(t)} e^{-jn\omega_0 t} \, dt.$$

Using the expression above for $\psi(t)$, we get

$$F_n = \frac{1}{2} \left\{ \text{Sa}\left[\frac{\pi}{2}(\beta_1 - n)\right] + (-1)^n \, \text{Sa}\left[\frac{\pi}{2}(\beta_1 + n)\right] \right\}$$

where

$$\beta_1 \triangleq \Delta\omega/\omega_0.$$

Then we can write

$$\phi_{FM}(t) = \mathcal{R}e \left\{ Ae^{j\omega_c t} \sum_{n=-\infty}^{\infty} F_n e^{jn\omega_0 t} \right\},$$

$$\phi_{FM}(t) = A \sum_{n=-\infty}^{\infty} F_n \cos(\omega_c + n\omega_0)t.$$

The magnitude line spectrum of $\phi_{FM}(t)$ is shown in Fig. 6.13 for three different values of β. It is interesting to compare these with the line spectra for the sinusoidal case shown in Fig. 6.10 (the same scaling has been used for both) and the limiting case of Drill Problem 6.3.1.

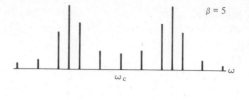

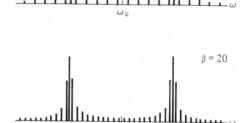

Fig. 6.13 Magnitude line spectra for FM waveforms with square-wave modulation (FSK).

6.3.3   Commercial FM Transmissions

As noted earlier, narrowband FM is linear and therefore much of the analysis for AM applies. Advantages in using narrowband FM over AM include the possibility of a response to zero Hz (important in telemetry and recording) and the rejection of large noise pulses (as a result of clipping, or limiting, the amplitude of the waveform) which may tend to saturate the receiver. Narrowband FM is used primarily in telemetry and mobile communications.

Provided that we are content with only the gross details, we can apply our knowledge of purely sinusoidal FM to more general waveforms also in the wideband case. For wideband FM we noted that the bandwidth depended mainly on the peak frequency deviation, Δf. This in turn depends, for a given modulator constant, on the amplitude of the modulating signal. Therefore some limit must be placed on the modulating signal to avoid excessive bandwidths even though the bandwidth of the modulating signal may be well-defined.

For commercial FM broadcasting, the Federal Communications Commission (FCC) in the U.S. assigns carrier frequencies spaced at 200 kHz intervals in the range 88–108 MHz and fixes the peak frequency deviation at 75 kHz.† The 200 kHz available to each station, in comparison with 10 kHz for AM broadcasting,

---

† See Appendix C.

allows the transmission of high-fidelity program material with room to spare, and wideband FM is used to fill the band. Suppose we take the modulating frequency f_m to be 15 kHz (typically the maximum audio frequency in FM transmissions). Use of Carson's rule then yields a bandwidth of $B \approx 2(\Delta f + f_m) = 180$ kHz, well within the required bandwidth. Our sinusoidal analysis indicates that $\beta = 5$ and the bandwidth occupied by significant sidebands is

$$2(8)(15 \text{ kHz}) = 240 \text{ kHz}$$

(see Fig. 6.9, 6.10, or 6.11). The discrepancy, of course, lies in the definition of bandwidth. However, we chose an extreme case as far as typical audio transmission is concerned, because we assumed that the 15 kHz tone was set at the maximum amplitude to produce a peak frequency deviation of 75 kHz. Typical program material does not contain as much at the higher frequencies. For lower audio frequencies the value of β increases and the bandwidth occupied by the significant sidebands approaches the wideband limit of $2\Delta f = 150$ kHz. (For audio signals with full maximum amplitude and frequencies below about 5 kHz all significant sidebands are within the 200 kHz bandwidth.) Note that if the amplitude weighting is uniform, it is the highest modulating frequency that governs the final bandwidth.

The transmission of one audio channel leaves room for additional program material within the bandwidth allocated. Stereo multiplexing and other auxiliary transmissions often occupy the higher frequency portions of the modulating spectrum. These were discussed in Chapter 5. To keep the bandwidth restricted, the maximum amplitude of these transmissions is reduced. The spectrum of a typical commercial transmission before the FM transmitter is shown in Fig. 5.12.

In the FM station, the left (L) and right (R) audio signals are derived from microphones, records, tapes, etc., and a preemphasis is applied to each channel (this is discussed later in this chapter). For stereo broadcasts, a pilot subcarrier at 19 kHz is permitted 10% of the total peak-frequency deviation (of 75 kHz). When there is a pause in program material (and no auxiliary transmissions), the modulation index is $\beta = (10\%)(75 \text{ kHz})/(19 \text{ kHz}) = 0.395$. This is approximately in the narrowband condition. Thus when there is a pause in program material, a stereo FM broadcast can be identified on a spectrum analyzer by a large carrier line plus two first-order sidebands each spaced 19 kHz from the carrier.

The Subsidiary Communications Authorization (SCA) system permits a commercial FM station to add another broadcasting channel in addition to the monophonic and stereo channels. The SCA transmissions carry no commercial messages and are intended for private subscribers who pay a fee for background music in stores, physicians' offices, etc. In contrast, the other FM transmissions are for general public use and are supported by commercial advertisements. The SCA channel uses narrowband FM. The subcarrier center frequency is usually set at 67 kHz, although this choice is not set by the FCC. A total peak-frequency deviation not exceeding 75 kHz is still required. For monophonic transmission only, this entire 75 kHz is available. If SCA is used with mono, the FCC limits the SCA portion to 30% of the maximum peak-frequency deviation, leaving 70% for the mono channel.

In stereo broadcasting with no SCA, 10% of the maximum peak deviation is used for the 19 kHz pilot subcarrier, leaving 90% to be divided between the (L + R) and (L − R) stereo channels. The average amplitudes of the L and R channels are normally kept equal. The maximum (L + R) amplitude is set to provide 90% modulation when the (L − R) amplitude is zero. Now if (L − R) is maximum, then (L + R) will go to zero, and if either L or R goes to zero, (L + R) and (L − R) will each take a maximum of 45% of the total peak frequency deviation. Thus there is a seesaw effect between the (L + R) and (L − R) channels such that the total does not exceed 90% of the peak frequency deviation capability allowed.

When used with stero multiplexing, the SCA channel is limited to 10% of the maximum peak frequency deviation. This leaves 80% for the stereo channels (i.e., 10% for SCA plus 10% for the 19 kHz pilot subcarrier). The system performance is the same as in the preceding paragraph except that now the stereo channels are allowed 80% of the maximum peak frequency deviation instead of 90%. Also, with the relatively low peak frequency deviation allowed, the SCA transmission does not have a very good signal-to-noise ratio and is used for only local coverage. Station muting (e.g., dropping the subcarrier to actuate an audio silencer circuit in the receiver) is often employed to silence the noise between records or tapes in the SCA transmissions.

Public FM broadcasting is an example of the use of DSB-SC and NBFM methods to frequency-multiplex several channels before using wideband FM for the final transmission. With the possibilities of transmitting compatible quadraphonic audio (e.g., see Drill Problem 5.1.5), the present status of the SCA transmission is an unresolved problem for the future.

Frequency modulation is also used for the audio in commercial television transmissions. The peak frequency deviation for this use is fixed at 25 kHz by the FCC. Assuming a maximum audio frequency of 15 kHz, use of Carson's rule gives a bandwidth of 80 kHz for the sound channel of a television receiver.

The relatively large bandwidth required for commercial FM, as compared with AM, is the penalty for obtaining substantial improvement in noise and interference rejection. This noise rejection increases with increasing Δf and, therefore, with increasing bandwidth. These topics will be discussed in a later section.

6.4 AVERAGE POWER IN ANGLE-MODULATED WAVEFORMS

For sinusoidal modulation, we can write [cf. Eq. (6.28)]

$$\phi_{FM}(t) = A \cos(\omega_c t + \beta \sin \omega_m t).$$

The mean-square value of this expression is

$$\overline{\phi_{FM}^2(t)} = A^2/2, \tag{6.42}$$

showing that the total average power in an FM waveform is a constant regardless of the modulation index. This is in contrast to AM where the total average power was proportional to the modulation index. This conclusion can be extended to any arbitrary band-limiting modulating waveform.

Equation (6.42) can be verified by writing $\phi_{FM}(t)$ in a series expansion [cf. Eq. (6.36)],

$$\phi_{FM}(t) = A \sum_{n=-\infty}^{\infty} J_n(\beta) \cos(\omega_c + n\omega_m)t.$$

As a result of the orthogonality of the cosine terms, the mean-square value of the sum is equal to the sum of the mean-square values and we get

$$\overline{\phi_{FM}^2(t)} = \frac{1}{2} A^2 \sum_{n=-\infty}^{\infty} J_n^2(\beta). \tag{6.43}$$

But from property (4) in Eq. (6.33),

$$\sum_{n=-\infty}^{\infty} J_n^2(\beta) = 1,$$

so that

$$\overline{\phi_{FM}^2(t)} = A^2/2.$$

The mean-square value of the unmodulated carrier is $A^2/2$. As the modulation index β is increased from zero and the sidebands are nonzero, the carrier component decreases. According to Eqs. (6.42) and (6.43), this takes place in such a manner as to always keep the total mean-square value constant. The mean-square value of each sideband is $\frac{1}{2} A^2 J_n^2(\beta)$ (recall also that sidebands occur in pairs). The mean-square value, of course, is identical to the average power if the resistance is one ohm and is related to the average power by a constant (i.e., the resistance) for all other cases so the conversion to units of power is straightforward.

It is possible to make any particular sideband, including the carrier, as small as desired by a proper choice of the modulation index β. From a table or graph of Bessel functions (e.g., Fig. 6.9), we see that the carrier term, $J_0(\beta)$, can be made zero for $\beta = 2.405, 5.52, \ldots$, and in these cases all of the average power is in the sidebands. These points are easy to read with a spectrum analyzer and serve as very convenient calibration points for β and Δf.

Example 6.4.1   A given FM transmitter is modulated with a single sinusoid. The output for no modulation is 100 watts into a 50-ohm resistive load. The peak frequency deviation of the transmitter is carefully increased from zero until the first sideband amplitude in the output is zero. Under these conditions determine (a) the average power at the carrier frequency; (b) the average power in all the remaining sidebands; and (c) the average power in the second-order sidebands.

Solution

a) Using Fig. 6.9 and Appendix I, we see that $J_1(\beta) = 0$ first occurs at $\beta \approx 3.8$ and that $J_0(3.8) \approx -0.40$. The average carrier power is then

$$P_c = J_0^2(3.8)(100 \text{ W}) = 16 \text{ W}.$$

b) The average power in the sum of the remaining sidebands is:

$$P_s = P_t - P_c = 100 \text{ W} - 16 \text{ W} = 84 \text{ W}.$$

c) $J_2(3.8) \approx 0.41$. The average power in the second-order sidebands is

$$2 J_2^2(3.8)(100 \text{ W}) = 34 \text{ W}.$$

Drill Problem 6.4.1   Determine the peak amplitude of (a) the total waveform and (b) the upper second-order sideband in Ex. 6.4.1.

Answer.   (a) 100 V; (b) 41V.

Drill Problem 6.4.2   Show that the rms value of Eq. (6.28) can be written as

$$\sqrt{\overline{\phi^2(t)}} = A \sqrt{\frac{J_0^2(\beta) + 2 \sum\limits_{n=1}^{\infty} J_n^2(\beta)}{2}}.$$

6.5   PHASE MODULATION

There is no basic difference between the mechanisms involved in the generation of phase modulation (PM) and frequency modulation (FM). In fact, the only difference is that the phase in the modulated waveform is proportional to the input signal amplitude in PM and to the integral of the input signal in FM. This introduces only a slight modification and we shall point that out here.

For an FM signal with the sinusoidal modulation $f(t) = a \cos \omega_m t$, the instantaneous frequency is

$$\omega_i(t) = \omega_c + ak_f \cos \omega_m t$$

$$= \omega_c + \Delta\omega \cos \omega_m t,$$

where $\Delta\omega$ is the peak frequency deviation (in radians per second) and k_f is the frequency-modulator constant (in radians per second per volt). The modulation index, $\beta = \Delta\omega/\omega_m$, is a dimensionless number and serves as a guide to the behavior of the carrier and sidebands.

For PM with the same modulating signal we have

$$\theta(t) = \omega_c t + ak_p \cos \omega_m t + \theta_0$$

$$= \omega_c t + \Delta\theta \cos \omega_m t + \theta_0,$$

where $\Delta\theta$ is the peak phase deviation (in radians) and k_p is the phase-modulator constant (in radians per volt). The instantaneous frequency is

$$\omega_i(t) = \frac{d\theta}{dt}$$

$$= \omega_c - ak_p\omega_m \sin \omega_m t$$

$$= \omega_c - \Delta\omega \sin \omega_m t.$$

Thus we see that the peak frequency deviation in PM is proportional not only to the amplitude of the modulating waveform but also to its frequency; that is,

$$\Delta\omega = \begin{cases} ak_f & \text{for FM} \\ ak_p\omega_m = (\Delta\theta)\omega_m & \text{for PM} \end{cases}. \tag{6.44}$$

This makes PM less desirable to transmit when $\Delta\omega$ is fixed (as in commercial FM). There are some advantages in the demodulation of PM, however, which make its use desirable. (These will become more evident later in this chapter.) The role of the modulation index β remains the same as in FM. Formally, then, we can compute $\Delta\omega = ak_p\omega_m = \Delta\theta\omega_m$ and then proceed as if the modulation were FM as far as bandwidth, sidebands, etc. are concerned. Note that the numerical value of β is the peak phase deviation, $\Delta\theta$, in the PM case.

Example 6.5.1   A carrier is phase modulated by a sinusoidal signal of 5 kHz and unit amplitude and the peak phase deviation is one radian. Calculate the bandwidth of the PM signal (a) using Carson's rule; and (b) using the definition of significant sidebands.

Solution

a) $\Delta f = (\Delta\theta)f_m = 5$ kHz and Carson's rule gives

$$B \approx 2(\Delta f + f_m) = 20 \text{ kHz}.$$

b) $\beta = \Delta\theta = 1$; using a Bessel function chart,

$$B \approx 2nf_m = 2(3)(5 \text{ kHz}) = 30 \text{ kHz}.$$

Drill Problem 6.5.1   Here we consider a PM system in which the phase may take on only two possible values—known as *phase-shift-keying* (PSK)—contrasted to the FSK system discussed in Ex. 6.3.3. Assume that a phase modulator is modulated by a periodic symmetric square wave of unit amplitude. Determine the required value of the peak phase deviation, $\Delta\theta(-\pi/2 \le \Delta\theta \le \pi/2)$, such that the average carrier is not present in the output.

Answer. $\pm 90°$.

6.6   GENERATION OF WIDEBAND FM SIGNALS

One method of generating wideband FM signals is to first produce a narrowband FM signal and then use frequency multiplication to increase the modulation index to the desired range of values. This is known as the indirect method of generating wideband FM signals. A second method—known as the direct method—is to vary the carrier frequency directly with the modulating signal. We shall now examine these two methods.

6.6.1   Indirect FM

We have seen (cf. Section 6.2) that the generation of narrowband PM is relatively easy and that narrowband FM can then be generated by first integrating the modulating signal. However, the modulation index obtainable by use of this method is restricted to very low values ($\beta < 0.2$ in theory; $\beta < 0.5$ in practice). To generate wideband FM, a method of increasing the modulation index must be used in this approach. The method used is that of the frequency multiplier.

A *frequency multiplier* is a nonlinear device designed to multiply the frequencies of the input signal by a given factor. For example, the input-output characteristic of an ideal square-law device is

$$e_o(t) = ae_i^2(t). \tag{6.45}$$

If the input signal is the FM signal,

$$e_i(t) = A \cos(\omega_c t + \beta \sin \omega_m t),$$

the output is

$$e_o(t) = aA^2\cos^2(\omega_c t + \beta \sin \omega_m t)$$
$$= (1/2)aA^2[1 + \cos(2\omega_c t + 2\beta \sin \omega_m t)]. \tag{6.46}$$

The first term in this result is simply a constant level and is easily removed with a filter. We conclude that both the carrier frequency and the modulation index have been doubled in this process. In a similar manner, use of an nth law device followed by a filter yields a carrier and a modulation index which have been increased by a factor of n.

In practice, very abrupt nonlinearities can be generated using special diodes (e.g., the varactor and the step-recovery diode) which yield many harmonic terms. It is possible to multiply by an order of magnitude or more in one step using these techniques. Limitations include the fact that losses incurred in the harmonic generation require additional amplification and small phase instabilities in the multiplication process accumulate and appear as noise in the output. With good design techniques, multiplication factors on the order of 10^3 are achievable with only a few degrees of phase noise.

Use of frequency multiplication increases the carrier of the FM waveform as well as the modulation index. This may result in very high carrier frequencies in order to achieve a given modulation index. To avoid this, frequency converters are often used to control the value of the carrier frequency. The frequency converter is essentially the same as discussed in connection with AM and translates the spectrum of a signal by a given amount but does not alter its spectral content. Block diagrams of the frequency multiplier and the frequency converter are shown in Fig. 6.14. Note carefully the differences between these two operations. In the frequency multiplier all spectral components of the input signal are multiplied by themselves (so that all cross-products are present), whereas in the frequency converter all spectral components of the input signal are multiplied by a sinusoid of a fixed frequency. The former operation spreads the spectral content (this can be verified using the frequency convolution property discussed in Chapter 3) and the latter translates the spectral content in frequency.

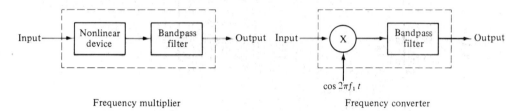

Fig. 6.14 Block diagrams of frequency multiplication and frequency conversion.

The method of obtaining a wideband FM waveform from a narrowband one using frequency multiplication is called the Armstrong indirect FM transmitter.[†] A block diagram of a typical Armstrong-type transmitter is shown in Fig. 6.15.

Example 6.6.1   A given angle-modulated signal has a peak frequency deviation of 20 Hz for an input sinusoid of unit amplitude and a frequency of 50 Hz. Determine the required frequency multiplication factor, n, to produce a peak frequency deviation of 20 kHz when the input sinusoid has unit amplitude and a frequency of 100 Hz, and the angle-modulation used is (a) FM; (b) PM.

Solution

a) $\Delta f_2 = 20$ kHz; $\Delta f_1 = 20$ Hz; $n = \Delta f_2/\Delta f_1 = 1000$

b) $\Delta f_2 = 20$ kHz; $\Delta f_1 = (100/50)(20$ Hz$) = 40$ Hz; $n = \Delta f_2/\Delta f_1 = 500$

---

† E. H. Armstrong was one of the first engineers to recognize the possible merits of FM broadcasting in the 1930's.

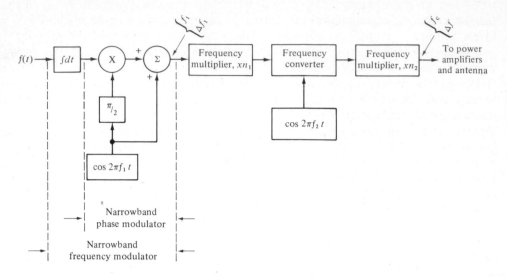

Fig. 6.15 Block diagram of an indirect (Armstrong) FM transmitter.

Drill Problem 6.6.1   Compute the carrier frequency f_c and the peak frequency deviation Δf of the output of the FM transmitter shown in Fig. 6.15 if $f_1 = 200$ kHz; $f_2 = 10.8$ MHz; $\Delta f_1 = 25$ Hz; $n_1 = 64$; $n_2 = 48$.

Answer.   96.0 MHz or 1132.8 MHz; 76.8 kHz.

6.6.2   Direct FM

In the direct method of generating FM the modulating signal directly controls the carrier frequency. An attempt is usually made to generate as wide a frequency deviation as possible and thus these systems often require less frequency multiplication, if any, than those using the indirect method. On the other hand, because the frequency is controlled by the modulating voltage, the long-term frequency stability is not as good as the crystal-stabilized oscillators generally used in the indirect method. Thus the direct FM methods usually employ some auxiliary methods for frequency stabilization.

A common method used for generating FM directly is to vary the inductance or capacitance of a tuned electronic oscillator. If L and C are the inductance and capacitance, respectively, of a simple tuned circuit, the frequency of oscillation is

$$\omega_0 = \frac{1}{\sqrt{LC}}.$$

If L or C is varied, the output frequency will also vary. For very small variations (that is, $\Delta\omega \ll \omega_c$, where ω_c is the carrier frequency), the square-root relationship can be approximated by a linear term and the conversion can be made quite linear.

There are various ways to make the capacitance or inductance of a tuned circuit dependent on the input signal. One common method at medium and high frequencies is to use a reverse-biased semiconductor diode as a voltage-variable capacitance. Although any semiconductor diode exhibits some capacitance change with change in reverse bias, the type of diode frequently used for this application is the varactor diode. The percentage frequency deviation which can be attained in this manner is quite small. To increase the percentage frequency deviation, the frequency modulation is performed at a high frequency and then heterodyned down to a lower frequency.

Other methods which are used successfully at high frequencies include the reflex klystron and the reactance-tube modulator. The latter consists of a pentode which is operated in such a manner as to produce a capacitance which is proportional to the grid voltage over a wide range. At lower frequencies the control of RC oscillators with FET's and similar devices has been used. Any oscillator whose frequency is controlled by the modulating-signal voltage is called a *voltage-controlled oscillator*, or VCO.

Example 6.6.2 A reverse-biased semiconductor diode can be used as a voltage-variable capacitance for frequency modulation. Assume that the capacitance of a given PN junction is given in terms of its reverse-bias voltage V by $C = C_0/\sqrt{1 + 2V}$. Such a diode is to be used as the capacitance in a parallel LC circuit tuned to a center frequency of 10 MHz when the reverse-bias voltage is 4 volts.

a) Determine the modulation constant k_f (i.e., the frequency-voltage slope near center frequency).

b) Determine the peak frequency deviation permissible for a maximum error of 1% from a linear frequency-voltage characteristic.

Solution

a) We can write the frequency f as

$$f = \frac{1}{2\pi\sqrt{LC}} = \frac{(1 + 2V)^{1/4}}{2\pi\sqrt{LC_0}}.$$

Letting $f = f_0$ when $V = V_0$ (i.e., at the operating point), we get

$$f = f_0 \frac{(1 + 2V)^{1/4}}{(1 + 2V_0)^{1/4}}.$$

Now let v be an incremental voltage about the operating point so that $V = V_0 + v$; also let $K = (1 + 2V_0)$ so that

$$f = f_0[1 + (2v/K)]^{1/4}.$$

Assuming that $v \ll K$, we can use the binomial expansion to obtain

$$f \approx f_0 \left[1 + \frac{1}{4} \left(\frac{2v}{K} \right) - \frac{3}{32} \left(\frac{2v}{K} \right)^2 + \cdots \right].$$

The modulation constant k_f is the slope of the linear frequency-voltage characteristic and is given by

$$k_f = \frac{1}{4} \left(\frac{2v}{K} \right) \frac{f_0}{v} = \frac{f_0}{2K} = \frac{f_0}{2(1 + 2V_0)}.$$

For the given operating point, $f_0 = 10$ MHz and $V_0 = 4$ volts so that

$$k_f = 0.56 \text{ MHz/V}.$$

b) Most of the error will arise from the second-order term in the series expansion so that we require

$$\frac{\frac{3}{32} \left(\frac{2v}{K} \right)^2}{\frac{1}{4} \left(\frac{2v}{K} \right)} \leq 0.01$$

which gives

$$v \leq \frac{4K}{300}.$$

The peak frequency deviation is then

$$\Delta f = k_f v_{max} = \left(\frac{f_0}{2K} \right) \left(\frac{4K}{300} \right) = \frac{f_0}{150}$$

$$= 66.7 \text{ kHz}.$$

In some cases it is not necessary that the output voltage be sinusoidal, or the output may be wave-shaped by nonlinear shaping circuits or by filtering. In these cases it is possible—and quite attractive—to generate wideband FM digitally. Perhaps the simplest way is to control the oscillation frequency of a relaxation oscillator or a multivibrator with the modulating-signal voltage. A more accurate and stable method is to generate the zero crossings of a PM waveform using digital techniques. Basically this involves sampling the input waveform and then using a precision ramp generator that resets at each sample point, followed by a voltage-variable threshold. The point in time at which the threshold is exceeded is used to generate a short pulse to signify a zero crossing. Applying this sequence of pulses to a bandpass filter results in a wideband PM signal. This method is stable and accurate and is capable of generating signals

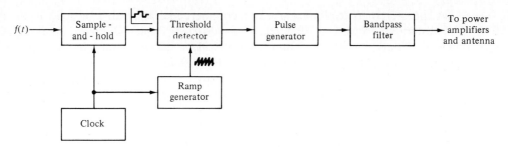

Fig. 6.16 Digital generation of wideband PM.

of very wide bandwidth. A simplified block diagram is shown in Fig. 6.16. These operations are discussed in more detail in Chapter 7.

6.6.3   FM Multiplexing

It is a common practice in data transmission to combine several channels of modulated signals using frequency multiplexing methods and then modulate a high-frequency carrier with the composite multiplexed signal. To do this, the individual data signals each modulate an assigned subcarrier. These subcarriers are arranged so that the channels occupy adjacent frequency bands with some frequency space between them, known as *guard bands*. The modulated subcarriers are used to angle-modulate a high-frequency carrier, as shown in Fig. 6.17.

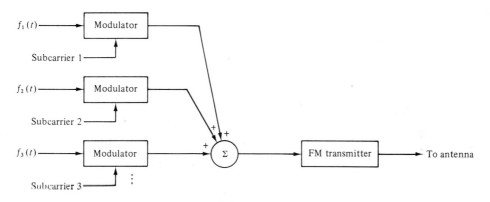

Fig. 6.17 A simplified composite modulation system.

If FM is used for the subcarrier modulation and for the main carrier modulation, the composite modulation is referred to as FM-FM; if AM is used, then it is referred to as AM-FM. The amplitude-modulation methods used for the subcarrier modulation are DSB-SC or SSB-SC. Large-carrier methods are avoided because too much of the peak frequency deviation would be used merely to send the AM carrier. Usually a pilot subcarrier is also sent for demodulation.

Note that the stereo multiplexing used in commercial FM is an example of an AM-FM system.†

Angle modulation (both FM and PM) is widely used in multichannel data transmission and telemetry systems. Standards exist for the assignment of subcarriers and guard bands for the latter.‡ To allow for realizable filter designs to separate adjacent channels, it is common to allow some frequency separation between channels.

6.7   DEMODULATION OF FM SIGNALS

There are a number of ways to recover the modulating signal from the FM waveform and we shall discuss only some of them. The overall characteristic must be the same, however—to provide an output signal whose amplitude is linearly proportional to the instantaneous frequency of the input waveform.

6.7.1   Direct Method

One method is to use some system which has a linear frequency-to-voltage transfer characteristic. Such a system is called a frequency *discriminator*. In our search for a simple discriminator, we need something with a linear amplitude vs. frequency characteristic. The simplest conceptually is that of the ideal differentiator, for we recall that its transfer function is given by $H(\omega) = j\omega$. (Certainly the magnitude characteristic is very linear!)

An expression for the general FM waveform is

$$\phi_{\text{FM}}(t) = A \cos \left[\omega_c t + k_f \int_0^t f(\tau) \, d\tau \right].$$

Assuming that A is a constant (a limiter can be inserted prior to the differentiator to ensure this), we have

$$\frac{d\phi}{dt} = -A[\omega_c + k_f f(t)] \sin \left[\omega_c t + k_f \int_0^t f(\tau) \, d\tau \right]. \tag{6.47}$$

If $k_f f(t) \ll \omega_c$, we see that Eq. (6.47) is in the form of an AM signal whose envelope is

$$A\omega_c \left[1 + \frac{k_f}{\omega_c} f(t) \right] \tag{6.48}$$

and whose carrier frequency is

$$\omega_c + k_f f(t). \tag{6.49}$$

---

† Applications of both amplitude modulation and angle modulation to proposed stereo AM transmissions are discussed in Appendix H.

‡ See, for example, E. L. Gruenberg (ed.), *Handbook of Telemetry and Remote Control*, New York: McGraw-Hill, 1967.

The differentiator has therefore changed FM into AM with only the slight difference that the new carrier frequency has some frequency variation. The resulting AM signal can be detected by an envelope detector and as long as $k_f f(t) \ll \omega_c$, the slight variation in the carrier frequency would not be detectable by the envelope detector. Figure 6.18(a) illustrates this type of discriminator.

The action of the ideal differentiator can be approximated by any device whose magnitude transfer function is reasonably linear within the range of frequencies of interest. In Fig. 6.18(b) an RL circuit approximation to a differentiator is used followed by an envelope detector. A bandpass version of this circuit is shown in Fig. 6.18(c). These discriminators are known as *slope detectors*. Although the slope detector is economical, it has a very limited linear range and its use is restricted to input signals with small frequency variations.

A more linear response can be obtained by taking the difference between two bandpass magnitude responses. The triple-tuned balanced discriminator shown in Fig. 6.18(d) uses this principle and has better sensitivity and linearity.

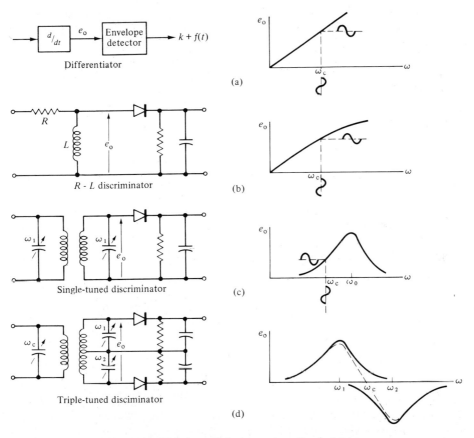

Fig. 6.18 FM demodulation using discriminators.

It also provides a zero output response at the carrier frequency which is an advantage in recording equipment design.

Another possibility is to use a time-delay/phase-shift approximation to the differentiator. It avoids the multiple tuning problems while retaining high sensitivity and good linearity. The most straightforward approach is the time-delay (phase-shift) demodulator shown in block diagram form in Fig. 6.19(a). The time-delay demodulator finds applications in demodulating FM microwave signals.

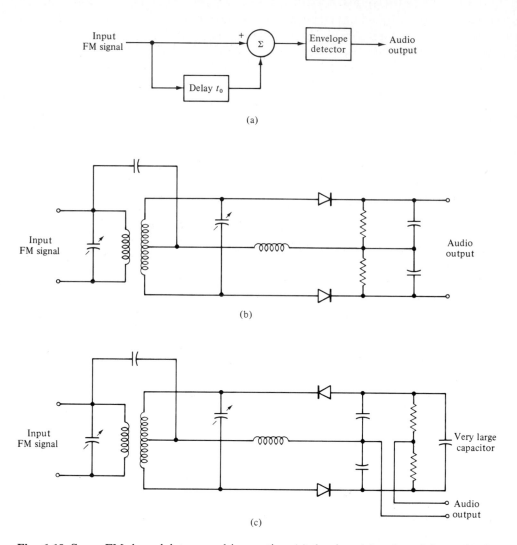

Fig. 6.19 Some FM demodulators used in practice: (a) the time-delay demodulator; (b) the Foster–Seeley discriminator; (c) the ratio detector.

At frequencies below the microwave region, one can take advantage of the nearly linear phase-shift characteristic of a tuned circuit near resonance. One circuit is the Foster–Seeley discriminator circuit shown in Fig. 6.19(b).† In the Foster–Seeley discriminator both tuned circuits are tuned to the carrier frequency and the output voltage varies with the frequency deviation as a result of the phase shifts in the secondary circuit. Circuit operation can best be traced using phasor diagrams; we shall not pursue this here.‡ The operation of the Foster–Seeley discriminator is very linear and produces a balanced output. It is amplitude-sensitive and requires prior amplitude limiting.

A more commonly used circuit in commercial receivers that is similar and yet requires no prior limiting is the ratio detector shown in Fig. 6.19(c). In contrast to the Foster–Seeley circuit, the diodes in the ratio detector are connected in series with the tuned secondary circuit and the output is taken across a bridge circuit. A very large capacitor placed across the opposite diagonal of the bridge keeps the total voltage relatively constant, suppressing the effects of amplitude variations. The ratio detector is used in many entertainment-type FM receivers.

Because the information about the modulating signal is contained in the zero crossings of the FM waveform, another approach is to severely clip (limit) the amplitude of the FM waveform and then detect the variation in the zero crossings of the resulting square wave. The spacing of the zero crossings can be measured by counting the number of zero crossings in a given time interval or by measuring the time interval for a given number of zero crossings.

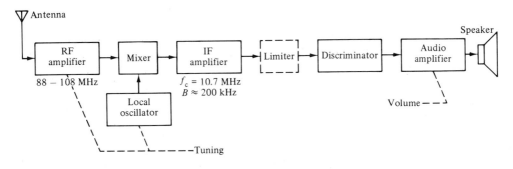

Fig. 6.20 A typical FM broadcast receiver.

---

† D. F. Foster and S. H. Seeley, *Proceedings of the IRE*, **25** (1937): 289.

‡ See, for example, J. J. DeFrance, *Communications Electronics Circuits*, Second Ed., San Francisco: Rinehart Press, 1972, p. 337; or A. B. Cook and A. A. Liff, *Frequency Modulation Receivers*, Englewood Cliffs, NJ: Prentice-Hall, 1968, Ch. 10.

A block diagram of a typical receiver for the commercial FM broadcast band covering 88–108 MHz is shown in Fig. 6.20. The receiver is similar to the AM superheterodyne receiver with the exception of the addition of a discriminator and possibly a limiter. The common choice of an intermediate frequency is 10.7 MHz.

★ **Example 6.7.1**   Suppose one attempts to convert a standard AM broadcast receiver to receive FM transmissions by only changing the frequency of the local oscillator. By tuning slightly to one side of the (nonideal) IF response, it is reasoned that the IF response and the envelope detector will then act as a slope detector. Choosing a simple parallel *RLC* network to model the IF response, estimate the performance of this receiver.

Solution.   For a parallel *RLC* network (see Fig. 6.21), we have

$$|H(\omega)| = \left| \frac{V(\omega)}{I(\omega)} \right| = \frac{1/C}{\sqrt{4\alpha^2 + [(\omega_0^2 - \omega^2)/\omega]^2}}$$

where

$$\omega_0 = 1/\sqrt{LC} \quad \text{and} \quad \alpha = 1/(2RC).$$

The -3-dB bandwidth is 2α; if $2\alpha \ll \omega_0$ (high-Q case), we can simplify things a little by translating to an equivalent low-pass case (that is, $\omega_0 = 0$) so that

$$|H_1(\omega)| = \frac{1/C}{\sqrt{4\alpha^2 + \omega^2}} = \frac{R}{\sqrt{1 + (\omega/2\alpha)^2}}.$$

A reasonable choice of operating point for maximum linear operating range is at an inflection point (see Fig. 6.21); i.e., the point where

$$\frac{d^2}{d\omega^2}|H_1(\omega)| = 0.$$

Solutions to this are at $\omega = \pm\sqrt{2}\alpha$.

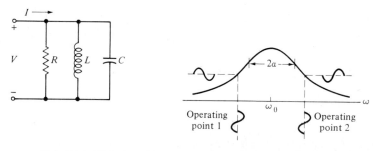

Fig. 6.21 FM slope demodulation using a tuned circuit.

Next we expand the tuning characteristic about the operating point. Taking the upper choice of operating points, let $\Delta\omega = \omega - \sqrt{2}\alpha$ and

$$|H_1(\omega)| = \frac{R}{\sqrt{1 + [(\Delta\omega + \sqrt{2}\alpha)/2\alpha]^2}} = \frac{R\sqrt{2}/\sqrt{3}}{\sqrt{1 + \frac{2}{3}[(\Delta\omega/\sqrt{2}\alpha) + (\Delta\omega/2\alpha)^2]}}$$

If we assume that $\Delta\omega \ll \alpha$, this can be approximated using the binomial expansion so that

$$|H_1(\omega)| \approx \sqrt{2/3}R[1 - 1/(3\sqrt{2}\alpha)\Delta\omega], \qquad \Delta\omega \ll \alpha.$$

Thus, within these assumptions, we have verified that $|H(\omega)|$ does have a linear slope component. Continuing our example, the bandwidth of this circuit for AM is on the order of 10 kHz so that $\alpha/2\pi = 5$ kHz and $\sqrt{2}\alpha/2\pi = 7$ kHz. Therefore $\Delta\omega$ should not exceed 0.7 kHz for our linearity assumption to hold and we conclude that although this scheme will work it is limited to very narrowband inputs. This approach is actually used in inexpensive converters to allow one to use an AM receiver to listen to police calls transmitted via narrowband FM.

6.7.2 Indirect Method—the Phase-Locked Loop

All of the preceding methods have some similarities and the choice of which one to use is governed by the required linearity, zero balance, ease of alignment, amplitude sensitivity, reliability, and economic factors. A different approach is to place a frequency modulator in the return branch of a feedback system. A feedback system with sufficient loop gain performs in its forward branch the inverse operation of that which is performed in its return branch and thus demodulates the signal. Among the detectors in this group are the FM demodulator with feedback (FMFB) and the phase-locked loop (PLL). Block diagrams of the FMFB and PLL demodulation systems are shown in Fig. 6.22. We shall center our discussion on the PLL. Its ease of alignment and efficient operation in the presence of noise makes the PLL increasingly more popular. The reader is referred to more advanced texts for details on the operation of the FMFB system.

A block diagram of the PLL is shown in Fig. 6.22(b). Both inputs to the phase comparator are assumed to be periodic and to have the same fundamental frequency. The phase comparator detects the timing difference between the two signals and produces an output voltage which is proportional to this difference. For sinusoidal inputs, this timing difference can also be expressed as a phase difference. One way to construct a phase comparator is to use a multiplier and low-pass filter, as shown in Fig. 5.13. The purpose of the low-pass filter in the phase comparator of Fig. 5.13 is to attenuate the second-harmonic frequency components which arise in the multiplication. We have added a loop filter in Fig. 6.22(b) whose function will be explained shortly.

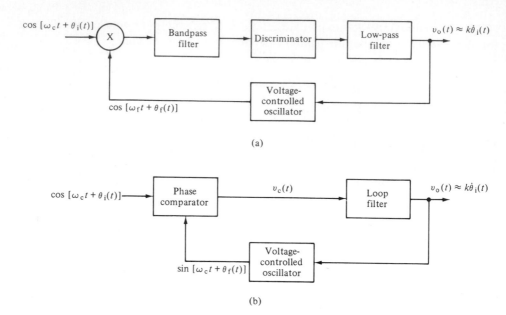

(a)

(b)

Fig. 6.22 Block diagram of (a) the FM demodulator with feedback (FMFB) and (b) the phase-locked loop (PLL).

In general, the output of the phase comparator is proportional to the average value of the product of the two inputs as a function of their relative time displacement. If the two inputs are $x(t)$, $y(t)$, this can be expressed as

$$\frac{1}{T} \int_0^T x(t)y(t + \tau)\, dt = R_{xy}(\tau) \tag{6.50}$$

where T is the period of the input waveform. Therefore the phase-comparator output is proportional to the crosscorrelation between the input waveforms. If both inputs are sinusoids, that is,

$$x(t) = A_1 \cos \omega_c t, \tag{6.51}$$

$$y(t) = A_2 \cos \omega_c t, \tag{6.52}$$

then Eq. (6.50) becomes (cf. Example 4.4.2)

$$R_{xy}(\tau) = \tfrac{1}{2}A_1 A_2 \cos \omega_c \tau. \tag{6.53}$$

Because the inputs are sinusoidal, this can also be expressed in terms of the phase $\theta = \omega_c \tau$:

$$R_{xy}(\theta) = \tfrac{1}{2}A_1 A_2 \cos \theta. \tag{6.54}$$

This is shown in Fig. 6.23(a).

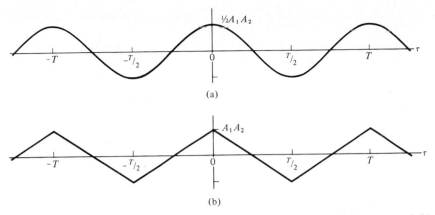

Fig. 6.23 Phase comparator characteristic for (a) two sinusoidal inputs and (b) two square-wave inputs.

We could also use square waves for both input signal waveforms. For example, the input could be a sinusoid which is amplitude-limited and the VCO could be designed for a square-wave output. In this case, the phase-comparator characteristic is a triangular characteristic (cf. Example 4.4.1), as shown in Fig. 6.23(b) (zero average value with peak values A_1 and A_2 assumed). Other cases (e.g., a sinusoid and a square wave) can be worked out using Eq. (6.50).

From the phase-comparator characteristics shown in Fig. 6.23, we observe that the output is zero at odd multiples of $\pm T/4$; these points are called *nulls*. The nulls form the desired operating points for our purposes and the remainder of the PLL is designed to keep the timing difference between input waveforms as near an operating point as possible. Note that the phase-comparator output goes through a null with a positive slope for $\tau = -T/4 \pm T$ ($\theta = -\pi/2 \pm 2\pi$ for sinusoids) and with a negative slope for $\tau = T/4 \pm T$ ($\theta = \pi/2 \pm 2\pi$ for sinusoids).

Now suppose that the loop is operating at the null point $\tau = T/4 \pm T$ and that the input is delayed by an increment $\Delta\tau$. This results in a small negative voltage out of the phase comparator (cf. Fig. 6.23). If the VCO frequency increases with input voltage and the gain in the feedback loop is noninverting, the loop will produce an incremental delay of opposite sign to the input which will tend to keep the loop near the desired operating point. On the other hand, if the feedback gain is inverting, the loop will produce an incremental delay which will add to $\Delta\tau$ and will drive the loop away from the operating point. Thus this null is a stable operating point for a noninverting gain and an unstable operating point for an inverting gain. In contrast, the slope is positive at the null $\tau = -T/4 \pm T$, and thus this point is a stable operating point for an inverting loop gain and an unstable operating point for a noninverting loop gain. Note that in either case the relationships are such that the timing difference between the two wave-forms is $\pm T/4$ (that is, $\pm \pi/2$ for sinusoids).

Both phase-comparator characteristics shown in Fig. 6.23 have linear slopes near the nulls. The square-wave characteristic has some advantage in that the slope is a constant over one-half a period whereas the slope decreases away from the null points for the sinusoidal case. This latter effect results in a nonlinear loop-gain factor.

Although the phase-comparator characteristic depends on the timing differ-ence between inputs, it is convenient to use a phase designation such that $\theta_i(t)$ is the input phase and $\theta_f(t)$ is the feedback phase. The timing difference is $[\theta_i(t) - \theta_f(t)]$ and the phase-comparator output voltage $v_c(t)$ about an operating point is then

$$v_c(t) = k_c[\theta_i(t) - \theta_f(t)].\qquad(6.55)$$

The proportionality constant k_c in Eq. (6.55) is the phase-comparator gain factor and has the dimension of volt per radian. From Fig. 6.23 we observe that k_c is proportional to the product of the amplitudes of the two input waveforms.

For the sinusoidal case, the two inputs must be 90° out of phase at a null. Using a multiplier and low-pass filter, the phase-comparator output is then

$$v_c(t) \propto A_1 \cos [\omega_c t + \theta_i(t)]A_2 \sin [\omega_c t + \theta_f(t)]_{\text{LP}}$$

$$\propto \tfrac{1}{2}A_1 A_2 \sin [\theta_i(t) - \theta_f(t)],$$

or

$$v_c(t) = k_c \sin [\theta_i(t) - \theta_f(t)].\qquad(6.56)$$

Equation (6.55) follows from Eq. (6.56) if $[\theta_i(t) - \theta_f(t)]$ is small. Of course, Eq. (6.55) must be restricted to an operating range $(-T/4, T/4)$ even for the square-wave case. The use of Eq. (6.55) in the analysis of the PLL is referred to as a "linearized" PLL.

The loop filter controls the dynamic response of the PLL. It has been sep-arated from the low-pass filter of the phase comparator to emphasize the different roles of the two filters. The purpose of the low-pass filter in the phase comparator is to suppress second-harmonic terms while the loop filter controls the dynamic response of the loop. Designating the impulse response of the loop filter by $h(t)$, the output voltage $v_o(t)$ is

$$v_o(t) = v_c(t) \circledast h(t).\qquad(6.57)$$

The voltage-controlled oscillator (VCO) generates a constant-amplitude pe-riodic waveform (e.g., a sinusoid, square-wave, etc.) whose fundamental fre-quency is proportional to the input voltage. Letting ω_f be the instantaneous frequency with respect to the fundamental frequency, we have

$$\omega_f(t) = k_f v_o(t),\qquad(6.58)$$

where the voltage $v_o(t)$ is the output of the loop filter (cf. Fig. 6.23b). The constant k_f has the dimensions of rad/sec per volt [cf. Eq. (6.5)]. The phase of the VCO is then, within an arbitrary constant,

$$\theta_f(t) = k_f \int_0^t v_o(\tau) \, d\tau. \tag{6.59}$$

Equations (6.55)[(6.56) for the nonlinear case], (6.57), and (6.59) describe the dynamic response of the PLL. The controlling parameters are the loop-gain factors k_c, k_f and the parameters of the loop filter.

Taking the derivative of Eq. (6.59), we get

$$v_o(t) = \frac{1}{k_f} \frac{d\theta_f}{dt}. \tag{6.60}$$

If the loop gain is high, $[\theta_i(t) - \theta_f(t)]$ is small, $\theta_f(t) \approx \theta_i(t)$ and Eq. (6.60) states that the output voltage is proportional to the instantaneous frequency of the input referred to the carrier. The PLL therefore accomplishes the desired demodulation of the input FM signal.

We have assumed that the input and the VCO are at the same frequency. Initially, this may not be true and the loop must go through a mode of acquisition to gain lock. If the frequency difference between the input and the VCO is less than the loop bandwidth (i.e., the closed-loop bandwidth of the PLL; this will be discussed in Section 6.7.3) the PLL will lock quickly. This is the usual case for FM detection.

If the range of expected frequency differences is much larger than the loop bandwidth, the VCO frequency can be swept at a suitable rate to search for the signal. In the absence of noise, this has similarities to the scan rate of a scanning spectrum analyzer for which (cf. Problem 3.13.3) the usual restriction is that

$$\dot{f} < B^2, \tag{6.61}$$

where $\dot{f}$ is the frequency scan rate and B is the bandwidth. Equation (6.61) is often used as an upper bound for the sweep rate of the PLL, but it must be lowered considerably in the presence of noise.

The frequency sweep can be obtained by applying a ramp voltage to the input of the VCO. The ramp is reset when a preset voltage level is attained and the sweep is then reinitiated. If the frequency sweep is slow enough, the loop will lock when the input and VCO frequencies coincide. Although the sweep voltage can often be left on and overcome by the locked loop, it is advantageous to have a lock indicator which automatically disengages the sweep when lock is attained.

The lock indicator widely used is the quadrature phase comparator shown in Fig. 6.24. In the unlocked condition, the outputs of the phase comparators have average values near zero. In the locked condition, the main phase comparator has an output proportional to sin $(\theta_i - \theta_f)$ (for sinusoidal inputs) while the quadrature output is proportional to cos $(\theta_i - \theta_f)$. Since $(\theta_i - \theta_f)$ is small in the locked condition, cos $(\theta_i - \theta_f) \approx 1$. The locked condition can be sensed by a threshold circuit which then disconnects the frequency sweep and connects the PLL for normal operation. Without the sweep acquisition circuitry, the lock indicator may be used to signal the lock condition (e.g., the stereo indicator in stereo FM detection).

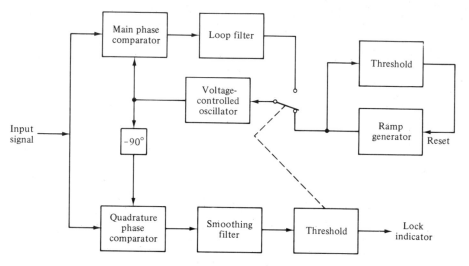

Fig. 6.24 PLL sweep acquisition.

Example 6.7.2 (a) Derive an integro-differential equation for $\theta_f(t)$ in terms of $\theta_i(t)$ for sinusoidal PLL inputs. (b) Derive a differential equation for the case of no loop filter and small phase errors.

Solution

a) Differentiating Eq. (6.59) and substituting Eqs. (6.56) and (6.57), we get

$$\frac{d\theta_f}{dt} = k_c k_f \sin [\theta_i(t) - \theta_f(t)] \circledast h(t).$$

b) For $h(t) = 1$ and sin $[\theta_i(t) - \theta_f(t)] \approx [\theta_i(t) - \theta_f(t)]$, we get the first-order linear differential equation,

$$\frac{d\theta_f}{dt} + k_c k_f \theta_f(t) = k_c k_f \theta_i(t).$$

Thus the linearized PLL for no loop filter is a first-order system with a time constant of $(k_c k_f)^{-1}$.

Drill Problem 6.7.1 (a) Rewrite the linear differential equation in Example 6.7.2(b) in terms of frequency. (b) Find ω_f for $t > 0$ if $\omega_i = u(t)$.

Answer.   (a) $d\omega_f/dt + k_c k_f \omega_f(t) = k_c k_f \omega_i(t)$; (b) $1 - \exp(-k_c k_f t)$.

★ 6.7.3   The Linearized Phase-Locked Loop†

To proceed farther into an analysis of the PLL, we linearize the phase-comparator characteristic and apply the Laplace transform to Eqs. (6.55), (6.57), and (6.59) to obtain, respectively,

$$V_c(s) = k_c[\theta_i(s) - \theta_f(s)], \tag{6.62}$$

$$V_o(s) = V_c(s)H(s), \tag{6.63}$$

$$\theta_f(s) = \frac{k_f}{s} V_o(s). \tag{6.64}$$

Combining Eqs. (6.62)–(6.64), we obtain the linearized PLL model shown in Fig. 6.25. Thus we can write

$$\theta_f(s) = k_c[\theta_i(s) - \theta_f(s)]H(s)k_f/s, \tag{6.65}$$

or

$$\frac{\theta_f(s)}{\theta_i(s)} = \frac{k_c k_f H(s)}{s + k_c k_f H(s)}. \tag{6.66}$$

Equation (6.66) is the basic result for the linearized PLL.

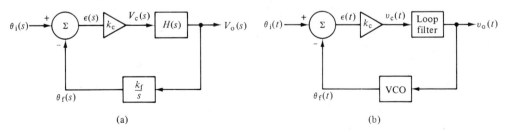

(a)   (b)

Fig. 6.25 The linearized PLL model in (a) frequency and (b) time.

Two quantities of interest to us are the loop phase error $\epsilon(t)$ and the input–output transfer function for phase and frequency demodulation. The former

---

† This section requires some knowledge of the Laplace transform.

can be found from Eq. (6.66) as follows (cf. Fig. 6.25a):

$$\frac{\epsilon(s)}{\theta_i(s)} = \frac{\theta_i(s) - \theta_f(s)}{\theta_i(s)} = 1 - \frac{\theta_f(s)}{\theta_i(s)}$$

$$= \frac{s}{s + k_c k_f H(s)} . \tag{6.67}$$

The latter can be found from (cf. Fig. 6.25a)

$$V_o(s) = k_c \epsilon(s) H(s),$$

or, using Eq. (6.67),

$$\frac{V_o(s)}{\theta_i(s)} = \frac{k_c s H(s)}{s + k_c k_f H(s)} . \tag{6.68}$$

We have thus expressed the desired output voltage $V_o(s)$ and the loop phase error $\epsilon(s)$ in terms of the input phase $\theta_i(s)$ and the loop filter $H(s)$. To proceed farther requires some choice of the loop-filter transfer function $H(s)$. We examine two cases.

The First-Order PLL

The simplest case is to let

$$H(s) = 1 \tag{6.69}$$

so that Eq. (6.68) becomes

$$\frac{V_o(s)}{\theta_i(s)} = \frac{k_c s}{s + k_c k_f} . \tag{6.70}$$

Because we are interested in the frequency-demodulation properties of the PLL, we apply a unit impulse in *frequency*. Then we have

$$\theta_i(s) = 1/s, \tag{6.71}$$

and Eq. (6.70) gives

$$V_o(s) = \frac{k_c}{s + k_c k_f} . \tag{6.72}$$

Because the PLL behaves like a first-order system, it is called a *first-order loop*. The -3-dB frequency is

$$\omega_{-3\,\text{dB}} = k_c k_f, \tag{6.73}$$

and therefore, to avoid attenuation of high-frequency content, we require that the modulation bandwidth W satisfy the restriction

$$W < k_c k_f. \tag{6.74}$$

For a step change, ω, in the input frequency, we have

$$\theta_i(s) = \omega/s^2, \tag{6.75}$$

and Eq. (6.70) gives

$$V_o(s) = \frac{k_c\omega}{s(s + k_ck_f)}, \tag{6.76}$$

or

$$v_o(t) = \frac{\omega}{k_f}[1 - e^{-k_ck_ft}]u(t). \tag{6.77}$$

The output voltage therefore attempts to follow the frequency step and, after the transients go to zero, it produces the desired voltage [cf. Eq. (6.60)]

$$v_o(\infty) = \frac{\omega}{k_f}. \tag{6.78}$$

The loop phase error for this step in frequency is found from Eqs. (6.67), (6.69), and (6.75):

$$\epsilon(s) = \frac{\omega}{s(s + k_ck_f)}, \tag{6.79}$$

or,

$$\epsilon(t) = \frac{\omega}{k_ck_f}[1 - e^{-k_ck_ft}]u(t). \tag{6.80}$$

The steady-state phase error is then

$$\epsilon(\infty) = \frac{\omega}{k_ck_f}. \tag{6.81}$$

This assumes, of course, that $|\epsilon(t)| < \pi/2$ for the square-wave inputs or that $|\epsilon(t)| \ll \pi/2$ for the sinusoidal inputs to remain within the linear range.

The product k_ck_f is the loop gain. A high loop gain is desirable for small phase errors [see Eq. (6.81)]. However, this product is also the loop bandwidth [cf. Eq. (6.73)] so that a high loop gain implies a large bandwidth. As a result of these restrictions, the first-order PLL is seldom used.

The Second-Order PLL

The loop-filter transfer function often used (or approximated) in the second-order PLL is†

$$H(s) = \frac{s\tau_2 + 1}{s\tau_1}, \tag{6.82}$$

---

† This is known as a *lead-lag* filter in the theory of feedback systems.

where τ_1, τ_2 are adjustable parameters. Using Eq. (6.82) in Eq. (6.68), we get

$$\frac{V_o(s)}{\theta_i(s)} = \frac{k_c s(s\tau_2 + 1)}{\tau_1(s^2 + 2\zeta\omega_n s + \omega_n^2)} \tag{6.83}$$

where the *natural frequency* ω_n and the *damping factor* ζ of the loop are parameters normally used in the theory of second-order systems and are given here by

$$\omega_n = \sqrt{k_c k_f / \tau_1}, \tag{6.84}$$
$$\zeta = \omega_n \tau_2 / 2. \tag{6.85}$$

For a unit impulse in frequency, we use Eq. (6.71) in Eq. (6.83):

$$V_o(s) = \frac{k_c(s\tau_2 + 1)}{\tau_1(s^2 + 2\zeta\omega_n s + \omega_n^2)}, \tag{6.86}$$

which exhibits the second-order behavior of the PLL. Usually the PLL is operated with slightly less damping than "critical damping" ($\zeta = 1$) to obtain a fast response. Therefore we are primarily interested in the range $\frac{1}{2} < \zeta < 1$($\zeta = 1/\sqrt{2} = 0.707$ is a popular choice). Within this range, the -3-dB bandwidth of Eq. (6.86) is approximately

$$\omega_{-3\text{ dB}} \approx (1 + \sqrt{2}\zeta)\omega_n \qquad \text{for} \qquad \tfrac{1}{2} < \zeta < 1. \tag{6.87}$$

A criterion often used for the modulating bandwidth W is the slightly more pessimistic choice,

$$W < 2\omega_n\zeta. \tag{6.88}$$

For a step change in frequency, Eqs. (6.75) and (6.83) give

$$V_o(s) = \frac{k_c\omega}{\tau_1} \frac{s\tau_2 + 1}{s(s^2 + 2\zeta\omega_n s + \omega_n^2)}, \tag{6.89}$$

whose time solution, for $\zeta^2 < 1$, is

$$v_o(t) = \frac{\omega}{k_f}\left[1 - \frac{1}{\sqrt{1 - \zeta^2}} e^{-\zeta\omega_n t} \cos(\omega_n\sqrt{1 - \zeta^2}t + \phi)\right]u(t), \tag{6.90}$$

where $\phi = \tan^{-1}(\zeta/\sqrt{1 - \zeta^2})$. After all transients have disappeared, Eq. (6.90) gives[†]

$$v_o(\infty) = \omega/k_f. \tag{6.91}$$

The loop phase error is found using Eqs. (6.67) and (6.82):

$$\frac{\epsilon(s)}{\theta_i(s)} = \frac{s^2}{s^2 + 2\zeta\omega_n s + \omega_n^2}. \tag{6.92}$$

[†] The final-value theorem of the Laplace transform may be used with Eqs. (6.89) and (6.84), and yields this result for general ζ.

As in the case of the first-order loop, this exhibits a high-pass characteristic and shows that the loop cannot keep up with rapid phase changes. Applying a frequency step, we use Eq. (6.75) in Eq. (6.92) to give

$$\epsilon(s) = \frac{\omega}{s^2 + 2\zeta\omega_n s + \omega_n^2},$$ (6.93)

whose time solution, for $\zeta^2 < 1$, is

$$\epsilon(t) = \frac{\omega}{\omega_n\sqrt{1 - \zeta^2}} e^{-\zeta\omega_n t} \sin(\omega_n\sqrt{1 - \zeta^2}\, t)u(t).$$ (6.94)

After all transients have disappeared, Eq. (6.94) gives

$$\epsilon(\infty) = 0.$$ (6.95)

In contrast to the first-order loop, the second-order PLL tracks out the static phase error resulting from a step change in frequency and returns the phase-comparator characteristic to the null point. This, together with the added versatility in setting the loop dynamic response, is why most PLL systems are operated as second-order systems.

The peak phase error is found by setting $d\epsilon/dt = 0$ and solving to give

$$\epsilon_p = \frac{\omega}{\zeta\omega_n\chi} \exp\left[-\frac{\tan^{-1}\chi}{\chi}\right] \sin(\tan^{-1}\chi),$$ (6.96)

where

$$\chi = \sqrt{1 - \zeta^2}/\zeta.$$

For a given value of ζ, Eq. (6.96) sets the maximum phase error for a step-frequency change.

Now let us investigate the sinusoidal case. Assume that the frequency modulation is sinusoidal with frequency ω_m and a peak frequency deviation $\Delta\omega$. The peak phase deviation is $\Delta\omega/\omega_m = \beta$ and from Eq. (6.92) the magnitude of the phase error is

$$|\epsilon(\omega)| = \frac{\beta\omega^2}{\sqrt{(\omega_n^2 - \omega^2)^2 + 4\zeta^2\omega_n^2\omega^2}} \quad \text{(sinusoidal case)}.$$ (6.97)

We let $\omega = \omega_m$ and keep $\Delta\omega$ fixed. Setting $d\epsilon/d\omega_m = 0$ and solving, we find that the maximum error occurs for $\omega_m = \omega_n$. Using this value in Eq. (6.97), we get

$$\epsilon_p = \left(\frac{\Delta\omega}{\omega_n}\right)\frac{1}{2\zeta} \quad \text{(sinusoidal case)}.$$ (6.98)

For no distortion in demodulating FM, this peak phase must be kept within the linear range of the phase-comparator characteristic.

The second-order PLL has the two parameters ω_n, ζ; these can be controlled by the choices of τ_2 and $k_c k_f / \tau_1$. The second-order PLL provides good performance and is widely used. Only in very special applications does the required complexity warrant a higher-order system.

Example 6.7.3 Determine the design parameters of a minimum-bandwidth second-order PLL to demodulate commercial FM. Base your design on a sinusoidal FM signal with $\Delta f = 75$ kHz, $0 \le f_m \le 75$ kHz. Assume a triangular phase-comparator characteristic, $\zeta = 0.707$ and $\tau_1 = \tau_2$.

Solution. The PLL must both demodulate the signal with uniform gain and stay within the linear phase-comparator characteristic. From Eq. (6.88), we have

$$f_n > f_m / (2\zeta) = 53 \text{ kHz}.$$

From Eq. (6.98), the peak phase error for $\zeta = 0.707$ is

$$\epsilon_p = \frac{1}{\sqrt{2}} \left(\frac{\Delta\omega}{\omega_n} \right) < \frac{\pi}{2},$$

or

$$f_n > \sqrt{2} \, \Delta f / \pi = 34 \text{ kHz}.$$

Therefore we choose $f_n = 53$ kHz; from Eq. (6.85),

$$\tau_2 = 2\zeta / \omega_n = 4.24 \ \mu\text{sec}.$$

Using Eq. (6.84) and noting that $\tau_1 = \tau_2$, we find

$$k_c k_f = \frac{2}{\omega_n} \zeta \omega_n^2 = 4.71 \times 10^5 / \text{sec}.$$

Drill Problem 6.7.2 A frequency step of 25 kHz is applied to the PLL of Example 6.7.3. If $k_c = 1$ V/rad, determine (a) the peak phase error, and (b) the steady-state voltage output.

Answer. (a) 12.3°; (b) 0.333 V.

6.8 SIGNAL-TO-NOISE RATIOS IN FM RECEPTION

Our purpose here is to examine the effects of band-limited white noise on the frequency demodulation process. The block diagram of an idealized FM receiver is shown in Fig. 6.26. We define the signal-to-noise ratio to be the ratio of the mean signal power without noise to the mean noise power in the presence of an unmodulated carrier.†

---

† In other words, we assume that the output noise power can be calculated independently of the modulating signal power. A justification for this choice for wideband FM in the presence of white noise is given in B. P. Lathi, *Communication Systems*, New York: John Wiley & Sons, 1968, p. 363.

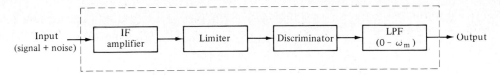

Fig. 6.26 The idealized FM receiver.

Assuming that the limiter is ideal and removes all amplitude variations, we can write the FM signal at the discriminator input as

$$s_i(t) = A \cos \theta(t),$$

$$s_i(t) = A \cos \left[\omega_c t + k_f \int_0^t f(\tau) \, d\tau \right]. \tag{6.99}$$

The discriminator output is proportional to the difference between the instantaneous frequency of $s_i(t)$ and the carrier frequency. Letting the constant of proportionality be 1 for convenience,

$$s_o(t) = \left(\frac{d\theta}{dt} - \omega_c \right) = k_f f(t). \tag{6.100}$$

The mean-square value of the output signal is

$$S_o = \overline{s_o^2(t)} = k_f^2 \overline{f^2(t)}. \tag{6.101}$$

Next we turn to calculating the mean output noise power in the presence of an unmodulated carrier. Using the bandpass representation for the band-limited noise [cf. Eq. (5.59)], we can write (see Fig. 6.27),

$$A \cos \omega_c t + n_i(t) = A \cos \omega_c t + n_c(t) \cos \omega_c t - n_s(t) \sin \omega_c t$$
$$= r(t) \cos[\omega_c t - \gamma(t)]. \tag{6.102}$$

Therefore the addition of noise introduces both amplitude noise [in $r(t)$] and phase noise [in $\gamma(t)$]. In the AM case we were interested in the effects of $r(t)$ but in the FM case we can assume amplitude limiting and are interested only in $\gamma(t)$.

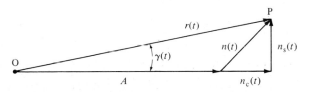

Fig. 6.27 Phasor diagram of the sum of a sine wave and narrowband noise, large S/N case.

The principal value of the phase noise angle is

$$\gamma(t) = \tan^{-1}\left[\frac{n_s(t)}{A + n_c(t)}\right].$$

(6.103)

Assuming that the noise is small, i.e., that $n_c(t), n_s(t) \ll A$, we have

$$\gamma(t) \approx \tan^{-1}\left[\frac{n_s(t)}{A}\right] \approx \frac{n_s(t)}{A}.$$

(6.104)

The discriminator output is proportional to the difference between the instantaneous frequency and the carrier frequency so that

$$n_o(t) = \frac{d\gamma}{dt} = \frac{1}{A}\frac{d}{dt}[n_s(t)].$$

(6.105)

The power spectral density corresponding to Eq. (6.105) is

$$S_{n_o}(\omega) = \frac{1}{A^2}S_{n_s}(\omega)|H(\omega)|^2,$$

(6.106)

where $\mathscr{F}^{-1}\{H(\omega)\} = h(t)$ is a time differentiator. But the Fourier transform equivalent of time differentiation is multiplication by $(j\omega)$ so that

$$S_{n_o}(\omega) = \frac{1}{A^2}\omega^2 S_{n_s}(\omega).$$

(6.107)

Thus those spectral components at the higher frequencies are emphasized—a general consequence of a differentiator.

The bandwidth of the discriminator output is limited by a low-pass filter with a cutoff frequency of ω_m radians per second. Using Eq. (5.65), we can express the power spectral density within this bandwidth by

$$S_{n_s}(\omega) = [S_n(\omega - \omega_c) + S_n(\omega + \omega_c)]_{LP}.$$

(6.108)

If the noise at the discriminator input is white,

$$S_n(\omega) = \eta/2,$$

Eq. (6.108) reduces to

$$S_{n_s}(\omega) = \eta,$$

and Eq. (6.107) becomes

$$S_{n_o}(\omega) = \eta\omega^2/A^2.$$

(6.109)

In other words, if one assumes a white noise spectral density for the discriminator input, the output noise spectral density will be parabolic (see Fig. 6.28).

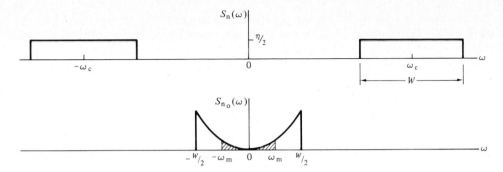

Fig. 6.28 Input-output noise spectral densities for the FM demodulator under high carrier-to-noise conditions.

Using Eq. (6.109), we find that the mean-square value of the output noise is

$$N_o = \overline{n_o^2(t)} = \frac{\eta}{\pi A^2} \int_0^{\omega_m} \omega^2 d\omega,$$

$$N_o = \frac{\eta \omega_m^3}{3\pi A^2}. \tag{6.110}$$

The mean carrier power is

$$S_c = A^2/2. \tag{6.111}$$

Comparing Eqs. (6.110) and (6.111), we see that the output noise power is inversely proportional to the mean carrier power in FM. This effect of a decrease in output noise power as the carrier power increases is called *noise quieting*.

Combining Eqs. (6.101) and (6.110), we have

$$\frac{S_o}{N_o} = \frac{3\pi A^2 k_f^2 \overline{f^2(t)}}{\eta \omega_m^3}. \tag{6.112}$$

The peak frequency deviation is proportional to k_f; for the wideband case, the bandwidth increases proportional to k_f. *For wideband FM, we conclude that the output signal-to-noise ratio increases as the square of the bandwidth.* In particular, if $f(t)$ is sinusoidal [that is, $f(t) = a \cos \omega_m t$ and $\Delta\omega = ak_f$], Eq. (6.112) becomes

$$\frac{S_o}{N_o} = \frac{3\pi A^2 (\Delta\omega)^2}{2\eta \omega_m^3}. \tag{6.113}$$

For a comparison between FM and AM, let the modulation be sinusoidal in each case and let us define the mean-square noise at the IF for the AM case by

$$N_c = \frac{1}{\pi} \int_{\omega_c - \omega_m}^{\omega_c + \omega_m} (1/2)\eta \, d\omega = \frac{\eta \omega_m}{\pi}. \tag{6.114}$$

Using Eqs. (6.111) and (6.114) and noting that $\beta = \Delta\omega/\omega_m$, Eq. (6.113) can be rewritten in terms of the carrier-to-noise ratio, we can write

$$\frac{S_o}{N_o} = 3\beta^2 \frac{S_c}{N_c}. \tag{6.115}$$

Under the most favorable conditions in AM the modulation index is 100% and for these conditions [cf. Eq. (5.90)],

$$\left(\frac{S_o}{N_o}\right)_{AM} = \frac{S_c}{N_c}. \tag{6.116}$$

Using Eq. (6.116) in Eq. (6.115), we get our final result:

$$\left(\frac{S_o}{N_o}\right)_{FM} = 3\beta^2 \left(\frac{S_o}{N_o}\right)_{AM}. \tag{6.117}$$

From this, we conclude that the output signal-to-noise ratio can be made much higher in FM than in AM by increasing the modulation index β. It must be remembered, of course, that we assumed that the noise was small in deriving this result. The effects of the noise in the output are reduced and the factor $3\beta^2$ is called the noise quieting of the system [this can be expressed in dB by taking $10 \log_{10} (3\beta^2)$]. But an increase in β also increases the bandwidth so that FM systems provide an improvement in signal-to-noise at the expense of an increase in bandwidth. For example, when $\beta = 5$ the output FM signal-to-noise ratio is 75 times that of an equivalent AM system but the bandwidth required is approximately 8 times larger (see Fig. 6.11). Thus the use of FM allows one to exchange bandwidth for signal-to-noise ratio.

To realize any signal-to-noise improvements in FM over AM we must have, from Eq. (6.117),

$$\beta > \frac{1}{\sqrt{3}} = 0.577.$$

However, this condition is approximately the transition point between narrowband and wideband FM. We conclude that narrowband FM provides no signal-to-noise improvement over AM. This is, of course, an expected result because NBFM does not use increased bandwidth to exchange for an improvement in output signal-to-noise ratio.

The exchange of bandwidth for signal-to-noise ratio in FM cannot be continued indefinitely. The reason for this is that the noise power increases with the increased receiver bandwidth necessary to demodulate the signal. Assuming a fixed signal power we find the noise power eventually becomes comparable to the signal power and the results of the above analysis no longer hold. As the noise power becomes large in angle modulation, the phase variations of the noise take over and the performance of the system becomes very poor. This transition depends heavily on the input signal-to-noise ratio, giving rise to a "threshold" effect.

Example 6.8.1   Develop a result similar to Eq. (6.117) for the improvement of PM over AM.

Solution.   For PM, we write

$$s_i(t) = A \cos [\omega_c t + k_p f(t)].$$

The phase detector output is proportional to the difference between the instantaneous phase of $s_i(t)$ and the carrier phase,

$$s_o(t) = [\theta(t) - \omega_c t] = k_p f(t).$$

Similarly, using Eq. (6.104), we get the output noise for high signal-to-noise conditions:

$$n_o(t) = n_s(t)/A,$$

$$S_{n_o}(\omega) = \frac{1}{A^2} S_{n_s}(\omega).$$

Assuming that the input noise spectral density is white and equal to $\eta/2$ watts per Hz, we write

$$\overline{n_o^2(t)} = \frac{1}{\pi} \int_0^{\omega_m} \frac{1}{A^2} \eta \, d\omega = \frac{\eta \omega_m}{\pi A^2}.$$

Then the output signal-to-noise ratio is

$$\frac{S_o}{N_o} = \frac{\overline{s_o^2(t)}}{\overline{n_o^2(t)}} = \pi A^2 \frac{k_p^2 \overline{f^2(t)}}{\eta \omega_m}.$$

For the sinusoidal case, $\overline{f^2(t)} = a^2/2$ and $\Delta\theta = ak_p$ so that

$$\frac{S_o}{N_o} = \frac{\pi A^2}{2\eta \omega_m} (\Delta\theta)^2.$$

Compared to the AM case for 100% modulation, this becomes

$$\left(\frac{S_o}{N_o}\right)_{PM} = (\Delta\theta)^2 \left(\frac{S_o}{N_o}\right)_{AM}.$$

Noting that $\Delta\theta = \beta$ for PM, we conclude that the signal-to-noise improvement in PM is similar to that in FM.

Drill Problem 6.8.1   The narrowband PM signal

$$\phi_{\text{NBPM}}(t) = \cos \omega_c t - \beta \cos \omega_m t \sin \omega_c t$$

is to be demodulated using a synchronous detector; i.e., multiplication by $\cos (\omega_c t + \theta)$ followed by a low-pass filter. Determine (a) if this will work satisfactorily; and (b) what value of θ is required for best results.
Answer.   (a) Yes; (b) $\pm \pi/2$.

Before leaving this discussion of S/N ratios, we digress briefly to complete the comparison of CW modulation systems that was started in Chapter 5. The results are summarized in Table 6.1. For convenience, the input S/N ratios have been normalized in this table for the case of band-limited white noise with power spectral density $S_n(\omega) = \eta/2$ W/Hz and a bandwidth f_m equal to the bandwidth of the input modulating signal. This places single-sideband and double-sideband modulation systems on an equal basis for S/N comparisons.

From Table 6.1, it is evident that SSB and VSB systems offer spectral efficiency, requiring only as much (or slightly more) bandwidth as the input

Table 6.1   A Comparison of CW Modulation Systems

Modulation	B/f_m	$\dfrac{S_o}{N_o}\left(\dfrac{S_i}{\eta f_m}\right)^{-1}$	$\dfrac{S_o}{N_o}\left(\dfrac{S_c}{\eta f_m}\right)^{-1}$	Typical Applications
DSB-SC	2	1		Analog instrumentation, multiplexing.
DSB-LC[1]	2	$\dfrac{\overline{f^2(t)}}{A^2 + \overline{f^2(t)}}$	$\dfrac{\overline{f^2(t)}}{A^2}$	Broadcast radio, point-to-point voice.
SSB-SC	1	1		Point-to-point voice, multiplexing.
SSB-LC[1]	1	$\dfrac{\overline{f^2(t)}}{A^2 + \overline{f^2(t)}}$	$\dfrac{\overline{f^2(t)}}{A^2}$	Point-to-point voice.
VSB-SC	1^+	1		Facsimile, digital data.
VSB-LC[1]	1^+	$\dfrac{\overline{f^2(t)}}{A^2 + \overline{f^2(t)}}$	$\dfrac{\overline{f^2(t)}}{A^2}$	Television video.
FM[2]	$\approx 2(1 + \beta)$	$\frac{3}{2}\beta^2$		Broadcast radio, mobile radio.
PM[2]	$\approx 2(1 + \beta)$	$\frac{1}{2}(\Delta\theta)^2$		Telemetry, digital data.

[1] Above-threshold envelope detection assumed; for sinusoidal modulation, $f(t) = mA\cos \omega_m t$ so that $\overline{f^2(t)} = m^2 A^2/2$.
[2] Above-threshold detection assumed; no deemphasis and sinusoidal modulation.

modulating signal. In contrast, angle modulation systems offer increased S/N at some expense in required bandwidth. Large-carrier amplitude modulation systems generally use envelope detection to minimize the complexity of the receiver. Angle modulation and SSB-SC systems require moderate receiver complexity while other types of suppressed-carrier transmissions require more substantial receiver complexity. Each method has its advantages and disadvantages, and these must be evaluated in each specific situation.

6.9 THRESHOLD EFFECT IN FM

The threshold effect turns out to be quite pronounced in FM. Because this effect occurs when the noise and signal levels are comparable, it is more difficult to analyze than the high signal-to-noise case.

To gain some insight into this, suppose we consider the following simplified problem. Two phasors, A and B, as shown in Fig. 6.29(a), are added to yield a resultant phasor C. For simplicity, we assume that the magnitudes of A and B are fixed and that the angle of the phasor B can take all values in the range $-\pi < \theta_B \leq \pi$ with equal weighting. The angle of phasor A remains fixed. Phasor A can be used to represent the signal and phasor B the noise. We wish to find the mean-square value of the phase angle θ_C of the resultant and this will represent the average phase noise present in the measurement of angle.

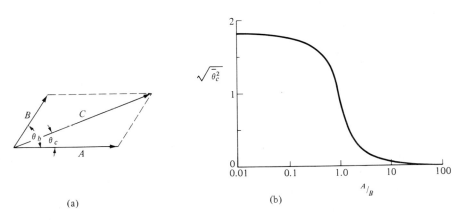

(a)   (b)

Fig. 6.29 Phasor diagram and determination of phase noise.

From Fig. 6.29(a), we can write

$$\theta_C = \tan^{-1}\left(\frac{B \sin \theta_B}{A + B \cos \theta_B}\right), \tag{6.118}$$

so that the mean-square value is:

$$\overline{\theta_C^2} = \frac{1}{2\pi} \int_{-\pi}^{\pi} \left[\tan^{-1} \left(\frac{B \sin \theta_B}{A + B \cos \theta_B} \right) \right]^2 d\theta_B. \tag{6.119}$$

This can be solved numerically and the results are shown in Fig. 6.29(b). Note that the ratio A/B may be regarded as a voltage signal-to-noise ratio in this simplified model. From these results, we see that the mean-square phase noise increases rapidly when the ratio A/B is smaller than about 3.

In actuality, the problem is more complicated because the amplitude of the noise varies, and the presence of angle modulation in the signal also has some effect. There is also another effect peculiar to FM that makes the threshold effect even more abrupt, and we turn our attention to that next.

Referring back to Fig. 6.27, we note that the tip of the resultant phasor, labeled point P, moves about in the neighborhood of the tip of carrier phasor A due to the random fluctuations in the additive noise $n(t)$. As long as $n(t) \ll A$, point P remains close to the tip of carrier phasor A, and the principal value of the angle $\gamma(t)$ is approximately $n_s(t)/A$, as given by Eq. (6.104). As the noise increases in magnitude, the wandering of point P about the tip of the carrier phasor increases, and the discriminator output noise increases accordingly.

As the noise is increased further, however, a condition is reached where not only may the magnitude of the phasor $n(t)$ exceed the magnitude of the carrier phasor, but the angles may be such that point P sweeps *around* the origin (point O). This latter condition is demonstrated in Fig. 6.30(a). The net effect is that the phase angle $\gamma(t)$ may change by $\pm 2\pi$ radians within a very short interval of time, as illustrated in Fig. 6.30(b). Because the output of the FM discriminator is proportional to $d\gamma/dt$, impulse-like noise spikes will occur in the output whenever the resultant phasor encircles point O. This is illustrated in Fig. 6.30(c). Heights and polarities of these noise spikes will vary, depending on the particular path taken by P around O, but the area of each is approximately $\pm 2\pi$ radians.

It is found experimentally that the occurrences of occasional noise spikes are heard as individual clicks in the FM discriminator output. Therefore the onset of spike noise in FM is sometimes referred to as click noise. As the signal-to-noise ratio is decreased a little further, these clicks rapidly merge into a crackling sound. As a result, the output signal-to-noise ratio decreases rapidly with only a relatively small decrease in the input S/N, and the results we described in the preceding section now fail to describe the performance adequately. This gives rise to a very pronounced FM threshold effect.

The presence of modulation varies the instantaneous angle of the carrier phasor in Fig. 6.30(a). The resulting behavior near the threshold condition is that, at a given time, there is a greater occurrence of encirclements about O in one direction than in the other. It turns out that the resulting decrease in encirclements in one direction does not quite offset the increase in encirclements in the other direction. Thus there is some dependence on modulation, although this effect is secondary.

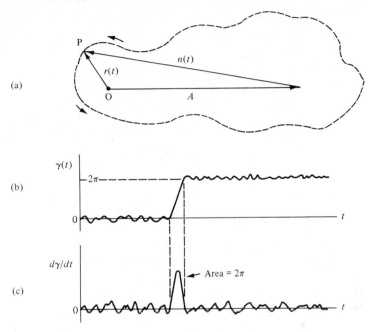

(a)

(b)

(c)

Fig. 6.30 (a) Phasor diagram of the sum of a sine wave and narrowband noise, small S/N case. (b) Example of $\gamma(t)$ for a case in which point P encircles the origin (point O). (c) Plot of resulting $d\gamma/dt$, exhibiting spike noise.

Figure 6.31 shows the signal-to-noise performance characteristics of a wideband FM system. The dependence of the threshold on the value of β is small and the input signal-to-noise ratio to achieve above-threshold operation is usually taken as about 10 dB for large β. Note that for small input signal-to-noise conditions the FM system may actually be inferior to an AM system. The amount of noise quieting above threshold increases with β, as noted earlier.

Improvement of the FM threshold has been the subject of considerable research, particularly for space communications where power is a main concern and signal-to-noise ratios may not be very large. One method used successfully to extend the FM threshold to lower values is that of the phase-locked loop (PLL). The occurrence of a noise spike in the PLL causes the phase error to exceed $\pm T/4$ ($\pm 90°$) in the phase-comparator characteristic (cf. Fig. 6.23). However, the dynamic range of the PLL output is limited by the phase-comparator characteristic and therefore the net energy due to the noise spike is decreased. It has been found that the overall threshold improvement is lowered by about 3 dB with the use of the PLL.

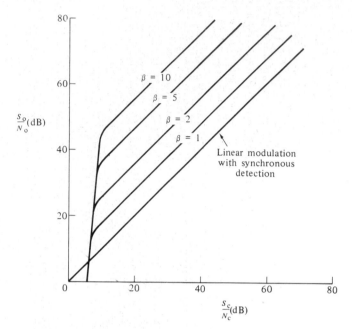

Fig. 6.31 Signal-to-noise performance of wideband FM.

Other methods used successfully to extend the FM threshold include the frequency-locked loop and FM with feedback (FMFB). A discussion of the lowered threshold of the PLL and FMFB can be found in Reference 6.

There is no reason, of course, why the second signal in the phasor example above could not have been a second FM or PM signal. In this case we are interested in the signal-to-interference ratio. The rather sharp transition between good and poor interference-to-signal ratios in FM results in a very definite suppression of small signals in the presence of larger ones. This small-signal suppression effect together with a higher carrier frequency results in more closely defined service areas for commercial FM than for AM.

Example 6.9.1   A given receiver for commercial FM has a noise temperature of 210 K. (a) Estimate the minimum required signal strength at the test generator terminals to obtain full quieting. The input resistance is 300Ω; assume that the signal generator is at 290 K and matched to the receiver. (b) What is the theoretical minimum required signal strength for full quieting under these conditions?

Solution

a) If we assume an IF bandwidth of 180 kHz, the available noise power is

$$P_n = kTB = (1.38 \times 10^{-23})(500)(180,000) = 1.24 \times 10^{-15} \text{ W}.$$

The assumed threshold is 10 dB so that

$$P_s = 1.24 \times 10^{-14} \text{ W},$$

$$\overline{e_s^2(t)} = P_s R = (1.24 \times 10^{-14})(300),$$

$$\sqrt{\overline{e_s^2(t)}} = 1.93 \ \mu\text{V}.$$

b) For no receiver-contributed noise, $T = 290$ K (from the signal generator) so that

$$P_s = 10 \ kT_0 B = 7.20 \times 10^{-15} \text{ W},$$

$$\overline{e_s^2(t)} = (300)(7.20 \times 10^{-15}),$$

$$\sqrt{\overline{e_s^2(t)}} = 1.47 \ \mu\text{V}.$$

★ **Drill Problem 6.9.1**   A telemetry system for VHF weather satellite transmissions that are compatible with facsimile (cf. Appendix E) uses a (low-pass) bandwidth of 1.6 kHz. This is modulated on a 2.4 kHz carrier using DSB-SC and then frequency-modulated (AM/FM).

a) Use Carson's rule to estimate the required receiver bandwidth if the peak frequency deviation is set at 20 kHz.

b) Estimate the maximum noise temperature of the receiver if the minimum signal level is 1 μV rms across 75 Ω and the expected antenna noise temperature is 1150 K.

Answer.   (a) 48 kHz; (b) 863 K.

6.10   SIGNAL-TO-NOISE IMPROVEMENT USING DEEMPHASIS

In commercial FM broadcasting it is found that signals arising from speech and music have most of the energy at the lower frequencies. In the output of an FM demodulator, however, the noise-power spectral density rises parabolically with frequency. Therefore the noise power spectral density is largest in the frequency range where the signal spectral density is smallest. This results in a rather inefficient communication system. To remedy this situation, we emphasize the high-frequency components in the input signal at the transmitter, *before the noise is introduced*. At the output of the FM demodulator in the receiver the inverse operation is performed to deemphasize the high-frequency components. The signal spectrum is restored to its original shape but the noise, which was added after the preemphasis, is now reduced.

The use of the preemphasis/deemphasis technique is an example of how a knowledge of the differences between characteristics of signal and noise can be applied to improve the performance of a communication system. This technique finds other applications, particularly in the audio recording area.†

Some care must be exercised in the selection of the preemphasis to be used. If the high-frequency components are emphasized too much, the bandwidth of the transmitted FM spectrum will increase unless β is lowered, offsetting the advantage. A simple and straightforward approach in choosing a preemphasis characteristic is to choose one which will yield a white noise spectral density after demodulation. Because the noise power spectral density at the FM demodulator output rises parabolically with frequency, we then use a matching preemphasis for the signal; this requires that $|H(\omega)|^2 = \omega^2$, suggesting the possibility $H(\omega) = j\omega$. This transfer function corresponds to a differentiator and we arrive at the surprising result that we now should use PM instead of FM! However, if we had used PM entirely, the bandwidth would be more difficult to control for commercial use. Obviously what we want here is the best combination of both modulation methods.

What is needed, then, is a filter whose transfer function is constant for low frequencies and behaves like a differentiator at the higher frequencies. An example of an RC network that approximates this type of response is shown in Fig. 6.32(a). The Bode magnitude plot for this network is also shown and a corresponding deemphasis network for the receiver is shown in Fig. 6.32(b).

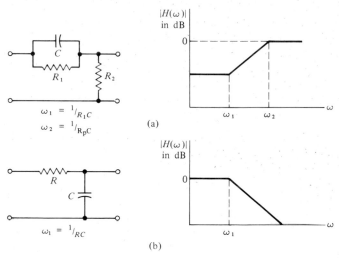

$$\omega_1 = \frac{1}{R_1 C}$$
$$\omega_2 = \frac{1}{R_p C}$$

(a)

$$\omega_1 = \frac{1}{RC}$$

(b)

Fig. 6.32 Example of a preemphasis/deemphasis network combination.

---

† Much interest has developed recently in applying combinations of preemphasis filtering and dynamic-range signal-amplitude compression to suppress the effects of noise in audio recordings. These are discussed in Appendix F.

The choice of ω_1 and ω_2 in Fig. 6.32 determines the component values to be used. A reasonable choice for $f_1 = \omega_1/2\pi$ is usually taken as that frequency at which the signal spectral density is down by 3 dB; for broadcasting this is taken as 2.1 kHz ($RC = 75\ \mu\text{sec}$) and it is assumed that the spectral density falls off very rapidly beyond this frequency.† The choice for $f_2 = \omega_2/2\pi$ is taken well above the highest audio frequency to be transmitted.

How much does the addition of the preemphasis-deemphasis improve the overall signal-to-noise ratio? Assuming that the two networks are chosen properly, there will be no net change in the signal. The noise spectral density at the demodulator output, however, is altered appreciably, as shown in Fig. 6.33.

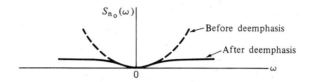

Fig. 6.33 Noise spectral density following deemphasis.

The improvement in the signal-to-noise ratio can be found by calculating the decrease in the noise power. From Eq. (6.109), the noise power spectral density at the output of the FM demodulator is

$$S_{n_0}(\omega) = \eta\omega^2/A^2.$$

The transfer function of the deemphasis filter can be written as

$$H(\omega) = \frac{1}{1 + j\omega/\omega_1}. \qquad (6.120)$$

The mean-square value of the noise after the deemphasis filter is

$$N_0' = \frac{1}{\pi}\int_0^{\omega_m} S_{n_0}(\omega)|H(\omega)|^2 d\omega,$$

$$N_0' = \frac{\eta}{\pi A^2}\int_0^{\omega_m} \frac{\omega^2}{1 + (\omega/\omega_1)^2} d\omega. \qquad (6.121)$$

---

† The high-frequency response is becoming so good in modern recordings that this assumption is no longer valid. Therefore stations have either had to decrease their modulation or purposely "color" their input spectrum to stay within the specified 75 kHz peak frequency deviation. Recently the FCC has permitted stations to use a 25 μsec time constant (6.4 kHz) and Dolby B compression (see Appendix F) instead of 75 μsec. This choice proves compatible with existing 75 μsec deemphasis and permits full program dynamic range with an increased S/N ratio.

Without the deemphasis filter, the noise would be

$$N_o = \frac{\eta}{\pi A^2} \int_0^{\omega_m} \omega^2 \, d\omega. \tag{6.122}$$

Defining a noise improvement factor,

$$\Gamma = N_o/N_o', \tag{6.123}$$

and evaluating Eqs. (6.121) and (6.122), we find that

$$\Gamma = \frac{1}{3} \frac{(\omega_m/\omega_1)^3}{(\omega_m/\omega_1) - \tan^{-1}(\omega_m/\omega_1)}. \tag{6.124}$$

For example, if $f_m = 15$ kHz, $f_1 = 2.1$ kHz, Γ is approximately 13 dB. A plot of the noise improvement factor Γ is shown in Fig. 6.34.

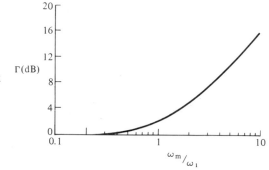

Fig. 6.34 Signal-to-noise improvement using deemphasis.

Note that the use of preemphasis/deemphasis has decreased the noise in the FM demodulator output but has left the signal unchanged. Therefore the value of Γ represents a net increase in the effective signal-to-noise ratio without any required increase in the transmitted power. As shown in Fig. 6.34, this increase can be appreciable. With the use of this preemphasis, the transmitter essentially sends out a signal that is frequency-modulated by the lower audio frequencies and phase-modulated by the higher audio frequencies.

Example 6.10.1   It was assumed, in the above discussion of preemphasis for commercial FM, that the higher spectral components in the modulating signal are negligibly small. Therefore Eq. (6.124) neglects any increase in the modulation power resulting from the preemphasis of higher spectral components. In modern recordings this assumption may lose some of its validity. For example, suppose that the power spectral density of the signal is $S_f(\omega) = (\text{constant})/[1 + (\omega/\omega_1)^2]$.

a) Determine the required change in modulation level with and without preemphasis for a fixed modulation power.

b) What is the net S/N improvement under these conditions?

Solution

a) Without preemphasis, the modulation signal power, within a constant, is

$$P = \frac{1}{\pi} \int_0^{\omega_m} \frac{1}{1 + (\omega/\omega_1)^2} \, d\omega = \frac{\omega_1}{\pi} \tan^{-1}\left(\frac{\omega_m}{\omega_1}\right).$$

With preemphasis, the modulation signal power increases unless an attenuation K is used so that, within a constant, the power is

$$P = \frac{1}{\pi} \int_0^{\omega_m} K \, d\omega = K\omega_m/\pi.$$

If this modulation power is to remain fixed, we have

$$K = \frac{\tan^{-1}(\omega_m/\omega_1)}{\omega_m/\omega_1}.$$

For $f_m = 15$ kHz, $f_1 = 2.1$ kHz, this gives $K = 0.200$ (-6.98 dB).

b) From Eq. (6.124), the S/N improvement as a result of preemphasis/deemphasis is 13.28 dB. However, now the modulation level must be decreased by 6.98 dB so that the net S/N improvement is only $(13.28 - 6.98) = 6.30$ dB. This gives some insight into why commercial FM stations are introducing alternative preemphasis/deemphasis methods as the quality of recordings continues to improve.

Drill Problem 6.10.1   Suppose that the preemphasis/deemphasis networks described in this section are proposed for use in commercial AM. Assuming that the input noise spectral density is white and that an envelope detector is used under high signal-to-noise conditions, calculate the improvement possible for $f_m = 5$ kHz. [*Hint*: Assume that the output noise is white.]

Answer.   3.07 dB.

6.11   SUMMARY

The variation of frequency of a continuous sinusoid in proportion to the amplitude of a modulating signal is called frequency modulation (FM). The variation of phase of a continuous sinusoid in proportion to the amplitude of a modulating signal is called phase modulation (PM). Both are examples of angle modulation. The instantaneous frequency is the time derivative of the phase so that PM and FM are closely related.

In the generation of FM there is an amplitude-to-frequency conversion and a frequency-to-frequency conversion. The peak frequency deviation is a measure of the former and the modulating frequency the latter. The ratio is called the modulation index, β.

The value of β determines the spectral characteristics of the FM waveform. For values of β below about 0.2, the spectral density of an FM waveform consists of two sidebands about a large carrier and this condition is called narrowband FM. The spectral details for larger values of β, called wideband FM, depend on each particular modulating signal because the FM generation is nonlinear. However, the overall bandwidth can be approximated by adding twice the peak frequency deviation to twice the modulating signal bandwidth (Carson's rule).

Most of the principles of FM also apply to PM. A difference is that the peak phase deviation is controlled by the modulating signal in PM and therefore the peak frequency deviation varies with the frequency of the modulating signal as well as its amplitude.

The total average power in an angle-modulated signal remains constant regardless of the modulation index.

Generation of angle-modulated signals is accomplished either by first generating a narrowband signal and then increasing the modulation index using frequency multiplication or by generating a wideband signal directly. Demodulation can be accomplished using a frequency-to-voltage converter circuit called a discriminator or by using feedback techniques around a voltage-to-frequency converter.

Wideband angle-modulated systems offer an improvement in output signal-to-noise over AM systems at the expense of increased bandwidths. This improvement can only be realized, however, when the input signal-to-noise ratio is above a threshold level. A preemphasis/deemphasis circuit is used in commercial FM broadcasting to partially suppress high-frequency noise in the demodulator output.

Selected References for Further Reading

1. J. J. DeFrance. *Communications Electronics Circuits*, Second ed. San Francisco: Rinehart Press, 1972.
   Chapter 9 contains a discussion of the circuits and operating characteristics of practical FM systems. Easy reading.

2. M. S. Roden. *Analog and Digital Communication Systems*. Englewood Cliffs, NJ: Prentice-Hall, 1979.
   Chapter 5 presents easily read explanations of angle modulation, particularly on the topics of instantaneous frequency and bandwidth approximation.

3. H. Stark and F. B. Tuteur. *Modern Electrical Communications*. Englewood Cliffs, NJ: Prentice-Hall, 1979.
   Chapter 7 presents angle modulation with brief explanations of some of the popular circuits used. For more detailed circuit descriptions, see Chapters 11, 12 in reference 5 of Chapter 5.

4. R. M. Gagliardi. *Introduction to Communications Engineering*. New York: John Wiley & Sons, 1978.

The topics of both receiver frequency acquisition, and phase referencing and timing are covered in Chapter 9.

5. F. M. Gardner. *Phaselock Techniques*, Second ed. Somerset, NJ: Wiley-Interscience, 1979.
   Contains a good discussion of the linearized second-order phase-locked loop, and its performance in the presence of noise; provides useful design criteria.

6. H. Taub and D. L. Schilling. *Principles of Communications Systems*. New York: McGraw-Hill, 1971.
   Chapters 4, 9, and 10 contain a well-written discussion of angle modulation at a more advanced level, stressing the noise suppression and threshold effects.

7. P. F. Panter. *Modulation, Noise, and Spectral Analysis*. New York: McGraw-Hill, 1965.
   Written at a more advanced level, this book is probably the most complete reference available on the subject of angle modulation, devoting ten chapters to the topic.

Problems

6.1.1 Determine the instantaneous frequency, in Hz, of each of the following waveforms at the time $t = 100$ seconds.

a) $\cos(100\,\pi t + 30°)$
b) $\cos[200\pi t + 200\sin(\pi t/100)]$
c) $10\cos[\pi t(1 + \sqrt{t})]$.
d) $2\exp(j\,200\pi\sqrt{t})$.

6.1.2 a) Find an approximation to the Fourier transform of the angle-modulated waveform $\psi(t) = \exp(j\beta\sin\omega_m t)$ for small β by using the MacLaurin series expansion for $\exp(x)$ and retaining only the first two terms in the expansion.

b) Repeat for the waveform: $\phi(t) = \psi(t)\exp(j\omega_c t)$.

6.2.1 A 1 GHz carrier is frequency-modulated by a 10-kHz sinusoid so that the peak frequency deviation is 1 kHz. Determine (a) the approximate bandwidth of the FM signal; (b) the bandwidth if the modulating signal amplitude were doubled; (c) the bandwidth if the modulating signal frequency were doubled; and (d) the bandwidth if both the amplitude and the frequency of the modulating signal were doubled.

6.2.2 Show that the error in using Eq. (6.20) first shows as third-harmonic distortion with a relative amplitude of approximately $(\beta/2)^2$.

6.2.3 The upper sideband of an AM waveform (DSB-LC) with sinusoidal modulation and a modulation index of 100% is attenuated by a factor α, where: $0 \leq \alpha \leq 1$.

a) Derive a relation for the resulting phase deviation from the carrier as a function of α.

b) Plot the peak (i.e., maximum) phase deviation as a function of α.

6.3.1 The voltage waveform $f(t)$ shown in Fig. P–6.3.1 is applied to an FM modulator with a modulation sensitivity of $k_f = 10$ Hz/volt and a center frequency of 11 Hz.

a) Sketch a diagram of the instantaneous frequency versus time.

b) Sketch the FM time waveform.

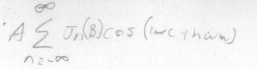

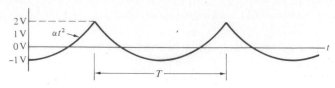

Fig. P-6.3.1

★ **6.3.2** Sketch an approximate magnitude spectral density of the waveform of Problem 6.3.1 if the period (T) is one second.

c★ **6.3.3** Using an FFT algorithm and $N = 256$, compute and plot the magnitude spectral density of a sinusoid whose instantaneous frequency varies linearly from $f_c - D$ to $f_c + D$ in the computation interval. [*Hint*: Let $f_c = N/4$, $D = N/8$.] Compare your result with that of Fig. 6.8.

6.3.4 Repeat Problem 6.2.1 with the exception that the peak frequency deviation is 20 kHz. Use both Carson's rule and Bessel functions and compare.

6.3.5 Using Appendix I and changing the criterion of significant sidebands to $|J_n(\beta)| \leq 0.10$, make a comparison between actual values and those given by Carson's rule for $\beta = 1, 2, \ldots, 8$.

6.3.6 A 10 kHz sinusoid is used to modulate two FM modulators whose unmodulated carriers are derived from a common oscillator and are at 3.000 and 3.040 MHz, respectively. The modulators are identical in all other respects and their outputs are added. The modulation index of each is $\beta = 2$.

a) Estimate the net bandwidth required for transmission of the composite FM signal.

b) Sketch to scale the resulting magnitude line spectrum.

6.3.7 A carrier is angle-modulated by the sum of two sinusoids:

$$\phi(t) = A \cos (\omega_c t + \beta_1 \sin \omega_1 t + \beta_2 \sin \omega_2 t).$$

Show that $\phi(t)$ has sidebands separated from the carrier not only at multiples of ω_1 and ω_2 but also has sidebands at separations of multiples of $(\omega_1 - \omega_2)$ and $(\omega_1 + \omega_2)$. [*Hint*: Express $\phi(t)$ in complex-valued notation and use Bessel functions for series representations.] Is this linear or nonlinear modulation?

6.3.8 a) DSB-LC modulation is used for commercial AM broadcasting. Why isn't it used to frequency-multiplex the stereo information in commercial FM broadcasting?

b) Suppose that the left (L) and right (R) channels of an FM stereo multiplex system are driven by the voltages $v_L(t) = \cos \omega_1 t$, $v_R(t) = \cos \omega_2 t$, where $\omega_1 \neq \omega_2$. Sketch the mono, stereo-multiplex, and pilot-carrier signals, indicating the correct proportions of each to meet FCC regulations. Assume that an SCA signal is being transmitted.

6.4.1 The FM signal (in volts),

$$\phi(t) = 20 \cos (2\pi \times 10^7 t + 10 \sin 2\pi \times 10^3 t),$$

is present across a 50-ohm resistive load.

$\Delta w = 10 \frac{rad}{sec}$

$w_m = 2\pi \times 10^3$

$P = \frac{2 \cdot 0^2}{2R} = 2\frac{98}{58} = 4 watt$

Power in Side Band
$= \frac{A^2}{2R} J_n^2(\beta)$
$n = 0$

a) What is the total average power?
b) What percentage of this power is at 10.000 MHz?
c) Find the peak frequency deviation.
d) Determine the approximate bandwidth of $\phi(t)$, using Carson's rule.
e) Can you determine from $\phi(t)$ whether this is FM or PM? Explain.

6.4.2 A certain sinusoid at a frequency of f_m Hz is used as the modulating signal in both an AM (DSB-LC) and an FM system. The unmodulated carrier powers are equal in both systems. When modulated, the peak frequency deviation of the FM system is set to four times the bandwidth of the AM system. The magnitudes of those sidebands spaced $\pm f_m$ Hz from carrier in both systems are equal. Determine (a) the modulation index of the FM system; and (b) the modulation index of the AM system.

6.4.3 A given FM transmitter is modulated with the sinusoidal input (in volts), $f(t) = 10 \cos 200\pi t$, and a modulation index of five. The unmodulated carrier power is ten watts across a 50-Ω resistive load. Determine (a) the modulator constant, k_f; (b) the peak amplitude of the first-order lower sideband and its phase relative to the unmodulated carrier; (c) the ratio of the average power in the sum of the third- and fourth-order sidebands to the power in all remaining sidebands, also excluding carrier; and (d) the bandwidth reduction factor if the input sinusoid peak amplitude is reduced to two volts (use Carson's rule).

6.5.1 A transmitter is angle-modulated with a 1-kHz sinusoid to generate the signal:

$$\phi(t) = 100 \cos (\omega_c t + 25 \cos \omega_m t).$$

If the modulation is FM, determine the modulation index and bandwidth when (a) ω_m is increased by a factor of 5; and (b) ω_m is decreased by a factor of 5. If the modulation is PM, determine the modulation index and bandwidth when (c) ω_m is increased by a factor of 5; and (d) ω_m is decreased by a factor of 5.

6.5.2 The bandwidths of three angle-modulated transmitting systems are compared, using the sinusoidal test signal, $f(t) = a \cos 2\pi f_m t$. The resulting approximate bandwidths are tabulated below:

$FM \quad \beta = \frac{\Delta w}{w_m}$

Test Results of Bandwidth

System	$a = 1$ V $f_m = 1$ kHz	$a = 2$ V $f_m = 1$ kHz	$a = 1$ V $f_m = 2$ kHz
A	2 kHz	2 kHz	4 kHz
B	40 kHz	80 kHz	80 kHz
C	50 kHz	100 kHz	50 kHz

Identify the type of angle modulation used (FM or PM; narrowband or wideband) for each of these three systems.

6.5.3 A carrier is phase-modulated by a sinusoidal signal $f(t)$. The peak phase deviation is one radian when the peak input amplitude is one volt. Find the ratio of the average power in the carrier to that in all sidebands excluding carrier for each of the following cases [$f(t)$ is expressed in volts] and the bandwidth in each case using Carson's rule:

a) $f(t) = 2 \cos 2500t$,
b) $f(t) = 3.8 \cos 200\pi t$,
c) $f(t) = 5.5 \cos 300\pi t$,
d) $f(t) = 7 \cos 8000\pi t$.

6.6.1 A 1-GHz signal is derived from a 10-MHz oscillator using frequency multipliers.

a) Find the maximum frequency inaccuracy in the 1-GHz signal if the accuracy of the 10-MHz oscillator is within one part in 10^9.
b) What is the frequency stability of the 1-GHz signal if the output of the 10-MHz oscillator varies ± 1 Hz; $\pm 5°$ per second?

6.6.2 An Armstrong-type FM system to handle inputs in the audio range: 100–3100 Hz first generates an FM signal with $\beta = 0.20$ (const.), $f_c = 200$ kHz. The desired output is required to have a carrier frequency of 40.0 MHz and a peak frequency deviation of 5 kHz. Using one frequency multiplier and one frequency mixer (plus required filters), devise two different system configurations to accomplish the desired objectives. Specify all frequencies in your diagrams.

6.6.3 A 200-kHz carrier signal is frequency-modulated by a 1-kHz sinusoid such that the peak frequency deviation is 150 Hz.

a) What is the bandwidth?
b) The above FM signal is applied to a $\times 16$ frequency multiplier. By what factor is the bandwidth increased? (Use Carson's rule.)
c) The FM signal of (b) is applied to a second $\times 16$ frequency multiplier. By what factor is the bandwidth increased over parts (b) and (a)?
d) Estimate the number of significant sidebands possible in the FM signal of part (c) above.

6.7.1 The angle-modulated voltage $\phi(t) = A \cos (\omega_c t + \beta \sin \omega_m t)$ is applied to the input of an RC high-pass filter.

a) If $RC \ll \omega^{-1}$ in the frequency band occupied by the signal $\phi(t)$, show that the output voltage across the resistor is amplitude-modulated and calculate the AM modulation index.
b) Estimate the maximum peak frequency deviation in terms of RC, ω_c, which can be used before the error in the voltage-frequency characteristic is 1% from a linear characteristic.

6.7.2 The FM signal $\phi(t) = A \cos (\omega_c t + \beta \sin \omega_m t)$ is applied to a certain demodulation system. This system generates two phasors $A_1 e^{j\theta_1}, A_2 e^{j\theta_2}$, which are adjusted to have equal magnitudes ($A_1 = A_2$) and complementary phase angles ($\theta_1 = -\theta_2$) at $\omega = \omega_c$. The sum of the phasor magnitudes is held constant and the difference between magnitudes is made proportional to the instantaneous frequency of the input from carrier. The output voltage of this demodulator is proportional to the phase angle of the resultant of the addition of the two phasors. How well (if at all) will this system demodulate $\phi(t)$?

c **6.7.3** Using a digital computer, determine the maximum peak frequency deviation as a percentage of the operating point given for the slope demodulator of Example

6.7.1 such that the total harmonic distortion in the output for a sinusoidally modulated FM input is less than 0.10%.

6.7.4 a) Sketch the phase-comparator characteristic for a symmetrical square-wave input
at a fundamental frequency f_c and a VCO square wave with a fundamental
frequency $3f_c$.
   b) Repeat if the VCO square wave had a fundamental $2f_c$.
   c) Generalize your results using the Fourier series.

6.7.5 a) Sketch the phase-comparator characteristic for a symmetrical square-wave input
at a fundamental frequency f_c and a VCO rectangular wave with one level
duration one-fourth the period and with a fundamental frequency f_c.
   b) Describe the effect of the rectangular wave on the comparator gain k_c and
compare with that of a square wave.

6.7.6 a) Using the result of Example 6.7.2(b), solve for $\theta_r(t)$ if $\theta_i(t) = u(t)$.
   b) Using Eq. (6.60), determine the output of the first-order PLL of (a) and compare
with the frequency of the input.

★6.7.7 Repeat Example 6.7.3 and Drill Problem 6.7.2 for a first-order PLL; also compute
the steady-state phase error.

★6.7.8 The sinusoidal input to a given PLL is amplitude-limited so that it is a square
wave with zero average value. The VCO output is a square wave, 0 to 1 V, at
the same fundamental frequency.

   a) Sketch the phase-comparator characteristic if the maximum output is $+1$ V.
   b) No loop filter is used and the loop time constant is 10 μsec. The loop is adjusted
so that the loop error is zero in the presence of unmodulated carrier. If the
input is frequency-modulated by a sinusoid, determine the largest peak frequency deviation which the PLL will follow without distortion.
   c) What is the highest modulating frequency?
   d) If the input signal is $\phi(t) = \cos(2\pi \times 10^6 t + 10 \sin 2\pi \times 10^3 t)$, determine
the peak amplitude of the output.

★6.7.9 Repeat Example 6.7.3 and Drill Problem 6.7.2 for a sinusoidal phase-comparator
characteristic. Linearize the characteristic with the restriction that the maximum
allowable percentage error from a linear characteristic be 5%.

6.8.1 Discuss the relative advantages and disadvantages of transmitting the color information for television using frequency modulation (cf. Appendix D).

6.8.2 Suppose the output of the discriminator is applied to a filter (ideal) with a passband
$\omega_1 - \omega_m$ rad/sec instead of $0 - \omega_m$ as assumed in Eq. (6.110). Determine the resulting
alteration in Eqs. (6.115) and (6.117) assuming that all of the signal power is within
the new passband.

6.8.3 A communication system operates in the presence of white noise with two-sided
power spectral density $S_n(\omega) = 0.25 \times 10^{-14}$ W/Hz and with total path losses
(including antennas) of 100 dB. The input bandwidth is 10 kHz. Calculate the
minimum required carrier power of the transmitter for a 10-kHz sinusoidal input
and a 40-dB output S/N ratio if the modulation is

   a) AM (DSB-LC), with $m = 0.707$ and $m = 1.0$;
   b) FM, with $\Delta f = 10$ kHz and $\Delta f = 50$ kHz;
   c) PM, with $\Delta \theta = 1$ radian and $\Delta \theta = \pi$ radians.

6.8.4 Repeat Problem 6.8.3 on the basis of the total transmitted power, including SSB-SC as a fourth alternative. Also calculate the bandwidth required for each system.

6.8.5 A frequency-division multiplexing system uses SSB-SC subcarrier modulation and FM main-carrier modulation. There are forty (40) equal-amplitude voice-input channels, each band-limited to 3.3 kHz. A 0.7-kHz guard band is allowed between channels and below the first channel.

    a) Determine the final transmission bandwidth if the peak frequency deviation is 800 kHz.

    b) Compute the degradation in signal-to-noise of input No. 40 when compared to input No. 1. (Assume a white input noise spectral density to the discriminator and no deemphasis.)

★ **6.9.1** Make a graph of the receiver noise figure (in dB) required for full quieting versus voltage input (in μV) across 300 Ω for a receiver designed for the commercial FM broadcast band.† State your assumptions.

★ **6.9.2** Two given commercial FM receivers are advertised at differing prices. Both have the same IF bandwidth and both have input resistances of 300 Ω. The more expensive receiver advertises an input of 1.8 μV for full quieting, whereas the cheaper one requires an input of 2.5 μV for full quieting.† Both receivers are connected to antennas; the effective antenna noise temperature in each case is 1400 K. How many dB of antenna gain can be sacrificed by the owner of the more expensive receiver to give the same S/N performance in both receivers when tuned to the same station?

6.10.1 A 13-kHz sinusoidal signal is to be transmitted using FM in the presence of additive white noise. If the S/N improvement at the demodulator output is required to be 20 dB, determine the required peak frequency deviation if (a) no preemphasis/deemphasis is used; and (b) the standard preemphasis/deemphasis is used (cf. Fig. 6.34).

6.10.2 An FM receiver is used to receive a PM transmission. Describe what will happen if (a) no deemphasis is used in the receiver; and (b) the standard commercial deemphasis is used in the receiver. Sketch some spectral plots to support your reasoning.

6.10.3 A given preemphasis/deemphasis system is shown in Fig. P–6.10.3. The power spectral density of the additive noise is $S_n(\omega) = \exp(10^{-4}|\omega|)$ μW/Hz. The frequency transfer function of the deemphasis filter is designed to yield a white output noise spectral density over the frequency range $0 < f < 10$ kHz.

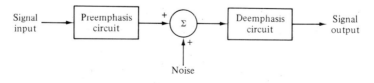

Fig. P–6.10.3

---

a) What is the magnitude frequency transfer function $H(\omega)$ of the preemphasis filter required to yield no overall net signal distortion?

b) Calculate the mean-square signal-to-noise improvement obtained using this system over the frequency range $0 < f < 10$ kHz if $H(0) = 1$.

6.10.4 In stereo FM receivers the deemphasis circuits are placed after the separation into the L and R signals.

a) Compare the resulting S/N ratio with that of a mono FM receiver if no preemphasis were used in either.

b) Show that stereo FM is 22 dB noisier than monophonic FM if the standard preemphasis circuits are used. [*Hint*: Use Eq. (5.65).]

CHAPTER 7

PULSE
MODULATION

In the two previous chapters we have considered applications of continuous-wave (CW) modulation. Recall that in CW modulation the parameters of a continuous sinusoid are varied in proportion to the modulating signal. In amplitude modulation, the amplitude of the sinusoid is varied; in angle modulation the angle and the angular rate are varied. Now we shall consider another category of modulation—that of pulse modulation.

From the sampling theorem, we know that to convey the information contained in a band-limited signal it is necessary to send only a finite number of discrete samples. A low-pass modulating signal that is band-limited to f_m Hz is completely specified by its values at intervals spaced no greater than $(2f_m)^{-1}$ seconds apart. We conclude that instead of transmitting the complete signal in analog form we really need to transmit only a discrete number of samples.

In pulse modulation, these discrete samples are used to vary a parameter of a pulse waveform. For example, one may vary the amplitude, width, or position of the pulse waveform in proportion to the sampled signal.

Consider a given analog signal $f(t)$ which is sampled at the uniformly spaced time intervals t_i, as shown in Fig. 7.1(a). A corresponding table of the sample values appears in Fig. 7.1(b).

This discrete data is to be transmitted in such a way that when the signal is demodulated at the receiver the first output pulse is a 4, the second a 3, the third a 1, etc., until the entire table of sample values is reconstructed correctly at the receiver. There is no reason, of course, why we must use the same time scale for the pulse train as was used for the modulating signal. For instance, one could sample the signal $f(t)$ using 1000 samples taken over a period of one second and then proceed to transmit this data at a rate of one pulse per minute for the next 1000 minutes. On the other hand, one could just as well transmit the entire 1000 sample values in one millisecond. (Such data slow-down and speed-up techniques have actually been used in satellite experiments.) However, unless we specify to the contrary, we shall not alter the time scaling in our discussion of pulse modulation and all signal transmissions will be in "real time."

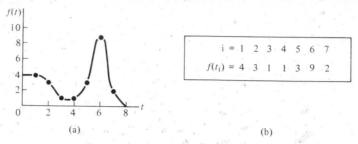

i =	1	2	3	4	5	6	7
$f(t_i)$ =	4	3	1	1	3	9	2

(a)   (b)

Fig. 7.1 A sampled signal and its table of sample values.

Note that we may or may not be able to reconstruct the original signal $f(t)$ exactly at the receiver. The sampling theorem provides the necessary conditions for recovering $f(t)$ from the $f(t_i)$. If the sampling permits the determination of $f(t)$ from the transmitting table, then we wish to make sure that it can be determined at the receiver. Therefore correct sampling is a problem to be solved *a priori* at the transmitter and no correction can be attempted at the receiver for an inadequate sampling rate. Our primary concern here will be to devise a system which will use no more bandwidth than necessary to permit the transmission of a table of sample values via pulse modulation and best reconstruction of the table at the receiver.

7.1   PULSE-AMPLITUDE MODULATION (PAM)

In pulse-amplitude modulation (PAM) the amplitude of a train of constant-width pulses is varied in proportion to the sample values of the modulating signal. The pulses are usually taken at equally spaced intervals of time. An example of a PAM signal is shown in Fig. 7.2. Because the generation of PAM has similarities to sampling, we shall quickly review sampling before proceeding.

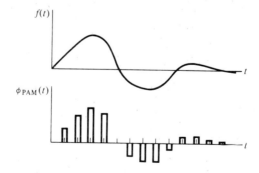

Fig. 7.2 Pulse-amplitude modulation.

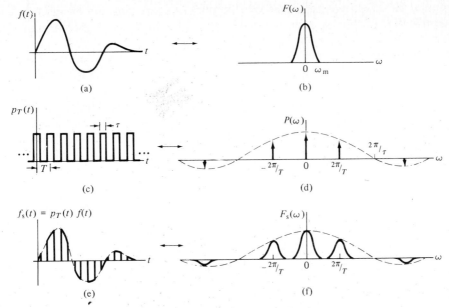

Fig. 7.3 Natural sampling of a band-limited signal.

Consider a low-pass signal $f(t)$ which is band-limited to f_m Hz and multiplied by a periodic train of rectangular pulses $p_T(t)$, as shown in Fig. 7.3. The sampling interval T is taken as the Nyquist interval $(2f_m)^{-1}$ seconds. The sampled signal $f_s(t)$ is the product of $f(t)$ and $p_T(t)$:

$$f_s(t) = f(t)p_T(t). \tag{7.1}$$

The spectral density of the sampled signal is found by taking the Fourier transform of Eq. (7.1):

$$F_s(\omega) = \frac{1}{2\pi}F(\omega) \circledast P(\omega). \tag{7.2}$$

The convolution is easy to perform in this case as a result of the impulse functions and is shown in Fig. 7.3(f). The equivalent analytical expression for this result is

$$F_s(\omega) = \frac{1}{2\pi}F(\omega) \circledast \frac{\tau}{T} \sum_{n=-\infty}^{\infty} \text{Sa}\,(n\pi\tau/T)2\pi\,\delta(\omega - n2\pi/T), \tag{7.3}$$

$$F_s(\omega) = \frac{\tau}{T} \sum_{n=-\infty}^{\infty} \text{Sa}\,(n\pi\tau/T)F(\omega - n2\pi/T). \tag{7.4}$$

We conclude from this brief review that the sampling of $f(t)$ results in the generation of spectral replicas at multiples of the periodic sampling rate. The sampling pulses do not have to be rectangular in shape and the particular choice of a pulse form will only alter the shape of the envelope of the spectrum of $F_s(\omega)$. What is important to us here is that each spectral replica generated by the sampling is an exact reproduction of the original spectral density $F(\omega)$ only displaced in frequency. The original signal $f(t)$ can be recovered from the sampled signal $f_s(t)$ by using an ideal low-pass filter.

In the case of natural sampling discussed above, the amplitudes of the pulses varied in proportion to the sampled values of the modulating signal $f(t)$. However, the pulse shapes also varied slightly, as a comparison of Figs. 7.2 and 7.3 will show. Specifically, the slopes of the pulse tops varied with the slopes of the modulating signal at the sample points in the case of natural sampling. In PAM, the pulse tops are flat. We shall investigate what effect this may have on the spectral characteristics.

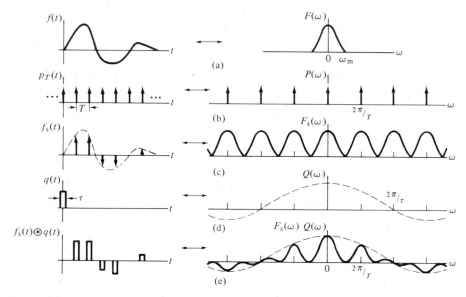

Fig. 7.4 Spectrum of a PAM signal.

First we let the sampling pulse width τ become very narrow so that essentially we have an impulse sampler, as shown in Fig. 7.4. Assuming unit area for each impulse, we have, as a special case of Eq. (7.4),

$$F_s(\omega) = \frac{1}{T} \sum_{n=-\infty}^{\infty} F(\omega - n2\pi/T).$$ (7.5)

The impulse-sampler time waveform can be written as

$$p_T(t) = \sum_{n=-\infty}^{\infty} \delta(t - nT).$$ (7.6)

The impulse-sampled waveform is, using Eq. (7.1),

$$f_s(t) = f(t) \sum_{n=-\infty}^{\infty} \delta(t - nT),$$

$$f_s(t) = \sum_{n=-\infty}^{\infty} f(nT) \, \delta(t - nT), \tag{7.7}$$

where the $f(nT)$ are the instantaneous sample values of $f(t)$. These impulse samples, weighted by the sample values of $f(t)$, are applied to a linear time-invariant filter with unit impulse response $q(t)$. The output of the filter is

$$f_s(t) \circledast q(t) = \sum_{n=-\infty}^{\infty} f(nT) \, \delta(t - nT) \circledast q(t)$$

$$= \sum_{n=-\infty}^{\infty} f(nT) q(t - nT). \tag{7.8}$$

Thus the impulse response of the filter, $q(t)$, can be chosen to approximate the desired output pulse shape. If $q(t)$ is the rectangular impulse response shown in Fig. 7.4(d), Eq. (7.8) represents the desired PAM pulse train. In this pulse train every pulse is rectangular in shape with an amplitude proportional to the sample value of $f(t)$ at the sampling points. The pulses are equally spaced in time.

Now that we are satisfied that we have generated a PAM pulse train, let us examine its spectral density. Using Eq. (7.5) and recalling that convolution in time is equivalent to a multiplication of spectral densities, we have

$$F_s(\omega)Q(\omega) = \frac{1}{T} \sum_{n=-\infty}^{\infty} F(\omega - n2\pi/T)Q(\omega). \tag{7.9}$$

This is illustrated in Fig. 7.4(e). But this spectral density is not the same as that which was obtained for the sampled waveforms in Fig. 7.3. In Fig. 7.3 the spectrum consisted of $F(\omega)$ repeated at multiples of the sampling frequency with only a gain variation of each spectral replica. This gain variation was determined by the shape of the sampling pulse, as expressed in Eq. (7.4). In contrast, Eq. (7.9) describes a point-by-point multiplication in frequency so that the spectral density $F(\omega)$ has lost its original shape. This distortion is dependent on the pulse shape; at low frequencies it is not severe if the pulse width is very narrow. Thus we conclude that there is a subtle difference between natural sampling with rectangular pulses and the generation of PAM with flat-topped pulses.

At this point, a question about why we are so interested in flat-topped pulses usually arises. The reason is that we do not need to use the shape of the pulses to convey information, and a rectangular pulse shape is an easy one to generate. When signals are transmitted over comparatively long distances, repeaters are often necessary to filter and amplify the signals before transmission to the receiver or next repeater. In analog CW modulation systems, as repeaters must amplify the signals faithfully, the effects of additive noise are compounded. In the type of pulse modulation system that we are considering here, the information is in the pulse amplitudes at the sample times only. Because pulse *shape* is not important, repeaters can regenerate rather than amplify the pulses. For example, a new pulse can be regenerated if we make the pulse amplitude proportional to the area of the pulse input detected over a fixed width or time interval. This pulse regeneration has some advantages in signal-to-noise ratio.

Next we shall consider the problem of recovering $f(t)$ from the PAM waveform. A possibility is to sample the PAM waveform with a periodic train of very narrow pulses (theoretically, impulses) and then use a low-pass filter to smooth the result. This sampling pulse train must be synchronized with the incoming PAM signal.

Another method of signal recovery, which was used in sampling, is to use low-pass filtering. However, because the low-frequency PAM spectrum is now given by $F(\omega) \times Q(\omega)$, it is not possible to recover $f(t)$ exactly using only a low-pass filter. The spectral weighting which has been introduced in the PAM process is known [that is, $Q(\omega)$] so the possibility of removing this effect does exist. Synthesizing a filter which has a transfer function of $Q^{-1}(\omega)$ will satisfy this goal. It generally is not possible to build such a filter over a wide range of frequencies, particularly if $Q(\omega)$ goes to zero. But because $f(t)$ is band-limited, it is sufficient to synthesize the inverse filter over only a very limited frequency range. We can combine the required low-pass filter with the inverse filter as shown in Fig. 7.5. The transfer function of the resulting filter is

$$H(\omega) = \begin{cases} Q^{-1}(\omega) & |\omega| < \omega_m \\ 0 & \text{elsewhere} \end{cases} \tag{7.10}$$

This technique of correcting the frequency response of a system for a known distortion is called *equalization*. Equalization is often used in correcting distortions which are known but over which one has little control. For example, telephone lines are known to introduce both amplitude and phase distortions and these become objectionable when the lines become quite long. Repeater amplifiers spaced along the lines are used not only to amplify the signals to offset signal losses but also to correct for known distortions.

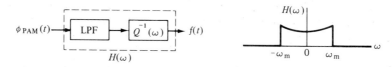

Fig. 7.5 Recovering $f(t)$ from a PAM waveform using equalization.

For a rectangular pulse shape, the spectral density $Q(\omega)$ is a $(\sin x)/x$ pattern, as shown in Fig. 7.6. Because the signal has been sampled at the Nyquist rate, the sampling period T is related to the maximum frequency f_m by $f_m = 1/(2T)$.

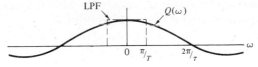

Fig. 7.6 Equalization for a rectangular pulse shape.

As the pulse width τ is made smaller, the zero crossings in the $(\sin x)/x$ pattern of $Q(\omega)$ move farther out in frequency and thus $Q^{-1}(\omega)$ tends to become more flat within the assigned bandwidth of the low-pass filter. Because T does not depend on τ, the ratio τ/T is a measure of the flatness of $Q(\omega)$ and $Q^{-1}(\omega)$ within the bandwidth of the low-pass filter. For a rectangular pulse, it turns out that as long as $\tau/T \leq 0.1$, the maximum difference between $Q^{-1}(\omega)$ and the ideal low-pass filter over the required range is less than 1%. In practice, then, the equalization for PAM can usually be neglected as long as $\tau/T \leq 0.1$.

7.2 TIME-DIVISION MULTIPLEXING (TDM)

The use of fairly short pulse widths in PAM signals leaves sufficient space between samples for insertion of pulses from other sampled signals. The method of combining several sampled signals in a definite time sequence is called *time-division multiplexing* (TDM). We shall discuss the principles of TDM here with particular reference to PAM although the principles apply as well to other types of pulse modulation.

Suppose we wish to time-multiplex two signals using PAM. Two alternative methods for accomplishing this are shown in Fig. 7.7. Usually digital logic circuitry is employed to implement the timing operations shown in these diagrams. The use of FET's at lower frequencies and the diode ring samplers discussed in Chapter 5 at higher frequencies is popular in realizing the sampling operations. The commutator determines the synchronization and sequence of the channels (signals) to be sampled. The pulse generator produces the narrow pulses desired to drive the sampler(s). The clock determines the timing of the overall system.

The two alternative methods shown in Fig. 7.7 basically differ only in that in the first system the commutator handles the analog signals before sampling and in the second system it handles the sampler control pulses. The second method is often preferred because it lends itself to digital logic circuitry even though it does require more samplers.

To illustrate the operation of time multiplexing it will be convenient to choose an example. Let us assume that both input signals $f_1(t)$, $f_2(t)$ are low-pass and band-limited to 3 kHz. The sampling theorem states that each must be sampled at a rate no less than 6 kHz. This requires a 12-kHz minimum clock rate for the two-channel system. The time-multiplexed PAM output might appear something like that shown in Fig. 7.8.

The time-multiplexed PAM signal could now be sent out on a line or used to modulate a transmitter (this will be discussed later in this section). But the bandwidth of this signal is very wide (theoretically infinite). Therefore we should decide just how much bandwidth is really necessary to transmit the information and to then correctly demodulate it at the receiver.

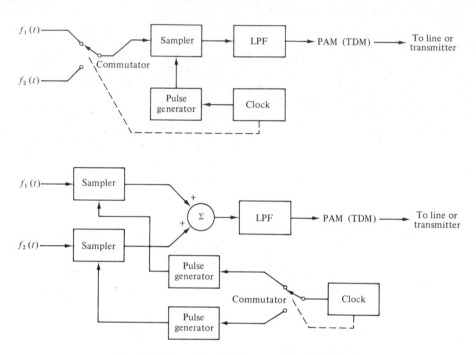

Fig. 7.7 Generation of time-multiplexed PAM.

First, let us define T_x as the time spacing between adjacent samples in the time-multiplexed signal waveform (see Fig. 7.8). If all input signals (channels) are sampled equally, it is fairly obvious that $T_x = T/n$, where n is the number of input signals and T is the sampling period for one signal. Now we ask the question: If only the amplitude information is important here (not reproduction of pulse shaping), what is the absolute minimum bandwidth required such that the information in each sampled channel remains independent of that in the other channels? To answer this question, we appeal to the arguments of the sampling theorem. The sampling theorem states that there are $2BT$ independent samples of information in a signal that is band-limited to B Hz and time-limited to T seconds. To prevent any irretrievable loss of information in the composite waveform then requires that the bandwidth B_x of the low-pass filter must satisfy the criterion

$$B_x \geq \frac{1}{2T_x}. \tag{7.11}$$

It should be emphasized that the use of this bound will not yield pulse shapes which closely resemble the pulses generated in the samplers. Pulse shape recognition requires additional bandwidth. What Eq. (7.11) does say is that independent amplitude information is present at the proper times if this condition is satisfied. Recall that the sampling theorem assumed the availability of ideal low-pass filters to satisfy the limiting condition.

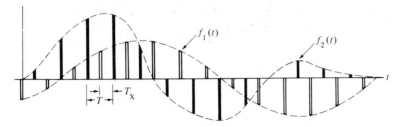

Fig. 7.8 Time multiplexing of two PAM signals.

A physical interpretation of Eq. (7.11) can be made as follows. After the two signals $f_1(t)$ and $f_2(t)$ in Fig. 7.8 have been sampled and combined, the resultant time-multiplexed signal can be considered as a new sampled signal, $f_3(t)$. The samples of the signal $f_3(t)$ are spaced at intervals of T_x seconds. The sampling theorem states that the highest frequency f_{max} which can be resolved in $f_3(t)$ without ambiguities is

$$f_{max} = \frac{1}{2T_x}.$$

In order to pass all of this possible range of frequencies, then, requires a system with a bandwidth satisfying Eq. (7.11).

We have preferred to find the bandwidth of the time-multiplexed PAM system in the time domain, just as we preferred to use the frequency domain for frequency multiplexing. At this point the unwary reader should be warned against some fallacious reasoning in computing minimum bandwidth for time-multiplexed PAM in the frequency domain. The line of reasoning could go something like this: Each input to the two-channel PAM system is band-limited to 3 kHz. The minimum bandwidth required for each channel is therefore 3 kHz, and if we add the two channels no new frequencies are generated so that we will need only 3 kHz. An extension leads to the ridiculous conclusion that no matter how many of these signals are multiplexed the composite waveform only requires 3 kHz! *This reasoning is wrong*—but why? A moment's reflection shows that this erroneous reasoning is based entirely on spectral magnitude considerations. However, *the phase relationships are essential to time-multiplexing and cannot be ignored*. It is generally more convenient to work in the time domain than to work with these phase relations.

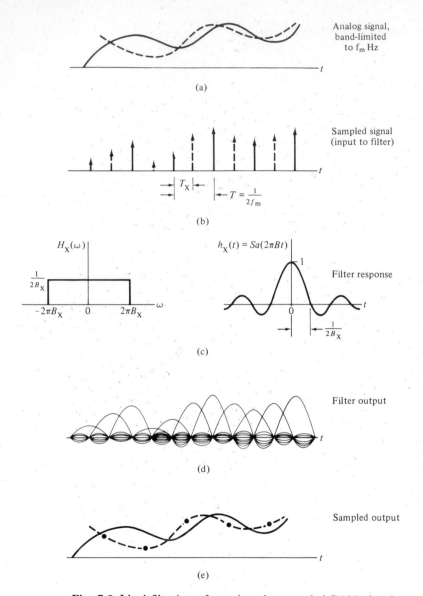

Fig. 7.9 Ideal filtering of two impulse-sampled PAM signals.

Some additional insight into the proper choice of bandwidth can be gained by assuming that the sample pulse widths are so narrow that they can be approximated by impulses. In addition, the filter is assumed to be an ideal low-pass filter with transfer function (see Fig. 7.9):

$$H_x(\omega) = \begin{cases} \pi/W & |\omega| < W, \\ 0 & \text{elsewhere.} \end{cases} \tag{7.12}$$

The impulse response of this filter is

$$h_x(t) = \mathcal{F}^{-1}\{H_x(\omega)\} = \text{Sa}(Wt).   \tag{7.13}$$

The effective width of this impulse response is $2\pi/W = 1/B$, as shown in Fig. 7.9(c).

The arrival of each impulse in the time-multiplexed waveform will produce a replica of the impulse response of the filter, weighted and delayed by the area and timing of the impulse. Using superposition, the overall response is a sequence of $(\sin x)/x$ terms, as shown in Fig. 7.9(d). Because we have chosen the spacing between successive samples to be $T_x = (2B)^{-1}$, contributions from all adjacent channels are exactly zero at the correct sampling instant. Therefore, by sampling the output at the correct instants, one can exactly reconstruct the original sampled values.

The equality sign in Eq. (7.11) then refers to the case in which impulse sampling and ideal filtering are used. Because neither of these conditions holds in practice, the requirement on the bandwidth must be relaxed somewhat. How much is a matter of engineering judgment and depends on the allowable crosstalk between channels. The amount of signal from an adjacent channel which spills over into the desired time slot gives rise to this *interchannel crosstalk*. Interchannel crosstalk can be minimized by allowing more bandwidth than the minimum required and, to a lesser extent, by operating the different channels at comparable voltage levels.

Let us apply the results of the preceding discussion to the two-channel PAM example that we started out to analyze. Each input signal was band-limited to 3 kHz so that $T = (6 \text{ kHz})^{-1} = 166.7 \ \mu\text{sec}$, $n = 2$, and therefore $T_x = 83.3 \ \mu\text{sec}$. Using Eq. (7.11), we find that the minimum bandwidth needed for the time-multiplexed signal transmission is

$$B_x \geq (166.7 \ \mu\text{sec})^{-1} = 6 \text{ kHz}.$$

From this result we learn that PAM is just as efficient in conserving bandwidth as SSB, at least when all input signals have the same bandwidth! This efficiency decreases when the bandwidths of the input signals differ. To conserve bandwidth, PAM time-multiplexed systems usually group input signals of comparable bandwidths.

At the receiver the composite time-multiplexed and filtered waveform must be resampled and separated into the appropriate channels. Once the pulses are separated, the normal sampling considerations apply and the analog reconstruction of the signals can be obtained by low-pass filtering. Block diagrams of two possible receivers are shown in Fig. 7.10. Operation of the sample-and-hold circuit shown will be discussed shortly.

Synchronization of the clock and the commutator in the time-multiplex receiver is necessary and can be achieved in various ways. One method used is to periodically send a pulse which is known to exceed the height of any other pulses in the transmission. These can be identified by the receiver and used for synchronization. Another possibility is to use a pilot tone which is shifted in frequency from the time-multiplexed transmission. A more elaborate scheme is to reserve a portion of the time-multiplexed transmission (e.g., one channel) to send some preassigned code which, when identified at the receiver, serves to synchronize the timing. This is considered in Section 7.11.

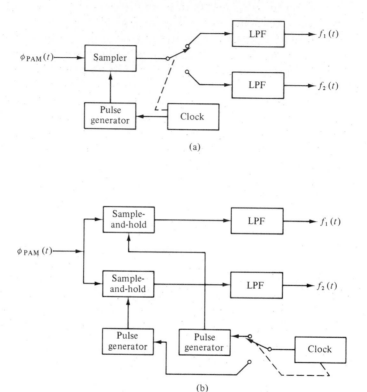

(a)

(b)

Fig. 7.10 Receivers for time-multiplexed PAM.†

When a large number of signals are time-multiplexed together in a PAM system, the width of the sampling pulses must be made very narrow in comparison to the sampling interval. Under these conditions the signal power at the final low-pass filter in the receiver decreases to the point where large amplification factors are needed. An attractive alternative to this is to use a sampling switch

---

† In the second system, the pulse generators can be connected to form a ring. Each pulse generator operates only when it receives simultaneous pulses from the preceding generator in the ring and the clock. This eliminates the need for a separate digital divider (commutator).

followed by a capacitor, known as a *sample-and-hold* circuit and shown in Fig. 7.11. The switch closes only when that particular channel is to be sampled. If the source impedance r is small, the capacitor voltage changes to the input voltage within the time τ that the switch is closed. The load impedance R is arranged to be high so that the capacitor retains the voltage level until the switch is closed again. Therefore the sample-and-hold circuit accepts only those values of the input which occur at the sampling times and then holds them until the next sampling time. A low-pass filter is still needed to smooth the output but the requirements on filtering are less severe with the sample-and-hold circuit. Use of the sample-and-hold circuit results in a very efficient, reliable, and relatively noise-free PAM demodulator without the need for high amplification. Note that the sample-and-hold circuit is an example of a linear time-varying system. Also note that, at least in this form, this circuit requires equalization to yield good signal fidelity. An alternative to the simple holding circuit is an interpolation algorithm that provides an accurate smoothed estimate of the correct output based on observations of past sample values.

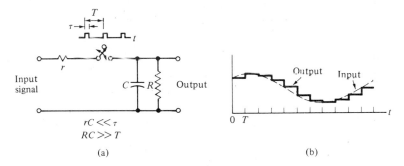

Fig. 7.11 The sample-and-hold circuit and its use as a PAM demodulator.

After time-multiplexing and filtering, the pulse-modulated waveform may be transmitted directly on a pair of wire lines. For long distance transmission, however, it is more convenient to use higher frequencies and transmit the signal using electromagnetic radiation. The PAM spectrum may be translated to these higher frequencies using the CW modulation techniques discussed in previous chapters. If amplitude modulation is used (DSB), the composite modulation is referred to as PAM/AM. If angle modulation is used, the systems are designated as PAM/FM or PAM/PM. A block diagram representation of a PAM/AM system is shown in Fig. 7.12.

As noted earlier, time-multiplexed PAM is most efficient when all input signals are band-limited to the same frequency. In large multiplexing systems where differing data rates are to be handled, it is fairly common practice to group the input data channels by bandwidths. Each group is pulse-modulated and time-multiplexed and then these groups are frequency-multiplexed using subcarrier frequencies. Finally this composite signal is modulated using CW modulation. An example of a PAM/AM/FM system is shown in Fig. 7.13.

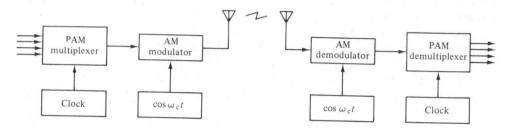

Fig. 7.12 A block diagram representation of PAM/AM.

Time multiplexing has been discussed here with particular emphasis on the use of PAM. The same principles apply to other types of pulse modulation.

Advantages in the use of pulse modulation with time-division multiplexing (TDM) include the fact that the circuitry required is digital, thus affording high reliability and efficient operation. This circuitry is simpler than the modulators and demodulators required in frequency-division multiplexing (FDM) systems. The multiplexing of many channels of relatively low-frequency data can be performed very efficiently using TDM when the inputs are all of comparable bandwidths.

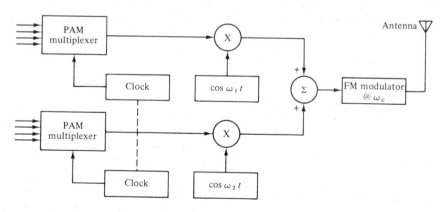

Fig. 7.13 Example of a PAM/AM/FM transmitter.

Another advantage of TDM systems is the relatively small interchannel cross-talk arising from nonlinearities in the amplifiers which handle the signals in the transmitter and receiver. These nonlinearities produce harmonics that affect higher-frequency channels in FDM systems. Thus the phase and amplitude linearity requirements of the amplifiers in FDM become more stringent as more channels are added. In contrast, in TDM systems the signals from different channels are not handled simultaneously but are allotted different time intervals. Hence the linearity requirements do not become more stringent as the number of channels is increased and the effects of distortion are shared equally by all channels.

Disadvantages of TDM include the fact that pulse accuracy and timing jitter become a major problem at high frequencies so that TDM systems normally operate at clock frequencies below 100 MHz. In addition, time synchronization is required between transmitter and receiver.

Example 7.2.1   Channel 1 of a two-channel PAM system handles 0–8 kHz signals; the second channel handles 0–10 kHz signals. The two channels are sampled at equal intervals of time using very narrow pulses at the lowest frequency that is theoretically adequate. The sampled signals are time multiplexed and passed through a low-pass filter before transmission. At the receiver the pulses in each of the two channels are passed through appropriate holding circuits (i.e., sample-and-hold) and low-pass filters.

a) What is the minimum clock frequency of the PAM system?

b) What is the minimum cutoff frequency of the low-pass filter used before transmission that will preserve the amplitude information on the output pulses?

c) What would be the minimum bandwidth if these channels were frequency-multiplexed, using normal AM techniques and SSB techniques?

d) Assume the signal in channel 1 is $\sin (5000\pi t)$ and that in channel 2 is $\sin (10,000\pi t)$. Sketch these signals; sketch the waveshapes at the input to the first low-pass filter, at the filter output, and at the output of the sample-and-hold circuit and output of the low-pass filter in channel 2.

Solution

a) In order to sample channel 2 adequately, we must take samples at a 20 kHz rate. Therefore the commutator clock rate is 40 kHz and the commutator must recycle at a 20 kHz rate.

b) For the composite (interlaced) signal, $T_x = 25 \ \mu$sec so that

$$B_x \geq \frac{1}{2T_x} = 20 \text{ kHz}.$$

c) For AM, min $B = 2(10+8) \text{ kHz} = 36 \text{ kHz}$.
   For SSB, min $B = (10+8) \text{ kHz} = 18 \text{ kHz}$.
   Note that SSB is more efficient in terms of the bandwidth required if the signals into the PAM system are not of equal bandwidth. For this reason,

low data rate signals are usually time-multiplexed separately from medium data rate signals.

d) See Fig. 7.14.

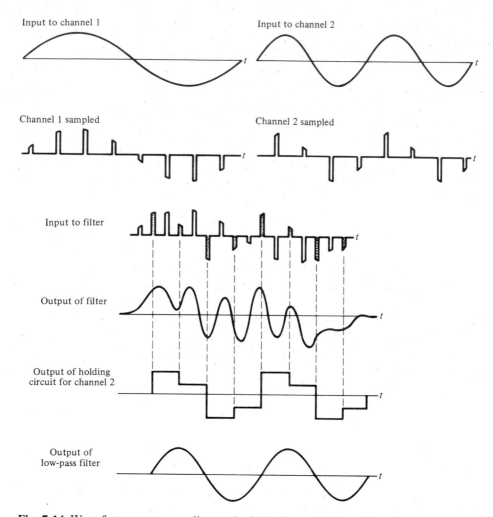

Input to channel 1

Input to channel 2

Channel 1 sampled

Channel 2 sampled

Input to filter

Output of filter

Output of holding circuit for channel 2

Output of low-pass filter

Fig. 7.14 Waveforms corresponding to the PAM time-multiplexed system of Example 7.2.1.

Drill Problem 7.2.1   Two low-pass signals, each band-limited to 4 kHz, are to be time multiplexed into a single channel using PAM. Each signal is impulse-sampled at a rate of 10 kHz. The time-multiplexed signal waveform is filtered by an ideal low-pass filter (LPF) before transmission.

a) What is the minimum clock frequency of the system?

b) What is the minimum cutoff frequency of the LPF?

c) In the receiver, the signal is impulse-sampled and demultiplexed using the system shown in Fig. 7.10(a). Determine the minimum and maximum acceptable bandwidth of the LPF used in retrieving the analog signals.

Answer.   (a) 20 kHz; (b) 10 kHz; (c) 4 kHz, 6 kHz.

7.3 PULSE SHAPING AND INTERSYMBOL INTERFERENCE

In the preceding discussion we discussed the use of narrow pulse widths and assumed an ideal low-pass filter with bandwidth B. As each pulse passes through this low-pass filter, the corresponding output is a $(\sin x)/x$ waveform, appropriately scaled and delayed by the pulse amplitude and position [Fig. 7.9(d)]. With pulses spaced by $T = 1/(2B)$ seconds, the received waveform, when observed at the correct sampling instants, results in zero interference between adjacent pulses.

Attainment of the zero-interference condition between adjacent pulses requires precise synchronization between transmitter and receiver. It also demands that the transmission medium (or channel) behaves like an ideal low-pass filter. However, this is physically unrealizable and difficult to approximate in practical systems. Use of a realizable filter characteristic causes the latter assumption to be invalid, and errors in transmitter-receiver synchronization are difficult to eliminate. The result is that each received pulse is affected somewhat by the presence of adjacent pulses. In the case of multichannel pulse modulation, overlap between pulses gives rise to interchannel crosstalk, which in general is called *intersymbol interference* (ISI).

ISI can be decreased by purposely widening the transmission bandwidth. However, this alternative may be unnecessarily wasteful of bandwidth if not carefully controlled. Therefore we try to design received signal waveforms (and hence the transmission filters) to minimize the ISI within as small a transmission bandwidth as possible.

We begin by representing a received PAM signal $y(t)$ as[†]

$$y(t) = \sum_{m} a_m x(t - mT) + n(t), \qquad (7.14)$$

where m is an integer and T is the sampling period. Here $x(t)$ represents the pulse waveform, a_m represents the sampled values of the input, and $n(t)$ is the additive noise. At a given time, $t = kT$, we have

$$y(kT) = a_k x(0) + \sum_{m \ne k} a_m x[(k-m)T] + n(kT). \qquad (7.15)$$

---

† We assume the single-channel case here for convenience, but the results are readily extended to the multichannel TDM case.

In Eq. (7.15), the first term represents the kth sampled value of the input that has been transmitted. The second term in Eq. (7.15) arises from the overlap of other pulses adding to the desired pulse $a_k x(t - kT)$ at the kth sampling time. Thus this term represents the ISI present. The last term in Eq. (7.15) represents additive noise at the time $t = kT$.

The ISI given by the second term in Eq. (7.15) can be eliminated by choosing the received pulse shape $x(t)$ to satisfy the following condition:†

$$x[(k - m)T] = \begin{cases} 1 & m = k \\ 0 & m \neq k \end{cases}. \tag{7.16}$$

Note that the ideal low-pass filter in the preceding section (see Fig. 7.9) satisfies this condition. The uniformly spaced zeros in the impulse response at all nonzero multiples of the sampling periods yield zero ISI in this case. Because that filter is not physically realizable and is even difficult to approximate in practice, we modify the sharp magnitude-cutoff characteristic to attain a more gradual "roll-off" in frequency that does not affect the desirable uniformly spaced zeros in the impulse response. This is possible if the adjusted filter magnitude response has odd symmetry about the low-pass cutoff frequency.‡ Satisfaction of this requirement gives rise to a class of waveforms known as *Nyquist waveforms*. In other words, all Nyquist waveforms have uniformly spaced zeros at all multiples of a basic interval except one (at the center).

One type of Nyquist waveform that is popular in both analytical work and practice is that of the raised cosine. A raised-cosine characteristic consists of a flat, or constant, magnitude at low frequencies and a roll-off portion that has a sinusoidal form with odd symmetry about the cutoff frequency. The raised-cosine characteristic can be expressed in equation form as§

$$X(\omega) = \begin{cases} T & 0 \leq |\omega| \leq (1 - \alpha)W \\ \dfrac{T}{2}\left\{1 - \sin\left[\dfrac{\pi}{2\alpha W}(|\omega| - W)\right]\right\} & (1 - \alpha)W \leq |\omega| \leq (1 + \alpha)W \quad (7.17) \\ 0 & |\omega| > (1 + \alpha)W \end{cases}$$

where $W = \pi/T$. The parameter α is the excess bandwidth used divided by the minimum Nyquist bandwidth. Thus the case for $\alpha = 0$ coincides with the ideal low-pass filter used in the preceding section. The corresponding impulse response is

$$x(t) = \left(\frac{\sin Wt}{Wt}\right)\left(\frac{\cos \alpha Wt}{1 - (2\alpha Wt/\pi)^2}\right). \tag{7.18}$$

---

† This assumes the convenient normalization: $x(0) = 1$.

‡ M. Schwartz. *Information Transmission, Modulation, and Noise*, Third ed. New York: McGraw-Hill, 1980, p. 181.

§ Note that this characteristic could be expressed in terms of $\cos^2(\cdot)$.

The first term in the right-hand side of Eq. (7.18) is the $(\sin x)/x$ waveform that we had used previously for the ideal low-pass filter, and it retains the original zero crossings of that waveform. The second term is a result of the more gradual spectral roll-off.

Plots of $x(t)$ and $X(\omega)$ are shown in Fig. 7.15 for three values of α. The case for $\alpha = 1$ is known as the full-cosine roll-off characteristic; its frequency transfer function is

$$X(\omega) = \begin{cases} \dfrac{T}{2}\left(1 + \cos\dfrac{\pi\omega}{2W}\right) & |\omega| \le 2W \\[2mm] 0 & \text{elsewhere} \end{cases} \qquad (7.19)$$

Note that for the $\alpha = 1$ case, $x(t)$ has zeros halfway between sampling times in addition to the sampling zeros. The amplitudes of the oscillatory tails of $x(t)$ are smallest when $\alpha = 1$ (i.e., the most gradual roll-off). Therefore this type of response results in lower ISI for a given timing error in sampling, even though it requires the transmission of some spectral components that are up to twice those required for the ideal low-pass case. The parameter α is called the *roll-off factor*.

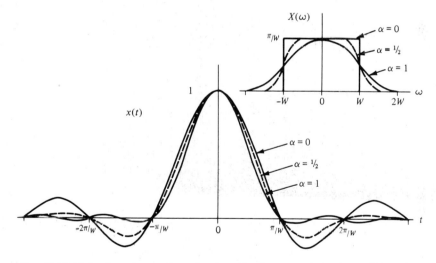

Fig. 7.15 Raised-cosine pulse shaping (in frequency) and resulting time responses.†

In summary, the bandwidth occupied for a raised-cosine-type transmission characteristic varies from a minimum of $B = 1/(2T)$ Hz ($\alpha = 0$) to a maximum

---

† The ratio π/W can be written as $1/(2B)$ to make this agree more closely with Fig. 7.9(c).

of $B = 1/T$ Hz ($\alpha = 1$). The larger values of α lead to faster decaying pulses so that receiver synchronization will be less critical and modest timing errors will not cause large amounts of ISI, but at the expense of added bandwidth. To control the amount of tolerable ISI, then, Eq. (7.11) should be modified to

$$B_X = \frac{(1 + \alpha)}{2T_X}. \tag{7.20}$$

So far we have considered those filters whose impulse response is the required pulse shape for zero ISI. In practice, the inputs are not impulses but finite-width pulses. Designating the Fourier transform of these finite-width pulses by $Q(\omega)$, we find that the magnitude of the frequency transfer function of the required shaping filter is

$$|H(\omega)| = \left|\frac{X(\omega)}{Q(\omega)}\right|. \tag{7.21}$$

As noted in Section 7.1, however, this correction is not significant if the pulse width is much narrower than the sample period.

Finally, note that, strictly speaking, none of the raised-cosine pulse waveforms are physically realizable. However, they can be approximated by causal filters if sufficient time delay (i.e., linear phase) is allowed.

Example 7.3.1   Twenty PAM signals, each band-limited to 3 kHz and sampled at 8 kHz, are time-multiplexed prior to transmission. Choose a raised-cosine filter characteristic that will permit transmission of this multiplexed signal within an absolute maximum bandwidth of 120 kHz.

Solution.   Using Eq. (7.11), we find that the minimum bandwidth with an ideal LPF is $(2T_x)^{-1} = 80$ kHz. Examination of the raised-cosine filter characteristic reveals that the frequency $(2T_x)^{-1}$ is set at the $\frac{1}{2}$-magnitude point. Therefore we set the $\frac{1}{2}$-magnitude point at 80 kHz and use $\alpha = (120 - 80)/80 = 0.50$.

7.4   OTHER TYPES OF ANALOG PULSE MODULATION: PWM AND PPM

Recall that to transmit a signal $f(t)$ that is band-limited to f_m Hz it is only necessary to transmit the information about its sample values at $(2f_m)^{-1}$-second intervals. We have discussed one way to accomplish this by using PAM and letting the amplitudes of a train of constant-width, uniformly spaced pulses vary in proportion to $f(t)$. An alternative modulation method is to vary some parameter in the timing of each pulse to convey the information. Pulse-timing modulation may be accomplished in a number of ways but the principles are basically the same and we only consider two specific types here.

One type of pulse-timing modulation uses constant-amplitude pulses whose width is proportional to the values of $f(t)$ at the sampling instants. This type is designated as *pulse-width modulation* (PWM).† Another possibility is to keep both the amplitude and the width of the pulses constant but vary the pulse positions in proportion to the values of $f(t)$ at the sampling instants. This is designated as *pulse-position modulation* (PPM). PAM, PWM, and PPM waveforms for a given $f(t)$ are shown in Fig. 7.16.

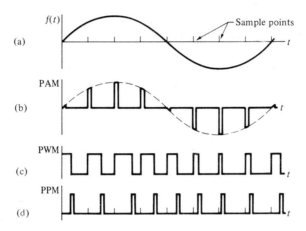

Fig. 7.16 An illustration of PAM, PWM, and PPM.

     The various types of pulse-timing modulation are closely related to each other and one type can be derived from another. In a more general way, the relation of pulse-timing modulation to angle modulation is analogous to the relation of PAM to AM. In fact, one of the methods to generate wideband PM or FM is to first generate PWM or PPM.

     In PWM, the signal $f(t)$ is sampled periodically at a rate fast enough to satisfy the requirements of the sampling theorem. At each sampling instant a pulse is generated with a fixed amplitude and a width that is proportional to the sample values of $f(t)$. A minimum pulse width τ_0 is assigned to the minimum value of $f(t)$. The variation of the pulse width from τ_0 is proportional to $f(t)$ and a modulation constant k_1 is defined in the proportionality. The pulse duration must be shorter than the time slot allocated to a particular sample and an additional guard time τ_g is usually allowed. The modulation in Fig. 7.16(c) is in the timing of the trailing edge of each pulse. Another version of PWM is to modulate both edges simultaneously (i.e., symmetrically).

     PWM is a popular choice where the remote proportional control of a position or a position rate is desired. The average value of a PWM waveform varies directly with the modulation and can be used to control a motor by employing efficient switching-type operations. Proportional control can be maintained rel-

---

† The designation *pulse-duration modulation* (PDM) is also used.

atively independent of signal strength over a wide range. FM is also used for remote position control, but PWM systems are generally easier to build and align. Disadvantages of PWM include the necessity for detection of both pulse edges and a relatively large guard time. Effects of signal transients introduced in the system may vary with pulse width causing nonuniform performance.

Only the trailing edges of the PWM waveforms discussed above contain the modulating information. Therefore one could convey the information by only sending these timing marks. In PPM these are sent as constant-width, constant-amplitude pulses, as shown in Fig. 7.16(d). The minimum pulse delay is used to designate the minimum value of $f(t)$ and the change in delay is proportional to the modulating signal. The constant of proportionality is the modulation constant k_1. While generally more efficient for communications purposes than PWM, the use of PPM does require a method of regenerating the clock timing. In contrast, both PAM and PWM are "self-clocking"; i.e., the clock timing is directly present in the modulated waveform.

Both PWM and PPM are nonlinear modulation methods and hence we cannot apply Fourier analysis directly. As in the case of angle modulation, however, we can obtain some helpful information.

In Chapter 3 we saw that band-limited systems have time responses with finite rise times. In pulse-timing modulation this finite rise-time constraint results in uncertainties in reconstructing the detailed structure of a signal. The uncertainty in specifying or reproducing a given signal amplitude is proportional to the rise time of the system and therefore inversely proportional to its bandwidth. This uncertainty is called the *resolution* of the system.

Without the deleterious effects of noise and distortion, the concept of resolution doesn't have much meaning; a given point on the rise time characteristic of the system can be chosen to indicate signal detection if the response is known exactly. In the presence of noise, however, the following criterion is quite generally used: *the time resolution of a pulse-timing system in the presence of noise is equal to the rise time of the system.*†

As a result of measurements on practical systems in the presence of noise, the rule that the two pulses must be spaced by at least the effective width of the system impulse response turns out to be a fairly good criterion. From our knowledge of linear time-invariant systems we recognize that the system rise time is a measure of the effective width of the impulse response.

---

† This rule is not quite as arbitrary as it may appear. The rise time of a system is the indefinite integral of the impulse response. The question we are really asking, then, is how close can two impulses of equal weighting be at the input of the system before one cannot tell that there were two and what their separation would be by observing the output. The problem is a classical one first studied in the separation of point images in optical telescopes. The Rayleigh criterion, as formulated in these studies, states that two point images are just resolvable when they are separated by the effective width of the impulse response of the observing system. What we have stated here is essentially the Rayleigh criterion.

This criterion can easily be demonstrated in practice by applying two narrow pulses to the input of a low-pass filter, as shown in Fig. 7.17. Here two pulses, each 0.3 μsec in width, are applied to the input of a fifth-order Butterworth low-pass filter with a -3-dB frequency of 80 kHz. (The rise time is about 6 μsec and the pulses are thus narrow enough to be considered impulses driving the system.) The input and output filter waveforms are shown for various pulse separations. The position of the first pulse is variable and the position of the second pulse is fixed. Some delay is noticeable in the filter response.

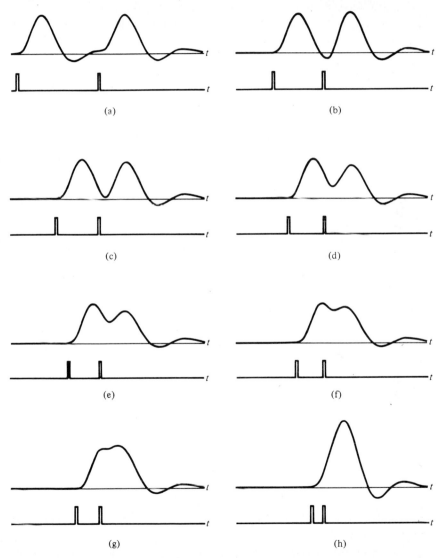

Fig. 7.17 Pulse resolution of a low-pass filter.

Starting with the waveforms in Fig. 7.17(a), note that the pulses and the corresponding responses are well separated. As the pulses are moved closer together, we approach a condition in Fig. 7.17(e) in which the pulse separation is equal to the width of the impulse response of the filter. For pulse separations less than this, it becomes difficult to detect from the output response that there are two pulses present. This would be particularly true if some noise were added before the measurement was made.

We have specifically stated the above rule for practical systems. Recall that for an ideal low-pass filter the relationship between rise time and bandwidth is $t_r = 1/B$, where t_r is the elapsed time (in seconds) between the minimum and maximum filter output and B is the filter bandwidth (in Hz). However, for practical filters the bandwidth is usually defined on the basis of the -3-dB criterion. The rise time is defined as the time elapsed for the leading edge of a pulse to go from 10% to 90% of its final value when a step function is applied. In terms of these quantities, a good rule of thumb for practical low-pass filters is $t_r \approx 1/(2B)$. Because our criterion for pulse-edge response used above is based on the response of practical systems, we shall use this latter result for pulse-timing systems. Therefore the rule-of-thumb criterion for minimum detectable timing accuracy $\Delta\tau$ in the presence of additive noise is

$$\Delta\tau \approx 1/(2B) \qquad \text{(low } S/N\text{)}, \tag{7.22}$$

where B is the bandwidth of the system.

The criterion in Eq. (7.22) turns out to be a pessimistic one for high signal-to-noise ratios. To derive an equation for this case, we assume that the system rise time can be approximated by a straight line. This is a fairly good approximation for an ideal filter (cf. Section 3.13). Using this approximation, we find that each pulse after filtering is trapezoidal in shape (cf. Fig. 7.18). For high signal-to-noise, the additive noise only shifts the trapezoidal waveform in amplitude, and the timing accuracy can be written from the geometry of Fig. 7.18:

$$\frac{\Delta\tau}{t_r} = \frac{n}{A},$$

or

$$\Delta\tau = t_r[\overline{n^2(t)}/A^2]^{1/2}.$$

For the ideal low-pass filter, $t_r \approx 1/B$ so that

$$\Delta\tau \approx \frac{1}{B\sqrt{S/N}} \qquad \text{(high } S/N\text{)}, \tag{7.23}$$

where S/N is the peak pulse power to average noise power ratio. Note that the "rule-of-thumb" result in Eq. (7.22) corresponds to $S/N = 4$ (that is, 6 dB). This value is often taken as the detection threshold for pulse timing modulation. We use the threshold value in system calculations unless otherwise specified.

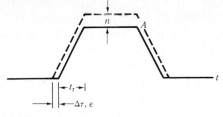

Fig. 7.18 Variation in pulse position
as result of additive noise.

Example 7.4.1   A signal which is band-limited to 1 kHz with an amplitude range
of -1 to $+1$ V is to be sent with a resolution of ± 0.5 mV using PWM. The
minimum pulse width is 1 μsec and the guard time used is equal to the minimum
pulse width. Estimate the bandwidth required for an input $S/N = 20$ dB.†

Solution.   The sampling time is $(2 \text{ kHz})^{-1} = 500 \ \mu\text{sec}$.

Time available for modulation: $500 - 2 = 498 \ \mu\text{sec}$.

Modulation constant: $k_1 = 498/2 = 249 \ \mu\text{sec/V}$.

Timing accuracy required: $\Delta\tau = k_1 \times$ (signal resolution)
$$= (249)(10^{-3}) = 0.249 \ \mu\text{sec}.$$

Bandwidth: $B \approx 1/(10 \ \Delta\tau) = 402$ kHz.

Drill Problem 7.4.1   A given five-channel radio-control telemetry system uses
PPM.‡ The time between consecutive samples of channel 1 is 100 msec; half of
this time is used for synchronization (see Fig. 7.19). The pulse width is 1 msec
and a 1 msec guard time is used. (a) What minimum receiver bandwidth would
you recommend if the desired resolution (of the servos) is to be $\pm 2\%$ of the
maximum for $S/N = 15$ dB? (b) What is the maximum frequency response of
each channel?

Answer.   (a) 556 Hz; (b) 5 Hz.

|←————50 msec————→|←————50 msec————→|

Fig. 7.19 The PPM system of Drill Problem 7.4.1.

---

† The reader may wish to investigate here the bandwidth-S/N trade-off for a given $\Delta\tau$
[cf. Eq. (7.23)].

‡ This type of system is commonly used for radio-controlled models. To simplify the
receiver design and avoid deriving a clock signal for channel synchronization, each channel
reference is triggered by the preceding one. This requires transmission of six pulses for
five channels, but otherwise all operating principles remain the same as for conventional
PPM.

Example 7.4.2   A pulse radar uses pulse timing to determine range. A short burst of high-frequency electromagnetic energy is radiated from an antenna. The receiver amplifies and detects returned energy reflected from objects. Measurement of the time delay between transmission and reception is an indication of the range to the reflecting object. If the reflecting object is relatively small the reflected waveform is a delayed replica of the transmitted waveform. Develop expressions for the range R and the range resolution ΔR if the measured time delay is τ seconds and the receiver bandwidth (low-pass) is B Hz.

Solution.   The transmitted pulse travels at the speed of light, c, to the reflecting object and back in the time τ so that

$$R = c\tau/2 = 150 \times 10^6 \tau \text{ m}.$$

Therefore one microsecond of time delay corresponds to 150 meters of range.

The range resolution is related to the pulse timing by $\Delta R = (\Delta\tau)c/2$. The pulse timing accuracy is limited by the rise time of the receiver and $B \approx 1/(2t_r)$ so that

$$\Delta R \approx c/(4B) \text{ m}.$$

Thus a receiver bandwidth on the order of 7.5 MHz is necessary to achieve a range resolution of 10 meters.

Generation of pulse-timing modulation commonly employs various combinations of a sample-and-hold circuit, a precision ramp voltage generator, and a comparator. The block diagram of a typical circuit for generating PWM and PPM is shown in Fig. 7.20(a). Operation of the sample-and-hold circuit has been discussed in a previous section. The ramp generator produces a precision ramp voltage which has a peak-to-peak amplitude slightly larger than the maximum amplitude range of the input signals. This ramp voltage is the basis for the amplitude-to-timing conversion and therefore must be accurately known. It is resettable by the clock command. The comparator is a high-gain amplifier intended for two-state operation. If the input signal is higher than a preset reference level, the output is held in one state (i.e., a given voltage level). Whenever the input signal level is less than the reference level, the output is held in the other state (voltage level). Which output state is present, then, depends upon whether the input is above or below the threshold (reference level) of the comparator. The transition between the two output states occurs very abruptly in good comparator designs.

Waveforms in the generation of PWM and PPM using this system are shown in Fig. 7.20(b). The voltage reference level of the comparator is adjusted so that there is always an intersection with the sum of the sample-and-hold circuit and the ramp voltage. In this system, the first crossing of the reference level indicates the clock timing and the second crossing generates the variable trailing edge.

The constant amplitude of the output pulses is determined by the output voltage levels of the comparator. If the ramp voltage is linear, the output pulse train is PWM.

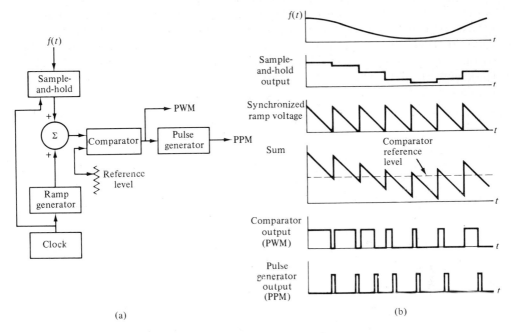

Fig. 7.20 A method of generating PWM and PPM.

    A convenient way to generate PPM is to use the PWM waveform generated above and then trigger a constant-width pulse generator on those edges of the PWM waveform with a negative slope. This is illustrated in Fig. 7.20(b). This pulse generator produces pulses of fixed width and amplitude whose time-of-occurrence relative to the clock is linearly related to the sampled values of the input signal.

    Differing versions of these circuits are used in practice. Sometimes the sample-and-hold operation is deleted if the sampling rate is high. Note that modulation of the leading edge of the PWM waveform can be effected by reversing the slope of the ramp voltage. If a symmetrical ramp voltage is used, both pulse edges in the PWM waveform are modulated symmetrically.

    A popular circuit for modulation of the trailing edges in PWM is shown in Fig. 7.21. Here the ramp generator and the pulse generator are both started by the clock. When the ramp voltage exceeds the input voltage level, the comparator initiates the reset command to the pulse generator. The resulting waveform is a train of pulses whose leading edges are governed by the clock and whose trailing edges are governed by the amplitude of the input signal. Again, PPM is easily derived from this PWM waveform.

Some means of synchronization must be provided when several channels of pulse-timing modulation are transmitted using time multiplexing. A common method of implementing this is to periodically send pulses which are wider than any used for modulation. Use of an integrator and a comparator in the receiver can then distinguish the synchronization pulses and use them to synchronize the commutators. This scheme is particularly advantageous when using PPM because the change in pulse width can be made large to easily distinguish the synchronization pulses.

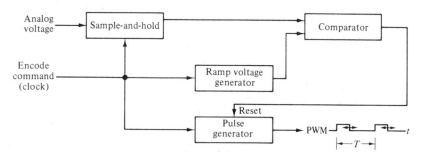

Fig. 7.21 A popular method for generation of PWM.†

7.5 SIGNAL-TO-NOISE RATIOS IN ANALOG PULSE MODULATION

The pulse modulations we have discussed in preceding sections employ a variation of some parameter (e.g., amplitude, position, width, etc.) in a train of pulses in response to the amplitude of an input signal. This variation is an analog one—i.e., continuous in amplitude—even though the information is sent at discrete time intervals. Thus we classify these generally as *analog pulse modulation*.

Here we investigate the performance of analog pulse modulation systems in the presence of additive noise. In many respects, the modulation performance of PAM is analogous to that of AM, while the performance of PWM and PPM is analogous to that of FM and PM. Therefore it is not surprising that the signal-to-noise performance of PAM is the same as that for SSB-SC and that of PWM and PPM are closely related to those obtained for FM and PM.

7.5.1 PAM

Noise is added in the transmission of the PAM signal, as illustrated in Fig. 7.22(a). But the noise occurring between pulses adds noise power to the trans-

---

† The pulse generator is not really necessary but is used as a buffer stage.

mission without any increase in signal power. To avoid this, a synchronized gating circuit is used in the receiver to accept samples only when the signal is known to be present.

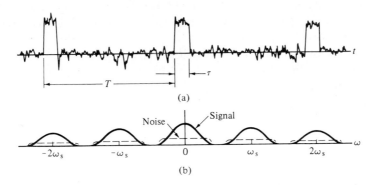

(a)

(b)

Fig. 7.22 (a) Time waveform of a PAM signal with additive transmission noise; (b) spectrum of PAM with band-limited input signal and noise.

We shall assume that the signal and the additive noise present in the input to the PAM receiver are band-limited and that the conditions of the sampling theorem are satisfied. Because the PAM receiver is linear, we can apply the signal and the noise separately, measure their powers, and then combine. The sampling and low-pass filtering at the receiver reproduce the band-limited signal and noise spectra within a constant, as shown in Fig. 7.22(b). Thus we can write

$$\overline{s_o^2(t)} = k\overline{s_i^2(t)},$$

$$\overline{n_o^2(t)} = k\overline{n_i^2(t)},$$

so that

$$\frac{S_o}{N_o} = \frac{S_i}{N_i}. \tag{7.24}$$

This can be extended to time-multiplexed PAM as long as the samples from adjacent channels are independent; i.e., if Eq. (7.11) is satisfied. Note that Eq. (7.24) is the same as that obtained earlier for SSB-SC systems [cf. Eq. (5.80)]. These results are on the basis of average power. The problem of maximizing the peak signal to rms noise ratio for each pulse will be considered in Section 7.9.

7.5.2 Pulse-timing Modulation

In pulse-timing modulation (such as PWM or PPM), the information is contained in the relative positioning of the pulse edges. While we restrict our discussion to PPM systems here, the same general conclusions hold for other pulse-timing systems.

Although the pulses to convey the information may be generated with extremely short (fast) rise times, after passage through a band-limited system they have rise times which are governed by the bandwidth of the system. We shall approximate this rise time by a linear ramp, as shown in Fig. 7.18, so that the pulse assumes a trapezoidal shape.

The position of the trapezoidal pulse is sensitive to additive noise. If the noise voltage is assumed to vary slowly compared to the rise time of the pulse, the variation in the pulse amplitude, n, may be represented by a shift ϵ in the pulse position, as shown in Fig. 7.18. In fact, this figure shows that the pulse positional error ϵ arising from the additive noise voltage n may be reduced by increasing the slope of the pulse edge. For a given noise power, this requires a decrease in the rise time t_r—thus increasing the system bandwidth—or an increase in the pulse amplitude A—thus increasing the transmitter power. Therefore an exchange of bandwidth and transmitted power toward improvement of the S/N ratio is possible here, much as it was for angle-modulated systems.

We can obtain a simple expression to exemplify these conclusions if we make assumptions similar to those used in our analysis of FM. Specifically, we assume that the average output noise power is measured in the absence of modulation and for a large signal-to-noise ratio. Also, the output signal power is measured under noise-free conditions.

From the geometry of Fig. 7.18, we have

$$\frac{\epsilon}{t_r} = \frac{n}{A},$$

or

$$\overline{\epsilon^2} = (t_r/A)^2 \overline{n^2}. \tag{7.25}$$

The output signal amplitude is proportional to the modulating signal $f(t)$ through a modulation constant k so that

$$s_o(t) = kf(t),$$

or

$$\overline{s_o^2(t)} = k^2 \overline{f^2(t)}. \tag{7.26}$$

The output noise, when we use Eq. (7.25), is

$$\overline{n_o^2(t)} = \overline{\epsilon^2} = (t_r/A)^2 \overline{n_i^2(t)}. \tag{7.27}$$

Also, we have

$$\overline{s_i^2(t)} = A^2(\tau/T), \tag{7.28}$$

and for the ideal filter (giving a nearly linear rise time),

$$B \approx 1/t_r. \tag{7.29}$$

Combining Eqs. (7.26)–(7.29), we have

$$\frac{S_o}{N_o} = \frac{k^2 \overline{f^2(t)}}{\tau/T} B^2 \frac{S_i}{N_i}, \tag{7.30}$$

or

$$\frac{S_o}{N_o} \propto B^2 \frac{S_i}{N_i}. \tag{7.31}$$

Therefore the S/N improvement in a PPM system is proportional to the square of the bandwidth, just as it was for angle-modulation systems.

As in the case of angle modulation, however, the output signal-to-noise ratio in pulse-timing systems cannot be improved indefinitely. The noise power accepted by the receiver increases with the bandwidth, eventually becoming comparable to the signal level, violating our assumption of a large signal-to-noise ratio, and taking over the system. This threshold effect is found to take place for input signal-to-noise ratios on the order of 6 dB.

It is instructive to make a comparison of the S/N expressions for FM and PPM. Recall that for FM we obtained [cf. Eq. (6.115)]

$$\frac{S_o}{N_o} = 3\beta^2 \frac{S_c}{N_c}.$$

For the wideband FM condition $\beta \approx B/(2f_m)$ so that

$$\frac{S_o}{N_o} \propto \left(\frac{B}{f_m}\right)^2 \frac{S_c}{N_c} \qquad \text{wideband FM.} \tag{7.32}$$

For PPM, the modulation constant k is proportional to the time allotted between pulses, which in turn is governed by the sampling theorem. Therefore $k \propto 1/f_m$ and we can rewrite Eq. (7.30) as

$$\frac{S_o}{N_o} \propto \left(\frac{B}{f_m}\right)^2 \frac{S_i}{N_i} \qquad \text{PPM.} \tag{7.33}$$

In both cases, the output voltage signal-to-noise ratio is proportional to the ratio of the transmission bandwidth B to the signal bandwidth f_m. The larger the ratio B/f_m, the greater the S/N improvement as long as operation is above the threshold level. Thus both FM and PPM are examples of analog signal transmission methods in which the rms noise improvement is linearly proportional to the ratio B/f_m.

7.6   PULSE-CODE MODULATION (PCM)

The preceding types of pulse modulation made use of the discrete time samples of analog signals. In these cases, the transmission is composed of analog information sent at discrete times. The variation of pulse amplitude or pulse timing is allowed to vary continuously over all values. A further refinement is to quantize the sampled analog signal into a number of discrete levels. This is sometimes done in PAM systems and is referred to as "*M*-ary PAM," where *M* designates the number of levels used. We now not only quantize the sampled analog signal into a number of discrete levels but also use a code to designate each level at each sample time. This type of modulation is called *pulse-code modulation* (PCM).

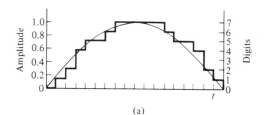

(a)

Fig. 7.23 Quantization of half-sinusoid:
(a) linear; (b) nonlinear.

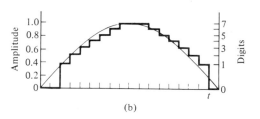

(b)

Suppose, for example, that we wish to quantize one-half a cycle of a one-volt (peak) sinusoid using eight discrete levels, as shown in Fig. 7.23. At each sample time we must decide which of these eight levels is the best approximation to the sinusoid. We choose the closest value (some other criterion can be chosen as well, depending on the particular circuit), and use this value until the next sample time, etc. Obviously, this process of *quantization* introduces some fluctuations about the true value; these fluctuations can be regarded as noise. An increase in the number of quantizing levels assigned will tend to decrease this quantization noise. The quantization may be linear (i.e., uniformly spaced levels), as shown in Fig. 7.23(a), or it may be nonlinear (i.e., nonuniform level spacings). Figure 7.23(b) illustrates a case in which the half-sinusoid is quantized using a square-law dependency for the level spacings. This type of quantization could be used to advantage, for example, in systems where the low levels are of less importance than the higher levels. Unless otherwise specified, quantization using equal level increments (i.e., linear) is usually assumed.

The next step is to assign a digit to each level in such a way that each level has a one-to-one correspondence with the set of real integers. This is called *digitization* of the waveform. The process of digitization has reduced the waveform to a set of digits at the successive sample times, resulting in a completely digital modulation system. The digits are expressed in a coded form. The most common code used for this purpose is a binary code (i.e., a code using only two possible pulse levels). A binary code commonly used for an eight-level system is illustrated in Fig. 7.24.

	Digit	Binary code	A binary pulse code
	0	000	
	1	001	
	2	010	
Fig. 7.24 A binary pulse code.	3	011	
	4	100	
	5	101	
	6	110	
	7	111	

Thus instead of transmitting the individual samples at the sample times, a pulse code or pattern is sent at each sample time to convey the information in quantized form. Systems making use of the transmission of digitized (i.e., quantized and coded) signals are called pulse-code modulation (PCM) systems.

Note that if the quantized samples were transmitted directly, the resulting system would simply be a quantized PAM (*M*-ary PAM) system. It is the assignment of the numbered signal levels which can be coded that gives PCM a distinct advantage over other types of modulation systems. Although we shall restrict our attention to the binary code here, one can also use pulse codes with more than two levels (called *M*-ary PCM).

In a binary code, each pulse can represent one of two possible states, e.g., a "1" or a "0". This basic quantum unit for conveying information is called a binary unit, or *bit*. For example, the eight-level binary code used above is referred to as a three-bit code. A four-bit code corresponds to sixteen quantization levels, etc.

Drill Problem 7.6.1   Determine the number of quantizing levels that can be used if the number of bits in a given binary code is (a) 5; (b) 8; (c) x.

Answer.   (a) 32; (b) 256; (c) 2^x.

PCM systems are becoming increasingly more important as a result of certain inherent advantages over other types of modulation systems. Some of these advantages are the following:

1. In long-distance communications, PCM signals can be completely regenerated at intermediate repeater stations because all the information is contained in the code. Essentially a noise-free signal is retransmitted at each repeater. The effects of noise do not accumulate and one need be concerned only about the effects of transmission noise between adjacent repeaters.

2. Modulating and demodulating circuitry is all digital, thus affording high reliability and stability, and is readily adapted to integrated-circuit logic design.

3. Signals may be stored and time scaled efficiently. For example, PCM data may be generated in an orbiting satellite once every minute during a 90-minute orbit and then retransmitted to a ground station in a matter of a few seconds. Digital memories can perform the required storage very efficiently.

4. Efficient codes can be utilized to reduce unnecessary repetition (redundancy) in messages. For example, if one wishes to send "A Merry Christmas and a Happy New Year" to a distant friend via Western Union, it is much more efficient to assign a code (i.e., a number) to this redundant message and then send the code (number). At the receiving station the decoder recognizes the code and types out the message.

5. Appropriate coding can reduce the effects of noise and interference. As we shall soon see, bandwidth can be exchanged for signal power; because PCM can be time scaled, time can also be exchanged for signal power. The communication-systems designer therefore has added versatility in the design of a PCM system to meet given performance criteria.

Offsetting these advantages is the fact that the complexity of a PCM system is greater than that required for other types of modulation systems. However, this added complexity varies little as the number of channels is increased. Thus the complexity of a PCM system can compare quite favorably with other types of modulation when the number of channels is very large. Advances in integrated-circuit logic-type circuitry have reduced size and power consumption to the point where a large number of channels can be multiplexed efficiently using PCM.

Now we shall devote some attention to the practical questions of generating PCM. The central operation is that of the analog-to-digital (A/D) converter; i.e., encoding analog signals into digital codes. We briefly describe several techniques for accomplishing this operation.

The first method is called the ramp encoder and uses the same basic principle as was used in the generation of PWM; i.e., the generation of a linear ramp which is compared with the input signal level. Instead of generating a pulse whose width is controlled by the time it takes for the ramp to reach the input-signal level as in PWM, however, we now operate a binary counter during this time and then read out the digital count. A simplified block diagram of the ramp encoder is shown in Fig. 7.25.

The ramp encoder uses relatively few precision components and is widely used. Its chief limitation is in the required linearity of the ramp voltage generated. Speed of operation is generally limited by the speed of the binary counter.

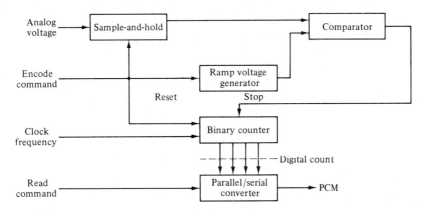

Fig. 7.25 A simplified diagram of the ramp encoder for generation of PCM.

A popular version of the ramp encoder in instrumentation-type systems is the dual-slope (dual-ramp) encoder. In the dual-slope encoder, an integrator is inserted between the sample-and-hold and the comparator. The ramp voltage generator is not needed. As the output of the sample-and-hold is integrated, the slope of the integrator output is proportional to the input voltage. When the integrator output reaches a preset threshold, the input to the integrator is switched from the sample-and-hold to a precision voltage source with polarity opposite that of the signal. Conversion is completed when the integrator output returns to its initial starting level. Advantages of this method include good linearity and accuracy, excellent noise rejection, simplicity, and economy. A major disadvantage is the relatively long conversion time required.

A second method is the feedback encoder, shown in simplified block diagram form in Fig. 7.26. The feedback encoder operates by successively comparing the analog voltage input with a series of trial voltages. Successive trials are governed by the outcomes of previous decisions as to whether the trial voltage is greater or less than the level of the analog input. Usually the routine is designed to take one-half the remaining voltage interval for each trial so that the trials converge rapidly, following the binary code sequence to describe the input level. (This procedure is analogous to that of an analytical balance used for weighing objects, with the trial weights in this case arranged in powers of two.) The trial voltages are generated from a series of voltage dividers (i.e., a "ladder" network) with switches controlled by a digital-logic configuration. The code corresponding to the input voltage level can be read from the position of the switches in the voltage-divider network.

The accuracy of the feedback encoder is dependent on the accuracy of the trial voltages; speed is limited by the speed with which the voltage-divider switches can be operated. The feedback encoder requires more precision components than the ramp encoder but it also allows a more accurate encoding.

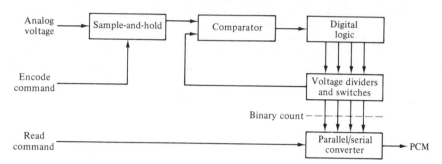

Fig. 7.26 A simplified diagram of the feedback encoder for generation of PCM.

A method of decreasing the conversion time is to apply the output of a sample-and-hold to many comparators operating in parallel, as illustrated in Fig. 7.27. Each comparator has its own reference. Outputs of all comparators are compared by digital logic circuits to determine the binary code output for each sample value. This parallel or *flash* encoder can be operated at extremely high speeds at the expense of providing $n - 1 = 2^m - 1$ comparators for m-bit quantization. The high parts count can be reduced somewhat by using a $(m/2)$-bit flash encoder feeding a high-speed D/A converter (DAC). The output of the DAC is subtracted from the input voltage, and the difference between the two is converted by a second $(m/2)$-bit flash encoder. This feedback arrangement increases the conversion time somewhat, which may be a reasonable compromise if extremely high speeds are not required. A fourth method of A/D conversion will be covered in Chapter 9, where delta modulation methods are discussed.

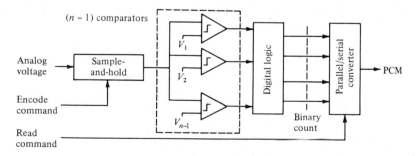

Fig. 7.27 A parallel or flash encoder for generation of PCM.

We shall postpone discussion of the optimum reception of PCM in the presence of noise until later. If transmission noise is not a problem, or if constraints on the receiving system complexity do not allow an optimum receiver, a suboptimal design can be used. The block diagram of a suboptimal PCM receiver is shown in Fig. 7.28 (also see Problem 7.6.3). It uses some of the same ideas as the feedback encoder without the necessity for the feedback. Resistive dividers are operated by switches that are controlled by the binary code. By weighting the resulting voltage levels in the same proportions as the weights of the code pulses, we produce a composite voltage, which reproduces the original quantized voltage. This receiver design is simple and widely used in instrumentation systems. It is generally not used on long-distance communication systems which attempt to maintain as low an error rate as possible.

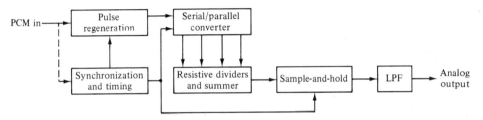

Fig. 7.28 A suboptimal PCM decoder.

There are several levels of synchronization necessary in PCM systems. The first one is the clock rate which can be sent as a separate frequency-multiplexed single-frequency tone or by using a phase-locked loop with a long time constant on the incoming PCM. Frame synchronization is necessary to establish when each pulse code is to begin. This is often accomplished by periodically sending a preassigned code which cannot be duplicated by any input level. When detected, this establishes the beginning of the frame. A divider on the clock frequency helps to maintain frame synchronization until the next synchronization arrives. A third level of synchronization is that needed to identify which channel is being sent when several channels of information are time-multiplexed. This is often combined with the frame synchronization.

In PCM telephone systems in the United States, 24 8-bit voice channels are time-multiplexed to give 192 bits per frame. One extra bit is inserted to give frame synchronization, yielding 193 bits/frame. Each input channel is sampled at an 8-kHz rate resulting in a frame interval of: $(8 \text{ kHz})^{-1} = 125 \text{ }\mu\text{sec}$ and a clock rate of: $(193) (8 \text{ kHz}) = 1.544 \times 10^6$ bits per second. Signaling (such as dial pulses) bits can be transmitted at a much slower rate and are either transmitted using one bit from each channel every sixth frame (yielding an effective $7\frac{5}{6}$-bit channel operation) or using a single data channel (yielding an effective 23-channel system).

This 24-channel multiplexer is used as a basic system, known as T1, in the modular T-carrier TDM/PCM telephony system designed by the American Telephone and Telegraph Company. The T1 system is designed primarily for short-distance, heavy usage in metropolitan areas. In this system two twisted wire pairs are terminated at each end in a "channel bank" which combines the 24 voice channels using TDM/PCM as described above. Each wire pair carries data transmissions at a rate of 1.544 megabits per second (Mbps). Digital repeaters are spaced at approximately one mile intervals (35-dB loss spacings) along the line.

Maximum length of the T1 system is limited to about 50 to 100 miles because of timing jitter. It is designed to be compatible with existing PAM telephone systems. Transmissions which are digital (e.g., digital computer links) can be handled by replacing the channel bank with a data terminal.

The overall T-carrier system is made up of various combinations of lower-order T-carrier subsystems in order to accommodate the requirements of voice channels, Picturephone® service, and commercial television network programming. A block diagram of the overall T-carrier system concept is shown in Fig. 7.29 and the operating specifications of the various levels are listed in Table 7.1.

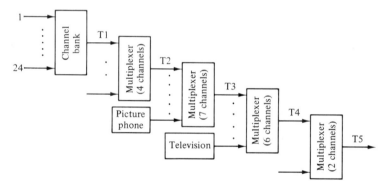

Fig. 7.29 The T-carrier TDM/PCM telephony system.

As in other types of pulse modulation, the maximum attainable rates for PCM are limited by switching speeds. After the data have been encoded, final transmission is often accomplished after a second modulation process. The PCM signal can be used to modulate a high-frequency carrier using AM, FM, or PM techniques. It is also possible to have a multichannel PCM system in which each of the pulse sequences modulates a subcarrier. These subcarrier signals are, in turn, frequency-multiplexed and then used to modulate a carrier.

Before the PCM signal is applied to a modulator, however, it may be put into several different forms. Choice of these different ways of conveying the binary code information will depend somewhat on the type of modulation and

Table 7.1 Specifications for the T-carrier TDM/PCM Telephony System

System*	Rate (Mbps)	MUX†	System capacity			Medium	Repeater Spacing (miles)	Maximum System Length (miles)	System Error Rate
			Voice Channels	Picture-phone	TV				
T1	1.544	T1	24	—	—	Wire pair	1	50	10^{-6}
T2	6.312	4(T1)	96	1	—	Coax	2.5	500	10^{-7}
T3	44.736	7(T2)	672	7	1	Coax	‡	‡	‡
T4	274.176	6(T3)	4032	42	6	Coax	1	500	10^{-6}
T5	560.160	2(T4)	8064	84	12	Coax	1	500	4×10^{-7}
WT4	18,500.0	58(T4)	233,000	2436	348	60 mm circ. waveguide	25	4000	10^{-8}

* The corresponding channel bank designation uses "D" in place of "T."
† Abbreviation commonly used for multiplexer.
‡ Multiplexing levels only; not intended for transmission.

demodulation employed and other constraints on bandwidth, receiver complexity, etc. Figure 7.30 illustrates some of the more commonly used PCM representations. We briefly describe each of those shown.

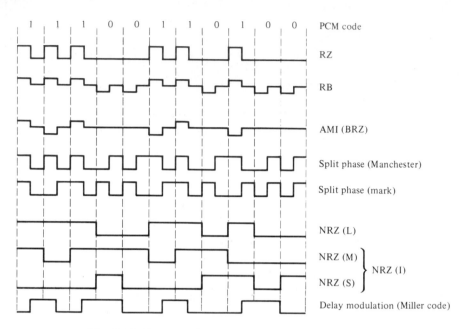

Fig. 7.30 Some methods of representing binary data.

The return-to-zero (RZ) method represents a 1 by a change to the 1 level for one-half the bit interval, after which the signal returns to the reference level for the remaining half-bit interval. A 0 is indicated in this method by no change, the signal remaining at the reference level. This is the representation used in the binary code shown in Fig. 7.24.

In the return-to-bias (RB) method, three levels are used — 0, 1, and a bias level. The bias level may be chosen either below or between the other two levels (it may be a zero reference, as shown in Fig. 7.30). The waveform returns to the bias level during the last half of each bit interval. The RB method has an advantage in being self-clocking; i.e., the clock frequency is easily found from the magnitude of the pulse code. However, the average value of the waveform depends on the particular proportion of ones and zeros present. The RB representation also takes more bandwidth than necessary and it uses three levels. The succeeding representations are designed to overcome some of these drawbacks at the expense of increased complexity.

In Alternate Mark Inversion (AMI) the first binary one is represented by $+1$, the second by -1, the third by $+1$, etc. The AMI representation is easily derived from an RZ binary code (and vice versa) by alternately inverting the 1's. It has zero average value and is widely used in telephone PCM systems. This is also referred to as a bipolar return-to-zero (BRZ) representation.

The split-phase binary representations eliminate the variation in average value using symmetry. In the Manchester split-phase method, a 1 is represented by a 1 level during the first $\frac{1}{2}$-bit interval, then shifted to the 0 level for the latter $\frac{1}{2}$-bit interval; a 0 is indicated by the reverse representation. In the split-phase (mark) method, a similar symmetric representation is used except that a phase reversal relative to the previous phase indicates a 1 (i.e., mark) and no change in phase is used to indicate a 0.

The nonreturn-to-zero (NRZ) representations reduce the bandwidth needed to send the PCM code. In the NRZ(L) representation a bit pulse remains in one of its two levels for the entire bit interval. In the NRZ(M) method a level change is used to indicate a mark (i.e., a 1) and no level change for a 0; the NRZ(S) method uses the same scheme except that a level change is used to indicate a space (i.e., a 0). Both of these are examples of the more general classification NRZ(I) in which a level change (inversion) is used to indicate one kind of binary digit and no level change indicates the other digit. The NRZ representations are efficient in terms of the bandwidth required and are widely used. Note that use of split-phase and NRZ representations require some added receiver complexity to determine the clock frequency.

In delay modulation (Miller code), a 1 is represented by a signal transition at the midpoint of a bit interval. A 0 is represented by no transition unless it is followed by another 0, in which case the signal transition occurs at the end of the bit interval. In this method, a succession of 1's and a succession of 0's each are represented by a square wave at the bit rate, but one is delayed a $\frac{1}{2}$-bit interval from the other. Delay modulation is insensitive to the initial 180° ambiguity that is present in NRZ and Manchester representations. It is efficient in terms of a required bandwidth without a required good low-frequency response.

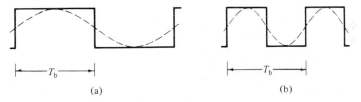

(a)　　　　　　　　　　　　　　　　　(b)

Fig. 7.31 Maximum PCM bit rates of (a) NRZ and (b) RZ codes.

Illustrations of sinusoids at the fundamental frequency for alternating levels in a PCM bit stream are shown in Fig. 7.31. The alternating-level sequence represents the most rapidly varying pattern possible in a binary code stream. In these diagrams, T_b represents the bit interval. Note that NRZ codes transmit one bit per level change while RZ codes transmit one bit per pair of level changes. It follows that the NRZ codes (and the Miller code) can transmit two bits per second per Hz (bps/Hz). In contrast, the limit for the other code representations depicted in Fig. 7.30 is one bps/Hz. We call the number of bits per second per Hz the *bandwidth efficiency*. Therefore the potential bandwidth efficiency of a binary PCM code is two bps/Hz using an NRZ (or Miller) code representation.

Let's next consider the minimum bandwidth required for a binary PCM system using n quantization levels. Let the number of bits per sample be designated by $[\log_2 n]$, where the use of the brackets indicates the next higher integer to be taken. Also, let the sample rate be $1/T$; then $[\log_2 n]/T$ is the number of bits per second to be sent.

From this discussion, we can see that the bandwidth efficiency for the NRZ code representation is 2 bps/Hz, so that the minimum bandwidth is

$$B \geq \frac{1}{2}\left(\frac{[\log_2 n]}{T}\right). \tag{7.34}$$

Extending this result to the time-multiplexed case, we have for the NRZ bandwidth

$$B_x \geq \frac{[\log_2 n]}{2T_x} \quad \text{(NRZ)}. \tag{7.35}$$

Because the bandwidth efficiency for the RZ code representation is 1 bps/Hz, the minimum bandwidth required for RZ is

$$B_x \geq \frac{[\log_2 n]}{T_x} \quad \text{(RZ)}. \tag{7.36}$$

A comparison of Eqs. (7.35), (7.36) with Eq. (7.11) reveals the following convenient rule: The minimum bandwidth required for a binary NRZ code representation is $[\log_2 n]$ times that required for a PAM system operating with the same sample rate and the same number of channels. The minimum bandwidth required for a binary RZ code representation is double this amount. These results can easily be extended to those systems that use more than two levels for allowable code representation levels. A logarithm to a different base accounts for the change, as shown in the following drill problem.

Drill Problem 7.6.2   Determine the minimum increase in bandwidth over that required by a PAM signal for the transmission of a PCM signal that has been quantized to 64 levels. Assume that each pulse in the code is allowed to take on the following number of levels: (a) 2; (b) 3; (c) x.

Answer.   (a) 6; (b) 4; (c) $\log_x 64$.

Several channels using PCM codes such as those shown in Fig. 7.30 may be time-multiplexed and the resulting bit stream sent on a transmission line. If electromagnetic propagation is to be used, the bit stream is used to modulate a high-frequency CW carrier. One choice of modulation is that of amplitude-shift keying (ASK) in which the amplitude of a CW carrier is switched between two (or three) values in response to the PCM code. Another choice is to shift the frequency of the CW carrier and this is called frequency-shift keying (FSK). A third possibility, called phase-shift keying (PSK), is that of shifting the phase of the CW carrier in response to the amplitude of the PCM code. These are illustrated in Fig. 7.32 for the PCM code used in Fig. 7.30. The S/N performance of these modulation methods will be discussed in Chapter 10.

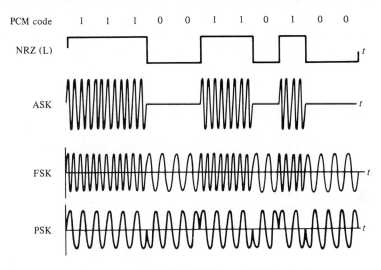

Fig. 7.32 Idealized ASK, FSK, and PSK waveforms.

7.7 USE OF PARITY AND REDUNDANCY IN PCM

In binary pulse-code modulation, each sample of the signal is represented by a codeword of, say, k bits. After these bits are transmitted, the receiver must recognize each codeword in order to reconstruct the samples. However, errors may occur in transmission as a result of noise. One way to increase the reliability of communication is to add extra bits, at the expense of added transmission time or bandwidth, in order to detect and, possibly, to correct the errors. We consider two simple and effective ways to accomplish greater reliability; more advanced methods are considered in Chapter 9.

One simple and effective error-detecting code can be formed by adding an extra binary digit at the end of each codeword. This extra bit is called a *parity-check* bit. For a binary code, the check bit is often chosen to make the number of 1's in each codeword an even number. (If the number of 1's is already even, a 0 is added to signify that fact.) Such a code is said to have *even* parity.

Similarly, for *odd* parity the added parity bit is chosen to make the total number of 1's in the codeword (including the parity bit) an odd number. Let's assume we are using even parity: In a seven-bit codeword (i.e., $k=7$), the message 0100110 becomes 01001101, and 0101101 becomes 01011010.

The extra bit added to each codeword is said to be redundant because it does not convey any information concerning the value of the sampled signal. It is used only to protect against possible errors by verifying the odd or even number of 1's. The code-rate efficiency of a system is defined as the ratio of the number of bits used to convey signal information divided by the total number of bits sent. Thus the code-rate efficiency decreases as more redundancy is added.

The single parity-check code has a code-rate efficiency of $k/(k + 1)$ and is widely used in digital computation and recording. It offers a simple and effective method for error detection when the possibility of making a bit error is low. When an error is detected, a retransmission can be requested. In other cases, those codewords with detected errors may simply be discarded. Note that an error in two bits in a codeword is not detectable using a single parity-check code.

We may wish to add enough redundancy to the code in order not only to detect single errors, but also to correct them. As an example of an error-correcting code, suppose that in transmitting binary PCM we transmit a sequence of three 0's for every 0 and a sequence of three 1's for every 1. Because the two added digits are completely redundant, this repetition code has a code-rate efficiency of $\frac{1}{3}$.

If the possibility of making more than a one-bit error in a codeword is low, we can use a "majority decision" rule to decode the received bit stream. Thus if we receive 001 or 010 or 100, we decode the message as 000; and if we receive 011, 101, or 110, we decode the message as 111. However, if two bit errors occur in a codeword, we would make the wrong decision. Thus the single-bit error correction code is most useful when the possibility of making a bit error in a codeword is low. Repetition codes have good error-detecting capabilities at the expense of transmitting many redundant symbols. The intentional use of redundancy to advantage, whether for error detection or for error correction, is known as *error-control coding*.

To appreciate some of the underlying principles of error-control coding, we introduce the concept of codeword separation or distance. To do this, we define the *Hamming weight* $w(s_i)$ of a codeword s_i as the number of ones in that codeword. The *Hamming distance* $d_{ij} = d(s_i, s_j)$ between codewords s_i, s_j is defined as the number of positions in which s_i, s_j differ. An equation for the Hamming distance is

$$d_{ij} = w(s_i \oplus s_j) \tag{7.37}$$

where $\oplus$ denotes modulo–2 addition. For example, the Hamming distance between the two codewords in the repetition code above is 3, and the decoder rule could have been stated as "choose that codeword with minimum Hamming distance from the received codeword."

It is obvious that a code having all distinct codewords must have a minimum Hamming distance of at least one. Not quite as obvious is the fact that a code which permits detection of up to e errors per word or correction of up to e errors per word must consist of codewords for which:

$$d_m = \begin{cases} e + 1 & \text{error detection} \\ 2e + 1 & \text{error correction} \end{cases} \tag{7.38}$$

where d_m is the minimum Hamming distance between codewords.

Drill Problem 7.7.1   A given code consists of the codewords: 0000000, 0011110, 0101101, 0111000, 1001100, 1011001, 1101010, 1110100. If 1011011 is received, what is the decoded codeword based on minimum Hamming distance?

Answer.   1011001.

Drill Problem 7.7.2   A single parity-check code used in computer listings of registration numbers, etc., is found by doubling alternate digits, beginning from the right, and adding the digits modulo–10. (Numbers greater than ten are counted as two digits.) The parity-check digit is found by taking the tens complement (why?) of this sum. (a) Determine the parity-check digit for the numbers 837412 and 834712. (b) In addition to single-error detection, what type of error is being detected?

Answer.   (a) 6, 2; (b) Interchange of two adjacent digits.

7.8   TIME-DIVISION MULTIPLEXING OF PCM SIGNALS

In the case of the time-division multiplexing of PAM signals, a number of analog signals are combined sequentially at a common sampling rate and the composite pulse stream is transmitted. This procedure can be extended to include PWM and PPM signals. Multiplexing of PCM signals presents some different problems that we will investigate in this section. Unless specified otherwise, we assume binary PCM.

Two basic types of transmission of digital data are known as *synchronous* and *asynchronous*. In synchronous transmission, the bit rate of each multiplexer input is fixed and is synchronized to a master clock. Thus within any given time period there is always a discrete specific number of bits, and the bits follow in a very regular predetermined progression. The multiplexer operates at a multiple of the master clock rate to sequentially format all data and then transmit the data block. Efficiency is high because one binary word follows the preceding one, with no special designators necessary to separate the words.

In asynchronous transmission, words are sent one at a time without necessarily having any fixed time relationship between one word and the next. Many data terminals—such as teletypewriters and other keyboard terminals—transmit alphanumeric symbols (i.e., letters of the alphabet, decimal numbers, and various control symbols) asynchronously, and we refer to the resulting data words as *characters*. In asynchronous transmission, the receiver has to reestablish synchronization for every character. This is accomplished by beginning each character with a start pulse. Thus the receiver clock is resynchronized at the beginning of each character, and small variations in the rate between transmit and receive clocks can be tolerated. A stop pulse is transmitted at the end of each character.

Although various codes could be used, the majority of data transmission systems use one of two international standardized codes, possibly with some minor deviations. These two codes were established by the Consultative Committee on International Telegraphy and Telephony (CCITT), a committee operating with the International Telecommunications Union and based in Geneva. The first of these is CCITT International Alphabet No. 2. This 5-bit code, also known as the Baudot code, is widely used for Telex transmissions around the world. The second is the CCITT International Alphabet No. 5. The U.S. ASCII code (American Standard Code for Information Interchange) is a modification of this 7-bit code (Ref. 1).

Most modern data terminals and associated devices in the U.S. have standardized a 10-bit word for each character, each word consisting of a start bit (always represented by 0), 7 data bits using the ASCII code, a parity-check bit, and a stop bit (represented by a 1). An example of a typical asynchronous terminal character is shown in Fig. 7.33. Because start and stop bits are not required in synchronous transmission, these extra bits are the overhead required for the asynchronous transmission.

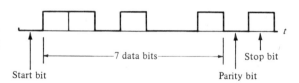

Fig. 7.33 A typical 10-bit asynchronous terminal character using the ASCII code.

7 data bits   Stop bit

Start bit   Parity bit

The purpose of the multiplexer (often abbreviated MUX) is to divide the information-handling capacity of a communications system between a number of terminals. In this respect, it is "transparent"—i.e., a MUX does not do anything to alter the data as it is handled.

A *concentrator* is a computer-based device with some similarities to the multiplexer in that it combines the data from a number of terminals onto a high-speed line for transmission. However, it can alter the forms of the data streams prior to merging them. The concentrator, acting as an intelligent terminal, often is capable of assembling complete messages or blocks of messages before transmission, performing code, speed, and format conversions, and improving on line utilization by statistically averaging the traffic from the inputs onto a high-speed line. It can also compress data so that fewer bits are required to convey the desired information, and automatically detect and correct errors. Recently some intelligent multiplexers, generally microprocessor controlled, have been designed to assume some of the functions of concentrators (e.g., statistical multiplexing, error correction, etc.). For our purposes here, we consider only the nonintelligent multiplexer.

There are three major classes of nonintelligent multiplexers in use. The first is used for completely synchronous transmissions. The object of such multiplexers is to combine a number of synchronous inputs and feed out the combined bit stream sequentially at a higher (synchronous) bit rate. The bit rates of all inputs and the multiplexing rate must be governed by a master clock. The efficiency of such multiplexing systems is high, but provisions must be made to supply a master clock signal to all portions of the network being serviced. Long-range plans for telephone service in the U.S. include plans for a completely synchronous system with a master clock located at Hillsboro, Missouri.

A second class includes those multiplexers designed for almost-synchronous (quasi-synchronous) systems, in which the clock rates are nominally the same for all inputs, but in which small variations about the clock frequencies may exist. In general, multiplexers designed for this type of service operate at high bit rates and are integral parts of larger data transmission services. They multiplex on a bit-by-bit basis and, therefore, are called bit-interleaved multiplexers. To keep high transmission efficiencies without elaborate control procedures, these multiplexers use synchronous timing methods that run slightly fast and then, when necessary, insert bits (a process known as *bit stuffing*) to keep the input rates up to the desired rate.

A third major class of multiplexers includes those designed to combine a number of low-speed asynchronous transmissions into one higher-speed multiplexed signal. Generally these multiplexers are designed to handle the outputs of data terminals and to transmit the data over voice-grade lines in a commercial telephone system. When used with modems to interface with the lines, they normally transmit at output rates of 1200, 2400, 3600, 4800, 7200, or 9600 bits per second. Choice of output bit rate depends on the complexity of the modem and whether the voice-grade line used is privately leased and specially conditioned, or dialed as in any ordinary telephone call.† Multiplexers in this class usually multiplex on a character-by-character basis; input rates may be as low as 75 bps per device.

Although we have grouped multiplexers into three broad categories, it is not always true that an asynchronous one is slow and a synchronous one is fast. There are terminals that operate over short distances at fairly high asynchronous rates. There are also some terminals in use that operate synchronously at 1200 bps.

---

† Some typical characteristics of both basic and conditioned telephone lines are given in Appendix E. The principles of modems are discussed in Chapter 10.

Another multiplexing method of increasing importance in communications satellite networks is that of time division multiple access (TDMA). In TDMA systems, each terminal samples and temporarily stores its user information and then transmits this information in short bursts at high bit rates. All terminals in the network transmit using the same carrier frequency (generally about 6 GHz for the up-link and 4 GHz for the down-link for the communications satellites). The bursts of each TDMA terminal are transmitted synchronously at assigned times relative to the transmissions of other terminals in the network. These bursts arrive at the satellite one at a time in a closely spaced sequence that never overlaps. The bursts are amplified by the satellite transponder, shifted to a different carrier frequency, and then transmitted back to the earth stations. A receiver accepts the burst, temporarily stores the bits and then reads out the bits at a slower rate and distributes the information to the users.

Implementation in TDMA terminals often uses a distributed-processing approach. The relatively slow-speed sampling and buffer-store operations are under the control of a conventional microprocessor (known as the background processor). A high-speed processor (known as the foreground processor) controls the high-speed bursts. Each processor has its assigned functions. Both share some functions to enable the background processor to load and read selected data into and from the foreground processor and to provide the proper synchronization and timing.

TDMA systems offer improved flexibility for the handling and multiplexing of digital data from sources with widely varying data rates. Their use is becoming more attractive as economical and reliable methods of digital storage and control become more readily available.

★ 7.8.1   Bit-interleaved Multiplexers and Bit Stuffing

In bit-interleaved multiplexers, input channels are sampled in succession on a bit-by-bit basis. Synchronization bits are added in a predetermined format to the resulting output data stream. After all inputs have been serviced at least once and the necessary synchronization bits have been added, the multiplexer is ready to begin its next cycle. Such a format is illustrated in Fig. 7.34 for a multiplexer with four inputs.

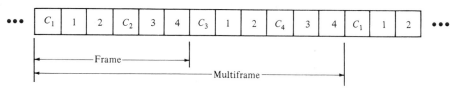

Fig. 7.34 A four-channel framing format.

The shortest period of time in which all multiplexer inputs are serviced at least once is called a *frame*. Location information of the channels is called framing information. The bits added for framing and synchronization are collectively known as control bits. The shortest period of time in which all inputs to be multiplexed and in which all control bits have been added to the frame is called a *multiframe* or *superframe*. In the simplified framing format of Fig. 7.34, there are four input channels plus two control bits per frame and two frames per multiframe. The control bits are merely overhead and decrease the information-handling capacity of the multiplexer. In this particular case, the throughput efficiency of the multiplexer is $\frac{4}{6} = 67\%$.

All inputs to the multiplexer are assumed to have the same nominal bit rate, R_i. Unless all inputs are completely synchronous, however, small fluctuations can be expected. If no provision is made for accommodating these fluctuations, some information could be lost in the multiplexing process. To prevent this, the multiplexer is operated at a rate slightly higher than the sum of the maximum expected rates of the inputs. Because the multiplexing rate will gradually gain, provision is made for the occasional adding of an arbitrary bit (0 or 1) to the bit stream in order to maintain the multiplexer rate. This operation is known as *bit stuffing* and is illustrated in Fig. 7.35. The stuffed bits carry no information and are added merely to maintain the higher bit rate of the multiplexer.

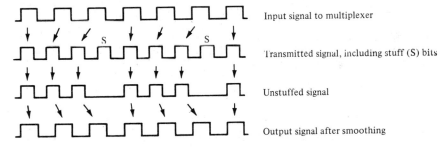

Fig. 7.35 Inputs and outputs of a bit-stuffed multiplexer/demultiplexer.†

To smooth out short-term fluctuations in the input bit rate and to signal the need for a stuff bit, each input is applied to a buffer memory, or *elastic store*, as shown in Fig. 7.36. This buffer acts as a reservoir whose fullness is monitored by the buffer control. Whenever the fullness of the buffer memory drops below a given threshold, a stuff bit is inserted into the data stream on command of the synchronizer.

An important function of the demultiplexer at the other end of the line is to remove the stuff bits from the data stream. This requires identification of the stuffed bits in as reliable a manner as possible without using excessive format overhead. Various techniques are used.

---

† From J. J. Spilker, Jr., *Digital Communications by Satellite*, © 1977, p. 118. Reprinted by permission of Prentice-Hall, Inc., Englewood Cliffs, New Jersey.

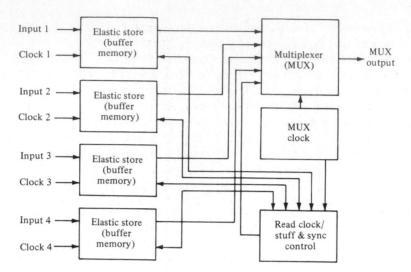

Fig. 7.36 A bit-interleaved multiplexer using bit stuffing.

Consider the Bell System M12 multiplexer as an example of framing and synchronization methods for bit stuffing.[†] This multiplexer is designed to combine four T1 carrier inputs into one T2 carrier output (cf. Fig. 7.29 and Table 7.1). The M12 multiplexer frame format is shown in Fig. 7.37. This multiplexer uses bit-by-bit interleaving of the four T1 carrier inputs until a total of 48 bits, 12 from each input, have been taken. Then a control bit is inserted and the sequence is repeated. Three types of control bits are used—the M-, F-, and C-series. These provide multiframe indication, frame synchronization, and identification of which of the four inputs has been stuffed. The first row of the format is transmitted first, then the second, etc. Each row represents a frame, and all four rows represent a multiframe. Only one stuff bit per input channel is allowed in each multiframe.

Order of transmission											
M_0	(48)	C_I	(48)	F_0	(48)	C_I	(48)	C_I	(48)	F_1	(48)
M_1	(48)	C_{II}	(48)	F_0	(48)	C_{II}	(48)	C_{II}	(48)	F_1	(48)
M_1	(48)	C_{III}	(48)	F_0	(48)	C_{III}	(48)	C_{III}	(48)	F_1	(48)
M_1	(48)	C_{IV}	(48)	F_0	(48)	C_{IV}	(48)	C_{IV}	(48)	F_1	(48)

Fig. 7.37 Frame format of the Bell System M12 multiplexer.

[†] Bell Telephone Laboratories, *Transmission Systems for Communications*, Fourth ed., 1970, p. 612.

The subscripts on the M- and F-series control bits in Fig. 7.37 denote the bit (0 or 1) transmitted. The F-series uses an alternating 0, 1 sequence that provides a fundamental frequency whose zero crossings give frame sync. Because the pattern is repetitive, a PLL in the demultiplexer can lock on to it. After doing so, the demultiplexer searches for the 0111 pattern of the M-series. Having found the M-series pattern, the demultiplexer has also identified the multiframe.

Bit stuffing information is supplied by the C-series, and subscripts are used in Fig. 7.37 to designate input channels; thus C_I stands for input channel 1, C_{II} for input channel 2, etc. The bit patterns in the M- and F-series are periodic, and detection is tolerant of transmission errors. In contrast, the C-series bit patterns are not repetitive, and they must be transmitted with sufficient redundancy to maintain bit integrity—i.e., without leaving in a stuff bit or removing a valid data bit, thus shifting the entire succeeding data stream one bit ahead or behind in time. This loss of bit integrity would cause the remaining bits in the frame to be read out erroneously—probably to the incorrect demultiplexer output port. Therefore this is a much more serious error than a data bit error.

To protect bit integrity, a $\times 3$ repetition code is employed in the M12 multiplexer. The insertion of a stuff bit in any one frame is denoted by setting all three C's in that frame to 1; similarly, three 0's denote no stuffing. Majority-logic decoding is used so that if two out of the three are detected as 1's, a decision is made that a stuff bit has been inserted. Conversely, if two out of the three are detected as 0's, a decision is made that no stuff bit was inserted.

Each multiframe contains $12 \times 6 \times 4 = 288$ bits for each of the four input channels. Because only one stuff bit per input channel is allowed in each multiframe, the maximum fractional change in input bit rate that can be accommodated is 1/288, or 0.35%. Good clocks can be maintained to within 60 ppm (parts per million) or 0.006%, well within the capabilities of this multiplexer. Note that this maximum fractional change represents the change from no stuff bits to all stuff bits. All practical situations lie between these extremes. Let the average number of stuff bits per channel per multiframe be designated as s; then $0 \leq s \leq 1$. To allow for plus or minus deviations from a nominal input clock rate R_i, we should choose s near $\frac{1}{2}$ so that bit stuffing will occur, on the average, 50% of the time.† For example, for the M12 multiplexer we have four input channels each at nominal rate R_i. Allowing for an overhead of one control bit in every 48 data bits and noting that there are 288 data bits in each multiframe, we find the multiplexer output rate R_o is

$$R_o = 4R_i\left(\frac{49}{48}\right)\left(1 + \frac{s}{288}\right).$$

---

† It might appear that one would always design for $s = \frac{1}{2}$, which gives the greatest frequency tolerance each way from the nominal rate. However, the rate at which the buffer memories require stuffing may vary, causing jitter in the demultiplexed bit stream. This jitter has a peak-to-peak value equal to s, and therefore, it is desirable to keep s small.

Using R_o = 6.312 Mbps and R_i = 1.544 Mbps (see Table 7.1), we have s = 0.335.

Drill Problem 7.8.1   Compute the average number of bits stuffed per channel in the Bell System M12 multiplexer if the accuracy on both the input data stream and the multiplexer clock is ± 90 ppm and (a) the input rate (R_i) is slow, but the output rate (R_o) is fast; (b) the input rate is fast, but the output rate is slow.

Answer.   (a) 0.387; (b) 0.283.

Example 7.8.1   A given bit-interleaved multiplexer operates at an output rate of 224 Mbps and 145 bits per frame—143 data bits plus two control bits, labeled the F-series and the S-series.† The frame format is shown in Fig. 7.38(a). The frame sync bit (F-series) simply alternates 0, 1. The synchronization bit (S-series) occupies the time slot normally used for channel 73, and the S-bit format is shown in Fig. 7.38(b). The marker word M is the 16-bit word 0110101001011010. Each channel stuff word C uses three-to-one redundancy: 000 not stuffed, 111 stuffed. Nominal input bit rate per channel is the T1 carrier rate of 1.544 Mbps. For this multiplexer, calculate (a) the throughput efficiency; (b) the number of frames in a multiframe; (c) the maximum fractional change in input rate that can be accommodated; (d) the average number of stuff bits per channel when the input is at the nominal T1 carrier rate.

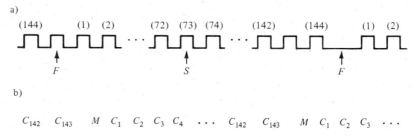

Fig. 7.38 (a) Frame format and (b) S-bit format for the multiplexer in Example 7.8.1.

Solution

a) Two bits out of 145 are used for overhead so that the throughput efficiency is (143/145) $\times$ 100 = 98.6%.

b) Because two channels (F and S) do not require stuff bits, transmission of one complete S-bit sequence requires M = 3(143) + 16 = 445 frames.

---

† This is a simplified version of the multiplexer described in F. J. Witt, "An Experimental 223 Mb/s Digital Multiplexer-Demultiplexer Using Pulse Stuffing Synchronization," *Bell System Technical Journal* (November 1965): 1843–1885; *see also* J. J. Spilker, *Digital Communications by Satellite*, Englewood Cliffs, N.J.: Prentice-Hall, Inc., 1977, p. 122.

c) The nominal multiplexed rate of 145 channels of T1 carrier signals is 145(1.544 Mbps) = 223.88 Mbps. Thus the 224 Mbps rate permits a 0.05% fractional change.

d) For this system, $R_o = 145R_i(1 + \delta)$, where δ is the maximum fractional increase in R_i. Because the S-bit sequence contains 445 bits, the maximum fractional change allowed is one bit in 445. But this would require a stuff bit in every multiframe. Letting s be the average number of stuff bits per channel per multiframe, we have

$$R_o = 145R_i\left(1 + \frac{s}{445}\right).$$

Using $R_o = 224$ Mbps and $R_i = 1.544$ Mbps, we find that $s = 0.239$.

★ 7.8.2   Character-interleaved Multiplexers and Word Stuffing

It is generally simpler to perform the time-division multiplexing of a number of asynchronous terminals on a character-by-character basis, rather than bit by bit. This is particularly true if flexibility to accommodate terminals with different character sizes is desired. Therefore, character-interleaved multiplexers have been adopted widely for the multiplexing of slow-speed asynchronous devices such as data terminals. Each character, usually with a parity check and possibly with an address code, is sent intact as a word and these are interleaved in the bit stream. Synchronous data streams, generally transmitted at higher speeds, are often multiplexed together with the asynchronous data.

The multiplexer rate for each input port is set to run slightly faster (say, 3%) than the maximum character rate expected from the asynchronous terminals. Because individual characters are sent out at a higher rate than received from the terminal, the multiplexer must be able to buffer at least one character from each terminal connected to it. Generally the start and stop bits are dropped from each character before buffering, and then reinserted by the demultiplexer at the other end of the line.

Each of the buffer memories is monitored; and when there is no character to transmit from a given input port, a predetermined control character is transmitted. This procedure is called *word stuffing*. Word stuffing is procedurally like bit stuffing, except that the data are handled on a character-by-character basis. Word stuffing is more tolerant of bit-rate changes than bit stuffing, and can accommodate bit-rate variations all the way from zero input rate to the maximum bit rate of the multiplexer. Throughput efficiency, however, is generally lower than for multiplexers using bit stuffing.

In word stuffing, a predetermined word is assigned to indicate a word stuff. This word must be at least a Hamming distance of one (preferably more) from every other word in the alphabet used. Usually this distance is accomplished by adding a parity bit in the multiplexer. Various techniques are possible; one relatively simple description follows, and a second is included as Problem 7.8.3.

To begin our example of a word stuffing procedure, we assume that an asynchronous terminal transmits 10-bit characters to a multiplexer input port. At the multiplexer, the start and stop bits are dropped, and parity-check bits for even parity are added. Thus each 10-bit character is transmitted as a 9-bit word. Assuming that one character in ten is needed for synchronization, the throughput efficiency of the multiplexer is then $(\frac{8}{9})$ $(\frac{9}{10})$ = 80%. The stuff word is assigned to one of the permissible characters in the alphabet; but to distinguish it from a data word, we use odd parity for the stuff word. This is sufficient to maintain a minimum Hamming distance of one between the stuff word and other words in the alphabet. The demultiplexer searches for this particular word with odd parity to detect the stuff word. One error, of course, is sufficient to cause an error in the output character.

Better protection against stuff word errors requires a greater distance between the stuff word and other words in the alphabet. Therefore a more sophisticated scheme is to note that in adding a parity-check bit to an m-bit word we have generated $2^{m+1} - 2^m$ new words, only one of which is the stuff word. The goal is to choose new data words out of this larger set in such a way that we maximize the distance from the stuff word.

Up to this point, we have assumed that the data format constitutes a frame when the multiplexer—either bit-interleaved or character-interleaved—has serviced all input ports at least once. This is what a conventional multiplexer does. Thus in a conventional multiplexer each input port has its own frame slot that regularly repeats, making available a constant and guaranteed transmission capacity. At any given time, however, some input ports may not have data to send, and the conventional multiplexer is inefficient in reserving time slots when there is no data to send. To remedy this, a version of time-division multiplexing known as *demand multiplexing* allocates available time according to input demand, keeping all slots filled with data as long as there is data waiting to be sent, and interchanging slot allocations, if necessary, to accomplish this.

In demand multiplexing, each input is applied to a buffer memory whose capacity is large enough to store at least several characters. The buffer memory control then monitors the fullness of the buffer memories and instructs the multiplexer to sample from those having the highest occupancy.† Every character or block of data to be transmitted is defined by an address attached to it by the multiplexer. Demand multiplexing can increase the efficiency of a transmission

---

† One version of demand multiplexing using long lines (e.g., submarine cables) or satellites is called TASI (Time Assignment Speech Interpolation). Telephone systems have found that subscribers speak only about 40% of the time, on the average, in normal conversations. Therefore some savings can be realized by detecting a talker's speech and then assigning it to an unused channel. That channel remains assigned until the talker is silent and the channel is needed by another speaker. During low traffic periods, the talker may keep the channel throughout a conversation. During high traffic periods, however, successive words may go over different routes.

system at the expense of higher bit overhead, some variable delays introduced in the data stream, and the risk of congestion if all terminals serviced should require transmission at the same time. If the latter situation should arise, a set of rules or procedures, with some hierarchy or priorities, must be assigned to handle the situation in an organized fashion.

A trend in modern time-division multiplexers is toward such devices as demand multiplexers, intelligent multiplexers, and concentrators. In fact, such devices can be viewed as intelligent terminals in an overall communication system that may not only multiplex and demultiplex data but also be users of data. As complexity increases, with added control features under the supervision of minicomputers or microprocessors, there are increasing requirements for development of a set of detailed operating procedures. Such a set of rules and procedures, usually involving a hierarchical set of instructions and priorities, is called a *protocol.*

Although the details of a protocol will differ from one system to another and will depend somewhat on the particular minicomputer or microprocessor used, there are some basic features that usually can be identified. These are designed to solve such operating problems as†

1. Framing—Determination of which bit groups constitute characters and which groups of characters constitute messages.

2. Error Control—Detection of errors, acceptance of correct messages and requests for retransmission of faulty messages.

3. Sequence Control—Identification of messages that are retransmitted by the error control system in order to avoid losing messages or duplicating them.

4. Line Control—Determination, in the case of a half-duplex line or a multipoint line, of which station is going to transmit and which station(s) is/are going to receive.

5. Time-out Control—Solution of the problem of what to do if message flow suddenly ceases entirely.

6. Start-up Control—Solution of the problem of what procedure should be followed to get transmissions started in a communication system that has been idle.

Several of these topics have been introduced, with specific reference to multiplexers, in the preceding discussion. It is beyond the purpose of this textbook, however, to pursue the subject further.‡

---

† J. E. McNamara, *Technical Aspects of Data Communication.* Maynard, MA: Digital Equipment Corporation, 1977, p. 191.

‡ See, e.g., D. W. Davies, D. L. A. Barber, W. L. Price, and C. M. Solomonides, *Computer Networks and their Protocols.* New York: John Wiley & Sons, 1977.

Drill Problem 7.8.2 A proposed character-interleaved multiplexer uses a standard 10-bit input word consisting of a start bit (0), 7 data bits using the ASCII code, a parity-check bit (even parity) and a stop bit (1). At the multiplexer a parity-check bit derived from the 8-bit data word is added using even parity, resulting in an 11-bit word. The stuff word uses one of the permissible characters in the input alphabet, inverts the start and stop bits, and uses odd parity for the multiplexer parity-check bit. (a) What is the minimum Hamming distance between the stuff word and other words in the alphabet? (b) Assuming that one character in ten is needed for synchronization, determine the throughput efficiency of the multiplexer. (c) Devise a stuff-bit decision rule.

Answer.   (a) 3; (b) 65%; (c) Two-of-three decision rule of (inverted) start and stop bits and the (odd) parity-check bit.

7.9 THE MATCHED FILTER

In both PAM and PCM systems we are interested in maximizing the peak pulse signal in the presence of additive noise. We are particularly interested here in the case in which the additive noise is white and the signal plus additive noise is passed through a linear time-invariant filter. Out of all the filters in this class which we could choose, which one will yield a maximum output in response to the input signal in the presence of additive white noise? This is the question to which we seek an answer in this section.

Let the input to the filter be $[f(t) + n(t)]$ where $f(t)$ is the signal and $n(t)$ is the additive noise. The output of the filter is $[f_o(t) + n_o(t)]$ and we wish to maximize the ratio $f_o(t_m)/\sqrt{n_o^2(t)}$, where $t = t_m$ is the best observation time (to be set). It turns out that it is more convenient to use the square of the amplitudes, so we seek to maximize the ratio:†

$$\frac{|f_o(t_m)|^2}{n_o^2(t)}. \tag{7.39}$$

Let the Fourier transform of $f(t)$ be $F(\omega)$ and let $H(\omega)$ be the transfer function of the desired optimum filter. Then we can write

$$f_o(t) = \mathscr{F}^{-1}\{F(\omega)H(\omega)\}$$

$$= \frac{1}{2\pi} \int_{-\infty}^{\infty} H(\omega)F(\omega)e^{j\omega t}\, d\omega,$$

$$f_o(t_m) = \frac{1}{2\pi} \int_{-\infty}^{\infty} H(\omega)F(\omega)e^{j\omega t_m}\, d\omega. \tag{7.40}$$

---

† Note that the mean-square value of $n_o(t)$ is independent of t.

The power spectral density of the noise is $S_n(\omega) = \eta/2$ so that

$$\overline{n_o^2(t)} = \frac{1}{2\pi} \int_{-\infty}^{\infty} \frac{\eta}{2} |H(\omega)|^2 \, d\omega. \tag{7.41}$$

Substituting Eqs. (7.40) and (7.41) into (7.39), we get

$$\frac{|f_o(t_m)|^2}{\overline{n_o^2(t)}} = \frac{\left| \int_{-\infty}^{\infty} H(\omega)F(\omega)e^{j\omega t_m} \, d\omega \right|^2}{\pi\eta \int_{-\infty}^{\infty} |H(\omega)|^2 \, d\omega}. \tag{7.42}$$

At this point we make use of the Schwarz inequality (cf. Example 2.5.2)

$$\left| \int_{-\infty}^{\infty} f_1(x)f_2(x) \, dx \right|^2 \leq \int_{-\infty}^{\infty} |f_1(x)|^2 \, dx \int_{-\infty}^{\infty} |f_2(x)|^2 \, dx. \tag{7.43}$$

The equality holds if, and only if,

$$f_1(x) = kf_2^*(x), \tag{7.44}$$

where k is an arbitrary constant.

Now we let the two functions in Eq. (7.43) be identified with $H(\omega)$ and $F(\omega)e^{j\omega t_m}$, respectively, so that Eq. (7.43) becomes

$$\left| \int_{-\infty}^{\infty} H(\omega)F(\omega)e^{j\omega t_m} \, d\omega \right|^2 \leq \int_{-\infty}^{\infty} |H(\omega)|^2 \, d\omega \int_{-\infty}^{\infty} |F(\omega)|^2 \, d\omega. \tag{7.45}$$

Substitution of this result into Eq. (7.42) gives

$$\frac{|f_o(t_m)|^2}{\overline{n_o^2(t)}} \leq \frac{1}{\pi\eta} \int_{-\infty}^{\infty} |F(\omega)|^2 \, d\omega,$$

or

$$\frac{|f_o(t_m)|^2}{\overline{n_o^2(t)}} \Bigg|_{\text{max}} = \frac{1}{\pi\eta} \int_{-\infty}^{\infty} |F(\omega)|^2 \, d\omega = \frac{E}{\eta/2}, \tag{7.46}$$

where E is the energy in $f(t)$ for a one-ohm load. The equality holds only if [cf. Eq. (7.44)]

$$H(\omega) = kF^*(\omega)e^{-j\omega t_m}, \tag{7.47}$$

or

$$\begin{aligned} h(t) &= \mathscr{F}^{-1}\{kF^*(\omega)e^{-j\omega t_m}\} \\ &= kf^*(t_m - t). \end{aligned} \tag{7.48}$$

The constant k is arbitrary and we assume $k = 1$ for convenience.

We conclude from this result that the impulse response of the optimum system is the mirror image of the desired message signal $f(t)$, delayed by an interval t_m. Hence the filter is matched to a particular signal, as conveyed by the terminology *matched filter*.

The result expressed in Eq. (7.47) makes good sense intuitively when applied to the magnitude characteristic of a filter so that: $|H(\omega)| = |F(\omega)|$. (Remember that this result was obtained for the case of additive white noise.) This result states that one should filter in such a way as to attenuate strongly those frequency components in frequency intervals having little relative signal energy while attenuating very little those components where the relative signal energy is high. Recall also that we are filtering for signal recognition in the presence of noise, not for signal fidelity (in which one usually desires a "flat" frequency response).

The phase response is also very important and Eq. (7.47) states that the phase shifts in $f(t)$ should be negated in such a way that all frequency components in $f(t)$ add *in phase* at exactly the time $t = t_m$. In contrast, the noise spectral components add with random phases so that the peak-signal-to-rms-noise ratio is maximized.

The signal representation $f(t)$ is assumed to have a finite duration $(0, T)$. The impulse response of the matched filter $f(t_m - t)$ can be obtained by folding $f(t)$ about the vertical axis and shifting it to the right by t_m seconds. Examples for $t_m < T$, $t_m = T$, and $t_m > T$ are shown in Fig. 7.39. Note that the impulse response of the matched filter for $t_m < T$ is noncausal (i.e., physically unrealizable). Both of the other systems are physically realizable but, to minimize the delay in obtaining a decision, the case for $t_m = T$ is chosen.

At the point $t = t_m$, the signal output of the matched filter is given by substituting Eq. (7.47) in Eq. (7.40) with $k = 1$:

$$f_o(t_m) = \frac{1}{2\pi} \int_{-\infty}^{\infty} |F(\omega)|^2 \, d\omega = E. \tag{7.49}$$

Thus the output of the matched filter at $t = t_m$ is independent of the particular waveform chosen and depends only on its energy! The mean-square noise output of the matched filter is [Eq. (7.49) in Eq. (7.46)]

$$\overline{n_o^2(t)} = E\frac{\eta}{2}. \tag{7.50}$$

The matched filter must be tailored to the signal waveform to achieve the maximum signal-to-noise ratio. One way in which this can be accomplished for the general case is by noting that the impulse response of a linear time-invariant filter can be approximated by a delay line that is tapped at various points and weighted by a set of fixed gains (cf. Section 3.10). Such a system is shown in Fig. 7.40, where the delay line has taps at the delays $k \, \Delta\tau$. For an N-tap delay line, the response $g(t)$ can be written as

$$g(t) = \sum_{k=0}^{N} f(t - k\,\Delta\tau)h(k\,\Delta\tau)\,\Delta\tau, \tag{7.51}$$

where the output of each tap is multiplied by the preset weight $h(k\,\Delta\tau)\,\Delta\tau$.

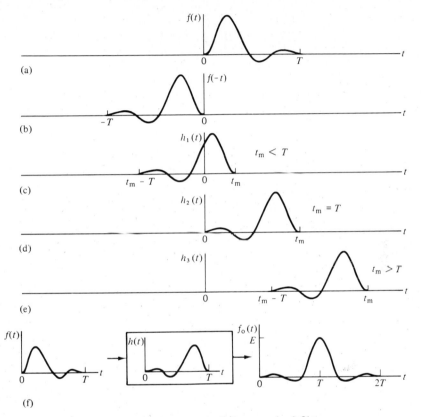

Fig. 7.39 Response of the matched filter.

For the specific case of a matched filter (real-valued case), we also have [cf. Eq. (7.48)]

$$h(t) = f(t_m - t),$$

so that the tap gains in Fig. 7.40 are given by

$$a_k = f(t_m - k\,\Delta\tau)\,\Delta\tau. \tag{7.52}$$

Thus once the signal $f(t)$ and the time t_m are known, the tap gains can be set to approximate the desired matched-filter characteristic.

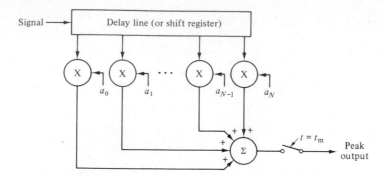

Fig. 7.40 Delay-line realization of the matched filter.

An alternative realization of the matched filter is as follows. Let the input to the matched filter be $y(t) = f(t) + n(t)$ and the corresponding output be

$$g(t) = y(t) \circledast f^*(t_m - t) = f^*(t_m - t) \circledast y(t)$$

$$= \int_{-\infty}^{\infty} f^*(t_m - \xi) \, y(t - \xi) \, d\xi. \tag{7.53}$$

We let $\zeta = t_m - \xi$, so that Eq. (7.53) becomes

$$g(t) = \int_{-\infty}^{\infty} f^*(\zeta) y(\zeta + t - t_m) \, d\zeta$$

$$= r_{fy}(t - t_m), \tag{7.54}$$

where

$$r_{fy}(\tau) = \int_{-\infty}^{\infty} f^*(\zeta) y(\zeta + \tau) \, d\zeta$$

is the time-crosscorrelation function for energy signals (cf. Chapter 4). Noting that $y(t) = f(t) + n(t)$, we find

$$g(t) = r_f(t - t_m) + r_{fn}(t - t_m), \tag{7.55}$$

where

$$r_f(\tau) = \int_{-\infty}^{\infty} f^*(\zeta) f(\zeta + \tau) \, d\zeta$$

is the time-autocorrelation function for energy signals. If the crosscorrelation between $f(t)$ and $n(t)$ is zero, then the peak output $g(t_m)$ is given by $r_f(0)$.

A block diagram of the time-crosscorrelator is shown in Fig. 7.41. The incoming waveform $y(t)$ is multiplied by $f(t)$. This requires that the signal $f(t)$ be known *a priori* and either stored in memory or supplied from another source.

The output of the multiplier is integrated to form $r_{fy}(\tau)$ in Eq. (7.54). To complete the operation, the switch is closed at $t = t_m$ to give the output $r_{fy}(0)$. Note that matched-filter detection is essentially a synchronous (i.e., coherent) detection.

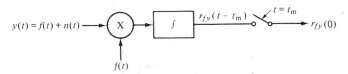

Fig. 7.41 The time correlator.

In some cases the realization of a matched filter may be much easier to accomplish than the above methods would indicate. For example, consider the matched filter for the rectangular pulse waveform shown in Fig. 7.42(a). The impulse response of the matched filter is given by $h(t) = f(T - t) = f(t)$ so that Fig. 7.42(a) also describes $h(t)$. The convolution of these rectangular waveforms yields the triangular waveform shown in Fig. 7.42(b).

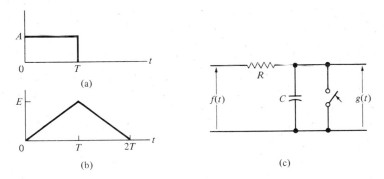

Fig. 7.42 Matched filtering of a rectangular pulse waveform: (a) input; (b) output; (c) integrate-and-dump circuit approximation.

At the time $t = T$, the output of the matched filter is maximum. A sample is taken for the output and the trailing portion of the triangular waveform for $t > T$ is not used. The linearly increasing portion of the triangular output waveform for $t < T$ can be realized by integrating an incoming rectangular pulse. The output is taken at $t = T$ and then the integrator is reset (dumped) and the procedure is repeated. A relatively simple circuit to accomplish this *integrate-and-dump* filtering is shown in Fig. 7.42(c). For $RC \gg T$, the circuit acts like an integrator on the incoming pulse waveform. The output $g(t)$ is sampled at $t = T$ and then the capacitor is shorted momentarily to reset the integrator. The integrate-and-dump filter thus forms a relatively simple realization of a matched filter for rectangular pulses. Note that synchronization is required for the sampling and reset operations. Another realization of a matched filter for rectangular pulse waveforms is discussed in Problem 7.1.2.

Drill Problem 7.9.1   Compute the matched-filter output over $(0, T)$ to the pulse waveform

$$f(t) = \begin{cases} \exp(-t) & 0 < t < T, \\ 0 & \text{elsewhere.} \end{cases}$$

Answer. $\exp(-T) \sinh t$.

7.10   MATCHED-FILTER CODE WORD DETECTION

In the previous section we discussed the matched-filter detection of pulse waveforms on a pulse-by-pulse basis. We can easily extend these concepts to the next level of detection—that of code words. This has application to the problem of an optimum detector of PCM in the presence of white noise.

Our interest in the use of the matched filter for PCM centers on the design of a filter which will yield a maximum output only for a particular code word. If we can design this filter, then we can use n such filters, each set for one of the n code words available in the code. These filters can be followed by a decision to determine which output is maximum. The appropriate signal level associated with the code word chosen is then generated and the receiver has performed its detection role at a particular sample point.

Both the tapped delay line and the time-correlator realizations of the matched filter can readily be extended to the case of code words. We shall restrict our consideration here to the binary case. This makes the required circuitry easy to build using digital-logic elements. As a result of the similarities between the two matched-filter methods for the binary case, we shall refer to both of them as correlators. Both methods require bit and word synchronization.

The design of an optimum PCM receiver now becomes evident. A code with n possible code words is chosen and then n correlators, each with a stored replica of one code word, are used. The output of the n correlators is supplied to decision logic that determines which correlator output is largest at the sample time $t = t_m$. The appropriate signal amplitude is then assigned to that particular sample point and the cycle is repeated. Such a receiver configuration is shown in Fig. 7.43. The reader will find it interesting to compare this diagram with that of Fig. 2.5.

Often these operations are performed digitally and the procedures are illustrated here with an example. Suppose that we decide to use a 16-level binary PCM code ($n = 16$) and for simplicity we assume that there is no noise present. At the receiver, each binary code symbol is stored in memory and the required multiplication and summation is performed digitally on command of a clock. For convenience, we designate a one by "$+$" and a zero by "$-$". Let us consider the correlator output before the sample switch for the code symbol of level 13 (that is, 1101, represented by $+ + - +$).† Upon command of the clock (which of course must be synchronized to the incoming bit stream), the incoming data are multiplied by the stored code symbols in each of the correlators and supplied to accumulators (i.e., integrators). At the next clock command, the incoming data are shifted by one bit interval and the operation is repeated.

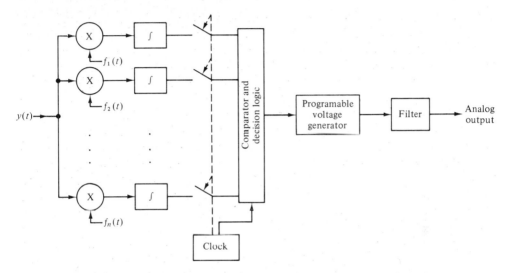

Fig. 7.43 An optimum receiver for PCM using correlation.

Let us assume that the incoming data are in the sequence 6, 13, 10, and that the correlator is set for the code symbol for 13. A listing of the successive correlator outputs before the sampling switch is shown in Table 7.2 and the results are graphed in Fig. 7.44. Note that the output is a maximum for $\tau = t_m$ and is proportional, at that point, to the signal energy. Closing the switch at that point assures us of the maximum output. However, this also requires receiver synchronization and the receiver must therefore maintain word synchronization as well as bit synchronization.

The necessity for word synchronization can be illustrated by considering the binary sequence for 3, 7 as the code enters the correlator shown in Table 7.2. The last two pulses of 3 and the first two pulses of 7 form the sequence $+ + - +$, matching the binary code word for 13. Thus there will be a maximum

---

† The usual (0, 1) binary notation can be used if the appropriate equivalence relation is defined for the multiplier ("EXCLUSIVE NOR"), although it will be shown in Section 10.3 that best performance is obtained if the binary signal states have equal energy and opposite signs.

correlator output midway between the processing for 3 and for 7. With word synchronization, however, this maximum will never occur at the correct time at which the output is taken unless the code word for 13 is present.

Table 7.2 Output of a Digital Correlator for the Binary Representation of 13 in the Sequence 6, 13, 10

Relative Shift, $\tau - t_m$	Incoming Data/Stored Replica			Accumulator Output $= r_{fy}(\tau - t_m)$
	10	13	6	
-3	$+ - + - + + - + - + + -$ $+ + - +$		$\begin{matrix} y \\ f \end{matrix}$	0
-2	$+ - + - + + - + - + + -$ $+ + - +$			$+2$
-1	$+ - + - + + - + - + + -$ $+ + - +$			-2
0	$+ - + - + + - + - + + -$ $+ + - +$			$+4$
$+1$	$+ - + - + + - + - + + -$ $+ + - +$			-2
$+2$	$+ - + - + + - + - + + -$ $+ + - +$			0
$+3$	$+ - + - + + - + - + + -$ $+ + - +$			$+2$

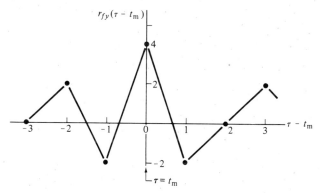

Fig. 7.44 Correlator output for the PCM sequence of Table 7.2.

Correlators can also be used to initially attain synchronization. In this case the receiver must search for the maximum correlator output. This search would be relatively easy if the correlator had an output only for $\tau - t_m = 0$. But this is not always the case, as can be seen from Fig. 7.44. One of the subjects of coding theory is to design codes which have minimum correlation for $\tau - t_m \neq 0$ (the maximum is proportional to the signal energy).

Drill Problem 7.10.1   A four-bit digital correlator similar to the one used for Table 7.2 is set for the binary code for 10 $(+ - + -)$. An infinite sequence of binary codes for a given number (0–15) is applied to the input. Determine the correlator output if the number is (a) divisible by 3; (b) divisible by 5 but not by 3.

Answer.   (a) 0; (b) ± 4.

7.11   PSEUDONOISE (PN) SEQUENCES

To aid in synchronization, particularly in initial synchronization, it is desirable to use a codeword whose autocorrelation is as large as possible at zero shift and as small as possible at all other shifts. Upon reception, the receiver can determine, upon recognition, that transmission is about to begin and establish an accurate timing reference for synchronization. This is particularly important in those transmission systems, such as satellites, that may not transmit data on a continuous basis.

   Although we can envision codewords that, to accomplish this goal, are stored in memory both at the transmitter and at the receiver, it turns out that some relatively simple linear systems are available to generate codewords with surprisingly good correlations. These codewords are generated using shift-register sequence generators.

   A shift register is a cascaded series of one-stage binary memories. The binary state of each memory is transferred to the next memory at the command of a clock. Thus the output of a three-stage shift register is the input delayed by three clock intervals; and at any one time, this shift register will contain the three successive most recent bits in the input bit stream. A shift-register sequence generator consists of an n-stage shift register with taps from certain stages and the output connected to logic combiners that feed back to the input. An example of a three-stage shift register sequence generator is shown in Fig. 7.45. In this example, the input to the shift register is the mod-2 (i.e., binary add without carry) sum of the content of stages S_2 and S_3, and the output is read from the content of S_3.

   Suppose we initialize the generator of Fig. 7.45 by loading all 1's into the shift registers. The two inputs to the mod-2 ADD operation are 1 and 1; the result, 0, is held by the buffer. In the next clock interval, the content of S_2 is transferred to S_3, the content of S_1 is transferred to S_2, and the content of the buffer is transferred to S_1. The mod-2 ADD operation again yields 0 and is held by the buffer for the next clock command. Continuing the sequence, we find that the three shift registers have the following contents:

$$111, 011, 001, 100, 010, 101, 110, 111, \ldots$$

Because this generator uses information from only the three most recent bits, once the initial state (111) repeats, the entire remaining sequence will repeat. The generator output is taken from the content of the last shift register to give the output codeword:

$$1\ 1\ 1\ 0\ 0\ 1\ 0 \ldots$$

The output codeword is periodic, with a period of 7 bits.

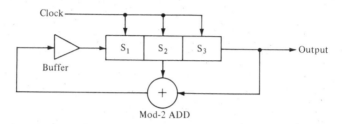

Fig. 7.45 A 3-stage shift register sequence generator for generation of a 7-bit PN sequence.†

In the progression of shift register contents above, every possible 3-bit combination occurs with the sole exception of the all-zero word. To the informed observer, this progression is fixed by the number of shift register stages, and the choice of taps to apply to the mod-2 ADD operation. To the uninformed observer, however, the progression (within a 7-bit period) appears to be entirely random. Therefore these sequences are sometimes called "pseudorandom" sequences.

The correlation properties of the above output sequence are of particular interest to us. Correlating the periodic sequence above with a 7-bit replica yields the correlation pattern shown in Fig. 7.46(a) for the on-off binary case (0, 1) using the EXCLUSIVE-NOR operation for multiplication, and the correlation pattern shown in Fig. 7.46(b) for the polar binary case $(-1, +1)$. In either case the maximum is given by the normalized energy (seven in this case) at zero shift. The correlation decreases to a plateau level for all other shifts up to one period away. As a result, the power spectral density approaches a white spectral density as the sequence length increases. Because these sequences are pseudorandom with white power spectral density, they are called *pseudonoise* (PN) sequences.

PN sequences are members of a class of codes known as maximal-length codes. By definition, maximal-length codes are the longest codes that can be

---

† Often the buffer is included as part of the mod-2 add operation. It is shown separately here for clarity.

generated by a shift register of given length. Some properties of maximal-length codes for the binary case are listed below.

1. The length of a maximal-length code that can be generated by n shift registers is $2^n - 1$.

2. With the exception of the all-zero word, every possible n-bit word exists in the n-stage shift register at some time during the generation of a complete code cycle, and only once during the cycle.

3. The number of 1's in an output sequence is one greater than the number of 0's.

4. The correlation of a code sequence with a one-period replica results in a peak equal to the sequence length $2^n - 1$ at zero shift, and at multiples of the sequence length. At all other shifts, the correlation is $2^{n-1} - 1$ for the on-off binary case $(0, 1)$, and -1 for the polar binary case $(-1, +1)$.

Not all tap combinations of an n-stage shift register will yield a maximal-length code. Proper feedback connections for several values of n are given in Table 7.3. Tables for many other feedback connections are available in the literature.[†]

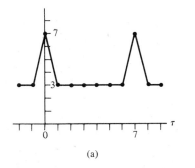 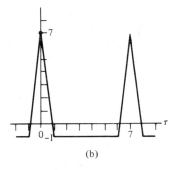

Fig. 7.46 Correlation of a 7-bit PN sequence for (a) on-off binary, and (b) polar binary output waveform.

The difference between the correlation value of a code at zero shift and the maximum correlation value at any nonzero shift (up to one period) is called the *index of discrimination*. The larger the index of discrimination the better the code is suited for synchronization. For a maximal-length code, the index of discrimination is $(2^n - 1) - (-1) = 2^n$ for the polar binary case and $(2^n - 1) - (2^{n-1} - 1) = 2^{n-1}$ for the on-off binary case. Some degradation may be expected in the presence of noise, and design procedures usually allow for an error of one or two bits in a codeword.

---

† See R. C. Dixon, *Spread Spectrum Systems*, New York: John Wiley & Sons, 1976, pp. 81–83.

Table 7.3   Feedback Connections for PN Code Generation

n	Sequence Length	Feedback Taps (for combination with output)	Output Sequence (initial state: all ones)
2	3	S_1	110
3	7	S_2	1110010
4	15	S_3	111100010011010
5	31	S_2	1111100110100100001010111011000

PN sequences are used widely in digital communication systems for synchronization. Often, if initial synchronization is the objective, several cycles of the sync codeword are sent and the receiver is required to identify two successive sync words. In a typical sync-search procedure, a bit (or group of bits) in the received bit stream is loaded into a receiver shift register and compared to the contents of a second register containing the sync codeword. If no match occurs, the process is repeated. If a match occurs, the receiver begins inputting additional bits and verifying. On verification of two successive sync words (within a given tolerance for error of, say, two bits), the receiver confirms that initial sync has been achieved. In conservatively designed systems, up to five sync words are sent to permit initial synchronization. A special codeword (meaning one that is distinct from the PN sequences) is placed after the PN sequences and becomes the signal that data transmission is about to begin.

PN sequences may be used also for frame synchronization. In this case, the sync word occurs periodically in the data transmission. A phase-locked loop (PLL) is used in the receiver to track this sync word and to provide accurate synchronization. The VCO in the PLL is a PN sequence generator.

PN sequences are used also in many other types of systems. Because of their good correlation properties, PN sequences find applications in ranging systems. In some systems, such as radar, the correlation of an aperiodic (one-shot) PN sequence is more important than that of a periodic sequence. Sequences with good aperiodic correlation properties are called Barker codes. The longest known Barker code is of length 13. (All of the known Barker sequences are listed in Table 7.4.) PN sequences may also be used as discrete white-noise generators that test systems and scramble signal transmissions.

Table 7.4   The Barker Sequences

```
+ −
+ + −
+ + − +
+ + + − +
+ + + − − + −
+ + + − − − + − − + −
+ + + + + − − + + − + − +
```

Drill Problem 7.11.1 The output of a six-stage PN sequence generator uses the polar binary representation $-1V$, $+1V$. (a) What is the (error-free) index of discrimination? (b) What is the average value of the sequence?

Answer. (a) 36.1 dB; (b) 15.9 mV.

7.12 SUMMARY

In pulse modulation, sample values are sent by varying a parameter in a train of pulses. The minimum sampling requirements are set by the sampling theorem.

In analog pulse modulation, the height, width, or position of the pulses is varied in direct proportion to the signal amplitude at the sampling instants. In pulse-code modulation, a code is changed in discrete steps in proportion to the quantized signal amplitude at the sampling instants.

Pulse-amplitude modulation (PAM) resembles conventional sampling except that the pulse tops are flat. Although this difference can be adjusted with equalizing filters, it is minimized by use of narrow pulse widths. This leaves space for interleaving pulses from other sampled signals. This method of combining several sampled signals in a definite time sequence is called time-division multiplexing (TDM). The minimum bandwidth to handle a TDM waveform must obey the requirements of the sampling theorem. PAM offers no signal-to-noise advantage.

Interference between pulses as a result of a finite bandwidth in a pulse modulation system is called intersymbol interference (ISI). Intersymbol interference can be controlled by widening the transmission bandwidth and by choosing the transmission spectral shaping. A good choice of the latter are the Nyquist waveforms, which have uniformly spaced zeros at multiples of a basic time interval governed by the bandwidth.

Pulse-timing modulation employs a variation in the timing of pulses to convey information. Examples of pulse-timing modulation are pulse-width modulation (PWM) and pulse-position modulation (PPM). Because pulse-edge response is important to the demodulation of pulse-timing modulation, it requires a much wider bandwidth than PAM. It also offers a signal-to-noise advantage similar to that for angle modulation.

Pulse-code modulation (PCM) uses a series of discrete code symbols to convey information. This type of modulation offers more versatility in signal design to combat effects of various types of noise. Different types of codes may be used, depending on system constraints on bandwidth, synchronization, etc.

The basic quantum unit for conveying information in a binary code is called the bit. Limiting bandwidth efficiency for a binary PCM system is two bits per second per Hz.

One way to increase the reliability of communication is to add extra bits, at the expense of added transmission time or bandwidth, in order to detect and, possibly, to correct the errors. The use of parity-check bits and repetition codes are simple and effective methods to accomplish this objective.

Two basic types of digital data transmission are synchronous and asynchronous. In synchronous systems, all sampling and multiplexing is governed by a master clock. Asynchronous multiplexers are operated slightly faster than would be necessary for comparable synchronous systems. Either extra bits (called bit stuffing) or extra words (called word stuffing) are inserted to keep the inputs up to the desired rate. Intelligent multiplexers that can alter the data-handling and routing characteristics are called concentrators.

Detection of PCM demands the recognition of a given code symbol in the presence of noise. The optimum linear time-invariant filter to accomplish this task in the presence of white noise is called a matched filter. A matched filter is the optimum linear time-invariant filter that gives the highest peak-signal–to–rms-noise ratio in the presence of white noise. The matched filter can be realized using tapped delay lines or crosscorrelation.

The principles of matched-filter detection can be extended to codewords. Codewords with good autocorrelation properties are desirable for synchronization in matched-filter detection. A good autocorrelation function for synchronization is one that is maximum at zero shift and zero (or nearly zero) for all other shifts. Pseudonoise (PN) sequences provide good periodic autocorrelation properties, and they are relatively easy to generate using shift registers and digital logic.

Selected References for Further Reading

1. J. Martin. *Telecommunications and the Computer*. Second Ed. Englewood Cliffs, N.J.: Prentice-Hall, 1976.
   Describes, in a qualitative manner, data transmission methods using telephone networks; includes a glossary for definition of terms used in telephony; easy reading.

2. H. S. Black. *Modulation Theory*. Princeton, N.J.: D. Van Nostrand, 1953.
   A good discussion of sampling theory and applications to pulse modulation.

3. M. Schwartz. *Information Transmission, Modulation and Noise*. Third Ed. New York: McGraw-Hill, 1980.
   Chapter 3 contains a good discussion of sampling, time-division multiplexing, and waveshaping for PCM, and uses typical operating systems as examples.

4. J. A. Betts. *Signal Processing, Modulation and Noise*. New York: American Elsevier Publishing Co., 1970.
   Chapter 6 describes some effects of inadequate band-limiting in TDM methods.

5. J. R. Pierce and E. C. Posner. *Introduction to Communication Science Systems*. New York: Plenum Press, 1980.
   An interesting approach to matched filters and correlation is presented in Chapter 7.

6. R. C. Dixon. *Spread Spectrum Systems*. New York: John Wiley & Sons, 1976.
   Chapter 3 contains a good discussion of maximal-length sequences and their generation. The remainder of this book makes interesting collateral reading.

7. P. Z. Peebles, Jr. *Communication Systems Principles*. Reading, MA: Addison-Wesley, 1976.
   Chapter 7 is an advanced discussion of sampling and S/N performance of PWM and PPM, including threshold effects.

Problems

7.1.1 The signals given below are not band-limited. Determine the minimum sampling rate, in terms of the number of -3-dB bandwidths, for each of these signals such that the magnitude of the largest alias frequency components introduced by sampling are down at least 10 dB from the largest magnitude of desired frequency components.

    a) $f(t) = e^{-at}u(t)$ b) $f(t) = \mathcal{F}^{-1}\{1/\sqrt{1 + \omega^{10}}\}$

7.1.2 The first-order hold circuit shown in Fig. P–7.1.2 is an alternative to the sample-and-hold circuit for recovery of PAM signals. The input $f_s(t)$ is a sampled version of $f(t)$ and the delay is equal to the sampling period T.

    a) Let the input be 16 equally spaced narrow rectangular functions (approximate with impulses) describing one cycle of a sinusoid. Sketch the output waveform.

    b) In contrast to the sample-and-hold circuit, this is a linear time-invariant system; determine its impulse response and frequency transfer function.

    c) Compare the magnitude frequency transfer function with that of an ideal low-pass filter.

Fig. P–7.1.2

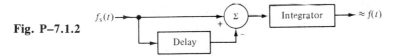

7.1.3 A certain signal $f(t)$ is band-limited to f_m Hz. A system is designed to transmit the sampled values of $f(t)$ by varying the heights of the train of triangular pulses shown in Fig. P–7.1.3, where $T \le (2f_m)^{-1}$.

    a) Sketch a block diagram of a system to accomplish this.

    b) Sketch a typical sampled signal and its spectrum.

    c) Determine the magnitude transfer function of an equalization filter for use at the receiver.

    d) Alter the system shown in Fig. P–7.1.2 appropriately to generate the required pulse train.

Fig. P–7.1.3

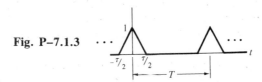

7.1.4 The signal $f(t) = 1000 \, \text{Sa}(1000 \, \pi t)$ is transmitted through a system with the frequency transfer function $H_1(\omega) = (1 - |\omega|/2000 \, \pi) \, \text{rect}(\omega/4000 \, \pi)$. The output $g(t)$ is sampled by a train of uniformly spaced impulse functions occurring at a 1 kHz repetition rate.

a) Sketch the spectral density of the sampled signal.
b) The sampled signal $g_s(t)$ is then transmitted through a second system as shown in Fig. P–7.1.4. Determine and sketch $H_2(\omega)$ if $x(t) = f(t)$.

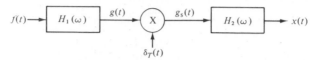

Fig. P–7.1.4

7.2.1 Two low-pass signals of equal bandwidth are to be time-multiplexed into a single channel using PAM. Each signal is impulse-sampled and the time-multiplexed signal is passed through an ideal low-pass filter with a bandwidth of 10 kHz before transmission.

a) What is the maximum frequency content allowable in each input signal to ensure that each signal can be recovered exactly at a receiver?
b) Sketch a block diagram of a receiver for recovering the two signals from the transmission.

7.2.2 Twenty-five audio input signals, each band-limited to 3.3 kHz and sampled at an 8 kHz rate, are time-multiplexed in a PAM system.

a) Determine the minimum clock frequency of the system.
b) Find the maximum pulse width for each channel.
c) The above PAM time-multiplexed signal is shifted in frequency by multiplying it by a carrier, $\cos \omega_c t$, thus forming a PAM/AM signal. What is the minimum bandwidth?

7.2.3 Five 10 kHz input signals are to be multiplexed and transmitted. Determine the minimum bandwidth required for each method if the multiplexing/modulation used is (a) FDM, SSB-SC; (b) FDM, DSB-LC; (c) TDM, PAM; (d) PAM/AM; (e) PAM/FM, $\beta = 40$.

7.2.4 A three-channel PAM system is shown in Fig. P–7.2.4; $f_1(t)$ is band-limited to 5 kHz and $f_2(t)$, $f_3(t)$ are each band-limited to 3 kHz. Determine the minimum and maximum values of the commutator clock rate such that all three signals can be reconstructed at a receiver without distortion.

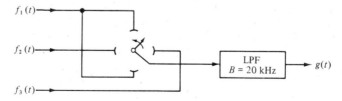

Fig. P–7.2.4

7.2.5 Eight input signals are sampled and time-multiplexed using PAM. The time-multiplexed signal is passed through an ideal low-pass filter. Six of the input signals are band-limited to 3.3 kHz and the other two are band-limited to 10 kHz.

a) What is the minimum sampling rate if all channels are sampled equally?
b) For the above sampling rate, what is the minimum bandwidth of the low-pass filter?
c) What would be the minimum bandwidth requirement if these input signals were frequency-multiplexed using SSB?
d) In the above sampling format, the low data rate inputs are sampled more than three times as often as necessary. A narrower transmission bandwidth can be obtained by sampling the two 10 kHz signals three times for each time that the 3.3 kHz signals are sampled. Devise a new sampling format to accomplish this; then repeat the calculations of parts (a) and (b) above for this new sampling format.

c 7.2.6 In the development of Eq. (7.11) we assumed that the samples were very narrow and could be approximated by impulses. Here we wish to investigate the effects of nonzero pulse width on this result. Suppose we have two equal-width rectangular pulses, each of one-volt amplitude with the pulse centers separated by 1 msec. These pulses are passed through an ideal low-pass filter with a bandwidth $B = 1/(2T) = 500$ Hz. Using the digital computer, determine the increase in pulse separation required such that an instantaneous sample of the peak amplitude of each pulse at the output of the filter is not contaminated by the presence of the other pulse. Compute for pulse widths from 1 μsec to 500 μsec.

7.3.1 Repeat Problem 7.2.2(c), assuming that a raised-cosine filter is used instead of an ideal LPF and that $\alpha = 0.25, 0.50, 1$.

7.3.2 Repeat Problem 7.2.1(a), assuming that a raised-cosine filter is used instead of an ideal LPF and that $\alpha = 0.25, 0.50, 1$.

7.3.3 Two unit impulses are applied to the input of an ideal low-pass filter. They are spaced such that the first zero crossings in the $(\sin x)/x$ output signal waveforms coincide.

a) Compute the peak-signal-to-highest-sidelobe ratio.
b) Repeat for the same pulse spacings if a raised-cosine filter ($\alpha = 1$) is used whose -6 dB bandwidth is the bandwidth of the ideal filter in part (a).

7.3.4 A number of 10 kHz channels are to be transmitted using PAM techniques. The sampling pulses are 10 μsec in width. The resulting time-multiplexed pulse train is passed through a low-pass RC filter with a time constant of $RC = 2$ μsec. The output of the filter is then shifted in frequency and transmitted. The problem here is to decide how many channels can be handled with an acceptable degree of fidelity. For this, define a "percent crosstalk" criterion as that amount of signal energy from a preceding pulse which "spills over" into the desired time slot divided by the desired signal energy (in the correct time slot) when both pulses are of equal magnitude. Using this somewhat arbitrary criterion, calculate the percent crosstalk for 5 channels.

7.4.1 A given PWM signal has fixed leading edges at the times $t = mT_s$. Sketch typical waveforms to show that the circuit in Fig. P–7.4.1 will approximately reconstruct the modulating waveform if the switch is closed momentarily whenever $t = mT_s$.

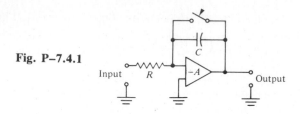

Fig. P–7.4.1

7.4.2 The surface of a rotating drum of axial length L is composed of half conductor and half dielectric along a diagonal cross section, as shown in Fig. P–7.4.2. A fixed contact makes an electrical connection with the conductor. A second contact can be moved along the axial length of the drum.

a) If the angular speed ω_0 of the drum is constant, show that the resulting electrical waveform is a PWM transmission of the position x of the movable contact.

b) Show that the average voltage at the output terminals is a linear function of x and hence can be used to linearly control the average speed of a dc motor.

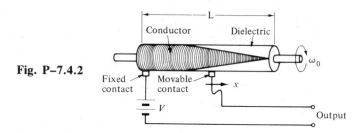

Fig. P–7.4.2

7.4.3 A given single-channel PPM system has the following specifications:

Input signal	$\|f(t)\| < 10$ V
Desired accuracy	± 0.05 V for $S/N = 10$ dB
Average pulse rate	10 kHz
Pulse width	1 μsec

Estimate the minimum bandwidth necessary for transmission in the presence of noise using a guard time of 1 μsec per channel.

7.4.4 Two channels of a PWM system are each sampled at a 10 kHz rate and then time-multiplexed. The width of each pulse, w, is given at the sample times t_i by:

$$w(t_i) = [5 + f(t_i)] \; \mu\text{sec}$$

where $f(t_i)$ is restricted to positive values and is measured in millivolts. The desired accuracy is specified as ± 0.5 mV.

a) What is the maximum frequency content allowable in each input signal?

b) Estimate the minimum bandwidth required for transmission.

c) Determine the maximum signal range, in millivolts, which can be handled by this system if a 5 μsec minimum guard time must be maintained between channels.

7.4.5 Twenty-five input signals are sampled and then time-multiplexed in a PPM system. The half-amplitude width of the pulses is 1 μsec, the rise and fall times are each 0.15 μsec, and the pulse repetition rate is 0.2 MHz.

a) What are the maximum frequency components allowable in each input if all inputs are sampled equally?

b) Estimate the minimum bandwidth required for transmission.

c) The above PPM time-multiplexed signal is shifted in frequency by multiplying it by a carrier, cos $\omega_c t$, thus forming a PPM/AM signal. What is the minimum bandwidth?

7.5.1 The S/N improvement in pulse-timing modulation is proportional to $k^2 B^2 \overline{f^2(t)}$, as shown in Eq. (7.30). What penalty must be paid for increasing k in order to obtain a better S/N improvement factor? Explain.

7.5.2 A 3 kHz sine wave with a peak amplitude of 2 volts is sampled at an 8 kHz rate. These sample values are transmitted using PPM. The pulse width is 25 μsec and the S/N improvement factor in detection is 30 dB.

a) Determine the minimum transmission bandwidth required and estimate the reconstructed signal amplitude resolution.

b) To obtain this S/N improvement using FM, how much bandwidth is required?

7.5.3 In the receiver, the signal pulses in a PPM system have a peak amplitude of 1 volt and the additive noise is 0.10 volt rms. The half-amplitude width of the pulses is 1 μsec, the rise and fall times are each 0.15 μsec, the pulse repetition rate is 10^5 $(sec)^{-1}$, and the rms pulse deviation is 1 μsec.

a) What is the minimum bandwidth required for transmission?

b) Compute the S/N ratio before and after demodulation. (Approximate the rise and fall times by linear line segments.)

7.6.1 An analog signal is sampled at the Nyquist rate, $1/T$, and quantized into n levels.

a) Show that the time duration τ of one bit of the binary encoded signal is $\tau \leq T/\log_2 n$.

b) When is the equality sign valid?

7.6.2 The sinusoidal signal (in volts) $f(t) = \cos 2\pi t$ is to be quantized into 16 uniformly spaced levels. The sampling rate is 8 Hz and the first sample occurs at $t = 0$; assume negligible sampler pulse width. Sampled voltages are to be quantized to the nearest quantizer level.

a) How many bits are required per sample?

b) Prepare a table showing the code number and its binary representation for each sample point if (1) the code numbers are arranged in order of increasing voltage; and (2) the code numbers are arranged in order of increasing voltage magnitude and the first bit is assigned to indicate the sign.

7.6.3 A four-bit D/A converter (such as could be used in Fig. 7.28) is shown in Fig. P–7.6.3.

a) Explain the operation of the circuit.

b) Draw a logic diagram to replace the switches.

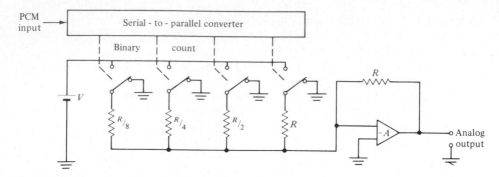

Fig. P–7.6.3

7.6.4 A computer outputs binary symbols at a rate of 56 kbps. Raised-cosine spectral shaping ($\alpha = 0.3$) is used prior to baseband transmission.

a) Determine the minimum bandwidth required.
b) Repeat if two successive digits are combined into one pulse with four possible amplitudes.
c) Repeat (a), assuming three successive digits are combined into one pulse with eight possible amplitudes.

7.6.5 Assume that the maximum usable bandwidth of a telephone channel is 2400 Hz and that DSB-SC at 1800 Hz carrier rate is used for modulation.

a) Find the maximum bit rates possible using binary pulses with raised-cosine shaping and $\alpha = 0.25, 0.50$.
b) Repeat (a), assuming that two successive digits are combined into one pulse having four possible amplitudes.

7.6.6 A series of high-density bipolar representations, known as HDB1, HDB2, HDB3, . . . , can be used to eliminate the long successions of zero-amplitude pulses that may occur in BRZ. The HDBn representations avoid the occurrence of more than n successive pulses of zero amplitude. They use the AMI rules but substitute special sequences for each succession of $(n + 1)$ 0's, and also use AMI rule violations to convey the extra information needed to replace the 0's. The most important of the HDB codes is the HDB3.

If the string 0000 occurs in HDB3, either of the special sequences 000D or 100D may be sent. The "D" is replaced by level $+1$ or -1 in such a way as to violate the AMI rule at that point. Choice of 000D or 100D is such that pulses violating the AMI rule take on the levels $+1$ and -1 alternately.

Write out the HDB3 pulse train for the following input: 100001100000000, assuming that 000D is the first special sequence. Then write out the HDB3 pulse train, assuming that 100D is the first special sequence.

7.6.7 Twenty-five input signals, each band-limited to 3.3 kHz, are each sampled at an 8 kHz rate and then time-multiplexed. Calculate the minimum bandwidth required to transmit this multiplexed signal in the presence of noise if the pulse modulation used is (a) PAM; (b) PPM with a required level resolution of 5%; or (c) binary PCM with a required level resolution of 0.5%. (This higher resolution requirement on PCM is normal for speech-type signals because the quantization noise is quite objectionable.)

7.6.8 One of the advantages of PCM is that maintenance of pulse shape is not essential for correct demodulation. As an example, a four-bit PCM system is designed to transmit through a communications channel which can be modeled by a *RC* low-pass filter with a -3-dB bandwidth of 100 kHz. Determine the maximum input bandwidth this system can handle if each received pulse must rise to at least 80% of the input pulse height to be correctly detected.

7.7.1 An *n*-bit binary codeword contains one parity bit. How many possible codewords are not usable in the set of *n*-bit codewords as a result of the parity bit? Compute numerical answers for the case where $n = 7$ and $n = 9$.

7.7.2 Verify that Eq. (7.38) is correct.

★ **7.8.1** Design an M12 bit-interleaved multiplexer (see Table 7.1) that has the same frame format as that given in Fig. 7.37, with the exception that only two bits per input channel are taken before a control bit is added. In designing this multiplexer, calculate the four quantities requested in Section 7.8.1.

★ **7.8.2** Design an M13 bit-interleaved multiplexer (see Table 7.1) that has the same frame format as the multiplexer in Example 7.8.1, with the exception of the added 7-bit marker word 1110010. In this design, calculate the four quantities requested in Example 7.8.1. Compare these results with those of Problem 7.8.1.

★ **7.8.3** A certain character-interleaved, time-division multiplexer with word stuffing uses an all-zero word as the stuff word. In the multiplexer, the *m*-bit input data word $\mathbf{A} = (a_1, a_2, \ldots, a_m)$ is encoded into the $(m + 1)$-bit output word $\mathbf{B} = (b_1, b_2, \ldots, b_m, b_{m+1})$ according to the following algorithm:[†]

$$b_{m+1} = 0 \quad \text{if} \quad w(\mathbf{A}) \geq \left[\frac{m}{2}\right] + 1$$

$$b_{m+1} = 1 \quad \text{if} \quad w(\mathbf{A}) \leq \left[\frac{m}{2}\right]$$

$$b_i = b_{m+1} \oplus a_i, \quad 1 \leq i \leq m,$$

where $w(\cdot)$ is the Hamming distance measured from the all-zero word, the notation $[m/2]$ represents the largest integer $\leq m/2$, and $\oplus$ indicates modulo-2 addition. At the demultiplexer, the bit order is reversed so that b'_{m+1} is detected first and the decoding algorithm is

$$a'_i = b'_{m+1} \oplus b'_i, \quad 1 \leq i \leq m,$$

where the primes indicate bits in the demultiplexer.

For such a multiplexer using $m = 8$-bit input data words,

[†] M. M. Buchner, Jr., "An Asymmetric Encoding Scheme for Word Stuffing," *Bell System Technical Journal* (March 1970): 379–398; *see also*, J. J. Spilker, Jr., *Digital Communications by Satellite*, Englewood Cliffs, N.J.: Prentice-Hall, 1977, pp. 124–128.

a) Make a table of several arbitrary data words and their encoded versions.
b) Estimate the minimum distance of the output words from the all-zero stuff word.

7.9.1 Let the power spectral density of $n_o(t)$ be $S_n(\omega)$. Derive the matched filter conditions and output for the case of nonwhite noise. [*Hint*: Multiply and divide the integrand of Eq. (7.42) by $S_n^{1/2}(\omega)$.]

7.9.2 A given real-valued signal waveform $f(t)$ exists over the time interval $(0, T)$ in the presence of white noise. A matched filter for $f(t)$ is chosen to be realizable with minimum delay. Show that (a) the output of the matched filter has even symmetry about the point $t = T$; and (b) the impulse response of the matched filter is equal to the signal waveform if $f(t)$ has even symmetry about the point $t = T/2$.

7.9.3 A given pulse signal $f(t)$ is shown in Fig. P–7.9.3, where $0 < \alpha < 1$.

a) Sketch the impulse response of a filter matched to this signal.
b) Sketch the matched-filter output as a function of time, and indicate the peak value of the output.

Fig. P–7.9.3

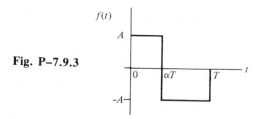

7.9.4 A given pulse signal waveform $f(t)$ is shown in Fig. P–7.9.4.

a) Determine the frequency transfer function of the matched filter for $f(t)$.
b) Develop a relation for the maximum peak-signal-to-rms-noise ratio at the output of this filter if the input noise is white with a (two-sided) power spectral density $\eta/2$ watts per Hz across one ohm.

Fig. P–7.9.4

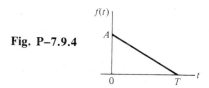

7.9.5 In a binary transmission in the presence of white noise, one of the message states is represented by a rectangular pulse of width τ and the other message state is transmitted by the absence of a pulse. It is decided to approximate the matched filter at the receiver with an RC low-pass filter. (a) Derive a relation for the S/N degradation factor (>1) expected using the RC filter instead of a matched filter. (b) What value of RC yields minimum degradation?

7.10.1 Three examples of codewords for synchronization are as follows:

$$N = 3: + + -; \qquad N = 5: + + + - +; \qquad N = 7: + + + - - + -,$$

Sketch the output of a correlator set for each of these codes under the following conditions:

a) The codeword is aperiodic (assume all 0's elsewhere).
b) The codeword is periodic with period N.

Assume that the stored replica is only for the length of the codeword.

7.10.2 Repeat Problem 7.10.1 using 0 and 1 to replace minus $(-)$ and plus $(+)$, respectively, and an EXCLUSIVE-NOR operation to replace multiplication in the correlator.

★7.10.3 To aid in the recognition of a binary PCM code of length T in the presence of a large amount of noise, each code word can be repeated m times. In the receiver, a delay line with attenuation is used, as shown in Fig. P–7.10.3.

a) In what sense can this be considered a matched filter? [*Hint*: Cf. Drill Problem 2.19.1.]
b) Suppose $K = 0.90$; what S/N improvement can be realized for $m = 3$? [*Hint*: Recall that the noise voltages add on a mean-square basis.]
c) What is the maximum S/N improvement attainable for a given value of K and $m \to \infty$?

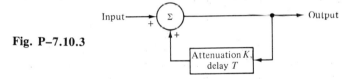

Fig. P–7.10.3

7.11.1 Show that the index of discrimination in a received PN sequence changes by at most -2 ± 2 for a one-bit error.

7.11.2 Consider a 15-bit maximal-length PN code generated by feeding back the last two stages of a four-stage shift register (see Table 7.3).

a) Assuming an initial state of 1111, find all the other possible shift-register states in order of their occurrence.
b) Determine and plot the correlation of the periodic PN sequence with a one-period replica.

7.11.3 In each of the following aperiodic code sequences, a polar binary representation is used and zeros are added beyond the codeword length.

a) Find and plot the aperiodic autocorrelation of the 13-bit Barker sequence of Table 7.4.
b) Find and plot the aperiodic autocorrelation of a 15-bit maximal-length PN code generated by feeding back the last two stages of a four-stage shift register (see Table 7.3). Compare the result with (a).

7.11.4 Verify that the discrimination index of the Barker codes (Table 7.4) is either k or $(k - 1)$, where k is the number of bits in the codeword.

CHAPTER 8

PROBABILITY AND RANDOM VARIABLES

Waveforms whose values can be specified exactly are said to be "deterministic." While deterministic signals cannot always be expressed explicitly, they can, within rather broad limitations, be expressed in terms of summations of explicitly known functions (e.g., the Fourier series). The use of deterministic signals has played a major role in our analysis of communication systems up to this point. Now we consider methods for describing nondeterministic, or "random," waveforms. To do this, we shall first introduce some basic concepts from probability theory.

Our approach to probability is heuristic. Because we are interested in some specific applications to communication systems, we cover only certain aspects of probability. For a more complete treatment of the subject of probability, the reader is advised to consult the references.

8.1 PROBABILITY

In signal analysis, it is common to use the concepts of signal energy and power as measures of the signal under consideration. In spatial coordinates, the use of distance and squared length are commonly employed. In a somewhat related manner, the concept of probability is used as a means by which we can measure (numerically) the favorable outcomes of a given experiment. To be a valid measure, we first need to discuss the basis for the measurement.

Suppose that some experiment is considered in which the outcome does not remain constant. One of the possible outcomes of the experiment is labeled A. Tossing a coin is such an experiment, the possible outcomes being heads and tails. If the experiment is repeated N times, suppose that the outcome A will occur N_A times. The *relative frequency of occurrence* of A is N_A/N. This ratio is not very predictable when N is small. If the experiment has statistical regularity, however, the relative frequency of a particular outcome may approach a limit as the number of repetitions N becomes very large. This limiting value of the relative frequency of occurrence is called the *probability* of the outcome A and is written as $P(A)$:

$$P(A) = \lim_{N \to \infty} \frac{N_A}{N}. \tag{8.1}$$

The limit in Eq. (8.1) is not a limit in the usual functional sense, but is used to indicate that if N is very large then the expected deviation of the ratio N_A/N from a constant becomes very small. This statement is sometimes called the "empirical law of large numbers" and agrees with our intuitive ideas of probability. For example, if an honest coin is tossed a large number of times, we expect that heads will show in about one-half the tosses and that this approximation becomes better as more tosses are attempted.

Example 8.1.1 An honest coin is tossed ten times and the outcomes of this experiment are: $\{H, T, T, H, H, H, H, T, T, T\}$. Let A be the occurrence of a head and let $x = N_A/N$. (a) Calculate x as the experiment progresses. (b) Also, as a measure of possible convergence to $x = 0.50$, calculate

$$y = \left[\sum_{i=1}^{N} (x_i - 0.5)^2/N \right]^{1/2}.$$

Solution.   The results of the above computation are shown below. Note the (slow) convergence as the number of trials is increased even though the given sequence is not a particularly "favorable" one.

N	H	x	y	N	H	x	y
1	1	1.000	0.500	6	4	0.667	0.229
2	1	0.500	0.354	7	5	0.714	0.227
3	1	0.333	0.304	8	5	0.625	0.217
4	2	0.500	0.264	9	5	0.556	0.206
5	3	0.600	0.240	10	5	0.500	0.195

In many cases, the experiments needed to determine the probability of an event can be done conceptually. For example, suppose that a die is rolled and we wish to calculate the probability that a "one" will appear. Assuming an honest die (so that all six outcomes are equally probable) and its symmetry, we can conclude that the probability of a one is 1/6 without ever actually running a practical experiment. This type of reasoning leads to the following alternative definition of the probability of occurrence of an outcome A:

$$P(A) = \frac{\text{number of possible outcomes favorable to the event } A}{\text{total number of possible equally likely outcomes}}. \tag{8.2}$$

Care is needed in a correct interpretation of Eq. (8.2) because the assumption that all outcomes are equally likely may be wrong and will lead to erroneous conclusions.

Drill Problem 8.1.1   What is the probability of an even integer showing on (a) a roll of one honest die? (b) the sum of two honest dice? (c) What would you expect from the sum of three honest dice?

Answer.   (a) 1/2; (b) 1/2; (c) 1/2.

A consideration of Eqs. (8.1) and (8.2) shows that our definition of the probability of occurrence of a given outcome is a positive real number between zero and one, or

$$0 \le P(A) \le 1. \tag{8.3}$$

Let us denote the fact that an outcome is an absolute certainty by S and the fact that it is not possible at all by $\varnothing$. Again, using Eqs. (8.1) and (8.2), we conclude that

$$\begin{cases} P(\varnothing) = 0, \\ P(S) = 1. \end{cases} \tag{8.4}$$

Two possible events are defined as being *mutually exclusive* (or disjoint) if they cannot possibly occur simultaneously: i.e., if the occurrence of one outcome precludes the occurrence of the other. In this case we can write†

$$P(A \text{ or } B) = P(A + B) = \lim_{N \to \infty} \frac{N_A + N_B}{N} = \lim_{N \to \infty} \frac{N_A}{N} + \lim_{N \to \infty} \frac{N_B}{N}, \tag{8.5}$$

or

$$P(A + B) = P(A) + P(B). \tag{8.6}$$

The additivity of probabilities as expressed by Eq. (8.6) permits us to assign probabilities to elementary events and then derive the probability of other events from these. In fact, in an advanced course in probability, Eqs. (8.3), (8.4), and (8.6) can be postulated and then the theory of probability is deduced from these three postulates.

It is convenient to represent A and B by closed plane figures, known as "Venn diagrams," which can be used as visual aids in the above operations. A Venn diagram for the two mutually exclusive events A and B is shown in Fig. 8.1. Because A and B are mutually exclusive, a given event cannot be represented as lying in both areas and therefore we conclude that the two areas do not overlap if the events are mutually exclusive. Because probability is a numerical measure of the outcomes of an experiment, Eq. (8.6) follows from this Venn diagram.

If all possible outcomes are included, an extension of Eq. (8.6) yields

$$P(A_1 + A_2 + A_3 + \cdots) = P(A_1) + P(A_2) + P(A_3) + \cdots, \tag{8.7}$$

---

† In mathematics, it is common to use the set notations of "union" ($A \cup B$) and "intersection" ($A \cap B$) instead of the sum ($A + B$) and product (AB) notations, respectively.

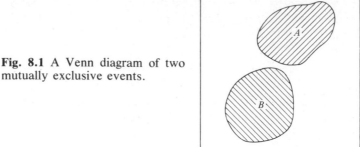

Fig. 8.1 A Venn diagram of two mutually exclusive events.

or, using Eq. (8.4),

$$\sum_{i=1}^{N} P(A_i) = 1. \tag{8.8}$$

If in addition all the outcomes have equal probability, then:

$$P(A_i) = 1/N. \tag{8.9}$$

In the latter case the outcomes are said to be *equally likely.*

Example 8.1.2   Consider the experiment of tossing two honest coins. What is the probability of obtaining two heads or two tails?

Solution I.   There are four possible outcomes of the experiment: $\{HH\}$, $\{HT\}$, $\{TH\}$, $\{TT\}$. The probability of each outcome [cf. Eq. (8.2)] is 1/4. Let the outcome $\{HH\}$ be A and $\{TT\}$ be B. As these outcomes are mutually exclusive,

$$P(A \text{ or } B) = P(A + B) = P(A) + P(B) = 1/4 + 1/4 = 1/2.$$

Solution II.   The number of favorable outcomes is two. A direct application of Eq. (8.2) then yields a probability of $2/4 = 1/2$.

Drill Problem 8.1.2   Consider the experiment of rolling two honest dice. The events of the experiment are the sums of the numbers showing up. What is the probability that this sum is less than five?

Answer.   1/6.

8.2   CONDITIONAL PROBABILITY AND STATISTICAL INDEPENDENCE

Consider the possible events A and B. These events may or may not occur together. (If they are mutually exclusive they cannot occur together.) The probability of the joint occurrence of A and B is $P(A \text{ and } B)$, or, more simply,

$P(AB)$.† Suppose that we repeat an experiment N times and let N_{AB} be the number of times that A and B occur together. Using the relative-frequency approach, we find that the joint probability of their occurrence is (the limit as $N \to \infty$ is implied)

$$P(AB) = \frac{N_{AB}}{N}. \tag{8.10}$$

Recall that N_A is the number of times that A occurs and N_B is the number of times that B occurs. Because A and B may not always occur together, we have $N_{AB} \leq N_A$ and $N_{AB} \leq N_B$.

It may be that the occurrence of the event B depends in some way on the occurrence of A. The probability of B given that A is known to have occurred is called the *conditional probability* of B given A and is written as $P(B|A)$. Using the relative-frequency approach for the N_A trials in which A occurs, we can write

$$P(B|A) = \frac{N_{AB}}{N_A} = \frac{N_{AB}/N}{N_A/N} = \frac{P(AB)}{P(A)}. \tag{8.11}$$

In a similar manner,

$$P(A|B) = \frac{N_{AB}}{N_B} = \frac{N_{AB}/N}{N_B/N} = \frac{P(AB)}{P(B)}. \tag{8.12}$$

Combining Eqs. (8.11) and (8.12), we can express the joint probability $P(AB)$ as

$$P(AB) = P(B|A)P(A) = P(A|B)P(B), \tag{8.13}$$

or

$$P(B|A) = \frac{P(B)P(A|B)}{P(A)}. \tag{8.14}$$

Equation (8.14) is known as *Bayes' theorem* in statistical decision theory.

Example 8.2.1   A standard deck of 52 playing cards is shuffled and two cards are drawn at random. After looking at the first card, what is the probability that the second is a heart if (a) the first card is a heart; (b) the first card is not a heart? (c) Calculate $P(AB)$ for part (a).

Solution.   Let A = event of a heart on the first card; B = event of a heart on the second card; C = event of no heart on the first card.

  a) Because there are 13 hearts in the original deck and one is known to have been removed,

$$P(B|A) = 12/51.$$

---

† This is equivalent to the set notation $P(A \cap B)$.

b) In this case, all 13 hearts are still distributed among $(52 - 1) = 51$ cards so that

$$P(B|C) = 13/51.$$

c) $P(A) = 13/52$; using Eq. (8.11) and the result of part (a) gives us

$$P(AB) = \frac{12}{51}\frac{13}{52} = \frac{3}{51}.$$

Drill Problem 8.2.1 Two honest dice are rolled. Event A is the sum which is divisible by two. Event B is the sum which is divisible by four. Calculate (a) $P(AB)$; (b) $P(B|A)$.

Answer. (a) 1/4; (b) 1/2.

Now suppose that B is independent of A so that the occurrence of A does not in any way influence the occurrence of B. If this is true, the conditional probability $P(B|A)$ is simply the probability $P(B)$; that is,

$$P(B|A) = P(B) \qquad \text{if independent.} \tag{8.15}$$

Using Eq. (8.15) in Eq. (8.13), we conclude that the two events A and B are *statistically independent* if

$$P(AB) = P(A)P(B). \tag{8.16}$$

Therefore when the possible outcomes of an experiment are statistically independent, the joint probability of the particular outcomes is equal to the product of the probabilities of the individual outcomes. This result may be extended to any arbitrary number of outcomes.

Note that the property of statistical independence is quite different from that of mutual exclusiveness. In fact, if A and B are mutually exclusive, the probability of their joint occurrence is zero by definition—that is, $P(AB) = 0$.

Example 8.2.2 Equation (8.6) gives the probability of two mutually exclusive events in terms of the sum of their probabilities of occurrence: $P(A + B) = P(A) + P(B)$. Derive a comparable result if the events are not mutually exclusive.

Solution. Let the two outcomes be A and B; a Venn diagram is shown in Fig. 8.2. Because A and B are not mutually exclusive, the joint probability $P(AB)$ is nonzero, as represented by an overlap in the Venn diagram. (If it were zero, then A and B would be nonoverlapping, or disjoint.)

From a consideration of the Venn diagram, it is quite obvious that simply adding A and B gives the correct result (A or B) only when A and B are disjoint (i.e., mutually exclusive). In the more general case, an addition of A and B includes (A and B) twice and to correct for this we write

$$P(A + B) = P(A) + P(B) - P(AB).$$

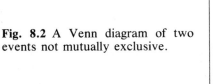

Fig. 8.2 A Venn diagram of two events not mutually exclusive.

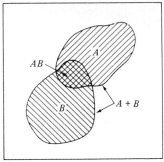

This is our desired result and is a more general version of Eq. (8.6). Note that when A and B are mutually exclusive, $P(AB) = 0$ and this result simplifies to that given in Eq. (8.6).

Drill Problem 8.2.2   A number of honest dice are rolled. Calculate the probability of every die rolled showing a "one" if the number of dice rolled is (a) 2; (b) 3; (c) N.

Answer.   (a) 1/36; (b) 1/216; (c) $(1/6)^N$.

8.3   THE RANDOM VARIABLE AND CUMULATIVE DISTRIBUTION FUNCTION

The concept of probability was introduced in the previous sections to measure numerically the possible outcomes of an experiment. The possible outcomes themselves are not necessarily numerical, however (e.g., heads and tails of a coin), and it becomes cumbersome trying to describe them. The concept of a random variable is introduced to signify a rule (or mapping) by means of which a real number is assigned to each of the possible outcomes of an experiment. A combination of these two concepts will enable us to make numerical graphs of probability versus events for a given experiment.

Let the events of an experiment having a finite number of possible outcomes be identified by the symbols λ_i. These symbols need not represent numbers but only the possible outcomes of the experiment. We establish some rule or functional relationship (mapping), $X(\lambda_i)$, which assigns a real number to each possible outcome. This rule is called a *random variable*.

As an example, we could assign the number -1 to "tails" and the number $+1$ to "heads" in the toss-of-a-coin experiment. Alternatively, we could just as well have associated $\sqrt{2}$ with "tails" and π with "heads." In either case, we have accomplished our purpose in assigning a number to each possible outcome. Note that in defining the random variable we have not assigned a measure to each outcome (i.e., whether a head occurs on the average in one-half of the tosses). The roles of probability and random variable are distinctly different.

Sometimes the identifying symbols λ_i turn out to be numerical (e.g., the six faces of a die—they are already assigned numbers in an arbitrary sequence). If we wish, we could assign new numbers to these events, e.g., squaring each number. Thus a random variable itself is not really random but is an arbitrarily established rule to assign numbers to events.

Although we have introduced the concept for a finite number of possible outcomes, a random variable may be discrete or continuous depending on the nature of the possible outcomes of the experiment. For example, the numbers assigned to the tossing of a coin (for a finite number of tosses) are discrete because the possible outcomes form a finite countable set. In contrast, the angular position of a pointer on a rotating wheel is a random variable whose values form a continuum over the range $0 < \theta \leq 360°$. To allow for both the discrete and continuous cases, we shall drop the subscript i and simply write $X(\lambda)$, realizing that we must supply the subscript when we are referring to the discrete case.

It is convenient and, in fact, quite advantageous to plot the outcome probabilities for a given experiment versus the random variable assigned. In order to make such a plot meaningful, we shall first define what we mean by the cumulative distribution function.

The *cumulative distribution function*, $F_X(x)$, associated with a random variable X is defined as†

$$F_X(x) \triangleq P\{X(\lambda) \leq x\}. \tag{8.17}$$

The cumulative distribution function is dependent on both the random variable X and the value of the argument x. It is for this reason that the subscript X is included in the notation.

Because the cumulative distribution function is based directly on the concept of probability, it has the following properties:

$$0 \leq F_X(x) \leq 1, \tag{8.18}$$

$$F_X(x_1) \leq F_X(x_2), \qquad \text{if} \quad x_1 < x_2, \tag{8.19}$$

$$\begin{cases} F_X(-\infty) = 0, \\ F_X(+\infty) = 1. \end{cases} \tag{8.20}$$

The property in Eq. (8.18) follows from the fact that $F_X(x)$ is a probability and therefore must obey Eq. (8.3). The property in Eq. (8.19) holds also as a result of Eq. (8.3) because probability is always nonnegative; for $x_1 < x_2$, $F_X(x_2)$ includes as many or more of the possible outcomes than are included in $F_X(x_1)$. The property in Eq. (8.20) follows from Eq. (8.4) because $F_X(-\infty)$ includes no events possible while $F_X(+\infty)$ includes all events possible.

---

† It is conventional to use a capital letter to denote a random variable and the corresponding lower-case letter to denote the value within the range.

For a discrete random variable $X(\lambda_i)$ with associated probabilities P_i, the cumulative distribution function can be expressed as

$$F_X(x) = \sum_i P_i u(x - x_i). \tag{8.21}$$

Therefore the cumulative distribution function of a discrete random variable has a series of finite ("step-wise") discontinuities occurring at the points $x = x_i$. The height of each step is given by: $P_i = P(X = x_i)$ and between the steps $F_X(x)$ is constant. At a point of finite discontinuity, $F_X(x)$ is taken to be continuous to the right.

Example 8.3.1   Consider the experiment which consists in the tossing of three honest coins. The random variable chosen is defined by assigning 0 to a "tail" and 1 to a "head," and then adding the numbers. Determine and plot the cumulative distribution function.

Solution.   The eight possible outcomes are

$$\begin{array}{ll}
000 & 100 \\
001 & 101 \\
010 & 110 \\
011 & 111
\end{array}$$

If we assume equally likely outcomes, the probability of each is 1/8. The outcomes are mutually exclusive so the probabilities are additive. A table of the random variable $X(\lambda_i)$ and the associated probabilities is shown below.

x	P
0	1/8
1	3/8
2	3/8
3	1/8

The cumulative distribution function is found by adding the probabilities for all outcomes below a given number and is plotted in Fig. 8.3(a). For example, the cumulative distribution function evaluated at $x = 2$ is

$$\begin{aligned}
F_X(2) &= P\{x \le 2\} \\
&= P(0) + P(1) + P(2) \\
&= 1/8 + 3/8 + 3/8 = 7/8.
\end{aligned}$$

In equation form, we can write our answer as [cf. Eq. (8.21)]

$$F_X(x) = \tfrac{1}{8}u(x) + \tfrac{3}{8}u(x - 1) + \tfrac{3}{8}u(x - 2) + \tfrac{1}{8}u(x - 3).$$

Note that $F_X(x)$ satisfies all of the properties given by Eqs. (8.18), (8.19), and (8.20).

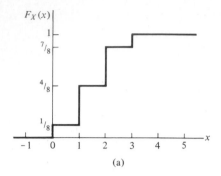
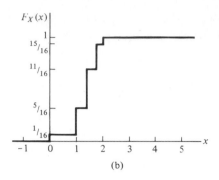

Fig. 8.3 Cumulative distribution function for the discrete random variable of (a) Example 8.3.1 and (b) Drill Problem 8.3.1.

Drill Problem 8.3.1   Four honest coins are tossed and X is designated as the square root of the number of heads which turn up. Determine and plot the cumulative distribution function, $F_X(x)$.

Answer.   The result is plotted in Fig. 8.3(b); the equation is

$$F_X(x) = \tfrac{1}{16}u(x) + \tfrac{4}{16}u(x - 1) + \tfrac{6}{16}u(x - \sqrt{2}) + \tfrac{4}{16}u(x - \sqrt{3}) + \tfrac{1}{16}u(x - 2).$$

8.4   THE PROBABILITY DENSITY FUNCTION

Instead of tossing coins or rolling dice, suppose we spin a wheel with an attached pointer. As a result of frictional losses, the wheel gradually slows and comes to rest at some angular position θ. Suppose that we mark off 360 equal divisions around the circumference. The stopping point of the wheel is the desired outcome of the experiment. There is an uncountably infinite number of possible outcomes and so we assign a continuous random variable θ to the experiment.

Consider the probability of stopping at points around the rim of the wheel. Because the wheel must stop somewhere within the range, the total probability of $0 < \theta \le 360°$ is one. Because we have divided this range uniformly, the probability of $0 < \theta \le 1°$ is 1/360. What, then, is the probability that the wheel comes to rest at a given point? There is an uncountably infinite number of points within each division (no matter how small we make the divisions), and the only reasonable answer is that this probability must be zero; otherwise we cannot possibly end up with a total probability of one.

Even though $P\{\theta = \theta_1\} = 0$, it makes good sense to speak of $P\{\theta \le \theta_1\}$, or $P\{\theta_1 < \theta \le \theta_2\}$ for $\theta_1 < \theta_2$. The concept of the cumulative distribution function is still valid, but it is more convenient in such cases to define a function whose *area* is the probability of occurrence within a specified range. Because area is being equated to probability, such a function is termed a *probability density function*.

Therefore we define the probability density function, $p_X(x)$, of a random variable X by

$$\int_{-\infty}^{x} p_X(x) \, dx = P\{X \le x\}. \tag{8.22}$$

However, we can also write [cf. Eq. (8.17)]

$$P\{X \le x\} = F_X(x), \tag{8.23}$$

so that

$$\int_{-\infty}^{x} p_X(x) \, dx = F_X(x). \tag{8.24}$$

Taking the derivative of each side of Eq. (8.24), we find that an equivalent definition of the probability density function (pdf) is

$$p_X(x) = \frac{d}{dx} F_X(x). \tag{8.25}$$

Thus the pdf of a random variable X is the derivative of the cumulative distribution function, $F_X(x)$.

It may happen that the derivative of the cumulative distribution function will not exist at all points. In this case we must resort to a limiting operation. The delta function notation used in Chapters 2 and 3 becomes useful in many cases.

The random variable subscript notation begins to get cumbersome at this point. When there is only a single random variable under discussion and no ambiguity can result, we shall omit the subscript in the probability density function for simplicity of notation.

From Eq. (8.24) and the properties of the cumulative distribution function [cf. Eqs. (8.18)–(8.20)], some properties of the probability density function are

$$\int_{-\infty}^{\infty} p(x) \, dx = F_X(\infty) = 1, \tag{8.26}$$

$$\int_{x_1}^{x_2} p(x) \, dx = F_X(x_2) - F_X(x_1) = P\{x_1 < X \le x_2\}. \tag{8.27}$$

Because Eq. (8.27) must always be nonnegative for any (x_1, x_2), we can also write

$$p(x) \ge 0 \qquad \text{for all} \quad x. \tag{8.28}$$

Example 8.4.1   A wheel with a pointer is rotated and allowed to slow down and stop. When the wheel comes to rest, the angular position of the pointer is θ. A scale, $0 < \theta \leq 360°$, is marked with uniform increments around the rim of the wheel. Determine the pdf of θ.

Solution.   Because all values in the range $0 < \theta \leq 360°$ are equally probable, the pdf is a constant over this range so that

$$p(\theta) = \begin{cases} K & 0 < \theta \leq 360°, \\ 0 & \text{elsewhere.} \end{cases}$$

The constant K can be evaluated because the total probability of the wheel stopping within the range is known to be one:

$$\int_{-\infty}^{\infty} p(\theta) \, d\theta = \int_0^{360} K \, d\theta = 360K = 1,$$

$$K = 1/360.$$

Note that the probability of the pointer stopping exactly at the point $\theta = 1°$ is, formally,

$$P(1) = \lim_{\epsilon \to 0} \int_{1-\epsilon}^{1+\epsilon} K \, d\theta = 0,$$

but the probability of the pointer stopping within the interval $0 < \theta \leq 1°$ is

$$\int_0^1 \frac{1}{360} \, d\theta = \frac{1}{360}.$$

A graph of the pdf is shown in Fig. 8.4.

Fig. 8.4 The probability density function of Example 8.4.1.

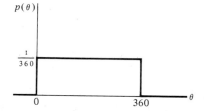

Drill Problem 8.4.1   In the reliability testing of small electronic component parts, the probability of the time-to-failure is often approximated using the exponential pdf, $p(y) = K \exp(-y/\alpha)u(y)$, where y is the equipment operating time in hours and the constant α is experimentally determined (and is called the *mean-time-before-failure*). (a) Compute the probability that a given part described by this pdf will fail within the first α hours of operation. (b) If $\alpha = 10,000$ hours, what is the probability that the part will fail during the first hour of operation?

Answer.   (a) 0.632; (b) 0.0001.

It is convenient to extend our definition of probability density function to include the case of discrete random variables. This follows readily if we recall that the cumulative distribution function is composed of a series of step functions and that the derivative of a step function, at least formally, is an impulse function. Using Eq. (8.21), we can write

$$p(x) = \frac{d}{dx}\left[\sum_i P_i u(x - x_i)\right],$$

or

$$p(x) = \sum_i P_i \delta(x - x_i). \tag{8.29}$$

Although our examples tend to fall into either of the two categories, there is no theoretical reason for a pdf to be purely discrete or purely continuous.

Example 8.4.2 Consider the experiment which consists in the rolling of two honest dice. The random variable X is assigned to the sum of the numbers showing up on the two dice. Determine and plot the cumulative distribution function and the pdf of X.

Solution. There are 36 possible outcomes because each die may show any of the numbers 1 through 6; the range of the sum then goes from 2 through 12. Each possible outcome is assumed equally likely and so its probability is equal to 1/36. The events are mutually exclusive so additivity applies. We find that $P(2) = P(12) = 1/36$, $P(3) = P(11) = 2/36$, $P(4) = P(10) = 3/36$, $P(5) = P(9) = 4/36$, $P(6) = P(8) = 5/36$, $P(7) = 6/36$. The resulting cumulative distribution function and pdf are shown in Fig. 8.5.

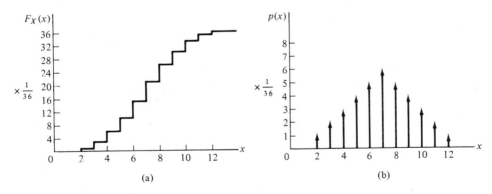

Fig. 8.5 (a) The cumulative distribution function and (b) the probability density function for Example 8.4.2.

Drill Problem 8.4.2   Four honest coins are tossed and X is designated as the number of heads which turn up. Determine the pdf of X.

Answer.† $p(x) = \frac{1}{16} \delta(x) + \frac{4}{16} \delta(x-1) + \frac{6}{16} \delta(x-2) + \frac{4}{16} \delta(x-3) + \frac{1}{16} \delta(x-4)$.

A comparison with Chapter 4 (cf. Section 4.2) reveals that our development of the cumulative power spectrum and power spectral density are practically identical to the above development of the cumulative distribution function and the probability density function. This analogy can be extended to the mathematical role of the periodic signal in power spectral density and the discrete random variable in the probability density function, since both give rise to series of impulse functions. Both density functions are nonnegative; both cumulative functions are therefore monotone nondecreasing with increasing argument. Each point in the power spectral density has zero power associated with it while the area under the density curve gives the power within a specified frequency range. Each point in the probability density has zero probability associated with it, while the area under the density curve gives the probability within a specified range of the random variable.

8.5   STATISTICAL AVERAGES

The probability density function is useful in evaluating the most probable values of desired quantities. A way to find the most probable value of X is to locate that point in the pdf of X which has the maximum value. This point is called the *mode*. A necessary condition for the determination of the mode (in the continuous case) is that $(d/dx)p(x) = 0$. The location of this point is extremely sensitive to minor changes in the pdf, however. Another measurement is to locate the point which divides the area into two equal parts. This is called the *median*. The determination of the median is improved over that of the mode but is still quite sensitive to minor changes in the pdf. Alternatively, we can use statistical averaging.

The statistical average of a random variable X (or a function of a random variable) is the numerical average of the values which X (or a function of X) can assume, weighted by their probabilities. As an example, consider N observations of a random variable X. If $X = x_1$ is observed n_1 times, $X = x_2$ is observed n_2 times, etc., the observed value is

$$n_1 x_1 + n_2 x_2 + n_3 x_3 + \cdots = \sum_{i=1}^{N} n_i x_i. \tag{8.30}$$

Dividing by N and then letting $N \rightarrow \infty$ to employ our relative-frequency approach

---

† A more compact notation for problems of this type is discussed in Section 8.6.2.

to probability, we have

$$M_1 = \lim_{N \to \infty} \sum_{i=1}^{N} x_i \frac{n_i}{N} = \sum_{i=1}^{N} x_i P_i = m_X, \tag{8.31}$$

where M_1 is called the first moment of X and m_X is the *average value* or *mean value* of X. Equation (8.31) is also called the *expected value* of X. Note that for the special case in which all events are equally likely [cf. Eq. (8.9)], Eq. (8.31) simplifies to the arithmetic average of X:

$$m_X = \frac{1}{N} \sum_{i=1}^{N} x_i. \tag{8.32}$$

In a similar manner the second moment M_2 is equal to the expected value of X^2:

$$M_2 = \sum_{i=1}^{N} x_i^2 P_i, \tag{8.33}$$

and is called the *mean-square* value of X. The square root of M_2 is called the *root-mean-square* (rms) value of X. Higher-ordered moments can be expressed in a similar way.

For the case of a continuous random variable, the summation in Eqs. (8.31) and (8.33) approaches an integration over the entire range of X so that the expected value of X, designated by $E\{X\}$, is

$$M_1 = E\{X\} = \int_{-\infty}^{\infty} x p(x)\, dx = m_X, \tag{8.34}$$

and the expected value of X^2, designated by $E\{X^2\}$, is

$$M_2 = E\{X^2\} = \int_{-\infty}^{\infty} x^2 p(x)\, dx. \tag{8.35}$$

By extension, the expected value of X^n is

$$E\{X^n\} = \int_{-\infty}^{\infty} x^n p(x)\, dx. \tag{8.36}$$

The expected value $E\{X^n\}$ is referred to as the "*n*th moment" of the random variable X. This is analogous to the concept of a moment in mechanics. If $p(x)$ were the mass density (normalized to one), the first moment [cf. Eq. (8.34)] corresponds to the center of gravity, the second [cf. Eq. (8.35)] to the moment of inertia about the origin, etc.

A knowledge of all the moments of a random variable, provided they are finite, specifies that random variable as completely as a knowledge of the pdf. Fortunately, we usually can get by with working only with the first two moments.

Often it is convenient to subtract out the first moment $M_1 = m_X$ before computing the other moments. This is analogous to mechanics, where moments are referred to the center of gravity rather than the origin of the coordinate system. Such moments are referred to as "central moments." The second central moment is of particular importance and is called the *variance* of the random variable X. Thus the variance, σ_X^2, of the random variable X is defined as

$$\sigma_X^2 = E\{(X - m_X)^2\} = \int_{-\infty}^{\infty} (x - m_X)^2 p(x)\, dx. \tag{8.37}$$

The square root of the variance is called the *standard deviation*, σ_X, of the random variable X. The standard deviation has the same units as X and is a convenient indicator of the effective width of the probability density function. For example, if two random variables, X, Y, both have average values of 1 but $\sigma_X = 1$ and $\sigma_Y = 2$, we can conclude that the effective width of the pdf for Y is more distributed (spread out) than that for X. Stated differently, the probability of X being close to 1 is greater than the probability that Y is close to 1.

Finally, the expected value of a function of a random variable $g(X)$ is defined as

$$E\{g(X)\} = \int_{-\infty}^{\infty} g(x)p(x)\, dx. \tag{8.38}$$

When $g(X) = X^n$, the expected value is the nth moment of X, as described above. If $g(X)$ is equal to a constant, then Eq. (8.38) gives that constant because $\int_{-\infty}^{\infty} p(x)\, dx = 1$. The expected value operator $E\{\cdot\}$ is linear so that

$$E\{a_1 X + a_2\} = a_1 m_X + a_2.$$

Example 8.5.1   Show that an equivalent expression to Eq. (8.37) is

$$\sigma_X^2 = E\{X^2\} - [E\{X\}]^2 = M_2 - M_1^2.$$

Solution.   Expanding the squared term and integrating term by term, we can rewrite Eq. (8.37) as

$$\begin{aligned}
\sigma_X^2 &= E\{X^2 - 2M_1 X + M_1^2\} \\
&= E\{X^2\} - 2M_1 E\{X\} + M_1^2 \\
&= M_2 - 2M_1^2 + M_1^2 \\
&= M_2 - M_1^2.
\end{aligned}$$

Note that if $M_1 = 0$, then $\sigma_X^2 = M_2 = E\{X^2\}$.

Drill Problem 8.5.1   Compute the mean m_X and the standard deviation σ_X of the random variable X described by the following probability density functions: (a) $p(x) = \delta(x - 1)$; (b) $p(x) = (8/9)x[u(x) - u(x - 3/2)]$; (c) $p(x) = \exp(-x)u(x)$.

Answer.   (a) 1, 0; (b) 1, $1/(2\sqrt{2})$; (c) 1, 1.

★ **Example 8.5.2**   The foregoing remarks about the variance of a random variable as a relative indicator of the spread of the probability density function can be supported by a very general result from statistics known as *Chebyshev's inequality*:

$$P(|X - m_X| > k\sigma_X) < \frac{1}{k^2} \qquad \text{for } k > 0.$$

This inequality states that the probability of X being outside $\pm k$ standard deviations of the mean is less than $1/k^2$ *regardless of p(x)*. Therefore the probability of finding X appreciably different from the mean is greater for a large standard deviation than for a small one. Because the total probability is unity, Chebyshev's inequality may also be written as

$$P(|X - m_X| \leq k\sigma_X) \geq 1 - \frac{1}{k^2}.$$

A proof of Chebyshev's inequality may be found in texts on probability and statistics.†

As an example, compute $P(X > 5)$ for the pdf of Drill Problem 8.5.1(c), and compare with the bound given by the Chebyshev inequality.

Solution.   Using a knowledge of the pdf, we have

$$P(X > 5) = \int_5^\infty e^{-x}\, dx = e^{-5} = 0.0067.$$

Using Chebyshev's inequality with $m_X = 1$, $\sigma_X = 1$, we have

$$P(X > 5) = P(|X - 1| > 4) < \frac{1}{4^2} = 0.0625.$$

Although this bound is almost an order of magnitude larger than the exact value in this case, note that the computation of this bound requires no knowledge of the pdf (only the first two moments) and hence is a very general result.

Although the equations for expected values have been written in terms of continuous random variables, they also hold for the case of discrete random variables when the impulse functions arising in the probability densities are handled correctly. Recall that the probability density for a discrete random variable can be written as [cf. Eq. (8.29)] $p(x) = \sum_{i=1}^{N} P_i \delta(x - x_i)$ so that the

---

† Chebyshev is spelled Tchebycheff in some texts.

expected value of X for the discrete case is, from Eq. (8.34),

$$m_X = E\{X\} = \int_{-\infty}^{\infty} x \sum_{i=1}^{N} P_i \delta(x - x_i) \, dx$$

$$= \sum_{i=1}^{N} x_i P_i, \tag{8.39}$$

and is in agreement with Eq. (8.31). In a similar manner,

$$E\{X^2\} = \int_{-\infty}^{\infty} x^2 \sum_{i=1}^{N} P_i \delta(x - x_i) \, dx = \sum_{i=1}^{N} x_i^2 P_i, \tag{8.40}$$

and the variance of X is

$$\sigma_X^2 = \sum_{i=1}^{N} (x_i - m_X)^2 P_i. \tag{8.41}$$

For the equally likely case, Eq. (8.40) simplifies to

$$E\{X^2\} = \frac{1}{N} \sum_{i=1}^{N} x_i^2, \tag{8.42}$$

and Eq. (8.41) becomes

$$\sigma_X^2 = \frac{1}{N} \sum_{i=1}^{N} (x_i - m_X)^2. \tag{8.43}$$

Example 8.5.3   An exam is taken by nine students. After grading, it is found that the percentage scores are as listed below. Compute the median and average grade (treat all students equally!). Also compute the standard deviation.

N	X	N	X
1	33	6	82
2	47	7	88
3	58	8	94
4	67	9	100
5	75		

Solution.   The student number which divides the distribution equally is No. 5 so the median grade is 75. Because all students are to be treated equally, $P_i = 1/9$ for all i. The average grade is [Eq. (8.32)]

$$E\{X\} = \sum_{i=1}^{9} x_i P_i = 644/9 = 71.6.$$

Using Eq. (8.43), we find the standard deviation is given by

$$\sigma_X = \sqrt{\sum_{i=1}^{9} (x_i - 71.6)^2/9} = 21.1.$$

Drill Problem 8.5.2 Compute the mean and the standard deviation of the random variable X in Example 8.4.2.

Answer. 7, 2.42.

The number of possible trials in an experiment with a discrete number of states is called a *population*. A *sample* is a finite number of trials which is less than the population. In statistical averaging, we make a distinction between a population average and a sample average. In the former, all possible trials are available to us for use in the statistical averaging, whereas in the latter we must estimate the statistical averages of the population based only on a finite number of samples.

Suppose that we wish to estimate the population average θ based on the N samples from the population which are available to us. The estimate of θ will be denoted by $\hat{\theta}$. We look for certain desirable qualities in our estimate, two of which we mention here.

First, we would like to have the estimate correct if the sample size were expanded to the entire population. Thus an estimator $\hat{\theta}$ is called an *unbiased estimator* of θ if $E\{\hat{\theta}\} = \theta$. A second desirable quality is that the estimate converge to the correct value as the number of samples becomes large. An estimator $\hat{\theta}$ is called a *consistent estimator* of θ (in a mean-square sense) if

$$\lim_{N \to \infty} E\{(\hat{\theta} - \theta)^2\} = 0.$$

It can be shown (cf. Example 8.5.4) that the sample mean [Eq. (8.32)] is an unbiased and consistent estimator of the population mean. The sample variance [Eq. (8.43)] is a consistent estimator; however,[†]

$$E\{\hat{\sigma}^2\} = \frac{N-1}{N} \sigma_X^2, \tag{8.44}$$

so that it is not an unbiased estimator. For this reason, the unbiased sample-variance estimator,

$$\sigma_X^2 = \frac{1}{N-1} \sum_{i=1}^{N} (x_i - m_X)^2, \tag{8.45}$$

is usually used. For large N, the difference between the two becomes negligible.

The concepts of estimation are not confined to discrete random variables. They can readily be extended to the case of continuous random variables.

Example 8.5.4 Show that the sample mean [Eq. (8.32)] is (a) an unbiased and (b) a consistent estimator of the population mean.

---

† See, for example, J. S. Bendat and A. G. Piersol, *Random Data: Analysis and Measurement Procedures*, New York: Wiley-Interscience, 1971, Chapter 4.

Solution

a)
$$E\{\hat{m}_X\} = E\left\{\frac{1}{N}\sum_{i=1}^{N}x_i\right\}$$

$$= \frac{1}{N}\sum_{i=1}^{N}E\{x_i\} = m_X.$$

b)
$$E\{(\hat{m}_X - m_X)^2\} = E\left\{\left[\frac{1}{N}\sum_{i=1}^{N}(x_i - m_X)\right]^2\right\}$$

$$= \frac{1}{N^2}E\left\{\sum_i\sum_k(x_i - m_X)(x_k - m_X)\right\}$$

$$= \frac{1}{N^2}\sum_i\sum_k E\{x_i x_k - m_X^2\}.$$

If we assume that the samples x_i, x_k are statistically independent, only those terms for $i = k$ are nonzero, and we have (using the result of Example 8.5.1),

$$E\{(\hat{m}_X - m_X)^2\} = \frac{1}{N^2}\sum_{i=1}^{N}[E\{x_i^2\} - m_X^2] = \frac{1}{N}\sigma_X^2.$$

Therefore the variance of the sample mean is equal to the population variance divided by the sample size. Because this approaches zero as $N \to \infty$, the sample mean is a consistent estimator of the population mean.

Drill Problem 8.5.3   Five resistors are selected at random from a box of 1-kΩ resistors. The measured values (in ohms) are 980, 1009, 1026, 963, and 997. (a) Estimate the mean and the standard deviation of the resistance values of the resistors in the box based on this sample. (b) Calculate the mean and standard deviation if these five were the only resistors in the box.

Answer.   (a) 995.0 Ω, 24.55 Ω; (b) 995.0 Ω, 21.95 Ω.

8.6   SOME PROBABILITY DISTRIBUTIONS

A few of the numerous examples of probability distributions, both discrete and continuous, which arise in physical problems and are the most useful to us in this book are described here.

8.6.1   The Uniform Distribution

A random variable which is equally likely to take on any value within a given range is said to be *uniformly distributed*. The random variable may be discrete or continuous. In the former case, the pdf is a series of equally weighted impulse functions. In the latter case, the pdf is a rectangular function, as illustrated in Fig. 8.6(a). Note that the height of the pdf must be chosen to give unit area. The corresponding cumulative distribution function is shown in Fig. 8.6(b).

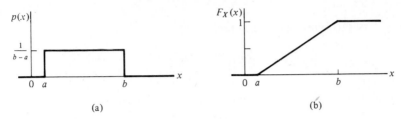

(a)   (b)

Fig. 8.6 The uniform probability density function and corresponding cumulative distribution function.

Example 8.6.1   Consider the random variable $X = \cos(\omega_0 t_1 + \theta)$, where θ is a random variable with a uniform probability density over $-\pi < \theta \le \pi$. Compute the mean and rms value.

Solution

$$E\{X\} = \int_{-\pi}^{\pi} \cos(\omega_0 t_1 + \theta)(1/2\pi) \, d\theta = 0,$$

$$E\{X^2\} = \int_{-\pi}^{\pi} \cos^2(\omega_0 t_1 + \theta)(1/2\pi) \, d\theta$$

$$= \int_{-\pi}^{\pi} \frac{1}{4\pi} \, d\theta + \int_{-\pi}^{\pi} \frac{1}{4\pi} \cos(2\omega_0 t_1 + \theta) \, d\theta = \tfrac{1}{2} + 0,$$

$$E\{X^2\}^{1/2} = 1/\sqrt{2} = 0.707.$$

That these results are exactly the same results as we obtain using time averages is not accidental. We shall return to the topic of statistical averages and time averages in a later section.

Drill Problem 8.6.1   Determine the mean and standard deviation of the uniform pdf shown in Fig. 8.6(a).

Answer. $(b + a)/2$; $(b - a)/2\sqrt{3}$.

8.6.2   The Binomial Distribution

Consider an experiment having only two possible outcomes, A and B, which are mutually exclusive. Let the probabilities be $P(A) = p$ and $P(B) = 1 - p = q$.

The experiment is repeated n times and the probability of A occurring i times is

$$P_i = \binom{n}{i} p^i q^{n-i}, \tag{8.46}$$

where $\binom{n}{i}$ is the binomial coefficient,

$$\binom{n}{i} = \frac{n!}{i!\,(n-i)!}. \tag{8.47}$$

The binomial coefficient enters here to account for the total number of possible ways to combine n items taken i at a time, with all possible permutations permitted. We let $0! = 1$ by definition. The derivation of Eq. (8.46) can be found in a text on probability.

When n, i are not too large, the coefficients can also be read from Pascal's triangle. Pascal's triangle contains the coefficients of the expansion of $(p + q)^n$ and is shown in Fig. 8.7 for $1 \le n \le 5$. The triangle is constructed by noting that each row begins and ends with the number 1. The other numbers in a given row are determined by adding those two numbers in the row immediately above the desired number. For example, the numbers in the fourth row (that is, $n = 4$) are 1, 4 = 1 + 3, 6 = 3 + 3, 4 = 3 + 1, 1; these numbers are the coefficients in the expansion of $(p + q)^4$. Note that each coefficient can also be expressed as $\binom{n}{i}$ where i is the power of one of the variables in the expansion of $(p + q)^n$. The binomial coefficients are useful because they enable us to find the number of possible ways to arrange n items taken i at a time.

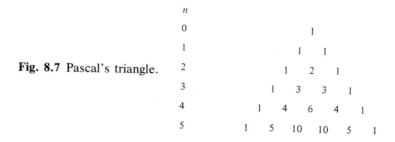

Fig. 8.7 Pascal's triangle.

n									
0					1				
1				1		1			
2			1		2		1		
3		1		3		3		1	
4	1		4		6		4		1
5	1	5		10		10		5	1

The binomial probability density function can be obtained by substituting Eq. (8.46) into Eq. (8.29):

$$p(x) = \sum_{i=0}^{n} \binom{n}{i} p^i q^{n-i} \delta(x - i), \tag{8.48}$$

where i is the integer number of times that event A occurs. The mean value of the binomial distribution is (np) and the variance is (npq).[†]

Example 8.6.2   A standard deck of 52 playing cards is shuffled and two cards are drawn at random. What is the joint probability of drawing two hearts? (Cf. Example 8.2.1.)

Solution.   We shall use the relative-frequency concept of probability [see Eq. (8.2)]. First we find the number of possible favorable outcomes and then divide by the total number of possible equally likely outcomes.

The favorable outcomes are the drawing of two hearts. There are 13 hearts in the deck; taken two at a time, the number of favorable outcomes is, using Eq. (8.47),

$$\binom{13}{2} = \frac{13!}{2!\,11!} = \frac{12 \cdot 13}{2} = 78.$$

The total number of possible equally likely outcomes is found by taking any two cards at a time out of a deck of 52 cards, or

$$\binom{52}{2} = \frac{52!}{2!\,50!} = \frac{51 \cdot 52}{2} = 1326.$$

The joint probability of drawing two hearts out of the deck is then

$$P(AB) = \frac{78}{1326} = \frac{3}{51}.$$

This agrees with the answer obtained in Example 8.2.1(c).

Example 8.6.3   Three honest dice are rolled. Write the pdf for the number of ones showing up.

Solution.   Let X be the number of ones; then $p = \frac{1}{6}$, $q = \frac{5}{6}$, $n = 3$, and Eq. (8.48) gives

$$p(x) = \sum_{i=0}^{3} \binom{3}{i} (\tfrac{1}{6})^i (\tfrac{5}{6})^{3-i}\, \delta(x - i),$$

$$p(x) = \tfrac{125}{216}\delta(x) + \tfrac{75}{216}\delta(x - 1) + \tfrac{15}{216}\delta(x - 2) + \tfrac{1}{216}\delta(x - 3).$$

Drill Problem 8.6.2   A painter has paints in six differing color hues. To obtain additional hues, he can combine any two in equal proportions. Assuming that all resulting hues are different, how many hues are obtainable?

Answer.   21.

[†] The random variable x_i may be substituted for the integers i in Eq. (8.48) if the mean and variance are appropriately modified.

Drill Problem 8.6.3   Three honest coins are tossed and X is designated as the number of heads which turn up. Determine the pdf of X, using Eq. (8.48); repeat for the case of six coins.

Answer.   See Fig. 8.8.

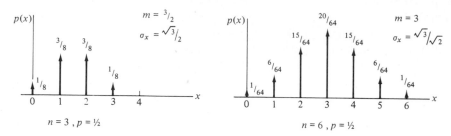

Fig. 8.8 The binomial probability density function for two different values of n.

8.6.3   The Poisson Distribution

When the number of trials n gets very large and p is very small, the binomial distribution becomes awkward to handle. However, if the mean value remains finite we can approximate the binomial distribution by the Poisson distribution,

$$P_i = e^{-v}\frac{v^i}{i!},\tag{8.49}$$

where v is the average value of the occurrence of the desired event.

The Poisson distribution is pertinent to the transmission of many data digits when the error rates are low. It also arises in some physical problems.

Example 8.6.4   Two honest dice are rolled and the desired outcome is the appearance of all ones. (a) Calculate the probability that this will occur twice in forty rolls, first using the binomial distribution and then approximating with the Poisson distribution. (b) Repeat for the case of four dice and 1000 rolls.

Solution

a) Using the binomial distribution, we have

$$P_2 = \binom{40}{2}\left[\frac{1}{36}\right]^2\left[\frac{35}{36}\right]^{38} = \frac{40!}{2!\,38!}\frac{35^{38}}{36^{40}} = 0.2063,$$

and using the Poisson distribution, we get

$$v = np = 40/36,$$
$$P_2 \approx e^{-v}(v)^2/2! = 0.2032.$$

The approximation becomes better with larger n for a given v. Note the comparatively simple calculation when contrasted with the binomial distribution; this is emphasized in part (b).

b) Using the binomial distribution, we have

$$P_2 = \binom{1000}{2}\left[\frac{1}{1296}\right]^2\left[\frac{1295}{1296}\right]^{998} = 0.13765,$$

and using the Poisson distribution, we get

$$v = np = 1000/1296,$$

$$P_2 \approx e^{-v}(v)^2/2! = 0.13761.$$

Drill Problem 8.6.4 A certain communication system transmits digital data with an average error rate of 1 digit in 10^6 digits. What is the probability that 3 errors will occur in a given transmission of 10^6 digits?

Answer. 6.13%.

8.6.4 The Gaussian Distribution

A widely known probability density function is the gaussian or "normal" density function. It arises when a large number of independent factors (within some fairly broad restrictions) contribute additively to an end result. This result, known as a "central-limit theorem," states that the sum of N independent random variables approaches a gaussian distribution as N becomes large. This result is not dependent upon the distribution of each random variable (within broad restrictions) as long as the contribution of each is small compared to the sum. This theorem is not easy to prove and interpret correctly, however, and we shall not pursue the topic further except to observe that electrical noise often results from summations of a large number of random effects and therefore tends to be gaussian-distributed.

The *gaussian pdf* is continuous and is defined by

$$p(x) = \frac{1}{\sqrt{2\pi}\sigma} e^{-(x-m)^2/2\sigma^2}, \tag{8.50}$$

where m, σ^2 are the mean and the variance respectively. The factor: $1/(\sqrt{2\pi}\sigma)$ is needed so that there is unit area. A graph of the gaussian pdf is shown in Fig. 8.9. Note that the gaussian pdf is uniquely specified by its first two moments. For convenience in tabulating numerical values, we define a *normalized gaussian pdf* having zero mean and unit variance,

$$p(z) = \frac{1}{\sqrt{2\pi}} e^{-z^2/2}. \tag{8.51}$$

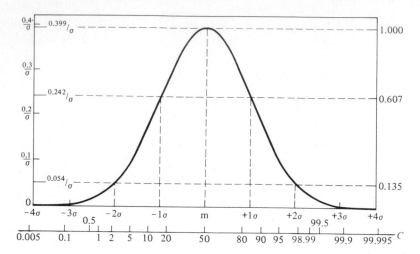

Fig. 8.9 Gaussian (normal) probability density function. The standard deviation is σ, and scale C is the cumulative distribution function in percent.

The cumulative distribution function corresponding to the gaussian pdf is

$$P\{X \le x\} = F_X(x) = \int_{-\infty}^{x} \frac{1}{\sqrt{2\pi}\sigma} e^{-(y-m)^2/2\sigma^2} \, dy. \tag{8.52}$$

This integral cannot be evaluated in closed form and requires numerical evaluation. The procedure is to first normalize and then refer to a table of tabulated values.

As an example, suppose we wish to calculate $P\{X \le m + k\sigma\}$ for a gaussian random variable X. Then we can write

$$P\{X \le m + k\sigma\} = \int_{-\infty}^{m+k\sigma} \frac{1}{\sqrt{2\pi}\sigma} e^{-(x-m)^2/2\sigma^2} \, dx. \tag{8.53}$$

To normalize, let $z = (x - m)/\sigma$ so that

$$P\{X \le m + k\sigma\} = \int_{-\infty}^{k} \frac{1}{\sqrt{2\pi}} e^{-z^2/2} \, dz. \tag{8.54}$$

As a result of the even symmetry in the integrand of Eq. (8.54), it is helpful to note that

$$\int_{0}^{\infty} \frac{1}{\sqrt{2\pi}} e^{-z^2/2} \, dz = \frac{1}{2} \int_{-\infty}^{\infty} \frac{1}{\sqrt{2\pi}} e^{-z^2/2} \, dz = \frac{1}{2}. \tag{8.55}$$

For convenience in referring to tabulated values, we define the *error function*, Erf (x), as†

$$\text{Erf}(x) \triangleq \frac{1}{\sqrt{2\pi}} \int_{-\infty}^{x} e^{-z^2/2} \, dz. \tag{8.56}$$

Therefore our answer to the above example can be expressed as

$$P\{X \le m + k\sigma\} = \text{Erf}(k). \tag{8.57}$$

For large values of x, the function Erf (x) approaches unity and it becomes convenient to define a *complementary error function*, designated by Erfc (x), as

$$\text{Erfc}(x) \triangleq 1 - \text{Erf}(x), \tag{8.58}$$

$$\text{Erfc}(x) = 1 - \int_{-\infty}^{x} \frac{1}{\sqrt{2\pi}} e^{-z^2/2} \, dz,$$

$$\text{Erfc}(x) = \int_{x}^{\infty} \frac{1}{\sqrt{2\pi}} e^{-z^2/2} \, dz. \tag{8.59}$$

Tabulated values of the complementary error function are included in Appendix G.

Example 8.6.5   The amplitude of a signal is gaussian-distributed with zero mean and a mean-square value σ^2. Find the probability of observing the signal amplitude above 3σ.

Solution

$$P\{x > 3\sigma\} = \int_{3\sigma}^{\infty} \frac{1}{\sqrt{2\pi}\sigma} e^{-x^2/2\sigma^2} \, dx$$

$$= \int_{3}^{\infty} \frac{1}{\sqrt{2\pi}} e^{-z^2/2} \, dz$$

$$= \text{Erfc}(3) = 0.0013 \, (\approx 10^{-3})$$

Drill Problem 8.6.5   A random variable X has a gaussian pdf. The mean value of X is $+2$ volts and the variance is 4 volts. Compute the following probabilities: (a) $P\{X < 6\}$; (b) $P\{X > 3\}$; (c) $P\{X < -2\}$; (d) $P\{2 < X < 3\}$.

Answer.   (a) 0.9772; (b) 0.3085; (c) 0.0228; (d) 0.1915.

---

† There are several different definitions used in current literature. They are all essentially equivalent, with minor differences in the choice of the constants.

★ 8.7   THE HISTOGRAM

The probability density function (pdf) provides much useful information about a random variable and allows us to determine statistical averages. But what if the pdf is not known? If experimental sample values of the random variable are available, we can estimate the pdf using a histogram.

A *histogram* is a tabulation of the frequency-of-occurrence of a random variable within preset ranges. The ranges for the tabulation are chosen to be adjacent and the total of the set of numbers gives the total number of events in the experiment. The histogram may be displayed by plotting the number of times that the random variable has been found in each range versus the value of the random variable. If the histogram is normalized by dividing by the total number of events, the height represents the relative frequency of occurrence within each range.

If the ranges are known or preset, the computation of a histogram is a straightforward number-tabulating routine and is easily implemented on a digital computer. This is illustrated in the following example.

---

c Example 8.7.1   Compute the histogram of the sum of the first two digits in the fractional part of the square root of the integers from 1 to 1000.

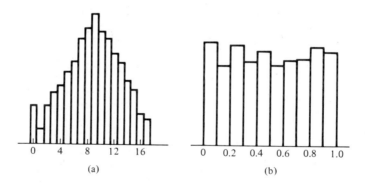

Fig. 8.10 Histograms of (a) Example 8.7.1 and (b) Drill Problem 8.7.1.

Solution.   The total range of possible outcomes is from 0 (0 + 0) to 18 (9 + 9). Event counters are set in 19 memory registers such that register No. 0 increments its count by one if $X_i = 0$, register No. 1 increments its count by one if $X_i = 1$, etc. The results are listed below and are plotted in Fig. 8.10(a). (*Note:* A two-place round-off was used in determining the first two integers.)

Register Number	Event Count		Register Number	Event Count
0	31		10	88
1	12		11	82
2	31		12	71
3	41		13	64
4	46		14	50
5	57		15	43
6	65		16	24
7	82		17	20
8	91		18	0
9	102		Total	1000

The relative frequency can be found by dividing each count by the total (1000). (Note that the relatively large contribution at zero is caused by the 31 perfect squares which occur within that range.) On the basis of this histogram data, the mean and standard deviations are 8.81 and 4.08, respectively. Also note that the histogram says nothing about the statistical independence (or lack of it) of the sample values.

---

c Drill Problem 8.7.1   A possible method of generating a "pseudorandom" number sequence within the range $0 \leq X < 1$ in small calculators is to use the recursive formula:

$$x_i = \text{fractional part of } \{(\pi + x_{i-1})^5\}.$$

Compute the histogram of the first 1000 sample values of X, using $x_0 = 0$, when each number is truncated to one decimal place.

Answer.   Histogram results for $0 \leq X < 0.1$, $0.1 \leq X < 0.2$, etc., are, respectively, 115, 88, 111, 93, 104, 88, 94, 96, 108, 103 (see Fig. 8.10b).

Computation of the histograms in Fig. 8.10 is straightforward because the ranges are preset. Each range interval for the tabulation in the histogram is often called a probability "cell." For instance, there are 19 cells in the above example problem and 10 in the drill problem. How many sample values should one take for each cell? This question is a typical one in applying a relative-frequency definition of probability and is not easy to answer concretely. "As many as possible" is desirable, of course, but a practical rule of about 100 per cell is reasonable.

   If the cell widths are not known or preset, the generation of a meaningful histogram becomes more complicated and involves some compromises in setting the cell widths. On the one hand, narrow cell widths are desirable to give as accurate an approximation to the pdf as possible. On the other hand, wider cell widths allow more averaging and the variance in the estimate of the cell height is decreased. If the sample values can be stored (e.g., on magnetic tape), the results of a histogram using one set of cell widths can be used to suggest different widths for a second histogram from the same data. The sample mean and variance can also be computed to aid in readjusting the cell widths and locations. The practical rule of tabulating at least 100 events per cell is applicable to retain some reliability in the estimate of the cell height. Obviously, the best histograms result from narrow cell widths and the tabulation of large amounts of data. It also helps to have some clues as to the nature of the pdf of the random variable before the tabulation is begun. For a more thorough discussion of these problems, the interested reader should consult the references.†

★ 8.8   TRANSFORMATIONS OF RANDOM VARIABLES

We are concerned with the transmission of signals through systems and are interested in how we can describe the effects of a system on a given input random variable. The approach here is to use the concept of a mapping of one random variable into another and to describe this in terms of probability density functions.

Let a random variable Y be a function $f(X)$ of a random variable X. We assume that $f(X)$ is known and has a continuous first derivative and that the pdf of X is known. Because probability is conserved in the mapping from X to Y, the probability of finding Y in a differential range dy is equal to the probability that X is in the corresponding range dx. To insure nonnegative pdf's, we use absolute magnitude signs and write:

$$p_Y(y)\,|dy| = p_X(x)\,|dx|, \tag{8.60}$$

or

$$p_Y(y) = p_X(x)\left|\frac{dx}{dy}\right|. \tag{8.61}$$

To obtain the transformation expressed in the new coordinate y, we assume that the inverse of $y = f(x)$ exists, so that our final result is

$$p_Y(y) = \left[p_X(x)\left|\frac{dx}{dy}\right|\right]_{x=f^{-1}(y)}. \tag{8.62}$$

Some intuitive insight into this procedure based upon a mapping of histograms is illustrated in Fig. 8.11. Here a histogram approximation to the output pdf $p_Y(y)$ is obtained from a histogram approximation of the input pdf $p_X(x)$ by a

---

† See, for example, M. Schwartz and L. Shaw, *Signal Processing: Discrete Spectral Analysis, Detection, and Estimation*, New York: McGraw-Hill, 1975.

mapping using the output-input gain characteristic of the system, $y = f(x)$. Note that $f(x)$ is not linear (however, it is linear with a slope of two for $x < 2$ and linear with a slope of one for $x > 6$).

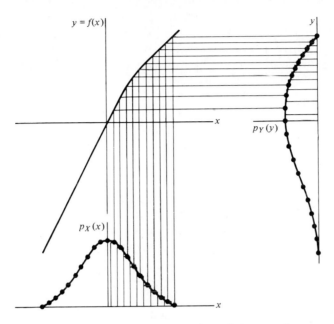

Fig. 8.11 A histogram interpretation of the transformation of random variables.

In the transformation, each probability cell in the histogram of the input is mapped into the corresponding cell in the histogram of the output. Each cell width in the output histogram is proportional to the width of the corresponding cell in the input histogram times the incremental slope (i.e., gain) of the output-input system gain characteristic. To conserve cell probability, each input histogram value must then be multiplied by the inverse of the magnitude of the incremental slope of the output-input gain characteristic. As the number of cells in the input histogram becomes very large, the transformation approaches the differential form expressed in Eq. (8.61). The final step taken in Eq. (8.62) is needed to express the result in the new coordinate y.

Example 8.8.1 A gaussian-distributed random variable X with zero mean and unit variance is applied to a "full-wave" rectifier whose output-input gain characteristic is $y = |x|/a$, $a > 0$. Determine the pdf of the output random variable Y.

Solution. The mapping is one-to-one for $x < 0$ and one-to-one for $x > 0$ (a sketch may help); in both cases, $y > 0$. Using Eq. (8.62) we have, for $x > 0$,

$$p_Y(y) = a\left[\frac{1}{\sqrt{2\pi}}e^{-x^2/2}\right]_{x=ay} = \frac{a}{\sqrt{2\pi}}e^{-a^2y^2/2}, \qquad y > 0,$$

and for $x < 0$

$$p_Y(y) = a\left[\frac{1}{\sqrt{2\pi}}e^{-x^2/2}\right]_{x=-ay} = \frac{a}{\sqrt{2\pi}}e^{-a^2y^2/2}, \qquad y > 0.$$

The solution is then

$$p_Y(y) = a\sqrt{\frac{2}{\pi}}e^{-a^2y^2/2}u(y).$$

It can be shown (see Appendix A) that the mean of Y is $(\sqrt{2/\pi})/a$ and the standard deviation is $\sqrt{1 - (2/\pi)}/a$.

Drill Problem 8.8.1   A random variable X is uniformly distributed over $(0, 2)$ and is applied to the input of a system whose output-input gain characteristic is $y = 2x + 1$. Determine the pdf of the output random variable Y.

Answer. $p_Y(y) = [u(y - 1) - u(y - 5)]/4$.

In cases in which the derivative of $f(X)$ has finite discontinuities, the above procedure can be altered somewhat to account for those probability cells which are all mapped into one. This can be handled conveniently by use of impulse functions and is demonstrated in the following example.

Example 8.8.2   The random variable X of Example 8.8.1 is applied to the "half-wave" rectifier whose output-input gain characteristic is $y = (x/a)u(x)$. Determine the pdf of the output.

Solution.   For $x > 0$, the solution proceeds in the same way as for Example 8.8.1. For $x < 0$, however, all points of the input pdf are mapped into zero in the output. To conserve probability we must add a contribution of $\int_{-\infty}^{0} p_X(x)\, dx = 1/2$ at the point $y = 0$ so that

$$p_Y(y) = \frac{a}{\sqrt{2\pi}}e^{-a^2y^2/2}u(y) + (\tfrac{1}{2})\,\delta(y).$$

The mean of Y is $(\sqrt{2\pi}\,a)^{-1}$, and the standard deviation is $\sqrt{\pi-1} \times (\sqrt{2\pi}\,a)^{-1}$.

Drill Problem 8.8.2   The random variable X of Drill Problem 8.8.1 is applied to the input of a system whose output-input gain characteristic is (a) $y = x - (x - 1)u(x - 1)$; (b) $y = xu(x - 1)$. Determine the pdf of the output.

Answer.   (a) $\frac{1}{2}[u(y) - u(y - 1)] + (\frac{1}{2})\delta(y - 1)$; (b) $\frac{1}{2}[u(y - 1) - u(y - 2)] + (\frac{1}{2})\delta(y)$.

8.9   JOINT AND CONDITIONAL DENSITY FUNCTIONS

In some cases we require two or more random variables for a description of a physical process. The same approach which was used for probability density functions of one dimension can be extended to more than one dimension. For simplicity we consider only the two-dimensional case.

The probability of finding X in the range $x_1 < X \le x_2$ while also finding Y in the range $y_1 < Y \le y_2$ is [cf. Eq. (8.27)]

$$P(x_1 < X \le x_2, y_1 < Y \le y_2) = \int_{y_1}^{y_2} \int_{x_1}^{x_2} p_{XY}(x, y) \, dx \, dy, \qquad (8.63)$$

where $p_{XY}(x, y)$ is the *joint probability density function* of the random variables X and Y. For the two-dimensional case, probability equals the volume under the surface described by $p_{XY}(x, y)$, as depicted in Fig. 8.12.

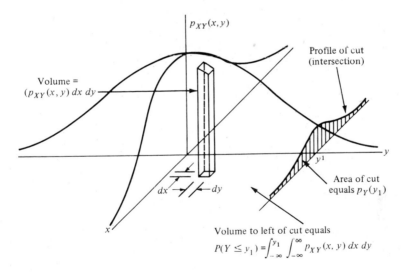

Fig. 8.12 A geometric interpretation of the joint probability density function.

Given the joint probability density function $p_{XY}(x, y)$, the probability density function $p_Y(y)$ can be found by obtaining the cumulative distribution function $F_Y(y)$ in the following manner. If we are interested only in Y, the value of X

does not matter, so that

$$P(Y \leq y) = P(X \leq \infty, Y \leq y),$$

or

$$F_Y(y) = F_{XY}(\infty, y),$$

yielding

$$F_Y(y) = \int_{-\infty}^{y} \int_{-\infty}^{\infty} p_{XY}(x, y) \, dx \, dy. \tag{8.64}$$

Differentiating both sides of Eq. (8.64) with respect to y, we obtain

$$p_Y(y) = \int_{-\infty}^{\infty} p_{XY}(x, y) \, dx. \tag{8.65}$$

Thus the probability density function of Y can be found by integrating the joint probability density function $p_{XY}(x, y)$ over all values of X. Similar arguments in the other dimension yield $p_X(x)$.

A geometrical interpretation of Eqs. (8.64) and (8.65) is shown in Fig. 8.12. Intersecting the surface $p_{XY}(x, y)$ by the plane $y = y_1$, we obtain the cross-sectional profile shown. From Eq. (8.65), we see that the cross-sectional area gives $p_Y(y_1)$, and $p_Y(y)$ can be determined by finding the dependence of the cross-sectional area on y. Note that the volume to the left of the intersection in Fig. 8.12 is equal to the integral of $p_Y(y)$ over all $y \leq y_1$, i.e.,

$$\int_{-\infty}^{y_1} p_Y(y) \, dy = \int_{-\infty}^{y_1} \int_{-\infty}^{\infty} p_{XY}(x, y) \, dx \, dy,$$

verifying Eq. (8.64). Similar arguments in the other dimension yield $p_X(x)$. The probability density functions $p_X(x)$ and $p_Y(y)$ are called *marginal densities*.

It is intuitively appealing to attempt to identify the profile formed by the intersection of the plane $y = y_1$ and the surface $p_{XY}(x, y)$ as the conditional probability density for X given $Y = y_1$. An investigation shows, however, that this profile function would not generally qualify as a probability density function because it does not necessarily have unit area. To achieve this requirement, we can normalize by dividing by the area. Thus we can write

$$p_X(x|y_1) = \frac{p_{XY}(x, y_1)}{\text{area shaded}},$$

$$= \frac{p_{XY}(x, y_1)}{p_Y(y_1)}.$$

Similar arguments hold for $p_Y(y|x)$ so that, in general, this result can be rewritten as†

$$p_{XY}(x, y) = p_X(x)p_Y(y|x) = p_Y(y)p_X(x|y). \tag{8.66}$$

In the particular case in which the two random variables X and Y are statistically independent, $p_X(x|y) = p_X(x)$, $p_Y(y|x) = p_Y(y)$, and Eq. (8.66) reduces to

$$p_{XY}(x, y) = p_X(x)p_Y(y). \tag{8.67}$$

In the case of two or more random variables, the concept of expected values can be extended by using multiple integration over the joint probability density function in the general form,

$$E\{g(x, y)\} = \int_{-\infty}^{\infty}\int_{-\infty}^{\infty} g(x, y)p_{XY}(x, y) \, dx \, dy. \tag{8.68}$$

Example 8.9.1   Let X and Y be two random variables with variances σ_X^2 and σ_Y^2, respectively. Let the sum be $Z = X + Y$, with variance σ_Z^2. Determine σ_Z^2 in terms of σ_X^2, σ_Y^2 if X and Y are statistically independent.

Solution.   For convenience, let the means of X and Y be zero so that

$$\sigma_Z^2 = \int_{-\infty}^{\infty}\int_{-\infty}^{\infty} (x + y)^2 p_{XY}(x, y) \, dx \, dy.$$

If X, Y are statistically independent, then

$$\sigma_Z^2 = \int_{-\infty}^{\infty}\int_{-\infty}^{\infty} (x + y)^2 p_X(x)p_Y(y) \, dx \, dy$$

$$= \int_{-\infty}^{\infty} x^2 p_X(x) \, dx + \int_{-\infty}^{\infty} y^2 p_Y(y) \, dy$$

$$+ 2\int_{-\infty}^{\infty} xp_X(x) \, dx \int_{-\infty}^{\infty} yp_Y(y) \, dy.$$

Because both factors of the last term of this equation are zero by definition, we have our desired result:

$$\sigma_Z^2 = \sigma_X^2 + \sigma_Y^2$$

if X, Y are statistically independent. It is left to the student to show that this result also holds for nonzero means. Therefore, we conclude that the variance

---

† The random variable subscript notation becomes cumbersome and is often deleted. Doing so streamlines the notation, but the random variable subscripts should be replaced if possible confusion could result in the notation.

of a sum is the sum of the variances if the random variables are statistically independent.

Example 8.9.2   A random variable Y is composed of the mean of N statistically independent random variables $X_1, X_2, \ldots, X_N$, each with a mean of m_X and variance σ_X^2. (a) Determine m_Y in terms of m_X. (b) Determine σ_Y^2 in terms of σ_X^2.

Solution.   We can write

$$Y = \frac{1}{N} \sum_{i=1}^{N} X_i = \sum_{i=1}^{N} X_i/N.$$

a) $m_Y = E\left\{ \frac{1}{N} \sum_{i=1}^{N} X_i \right\} = \frac{1}{N} \sum_{i=1}^{N} E\{X_i\},$

$m_Y = m_X.$

b) Treating X_i/N as new random variables and using the result of Example 8.9.1, we have

$$\sigma_Y^2 = \sigma_1^2/N^2 + \sigma_2^2/N^2 + \cdots + \sigma_N^2/N^2 = N\sigma_X^2/N^2,$$

$$\sigma_Y^2 = \sigma_X^2/N.$$

Combining these two results, we conclude that if an experiment having statistical regularity is repeated, then our confidence that the averaged results are closer to the true results is improved. Stated more precisely, the width of the pdf of the random variable describing the experiment narrows proportional to $\sqrt{N}$.

Drill Problem 8.9.1   Let X and Y be two random variables with means m_X and m_Y respectively. Let the sum be $Z = X + Y$, with mean m_Z. Determine m_Z in terms of m_X and m_Y.

Answer. $m_Z = m_X + m_Y$ (i.e., "the mean of a sum is the sum of the means.") Note that this result holds regardless of whether or not the random variables are statistically independent and is a special case of the more general statement that the expected value operation is linear.

★ 8.10   CORRELATION BETWEEN RANDOM VARIABLES

In an analogy to the variance of one random variable, the *covariance* of two random variables X and Y is defined as

$$\mu \triangleq E\{(X - m_X)(Y - m_Y)\}. \tag{8.69}$$

If X and Y are statistically independent, use of Eq. (8.67) in Eq. (8.69) gives

$$\mu = \int_{-\infty}^{\infty} \int_{-\infty}^{\infty} (x - m_X)(y - m_Y)p_{XY}(x, y) \, dx \, dy$$

$$= \int_{-\infty}^{\infty} (x - m_X)p_X(x) \, dx \int_{-\infty}^{\infty} (y - m_Y)p_Y(y) \, dy$$

$$= (m_X - m_X)(m_Y - m_Y) = 0. \tag{8.70}$$

Thus the covariance is zero if X and Y are statistically independent. Next suppose that X and Y are completely dependent so that either $X = Y$ or $X = -Y$. In these cases, we find that Eq. (8.69) gives

$$\mu = \sigma_X^2 = \sigma_Y^2 = \sigma_X\sigma_Y, \tag{8.71}$$

and

$$\mu = -\sigma_X^2 = -\sigma_Y^2 = -\sigma_X\sigma_Y. \tag{8.72}$$

At this point it is convenient to define a normalized covariance which is called the *correlation coefficient*:

$$\rho \triangleq \frac{\mu}{\sigma_X\sigma_Y}. \tag{8.73}$$

Because Eqs. (8.71) and (8.72) are the extreme cases, we conclude that

$$-1 \le \rho \le 1. \tag{8.74}$$

The correlation coefficient ρ serves as a measure of the extent to which X and Y are dependent. When $\rho = 0$, the random variables X and Y are said to be *uncorrelated*.

From Eqs. (8.70) and (8.73) we conclude that if two random variables are statistically independent then they are uncorrelated. However, *the converse is not necessarily true.*

In statistical testing, usually the joint probability function is unknown and a uniform distribution is assumed. Thus Eqs. (8.69) and (8.73), for the discrete case, become

$$\rho = \frac{\Sigma_i(x_i - m_X)(y_i - m_Y)}{[\Sigma_i(x_i - m_X)^2\Sigma_i(y_i - m_Y)^2]^{1/2}}. \tag{8.75}$$

Equation (8.75) is readily calculated from numerical data.

The correlation coefficient, as defined here, refers to a linear correlation between the random variables X, Y. Thus the correlation coefficient is not so much a measure of how dependent the random variables X, Y are in a general sense but is a measure of the linearity of that relationship. In fact, the square of the correlation coefficient can be used as a relative measure, on a mean-square basis, of how well a straight line can be fit to the sample values of X, Y. For example, the two random variables X, Y may have perfect dependence such as in the relation $X^2 + Y^2 = 1$, yet may have a zero correlation coefficient [as they do if uniformly distributed over $(-1, 1)$]. This behavior is also illustrated in Example 8.10.1.

Computation of the correlation coefficient is a convenient indicator of how dependent two random variables are on each other within the above restriction. For example, the correlation coefficient determined between adjacent sample values (that is, x_i, x_{i-1}) in Example 8.7.1 is $\rho = 0.391$, while for Drill Problem 8.7.1, $\rho = 5.96 \times 10^{-4}$. Thus the successive sample values of the latter number generator are much more uncorrelated than the former.

An expansion of Eq. (8.69) gives

$$\mu = E\{XY\} - m_X m_Y. \tag{8.76}$$

The *correlation* of the two random variables X and Y is defined as the first term in the right-hand side of Eq. (8.76):

$$E\{XY\} = \int_{-\infty}^{\infty}\int_{-\infty}^{\infty} xy p_{XY}(x, y)\, dx\, dy. \tag{8.77}$$

For zero mean values of either X or Y, the correlation is equal to the covariance of X and Y.

Example 8.10.1   Let the joint probability density function of the random variables X, Y be uniform over $(-1 < X < 1, -1 < Y < 1)$ and suppose $Y = X^2$. Show that even though X and Y are not statistically independent, they are uncorrelated.

Solution.   Using Eq. (8.66), we have

$$p_X(x) = \int_{-\infty}^{\infty} p_{XY}(x, y)\, dx = \int_{-1}^{1} (\tfrac{1}{4})\, dy = \tfrac{1}{2},$$

so that

$$E\{X\} = \int_{-1}^{1} x(\tfrac{1}{2})\, dx = 0.$$

Equation (8.77) then gives

$$E\{XY\} = \int_{-1}^{1} x^3(\tfrac{1}{4})\, dx = 0.$$

Therefore $\rho = 0$ [cf. Eqs. (8.73) and (8.76)] and X, Y are uncorrelated.

---

c Drill Problem 8.10.1   Compute the correlation coefficient of the first six successive digits in the fractional part of π.

Answer. -0.194.

Example 8.10.2   (Linear Regression) Suppose that a group of data points described by the random variables X, Y are to be approximated by the straight line $Y = aX + b$. The best straight-line fit is defined here by minimizing the mean-square error $E\{[Y - (aX + b)]^2\}$.

 a) Find a and b in terms of the statistical averages of X, Y.

 b) Show that ρ^2 is a relative measure of how well the straight line fits the data.

Solution

a) $\begin{aligned} E\{[Y - (aX + b)]^2\} &= E\{Y^2 - 2aXY - 2bY + a^2X^2 + 2abX + b^2\} \\ &= E\{Y^2\} - 2aE\{XY\} - 2bE\{Y\} + a^2E\{X^2\} \\ &\quad + 2abE\{X\} + b^2. \end{aligned}$

Taking a partial derivative with respect to b and setting the result equal to zero, we obtain

$$-2E\{Y\} + 2aE\{X\} + 2b = 0$$

or

$$b = m_Y - am_X.$$

Using this value of b in the equation above, we get

$\begin{aligned} E\{[Y - (aX + b)]^2\} &= E\{[Y - (aX + m_Y - am_X)]^2\} \\ &= E\{[(Y - m_Y) - a(X - m_X)]^2\} \\ &= E\{(Y - m_Y)^2 - 2a(X - m_X)(Y - m_Y) + a^2(X - m_X)^2\} \\ &= \sigma_Y^2 - 2a\mu + a^2\sigma_X^2. \end{aligned}$

Taking a partial derivative with respect to a and setting the result equal to zero, we obtain

$$-2\mu + 2a\sigma_X^2 = 0$$

or

$$a = \mu/\sigma_X^2 \qquad \text{and} \qquad b = m_Y - \frac{\mu}{\sigma_X^2}m_X.$$

b) $$E\{[Y - (aX + b)]^2\} = \sigma_Y^2 - 2a\mu + a^2\sigma_X^2$$

$$= \sigma_Y^2 - 2\frac{\mu^2}{\sigma_X^2} + \frac{\mu^2}{\sigma_X^2}$$

$$= \sigma_Y^2 - \frac{\mu^2}{\sigma_X^2 \sigma_Y^2}\sigma_Y^2$$

$$= \sigma_Y^2(1 - \rho^2).$$

The value of ρ^2 is called the *coefficient of determination*. Note that for $\rho = 1$, there is a perfect fit of the data to the straight line calculated from the results above. Many modern electronic calculators are preprogrammed to perform these calculations.

Drill Problem 8.10.2   Find the best straight-line fit to the following six (X, Y) data points and calculate the coefficient of determination: (1.0, 1.1), (2.0, 1.4), (3.0, 3.1), (4.0, 2.5), (5.0, 5.9), (6.0, 3.2).

Answer. $y = 0.67x + 0.53$; $\rho^2 = 0.53$.

★ 8.11   THE BIVARIATE GAUSSIAN DISTRIBUTION

A random variable or "variate" representing the outcome of a given experiment may require more than one dimension for its description of all possible outcomes. In this case we can speak of a "multivariate" probability distribution. Such a multivariate probability distribution may assume any form, as long as the integration or summation over all possible outcomes is normalized to one. However, we only discuss one very important case here—that of the bivariate (i.e., two-dimensional) gaussian distribution.

Two random variables X and Y are said to be jointly gaussian if their joint probability density is

$$p_{XY}(x, y) = \frac{1}{2\pi\sigma_X\sigma_Y\sqrt{1 - \rho^2}}$$
$$\exp\left\{-\frac{\left[\dfrac{(x - m_X)^2}{\sigma_X^2} + \dfrac{(y - m_Y)^2}{\sigma_Y^2} - \dfrac{2\rho(x - m_X)(y - m_Y)}{\sigma_X\sigma_Y}\right]}{2(1 - \rho^2)}\right\}. \tag{8.78}$$

We can make the following conclusions for the bivariate gaussian distribution based on Eq. (8.78) and some integrations (which we do not include here).

1. *The bivariate gaussian distribution depends only on the first- and second-order moments of the random variables.* Thus Eq. (8.78) requires only the means, variances, and covariances of the random variables.

2. *All individual (one-dimensional) pdf's are gaussian.* For example, an application of Eq. (8.65) to Eq. (8.78) gives

$$p_X(x) = \int_{-\infty}^{\infty} p_{XY}(x, y)\, dy = \frac{1}{\sqrt{2\pi}\sigma_X} e^{-(x - m_X)^2/2\sigma_X^2}. \tag{8.79}$$

3. *All conditional pdf's are gaussian.* This follows from a use of Eq. (8.66):

$$p(x|y) = \frac{p_{XY}(x, y)}{p_X(x)},$$

and Eqs. (8.78) and (8.79).

4. *Uncorrelated gaussian variates are also statistically independent.* (The converse is true for any distribution.) This is easily shown by setting $\rho = 0$ in Eq. (8.78) so that [cf. Eq. (8.79)]

$$p_{XY}(x, y) = p_X(x)p_Y(y).$$

5. *A linear combination of jointly gaussian variates has a gaussian distribution.* For example, if

$$Z = \alpha X + \beta Y,$$

where X and Y are jointly gaussian, then Z is gaussian with

$$m_Z = \alpha m_X + \beta m_Y,$$

$$\sigma_Z^2 = \alpha^2 \sigma_X^2 + \beta^2 \sigma_Y^2 + 2\rho\alpha\beta\sigma_X\sigma_Y.$$

As a result of these properties and the central limit theorem, the gaussian random variable is a popular choice for a probability model in engineering problems. The above concepts (and conclusions) can be extended to more than two variates. The resulting expressions become unwieldy, however, unless matrix notation is adopted.

8.12   RANDOM PROCESSES

To determine the probabilities of the various possible outcomes of an experiment, we repeat the experiment many times. For example, suppose we are interested in establishing the statistics associated with the tossing of a die. We might proceed by either of the two following methods. On the one hand, we might toss a large number of dice simultaneously. Alternatively, we could toss a single die many times. Intuitively, we would expect that both methods would yield the same results if all dice were identical and remained identical throughout the experiment. Assuming honest dice, we would expect, for example, that a "one" would show on $\frac{1}{6}$ of the dice in the first method and would show $\frac{1}{6}$ of the time in the second method.

Up to this point, we have conducted our experiments using the first method. However, our interests in the application of probabilistic concepts to communication systems actually are closer to the second method. Under what conditions can we expect the same results? We shall investigate this question in this section. But first we must specify more precisely just what we mean by the two methods above.

Conceptually, we could consider the second method as a random number generator which puts out a number between one and six at the command of a clock. Now suppose we have many of these random number generators and that they are all operated from commands from a common clock. Such a collection of waveform sources is called an *ensemble*. The individual random number generator waveforms (as a function of time) are called *sample functions*. The sample functions can be written as $X(t, \lambda_i)$ when there are i discrete outcomes possible. The ensemble of possible sample functions is called a *random process*. This is illustrated in Fig. 8.13.

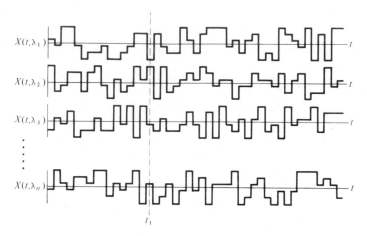

Fig. 8.13 An ensemble of random waveforms.

The statistics of the waveforms in our ensemble may be determined from measurements made at some fixed time $t = t_1$ on all the sample functions. Thus we can write $X(t_1, \lambda_i)$, which is just another way of writing $X(\lambda_i)|_{t_1}$. Therefore a random process $X(t, \lambda_i)$ at a fixed time is simply the random variable X. A statistical average can be determined from the measurements made on $X(\lambda_i)|_{t_1}$ and the average so determined is called an *ensemble average*. For example, we could toss one hundred coins simultaneously at $t = t_1$ and determine the statistics on the basis of this experiment. Note that the statistics and averages obtained in this method are those which we have been discussing in previous sections.

On the other hand, we could choose to observe just one sample function for a period of time and determine its statistics. (In the above example, this would correspond to taking one coin and tossing it one hundred times in succession.) From these observations, we can compute averages which we call *time averages*.

We can illustrate some of these ideas by plotting the output of a number of identical random waveform generators as shown in Fig. 8.13. Such a collection of waveforms is called an ensemble and is described by the random process $X(t, \lambda_i)$. Each generator output as a function of time is one sample function of the ensemble. Values of all waveforms in the ensemble observed at a particular time $t = t_1$ are described by the ensemble statistics and designated by the random variable X.

An ensemble average is found by using the ensemble statistics at a particular time. A time average is found by using time averaging on a particular sample function. The question then arises: Are time averages equal to ensemble averages?

In general, ensemble averages and time averages are not equal. Suppose, for example, that the statistical characteristics of the sample functions in the ensemble varied with time. Then the ensemble averages would be different at different times but the time averages would not reflect this variation. When the statistical characteristics of the sample functions do not change with time, a random process is said to be *stationary*. However, even the property of being stationary does not ensure that the ensemble and the time averages are the same. For it may happen that while each sample function is stationary, the individual sample functions may differ statistically from one another. In this case the time averages will depend on the particular sample function that is used to form the average.

When the nature of a random process is such that ensemble averages and time averages are equal, the process is referred to as being *ergodic*. Hence if $X(t, \lambda_i)$ is ergodic, then all its statistics can be determined from a single sample function of the random process. An ergodic random process is stationary but a stationary random process is not necessarily ergodic. For example, the random process $X(t) = C$, where C is a random constant, is stationary but not ergodic.

The foregoing reasoning can be applied to the continuous as well as to the discrete case. Suppose, for example, that the ensemble is composed of many random noise generators. Each generates a sample function and at a particular instant of time the outputs of all noise generators may be observed to find an ensemble average. If the random process describing the outputs of the noise generators is ergodic, then observations made on one sample function over a period of time are sufficient to obtain the statistics of the random process.

We shall consider only those random processes which are ergodic. Thus we can equate ensemble averages [cf. Eqs. (8.34) and (8.35)] with the time averages [cf. Eqs. (4.23) and (4.24)]:

$$E\{X\} = \int_{-\infty}^{\infty} xp(x)\ dx = \lim_{T\to\infty} \frac{1}{T}\int_{-T/2}^{T/2} x(t)\ dt, \tag{8.80}$$

$$E\{X^2\} = \int_{-\infty}^{\infty} x^2 p(x)\ dx = \lim_{T\to\infty} \frac{1}{T}\int_{-T/2}^{T/2} x^2(t)\ dt. \tag{8.81}$$

Also, the cumulative distribution function and the probability density function can now be determined from observations of one sample function just as well as from observations of an ensemble of sample functions at one particular instant in time.

A consideration of Eqs. (8.80) and (8.81) for ergodic random processes allows us to make the following statements:

1. The average or mean value, $E\{X\}$, is equal to the dc level of the signal.
2. The square of the mean, $[E\{X\}]^2$, is equal to the power in the dc component across one ohm.

3. The mean-square value, $E\{X^2\}$, is equal to the total average power of the signal $x(t)$ across one ohm.

4. The square root of the mean-square value is the rms value.

5. The variance, σ_X^2, is equal to the average power in the time-varying or ac component of the signal across one ohm.

6. The standard deviation, σ_X, is the rms value of the time-varying (ac) component of the signal.

7. If $E\{X\} = 0$, then σ_X is the rms value of the signal.

These relations tend to make us feel much more comfortable in the use of random signals. However, it must be remembered that these equalities hold only for ergodic random processes. As a result of the close tie with signal analysis, as indicated above, we shall designate a random process by the same type of notation that we use for deterministic waveforms so that $X(t, \lambda)$ will be written simply as $x(t)$.

It is instructive to review the relationship of the cumulative distribution function and the probability density function of an ergodic random process $x(t)$. Consider, for example, the sample function of an ergodic random process as shown in Fig. 8.14(a). This waveform is applied to the input of a comparator which in turn activates a counter connected to a clock. The proportionate amount of time that the waveform spends below the preset threshold of the comparator, x_0, is then recorded versus the setting x_0 and is shown in Fig. 8.14(b). Because we have assumed ergodicity, we now recognize that this describes the cumulative distribution function of the random variable X associated with the random process $x(t)$. Taking the derivative of the cumulative distribution function yields the pdf as shown in Fig. 8.14(c).

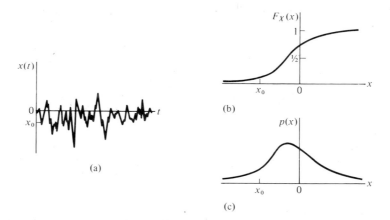

Fig. 8.14 A sample function of an ergodic random process and its cumulative distribution function and pdf.

Another possibility in making a measurement of the pdf of an ergodic random process directly is to use a series of threshold devices each of which detects whenever the input waveform is within a specified amplitude range. When the amplitude of the input waveform is within the specified range for a given threshold detector, a constant-level gate pulse is produced and integrated. At the end of the desired measurement interval, the outputs of the integrators are sampled and displayed, as shown in Fig. 8.15. As the time interval used becomes large, the resulting samples approximate the probability density function if they are normalized by the length of the time interval. A device available in the laboratory for performing this measurement is called a "multichannel analyzer." Note the similarities with the multichannel spectral analyzer shown in Fig. 4.2.

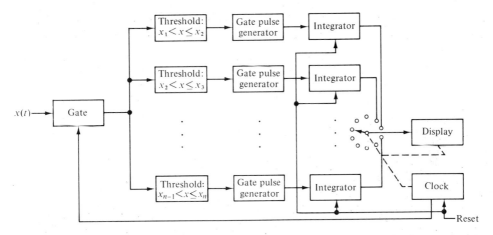

Fig. 8.15 Relative measurement of the probability density function of an ergodic random process.

Example 8.12.1   Suppose we wish to trigger an alarm system whenever a given signal voltage is present. There is also additive noise present; this noise voltage is gaussian-distributed with zero mean value. The alarm has a preset threshold such that whenever the total input voltage exceeds the threshold the alarm triggers and rings a bell.

a) It is desired to set the threshold at the lowest voltage level possible and yet have the alarm ring the bell no more than 1% of the time, on the average, when there is no signal present. Determine the required threshold voltage if the rms value of the noise is one-half volt.

b) Assume that when the signal is present, this signal is a positive two volts. With the threshold set as in part (a), what is the probability that the bell will not ring even when the signal is present?

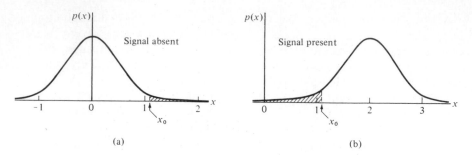

Fig. 8.16 The probability density functions of Example 8.12.1.

Solution

a) When no signal is present, $m_x = 0$ and $\sigma_x = \frac{1}{2}$. We determine the threshold x_0 such that (see Fig. 8.16a)

$$\int_{x_0}^{\infty} \frac{2}{\sqrt{2\pi}} e^{-2x^2} \, dx = 0.01.$$

Changing the variable, let $z = 2x$ so that this becomes

$$\int_{2x_0}^{\infty} \frac{1}{\sqrt{2\pi}} e^{-z^2/2} \, dz = 0.01,$$

or

$$\text{Erfc} \, (2x_0) = 0.01.$$

Using the tables in Appendix G, we find that the required threshold value is $x_0 = 1.163$ volts.

b) When the signal is present, $m_x = 2$ and $\sigma_x = \frac{1}{2}$. The probability that the bell will not ring even if the signal is present is (see Fig. 8.16b)

$$P(x \le x_0) = \int_{-\infty}^{1.163} \frac{2}{\sqrt{2\pi}} e^{-2(x-2)^2} \, dx$$

$$= \int_{-\infty}^{-1.674} \frac{1}{\sqrt{2\pi}} e^{-z^2/2} \, dz$$

$$= \int_{1.674}^{\infty} \frac{1}{\sqrt{2\pi}} e^{-z^2/2} \, dz$$

$$= \text{Erfc} \, (1.674) = 0.0471.$$

Drill Problem 8.12.1   (a) Determine a new alarm threshold setting for Example 8.12.1 such that the probability of the bell ringing with no signal present (i.e., a false alarm) is exactly twice that of the bell not ringing when a signal is present (i.e., a miss). (b) What is the threshold which will make these equal?

Answer.   (a) 0.927 volt; (b) 1.000 volt.

8.13   AUTOCORRELATION AND POWER SPECTRA

Suppose that we have a random process $x(t)$, such as the one for which several sample functions are exhibited in Fig. 8.17. At a time $t = t_1$, the random process $x(t)$ is characterized by the random variable $X_1 = x(t_1)$. Similarly, at a time $t = t_2$, the random process $x(t)$ is characterized by the random variable $X_2 = x(t_2)$. The possibility now arises that we characterize the random process $x(t)$ by use of a joint probability density function of these random variables. If we were to completely describe the random process in this way we would, of course, have to use an infinite number of random variables and an infinite-dimensional joint density. This is not practical and therefore we limit ourselves to a two-dimensional joint density.

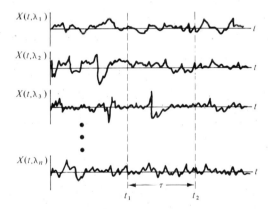

Fig. 8.17 An ensemble of continuous random waveforms.

The first moment of the random process under consideration is

$$E\{x(t)\} = m_x(t). \tag{8.82}$$

If the random process $x(t)$ is stationary, then its statistics do not change as a function of time and the first moment becomes a constant:

$$E\{x(t)\} = m_x \qquad x(t) \text{ stationary.} \tag{8.83}$$

We characterize the random process $x(t)$ in terms of a two-dimensional joint density $p_{X_1 X_2}(x_1, x_2)$, where $X_1 = x(t_1)$ and $X_2 = x(t_2)$ (such as depicted in

Fig. 8.17). There are three different second (joint) moments possible: $E\{X_1^2\}$, $E\{X_2^2\}$, and $E\{X_1 X_2\}$. This last joint moment is of particular importance and is given the special notation,†

$$E\{x(t_1)x(t_2)\} \triangleq R_{xx}(t_1, t_2), \tag{8.84}$$

or, writing this out,

$$R_{xx}(t_1, t_2) = \int_{-\infty}^{\infty}\int_{-\infty}^{\infty} x_1 x_2 p_{X_1 X_2}(x_1, x_2)\, dx_1\, dx_2.$$

The function $R_{xx}(t_1, t_2)$ is called the *autocorrelation function* of the random process $x(t)$ evaluated at $t = t_1, t_2$. The term *autocorrelation* has been used previously (cf. Chapter 4) and we shall soon see why the same word is used again here.

If the random process $x(t)$ is stationary, the second joint moment is independent of the absolute time reference and is only dependent on the time difference between t_1 and t_2 (see Fig. 8.17) so that we can write

$$R_{xx}(t_1, t_2) = R_{xx}(t_2 - t_1) \qquad x(t) \text{ stationary.} \tag{8.85}$$

In this case, we can let $\tau = t_2 - t_1$, or $t_2 = t_1 + \tau$, and then delete the subscript on t_1 to obtain the simpler notation

$$R_{xx}(\tau) = E\{x(t)x(t + \tau)\} \qquad x(t) \text{ stationary.} \tag{8.86}$$

The autocorrelation function is the second joint moment and therefore does not completely describe or define the random process $x(t)$ (i.e., it is not unique), but it does give us much information about $x(t)$. In particular, it gives us a measure of how dependent a particular value of a sample function is on another value that is removed τ units of time. For example, if t_1 and t_2 are sufficiently far apart such that $x(t_1)$ and $x(t_2)$ are statistically independent, the autocorrelation function reduces to

$$\begin{aligned} R_{xx}(\tau) &= E\{x(t)x(t + \tau)\} \\ &= E\{x(t)\}E\{x(t + \tau)\} \\ &= m_x^2. \end{aligned} \tag{8.87}$$

If the mean value of $x(t)$ is zero, that value of τ over which $R_{xx}(\tau)$ first goes to zero represents the time separation required for the random process to become "uncorrelated." In other words, the shape of $R_{xx}(\tau)$ gives us some idea as to how past values affect present values of $x(t)$ and hence how fast a particular sample function can vary with time.

---

† This definition is for a real-valued random process; for the complex-valued case, the second term in the expectation has a complex conjugate operation.

Thus for $m_x = 0$ we say that $x(t)$ and $x(t + \tau)$ are *uncorrelated* if $R_{xx}(\tau) = 0$. If $x(t)$ and $x(t + \tau)$ are statistically independent, they are uncorrelated. (However, if $x(t)$ and $x(t + \tau)$ are uncorrelated, they are not necessarily independent.)

In addition, if the random process $x(t)$ is ergodic (recall that ergodicity implies stationarity but not vice versa), the autocorrelation function can be found from any one sample function by use of time-averaging, or

$$R_{xx}(\tau) = E\{x(t)x(t + \tau)\}, \tag{8.88}$$

$$R_{xx}(\tau) = \lim_{T \to \infty} \frac{1}{T} \int_{-T/2}^{T/2} x(t)x(t + \tau)\, dt.$$

At this point we can see why the functional notation $R(\tau)$ and the name "autocorrelation" were used in Chapter 4. In fact, we can now tie our results together so that the relations for deterministic signals become special cases of the relations developed for random processes. For example, the power spectral density of a stationary random process $x(t)$ is given by

$$S_x(\omega) = \mathcal{F}\{R_{xx}(\tau)\} = \int_{-\infty}^{\infty} R_{xx}(\tau)e^{-j\omega\tau}\, d\tau. \tag{8.89}$$

If $x(t)$ is ergodic, then $R_{xx}(\tau)$ is given by Eq. (8.88), agreeing with our earlier result in Chapter 4. Conversely, the autocorrelation function is given by the inverse Fourier transform of the power spectral density:

$$R_{xx}(\tau) = \mathcal{F}^{-1}\{S_x(\omega)\} = \frac{1}{2\pi} \int_{-\infty}^{\infty} S_x(\omega)e^{j\omega\tau}\, d\omega. \tag{8.90}$$

The average power in $x(t)$ is (for a one-ohm load)

$$P = \lim_{T \to \infty} \frac{1}{T} \int_{-T/2}^{T/2} x^2(t)\, dt = E\{x^2(t)\} = R_{xx}(0),$$

and from Eq. (8.90),

$$R_{xx}(0) = \frac{1}{2\pi} \int_{-\infty}^{\infty} S_x(\omega)\, d\omega,$$

agreeing with our concept of a power spectral density as developed in Chapter 4.

In Chapter 4 the concept of a power spectral density was introduced as a limiting case of an energy spectral density measured over an interval and divided by that interval [see Eq. (4.17)]. This approach served our purposes for applying deterministic functions, and it can be applied with care to some other situations. In general, however, the approach is not always correct. To be valid, the expected value of the quantity $|F_T(\omega)|^2/T$ should tend toward the true value $S_f(\omega)$ over all sample functions, and its variance should tend toward zero with $T \to \infty$. To remedy this possible defect, we use the following definition for the power spectral density of random waveforms:

$$S_f(\omega) = \lim_{T \to \infty} \left[E\left\{ \frac{|F_T(\omega)|^2}{T} \right\} \right]. \tag{8.91}$$

Some implications of using Eq. (8.91) in the computation of power spectral density are discussed in Section 8.14.

All of the relations developed in Chapter 4 for the transmission of power spectra through linear systems can now be applied here. The methods for using random inputs form a very powerful analytical tool in general systems theory.

Example 8.13.1   The stationary random process $x(t)$ has a power spectral density $S_x(\omega)$. Find the power spectral density of $y(t) = x(t) - x(t - T)$.

Solution I.   Using the autocorrelation function, we have

$$\begin{aligned}
R_{yy}(\tau) &= E\{[x(t) - x(t - T)][x(t + \tau) - x(t + \tau - T)]\} \\
&= E\{x(t)x(t + \tau)\} - E\{x(t)x(t + \tau - T)\} \\
&\quad - E\{x(t - T)x(t + \tau)\} + E\{x(t - T)x(t + \tau - T)\} \\
&= R_{xx}(\tau) - R_{xx}(\tau - T) - R_{xx}(\tau + T) + R_{xx}(\tau),
\end{aligned}$$

$$\begin{aligned}
S_y(\omega) &= 2S_x(\omega) - e^{-j\omega T}S_x(\omega) - e^{j\omega T}S_x(\omega) \\
&= 2(1 - \cos \omega T)S_x(\omega).
\end{aligned}$$

Solution II.   Using the concept of a system transfer function, we can write

$$y(t) = x(t) \circledast h(t),$$

where

$$\begin{aligned}
H(\omega) &= \mathcal{F}\{h(t)\} = 1 - e^{-j\omega T}, \\
S_y(\omega) &= S_x(\omega)|H(\omega)|^2 \\
&= |1 - e^{-j\omega T}|^2 S_x(\omega) \\
&= (2 - e^{-j\omega T} - e^{j\omega T})S_x(\omega) \\
&= 2(1 - \cos \omega T)S_x(\omega).
\end{aligned}$$

The results must, of course, agree. In most cases the former approach is the more powerful, though not always the most direct.

Drill Problem 8.13.1   White noise with a (two-sided) power spectral density of $\eta/2$ watts per Hz is applied to the input of an ideal low-pass filter with a cut-off frequency of B Hz.

a) Find the autocorrelation function of the output of this filter.

b) What is the minimum time interval over which the output becomes uncorrelated?

Answer.   (a) $\eta B \text{Sa}(2\pi B\tau)$; (b) $\tau = 1/(2B)$ sec.

The autocorrelation function and power spectral density, together with the probability density function, are widely used in both theoretical and applied signal and system analysis. These functions are sketched in Fig. 8.18 for some of the more commonly used waveforms.

Example 8.13.2   (Random Binary Waveform) Consider a sample function $x(t)$ shown in Fig. 8.19 of the *random binary waveform* that has the following properties: (a) each pulse is of duration T_b; (b) the two possible pulse levels $-A$ and $+A$ are equally likely, and the presence of $-A$ or $+A$ in any one pulse interval is statistically independent of that in all other intervals; (c) the starting time T of $x(t)$ is uniformly distributed over $(0, T_b)$. Determine the autocorrelation function and the power spectral density of this random waveform.

Solution.   Let t_1, t_2 be two arbitrary values of time and assume that $0 < t_1 < t_2 < T_b$ and that $|t_1 - t_2| < T_b$. Then $X(t_1)$ and $X(t_2)$ are in the same pulse interval if (see Fig. 8.19)

$$t_2 \leq T \leq t_1 + T_b.$$

The probability that T is within this range is

$$P(t_2 \leq T \leq t_1 + T_b) = \int_{t_2}^{T_b + t_1} p_T(\xi)\, d\xi = \frac{T_b + t_1 - t_2}{T_b}.$$

Letting $\tau = t_2 - t_1$, we can rewrite this result as

$$P(t_2 \leq T \leq t_1 + T_b) = 1 - \frac{\tau}{T_b}.$$

Using absolute magnitude signs to permit the case for negative arguments, we have

$$E\{x(t)x(t + \tau)\} = A^2\left(1 - \frac{|\tau|}{T_b}\right) \qquad |\tau| < T_b.$$

For $|\tau| > T_b$ the random variables $X_t X_{t+\tau}$ are statistically independent (that is, the pulses are in different intervals), and we can write [cf. Eq. (8.87)]

$$E\{x(t)x(t + \tau)\} = m_x^2 = 0 \qquad |\tau| > T_b.$$

Combining the results above, we can write the autocorrelation function as [cf. Eq. (3.49)]

$$R_{xx}(\tau) = A^2 \Lambda \frac{\tau}{T_b}.$$

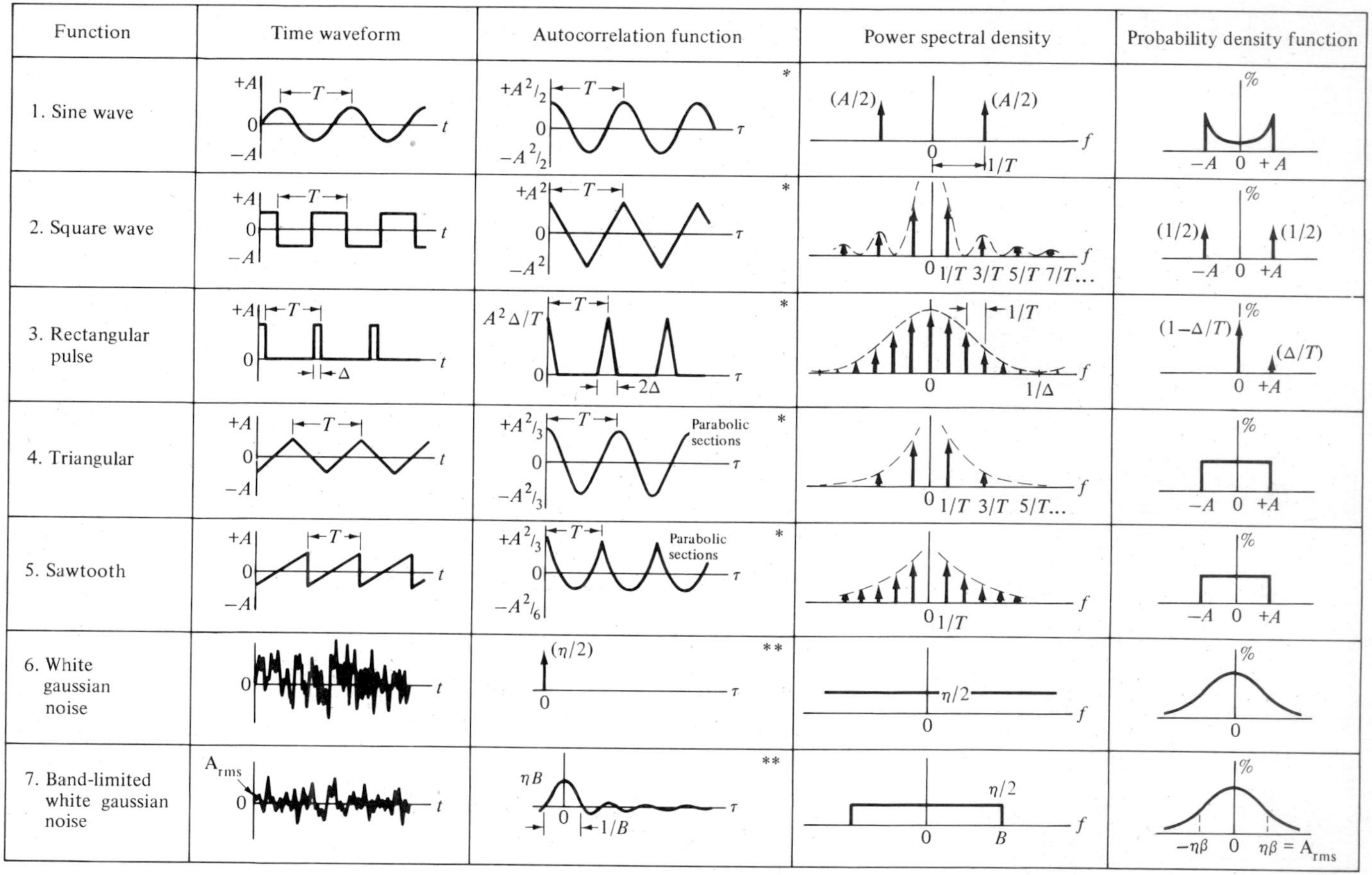

Fig. 8.18 Autocorrelation, spectra, and probability densities for some commonly used waveforms (* = time, ** = statistical).

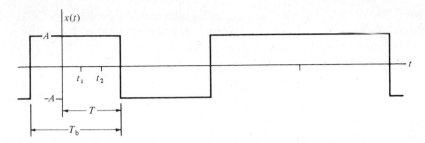

Fig. 8.19 A sample function of the random binary wave in Example 8.13.2.

Using Fourier transform pair #19 in Table 3.1, we find the corresponding power spectral density is

$$S_x(\omega) = A^2 T_b[\text{Sa}^2(\omega T_b/2)],$$

which can be written as

$$S_x(\omega) = \frac{|X(\omega)|^2}{T_b}.$$

★ **Example 8.13.3**   A given random process $x(t)$ is applied to the input of an N-stage transversal filter (Fig. 3.14); the output of the filter is designated here by $y(t)$. Assume that $x(t)$ can be written in terms of a pulse waveform $f(t)$ of duration T_b and a stationary sequence of discrete random amplitudes $\{b_k\}$ to form the random pulse train†

$$x(t) = \sum_{k=-\infty}^{\infty} b_k f(t - kT_b - T),$$

where T is a random time delay with a uniform pdf over $(0, T_b)$. The random variable b_k governs the choice of amplitude over the interval $(0, T_b)$. The choice in each interval is statistically independent of the choices in all other intervals. For convenience, we assume that the random variable b_k has zero mean and unit variance. For example, $x(t)$ can be envisioned as a binary random waveform, with b_k equaling $+1$ with probability $\frac{1}{2}$, and -1 with probability $\frac{1}{2}$.

The transversal filter introduces a controlled amount of dependence in $y(t)$ between intervals. In general, this dependence will extend over N adjacent intervals; therefore N is called the *constraint length*. Also, the maximum number of discrete output levels in $y(t)$ is increased by N over the number of levels in $x(t)$. Therefore if $x(t)$ is a binary random waveform, $y(t)$ is a ternary random

---

† Note that although the $\{b_k\}$ can be considered to be random variables here, they also represent a random process as a function of the discrete parameter k and, therefore, we use the lower-case notation.

waveform for $N = 1$. Dependence between adjacent intervals introduced by the transversal filter alters the power spectral density of $x(t)$, and such filters can be used to alter the spectral density of a random waveform.

Designating the filter output $y(t)$ as

$$y(t) = \sum_{k=-\infty}^{\infty} c_k f(t - kT_b - T),$$

determine

a) the autocorrelation function $R_{cc}(k)$, and

b) the power spectral density of $y(t)$ in terms of the power spectral density of $x(t)$.

Solution

a) The c_k's are statistically independent beyond the constraint length of the transversal filter so that

$$R_{cc}(k) = 0 \qquad \text{for } k > N.$$

Writing out a few terms for $k \leq N$, we find that

$$c_k = a_0 b_k + a_1 b_{k-1} + a_2 b_{k-2} + \cdots + a_N b_{k-N}$$

$$= \sum_{i=0}^{N} a_i b_{k-i}.$$

Using this result and adopting k for the shift parameter, we find the auto-correlation function is

$$R_{cc}(k) = E\{c_\ell c_{\ell+k}\} = \sum_{i=0}^{N} a_i \sum_{j=0}^{N} a_j E\{b_{\ell-i} b_{\ell+k-j}\}$$

$$= \sum_{i=0}^{N-k} a_i a_{i+k}.$$

Combining the functions above results in our desired result,

$$R_{cc}(k) = \begin{cases} \displaystyle\sum_{i=0}^{N-k} a_i a_{i+k} & k \leq N \\ 0 & k > N \end{cases}.$$

b) Development of the desired result is facilitated by first choosing $N = 2$ and then generalizing for larger N. Noting that what is requested is the magnitude squared of the frequency transfer function of the filter [cf. Eq. (4.21)], we find the most straightforward way to proceed is to determine the impulse response, take a Fourier transform, and then take the magnitude squared. However, we follow a statistical approach here to illustrate the methods developed in this chapter. It is left to the reader to verify that both approaches yield the same result.

For the case of $N = 2$,

$$
\begin{aligned}
y(t) &= a_0 x(t) + a_1 x(t - T_b) + a_2 x(t - 2T_b) \\
R_{yy}(\tau) &= E\{[a_0 x(t) + a_1 x(t - T_b) + a_2 x(t - 2T_b)] \\
&\quad \times [a_0 x(t + \tau) + a_1 x(t + \tau - T_b) + a_2 x(t + \tau - 2T_b)]\} \\
&= (a_0^2 + a_1^2 + a_2^2) R_{xx}(\tau) + (a_0 a_1 + a_1 a_2) \\
&\quad \times [R_{xx}(\tau - T_b) + R_{xx}(\tau + T_b)] + (a_0 a_2)[R_{xx}(\tau - 2T_b) + R_{xx}(\tau + 2T_b)].
\end{aligned}
$$

From part (a) above and $N = 2$,

$$
\begin{aligned}
R_{cc}(0) &= a_0^2 + a_1^2 + a_2^2, \\
R_{cc}(1) &= a_0 a_1 + a_1 a_2, \\
R_{cc}(2) &= a_0 a_2, \\
R_{cc}(k) &= 0 \qquad \text{for } k > 2,
\end{aligned}
$$

so that

$$
\begin{aligned}
R_{yy}(\tau) &= R_{cc}(0) R_{xx}(\tau) + R_{cc}(1)[R_{xx}(\tau - T_b) + R_{xx}(\tau + T_b)] \\
&\quad + R_{cc}(2)[R_{xx}(\tau - 2T_b) + R_{xx}(\tau + 2T_b)].
\end{aligned}
$$

Taking the Fourier transform of both sides and using the delay theorem yield

$$
\begin{aligned}
S_y(\omega) &= R_{cc}(0) S_x(\omega) + R_{cc}(1) S_x(\omega)[e^{-j\omega T_b} + e^{j\omega T_b}] \\
&\quad + R_{cc}(2) S_x(\omega)[e^{-j2\omega T_b} + e^{j2\omega T_b}]
\end{aligned}
$$

or

$$
S_y(\omega) = S_x(\omega)\left[R_{cc}(0) + 2\sum_{k=1}^{2} R_{cc}(k) \cos(k\omega T_b)\right].
$$

Generalizing this result, we have

$$
S_y(\omega) = S_x(\omega)\left[R_{cc}(0) + 2\sum_{k=1}^{N} R_{cc}(k) \cos(k\omega T_b)\right].
$$

Note that the terms within the brackets represent the spectral shaping (filtering) that results from the dependency between N adjacent pulses.

★ **Example 8.13.4** Consider a random binary waveform $x(t)$ that has the following properties: (a) each pulse is of duration T_b; (b) the two possible states in each interval are represented by the waveforms $f_1(t)$, $f_2(t)$ with corresponding Fourier transforms $F_1(\omega)$, $F_2(\omega)$; (c) the probability that $f_1(t)$ is selected in any interval is p and the probability that $f_2(t)$ is selected is $q = (1 - p)$; (d) the choice in any interval is statistically independent of that in all other intervals. Determine the power spectral density of this random binary waveform.†

Solution.   As a first step, we exhibit the possible periodic content in $x(t)$ as $v(t)$:

$$v(t) = \sum_{n=-\infty}^{\infty} [pf_1(t - nT_b) + qf_2(t - nT_b)].$$

Checking, we observe that $v(t + T_b) = v(t)$ and therefore $v(t)$ is periodic with period T_b. Therefore it can be expressed in terms of a Fourier series whose coefficients are the V_n. The V_n can be expressed in terms of the Fourier transforms $F_1(\omega)$, $F_2(\omega)$ as [cf. Eq. (3.15)]

$$V_n = \frac{1}{T_b} [pF_1(n\omega_b) + q F_2(n\omega_b)],$$

where $\omega_b = 2\pi/T_b$. The resulting power spectral density is [cf. Eq. (4.20)]

$$S_v(\omega) = \frac{1}{T_b^2} \sum_{n=-\infty}^{\infty} |pF_1(n\omega_b) + q F_2(n\omega_b)|^2 2\pi \, \delta(\omega - n\omega_b).$$

To consider the remaining (nonperiodic) content, we truncate the waveform to an interval T and then proceed to use Eq. (8.91). Let the truncated waveform $x_T(t)$ be represented by

$$x_T(t) = \sum_{n=-N}^{N} x_n(t),$$

where

$$x_n(t) = \begin{cases} f_1(t - nT_b) & \text{with prob } p \\ f_2(t - nT_b) & \text{with prob } q \end{cases}.$$

Subtracting out the periodic content, let $u_T(t) = x_T(t) - v_T(t)$, or

$$u_T(t) = \sum_{n=-N}^{N} u_n(t),$$

where

$$u_n(t) = \begin{cases} f_1(t - nT_b) - pf_1(t - nT_b) - qf_2(t-nT_b) & \text{with prob } p \\ f_2(t - nT_b) - pf_1(t - nT_b) - qf_2(t - nT_b) & \text{with prob } q \end{cases}.$$

This can be rewritten in the more compact notation

$$u_n(t) = a_n[f_1(t - nT_b) - f_2(t - nT_b)],$$

---

† W. R. Bennett and J. R. Davey, *Data Transmission*, New York: McGraw-Hill, 1965, 317–319; *see also*, W. C. Lindsey and M. K. Simon, *Telecommunication Systems Engineering*, Englewood Cliffs: Prentice-Hall, Inc., 1973, 17–19. The derivation in this example follows the approach used in Bennett and Davey.

where

$$a_n = \begin{cases} q & \text{with prob } p \\ -p & \text{with prob } q \end{cases}.$$

The $\{a_n\}$ form a sequence of random amplitudes with properties

$$E\{a_n\} = 0; \qquad E\{a_m a_n\} = \begin{cases} 0 & \text{for } m \neq n \\ pq & \text{for } m = n \end{cases}.$$

Allowing an interchange in integration and summation, and using the delay property, we find the Fourier transform of $u_T(t)$ to be

$$U_T(\omega) = \sum_{n=-N}^{N} a_n e^{-jn\omega_b} [F_1(\omega) - F_2(\omega)].$$

Proceeding, we write

$$E\{|U_T(\omega)|^2\} = E\left\{ \sum_{m=-N}^{N} \sum_{n=-N}^{N} a_m a_n e^{j(n-m)\omega_b} [F_1(\omega) - F_2(\omega)][F_1^*(\omega) - F_2^*(\omega)] \right\}.$$

To evaluate the expectation, we note that the only variables are the a's, and this expectation is nonzero only for $n = m$, so that

$$E\{|U_T(\omega)|^2\} = \sum_{n=-N}^{N} pq |F_1(\omega) - F_2(\omega)|^2$$

$$= (2N + 1)pq |F_1(\omega) - F_2(\omega)|^2.$$

Using Eq. (8.91) and noting that $T = (2N + 1)T_b$, we find the power spectral density of $u(t)$ is

$$S_u(\omega) = \lim_{N \to \infty} \frac{(2N + 1)pq |F_1(\omega) - F_2(\omega)|^2}{(2N + 1) T_b}$$

$$= pq \frac{|F_1(\omega) - F_2(\omega)|^2}{T_b}.$$

Combining the results for both nonperiodic and periodic content, we have

$$S_x(\omega) = p(1 - p) \frac{1}{T_b} |F_1(\omega) - F_2(\omega)|^2$$

$$+ \frac{2\pi}{T_b^2} \sum_{n=-\infty}^{\infty} |pF_1(n\omega_b) + (1 - p)F_2(n\omega_b)|^2 \, \delta(\omega - n\omega_b).$$

Note that if the Fourier transforms of both $f_1(t)$ and $f_2(t)$ vanish at a harmonic of ω_b, or if the weighted contributions from the Fourier transforms are equal but opposite signs, then no discrete component will appear in the power spectral density at the frequency of that harmonic. For the special case in which

$p = \frac{1}{2}$ and $f_1(t) = -f_2(t)$, this result simplifies to

$$S_x(\omega) = \frac{|F(\omega)|^2}{T_b}.$$

★ 8.14   NUMERICAL COMPUTATION OF POWER SPECTRA

In the preceding section the concept of a power spectral density was introduced as the limiting case of the ensemble average of an energy spectral density [Eq. (8.91)]

$$S_f(\omega) = \lim_{T \to \infty}\left(E\left\{\frac{|F_T(\omega)|^2}{T}\right\}\right).$$

In this equation, $S_f(\omega)$ is the power spectral density of the signal $f(t)$ having a Fourier transform $F_T(\omega)$ over $(-T/2, T/2)$. It is not practical to take the limit as $T \to \infty$, however, and we wish to examine some of the consequences of using a finite observation interval. Our emphasis here is on the numerical methods to compute power spectral density.

We use the notation $S_T(\omega)$ for the power spectral density estimate of $f(t)$ based on an observation interval of T units. Thus, for a given sample function, Eq. (8.91) becomes

$$S_T(\omega) = \frac{|F_T(\omega)|^2}{T}. \tag{8.92a}$$

The discrete case is often of interest in which the Discrete Fourier Transform (DFT) is used to approximate the Fourier transform based on N discrete data samples. Thus N replaces T and $F_N(\omega_n)$ replaces $F_T(\omega)$, where $F_N(\omega_n)$ is the DFT of $f(kT/N)$, $k = 0, 1, 2, \ldots, (N - 1)$. With these changes, we can rewrite Eq. (8.92a) as

$$S_N(\omega_n) = \frac{|F_N(\omega_n)|^2}{N}. \tag{8.92b}$$

For convenience we shall treat both the continuous and the discrete cases and it will be obvious from the notation which one is being considered.

The spectral estimate given by Eq. (8.92) is often called a *periodogram* because it was first used to search for possible periodicities in data records. Its use in power spectral estimation from numerical data is becoming more popular as a result of the efficiency and convenience of the FFT algorithm for finding $F_N(\omega_n)$.

Because the data are assumed to arise at least partially from random sources (e.g., additive noise), different sample functions $f(k)$, $k = 0, 1, 2, \ldots, (N - 1)$ will result in different values of $S_N(\omega_n)$ for each ω_n and the collection of these values forms an ensemble. A desirable characteristic in the estimator is that the mean of this ensemble should be the true value at that frequency, i.e., that $E\{S_N(\omega_n)\} = S_f(\omega_n)$. If this is true, then we say that the estimator is *unbiased*.

Using Eq. (8.92) and taking the expected value, we have

$$E\{S_T(\omega)\} = E\left\{ \mathcal{F}\left\{ \frac{1}{T} \int_{-\infty}^{\infty} f(t) \text{ rect } (t/T) f(t + \tau) \text{ rect } [(t + \tau)/T] \, dt \right\} \right\}$$

$$= \mathcal{F}\left\{ \frac{1}{T} \int_{-\infty}^{\infty} E\{f(t)f(t + \tau)\} \text{ rect } (t/T) \text{ rect } [(t + \tau)/T] \, dt \right\}$$

$$= \mathcal{F}\left\{ R_{ff}(\tau) \frac{1}{T} \int_{-\infty}^{\infty} \text{ rect } (t/T) \text{ rect } [(t + \tau)/T] \, dt \right\}$$

$$= \mathcal{F}\{R_{ff}(\tau)[1 - |\tau|/T]\}$$

$$= \mathcal{F}\{R_{ff}(\tau)\Lambda(\tau/T)\}. \qquad (8.93a)$$

In the discrete variable notation, this can be rewritten as

$$E\{S_N(\omega_n)\} = \text{DFT}\{R_{ff}(k)\Lambda(k/N)\}. \qquad (8.93b)$$

The second term in the braces in Eq. (8.93) arises as a result of the data truncation and is called a *window* function. As a result of the presence of the window function, the spectral estimator $S_N(\omega_n)$ provides a *biased* estimate of $S_f(\omega)$. Note that as $N \to \infty$ $(T \to \infty)$, $S_N(\omega_n)[S_T(\omega)]$ becomes *asymptotically unbiased*.

The effects of the window function are not unlike those discussed in Chapter 3 for the DFT and can be exhibited by using the frequency convolution property in Eq. (8.93):

$$E\{S_T(\omega)\} = S_f(\omega) \circledast T \text{ Sa}^2 (\omega T/2). \qquad (8.94)$$

The window function limits the frequency resolution to $1/T$ (or $1/N$) and also has some effect on the bias in the spectral estimate.

Another desirable property of an estimator is that the variance of the estimate should become small as T (or N) becomes large. An estimator with this property is said to be *consistent*. To derive a relation for the variance of the spectral estimator involves fourth-order moments and it is difficult to obtain a general result. If the data consists of sample values of a gaussian random process it can be shown that the standard deviation of the spectral-density estimate is on the order of that of the estimate and is not dependent on N.[†] This conclusion also holds approximately for nongaussian data and we conclude that the estimator described by Eq. (8.92) produces an *inconsistent* estimate of the power spectral density.

[†] J. S. Bendat and A. G. Piersol, *Random Data: Analysis and Measurement Procedures*, New York: Wiley-Interscience, 1971.

In practice, the variance of the spectral estimator can be reduced by smoothing. One way is to smooth over an ensemble. Given N data samples, we can use $N' = N/K$ data points for each of K periodograms and then the periodograms are averaged for each frequency. Assuming statistical independence, the variance decreases by the factor K (cf. Example 8.9.2). Note that the frequency resolution has also been decreased by the same factor (that is, $1/N'$ instead of $1/N$).

A second way is to smooth over frequency. In this method a periodogram is computed from the N data samples and then several spectral estimates adjacent to each other in frequency are averaged. This is analogous to low-pass filtering on the periodogram and decreases the variance at the expense of decreased frequency resolution.

In practice, the proper choices for the computation of a power spectral density can be made more easily if some information is known about the characteristics of the spectral density. If little is known, a few trial spectral plots should be formed before the best compromises can be made as to what windowing, averaging, and data record lengths should be used.

Drill Problem 8.14.1   A possible estimate of the autocorrelation function $R_{ff}(\tau)$ based on finite observation data over $(-T/2, T/2)$ is

$$\hat{R}_{ff}(\tau) = \frac{1}{T} \int_{-T/2}^{T/2} f(t)f(t + \tau)\, dt.$$

(a) Is $\hat{R}_{ff}(\tau)$ an unbiased estimator of $R_{ff}(\tau)$? (b) Compare $\hat{R}_{ff}(\tau)$ and $R_{ff}(\tau)$ used in Eq. (8.93).

Answer.   (a) Yes; (b) In $\hat{R}_{ff}(\tau)$ the overlap of $f(t)$ and $f(t + \tau)$ is truncated to $(-T/2, T/2)$; in Eq. (8.93) *each function* is truncated to this interval.

8.15   SUMMARY

Probability is a way to measure the number of favorable outcomes of an experiment relative to the total number of possible outcomes. The probability of an outcome is positive, real-valued, and bounded between 0 and 1. If the occurrence of one outcome precludes the occurrence of another, the two outcomes are mutually exclusive. When the outcomes are mutually exclusive, their probabilities are additive.

The probability of a given outcome based on knowledge of a second outcome is called a conditional probability. Two outcomes are statistically independent if the conditional probability is equal to the probability of occurrence. The joint probability of several outcomes is the probability of their joint occurrence. The joint probability of statistically independent outcomes is equal to the product of the probabilities of the individual outcomes.

A random variable is a rule (mapping) which assigns a number to each possible outcome. A random variable may be continuous or discrete.

The cumulative distribution function of a random variable is the probability of the random variable being less than a specified value (which then forms the argument) for a given experiment. The probability density function (pdf) is the derivative of the cumulative distribution function.

An expected value is a statistical average with a pdf weighting function. The expected value of X^n is called the nth moment of the random variable X. The first and second moments are called the mean and the mean-square value, respectively. The second moment about the mean is called the variance and its square root is the standard deviation.

Examples of continuous random variables include the uniform, Poisson, and gaussian distributions. The binomial distribution is an example of a discrete random variable.

A histogram is a tabulation of the frequency of occurrence of a random variable within preset ranges. The histogram can be used to approximate the pdf of the random variable. The transformation of a random variable through a system can be handled as a mapping of the pdf using the differential slope of the system output-input gain characteristic.

Bivariate probability distributions describe two-dimensional random variables. The statistical dependence of two random variables is measured by the correlation between them. If the correlation coefficient between two random variables is zero, the random variables are said to be uncorrelated.

A random process is a collection or "ensemble" of random variables arranged in a time progression. Each member waveform versus time is called a sample function. The random process is stationary if its statistics do not vary with time. It is ergodic if all its statistics can be computed from a single sample function just as well as from the ensemble.

A joint moment is the expected value of the joint occurrence of two or more outcomes. The autocorrelation function is the second joint moment of two random variables. For a random process, these random variables are specified at two instants of time. If the random process is stationary, only the time difference between them is important.

The power spectral density of a stationary random process is given by the Fourier transform of the autocorrelation function. It may be estimated numerically by computing the squared magnitude of the Discrete Fourier Transform and dividing by the number of data samples.

Selected References for Further Reading

1. E. Parzen. *Modern Probability Theory and its Applications*. New York: John Wiley & Sons, 1960.
   This well-written book is a standard undergraduate textbook on the mathematical theory of probability.

2. P. Beckmann. *Elements of Applied Probability Theory*. New York: Harcourt, Brace & World, 1968.

A less mathematical treatment of probability, random variables, and random processes with a number of interesting examples.

3. G. R. Cooper and C. D. McGillem. *Probabilistic Methods of Signal and System Analysis*. New York: Holt, Rinehart & Winston, 1971.
An introduction to probability written for undergraduate engineering students and stressing applications to engineering problems; additional references are given with comments at the end of the first chapter.

4. P. Z. Peebles, Jr. *Probability, Random Variables, and Random Signal Principles*. New York: McGraw-Hill, 1980.
An introduction to probability written with numerous examples and problems for undergraduate engineering students.

5. W. B. Davenport, Jr. *Probability and Random Processes*. New York: McGraw-Hill, 1970.
A more detailed treatment of probability from a mathematical point of view.

6. R. B. Blackman and J. W. Tukey. *The Measurement of Power Spectra*. New York: Dover Publications, 1958.
Discusses the techniques of estimating power spectra from numerical data. (Also see footnotes in this chapter.)

7. A. Papoulis. *Probability, Random Variables and Stochastic Processes*. New York: McGraw-Hill, 1965.
A well-written graduate treatment of probability with some engineering applications. Chapter 1 has an interesting and easily-read discussion of the various concepts of probability.

Problems

8.1.1 A card is drawn from a deck of 52 cards. What is the probability that the card is (a) a diamond? (b) a two? (c) the two of diamonds?

8.1.2 A number is selected "at random" from the interval [0, 1). What is the probability that in the decimal expansion of the selected number (a) the first digit is odd? (b) the second digit is five? (c) the sum of the first two digits is five or less?

8.1.3 What is the probability that the last digit of a telephone nunber is (a) either odd or divisible by four; (b) either odd or divisible by three? (Assume all digits equally likely.)

8.1.4 A box filled with black and white marbles is known to contain twice as many white marbles as black marbles. Assume they are mixed thoroughly and are indistinguishable except by color. Suppose that twelve (12) marbles are drawn from the box without replacement and that all of them are found to be black. If the probability of drawing one additional black marble is 7/27, how many marbles were in the box originally?

8.2.1 A card is drawn from a 52-card deck and then replaced. The deck is shuffled and a second card is drawn. What is the probability of (a) drawing a 2 both times? (b) drawing the same card twice? (c) first drawing a 2 of clubs and then drawing a 3 of clubs?

8.2.2 A certain transmitter is programmed to shut down if both events A and B (such as certain voltages out of tolerance, etc.) occur simultaneously. The probability that A will occur is 0.001, and the probability of B occurring is 0.002. In addition, it is known that $P(B|A) = 0.01$.

a) What is the probability of transmitter shut-down?
b) Are the events A and B statistically independent?
c) What is $P(A|B)$?

8.2.3 Find the probability of hitting an aircraft if four missiles are fired in succession, each having a probability of 25% of scoring a hit. Assume that the aircraft goes down when hit.

8.2.4 A certain binary PCM system transmits ones and zeros with equal probabilities. However, as a result of noise the receiver makes some errors in recognizing the ones and zeros. Suppose that $P(1|0) = 0.20$ and $P(0|1) = 0.10$.

a) What are the probabilities of ones and zeros at the receiver?
b) If the receiver makes a decision for a one, what is the probability that a one was actually sent?

8.3.1 Consider the experiment which consists in the tossing of four honest coins. Assign 0 to a "tail," 1 to a "head," and then add to form the number N. The random variable X chosen is defined by assigning $X = \log(1 + N)$. Determine and plot the cumulative distribution function of X over $(0, 1)$.

8.3.2 The cumulative distribution function for a certain random variable X is shown in Fig. P–8.3.2. Determine (a) $P(X \le 3)$; (b) $P(1 < X \le 3)$.

Fig. P–8.3.2

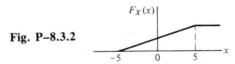

8.3.3 The integer numbers 1–100 are selected at random. The random variable X is defined as

$$X = \begin{cases} 2 & \text{if the number is divisible by 3, but not by 2,} \\ 1 & \text{if the number is divisible by 2,} \\ 0 & \text{if the number is not divisible by 2 or 3.} \end{cases}$$

Determine and plot the cumulative distribution function of X.

8.4.1 Consider the triangular pdf shown in Fig. P–8.4.1. Determine (a) the constant b in terms of a; (b) $P(X > a/2)$; (c) $P[(a/3) < X \le (a/2)]$.

Fig. P–8.4.1

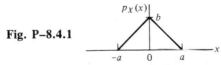

8.4.2 A random number generator produces the digits: $N = 0, 1, 2, \ldots, 9$ uniformly. Plot the pdf of (a) the random variable Y defined by $Y = -1$ if $N \le 3$; $Y = +1$ if $N \ge 7$; $Y = 0$ otherwise; (b) the random variable $Z =$ integer part of $\{N/2\}$.

8.4.3 A certain random variable X has the following pdf: $p_X(x) = K \exp(-|x|)$.

a) What is the required value of K?
b) What is $P(|X| > 1)$?
c) Calculate and sketch the cumulative distribution function of X.

8.4.4 The Rayleigh pdf is $p_X(x) = x \exp(-x^2/2)u(x)$.

a) Determine the cumulative distribution function.
b) What is $P(X \le 2)$?
c) Determine the constant a such that $P(X > a) = 0.01$.

8.4.5 Suppose a certain random variable X has the following cumulative distribution function:

$$F_X(x) = K[x^2 u(x) + x(1 - x)u(x - 1) + (2 - x)u(x - 2)].$$

a) Determine the required numerical value of K.
b) Determine the pdf of X. [*Hint:* First regroup in descending powers of x.]

8.5.1 Determine (a) the mode of the Rayleigh distribution in Problem 8.4.4; (b) under what conditions any two of the following three quantities are the same for a given random variable: mode, median, mean; sketch an example for each case.

8.5.2 Five sampled observations of a random variable X yield the following numbers: {24, 25, 27, 21, 23}. Determine the sample mean and sample variance of X.

8.5.3 A random variable X has a mean m and a variance σ^2. The random variable Y is related to X by $Y = aX + b$, where a, b are constants. Find the mean and variance of Y.

8.5.4 Find the first and second moment of the random variable X whose pdf is shown in Fig. P–8.5.4.

Fig. P–8.5.4

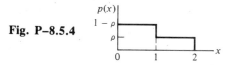

8.5.5 Compute the variance of the random variable described in (a) Problem 8.4.3; (b) Problem 8.4.1.

8.5.6 The standard deviation can be used as a measure of the effective width of a pdf. Show that a comparable result for power spectral density over positive frequencies gives a measure of bandwidth W; that is [cf. Eq. (3.81)]

$$W^2 = \frac{\int_0^\infty (\omega - \omega_0)^2 S_f(\omega)\, d\omega}{\int_0^\infty S_f(\omega)\, d\omega}.$$

In your answer include an expression for the "center frequency," ω_0.

8.6.1 A periodic waveform $x(t)$ is shown in Fig. P–8.6.1.

a) Sketch the cumulative distribution function and probability density function of the amplitude X.

b) Compute the average and rms value using the results of part (a). Compare with the corresponding time-averaged quantities.

c) Repeat part (a) if the peak amplitude were 2 instead of 1.

Fig. P–8.6.1

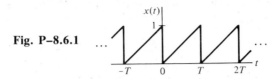

8.6.2 A randomly selected point lies within two concentric circles centered on the origin of radii r_1, r_2, where $r_1 < r_2$. The pdf of the radial distance r of the point from the origin is uniform.

a) Find the expected value of r.

b) Find the expected value of the area enclosed by the circle centered on the origin and passing through the point.

8.6.3 a) Using the binomial distribution, find the probability that there will be exactly four heads when eight honest coins are tossed.

b) What is the probability that there will be fewer than three heads?

8.6.4 A gambler offers to pay even odds (i.e., a one-to-one advantage) if on the throw of 3 honest dice any pair shows. Should you take the bet?

8.6.5 A two-dimensional cross-section of five rows of pins with slots below is shown in Fig. P–8.6.5. Marbles are entered through the funnel at the top. Each time a marble hits a pin, it deflects either to the right or to the left by 45° downward. The probability of either deflection is 50%, regardless of previous route. Determine how many marbles are in each slot, on the average, after 960 marbles have been poured into the funnel.

Fig. P–8.6.5

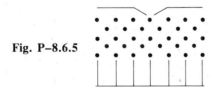

8.6.6 The average time that a customer must wait at a check-out in a given supermarket is found to be one minute. What is the probability that a customer selected at random on entering the store will have to wait at least four minutes at the check-out? (Assume a Poisson process.)

8.6.7 A certain data transmission system has an average error probability of 10^{-6} per digit. What is the probability that there will be more than two errors in a message of 2×10^6 digits?

8.6.8 Determine the constant K and the mean of the gaussian random variable X whose pdf is $p_X(x) = \exp[-(x - 1)^2/3]/\sqrt{12} + K\,\delta(x)$. [*Hint*: Put into the standard form and then find answers without integration.]

8.6.9 A gaussian-distributed signal has zero mean and mean-square value of σ^2. Find the probability of observing the signal amplitude (a) above σ; (b) above σ or below $-\sigma$; (c) within $\pm k\sigma$ of the mean; evaluate for $k = 1, 2, 3$.

8.6.10 A rule-of-thumb measure for acceptable reception of PPM is that the probability of the noise exceeding the pulse amplitude after detection should not exceed 1%. Assuming that the pulses go from 0 to A volts and that the noise is gaussian with zero mean, determine the required peak-signal-to-mean-square noise (power) ratio in dB.

8.6.11 The output noise from a high-gain amplifier is gaussian-distributed with an average value of one volt. An average power of 100 mW is dissipated in a 50-Ω resistive load connected across the output terminals. What is the probability of observing the output noise amplitude less than zero volts?

8.6.12 A manufacturer reports that a certain item is expected to be shipped in six weeks. From past experience, you interpret this to mean that there is a 95% probability of his shipping in six weeks. Using a zero-mean gaussian distribution which is then truncated for positive values only, estimate the probability that the item will be shipped after eight weeks.

8.6.13 In the quality control of manufacturing 100-Ω resistors, the measured resistance values are approximately gaussian-distributed with a mean of 100 and a standard deviation of 5 Ω. The resistors are classed and their net worth is in the ratio 1:2:5 for $\pm 10\%$, $\pm 5\%$, $\pm 1\%$, respectively. Assume that a $\pm 5\%$ resistor is in the range $95 < R < 99$ or $101 < R < 105$, etc., and that resistors for which $R < 90$ or $110 < R$ are rejects and worthless. If the $\pm 5\%$ resistors are sold at cost, determine the average profit on all resistor sales.

c★ 8.7.1 The subroutine library for most digital computers contains a random number generator which produces the numbers x_i uniformly distributed over $(0, 1)$. Compute and plot a histogram over $(0, 1)$; use 100 points and tabulate 100 sample values per point.

c★ 8.7.2 For the x_i in Problem 8.7.1, let $y_i = 1/\sqrt{12N} \sum_{i=1}^{N}(x_i - \frac{1}{2})$. Compute the y_i for $N = 10$ and plot as a 100-pt histogram over $(-0.2, 0.2)$. This plot should resemble the pdf of a gaussian random variable.

★ 8.8.1 The random variable X representing a waveform $x(t)$ has a pdf as shown in Fig. P–8.8.1(a).

a) Find the variance of X.
b) The waveform $x(t)$ described by X is amplified by an amplifier whose output-input gain characteristic is shown in Fig. P–8.8.1(b). Letting Y be the amplifier output, find the pdf of Y.
c) Find the variance of Y.

Fig. P–8.8.1

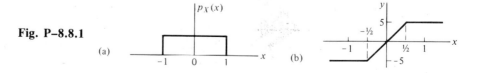

★ **8.8.2** A random variable X, uniformly distributed over (0, 1), is applied to the input of a system which has an output-input gain characteristic $y = 1 - (x - 1)^2$. Determine and sketch the pdf of the output Y.

c★ **8.8.3** For the x_i in Problem 8.7.1, let $y_i = 1 - (x_i - 1)^2$; compute the y_i for $1 \le i \le$ 10,000 and plot as a 100-pt histogram over (0, 1). Compare your plot with the results of Problem 8.8.2.

8.9.1 The joint pdf of the random variables X and Y is

$$p_{XY}(x, y) = a \exp[-(x + y)]u(x)u(y).$$

a) Are X, Y statistically independent?
b) Determine the value of the constant a.
c) Find $P(X < 2, Y < 2)$.
d) Find $p_X(x)$.

8.9.2 The joint pdf of the random variables X, Y is given as

$$p_{XY}(x, y) = x \exp[-x(1 + y)]u(x)u(y).$$

a) Are X, Y statistically independent?
b) Find $p_X(x)$.
c) Compute the probability that $X > 1$.
d) What is the pdf of Y given that $X = 1$?

★ **8.10.1** Show that it is a sufficient condition that X and Y are uncorrelated in Example 8.9.1.

★ **8.10.2** Find the correlation of X, Y in Problem 8.9.2.

★ **8.10.3** Obtain a listing of the total exam scores (Y) and corresponding total homework scores (X) for this course last term. Normalize both scales to 100%. Find the best straight-line fit and calculate the coefficient of determination. Did it pay to do the homework?

c★ **8.10.4** a) Compute the correlation coefficient of successive sample values of the fractional part of $\sqrt{N}$ for $1 \le N \le 1000$.
b) Repeat for the fractional part of $10\sqrt{N}$. Does the removal of the first decimal make the successive values more uncorrelated?
c) What would you expect if $\sqrt{N}$ were used? (You may wish to try it!)

★ **8.11.1** Write out the bivariate gaussian pdf, in as simplified a form as possible, for $m_X = m_Y = 1$; $\sigma_X = \sigma_Y = 1$; $E\{XY\} = 1 + 1/\sqrt{2}$.

★ **8.11.2** a) Using a table of integrals (e.g., Appendix A), show how Eq. (8.79) follows from Eqs. (8.65) and (8.78).
b) Use the result of part (a) to find an expression for $p(x|y)$ [cf. Eq. (8.66)] for the special case where $\rho = 0$. Does $\rho = 0$ imply statistical independence for the gaussian distribution?

8.12.1 Sunspot activity varies over intervals of several hours or more and affects high-frequency communications. This activity is measured and recorded daily. Would you classify this random process as being stationary (a) over a week? (b) over a minute? Explain.

8.12.2 Show how the output of a multichannel analyzer is related to the histogram (and to the pdf) of a random variable X if the input waveform $x(t)$ is ergodic.

8.12.3 Draw a block diagram of a system which will implement the operations depicted in Fig. 8.14 if $x(t)$ is ergodic and if only one threshold device is permitted in the design.

8.12.4 In the random process $z(t) = X \cos \omega_0 t - Y \sin \omega_0 t$, X and Y are independent gaussian random variables each with zero mean and variance σ^2.

a) Determine m_z and σ_z^2.
b) Is your result valid if it is only known that X, Y are uncorrelated and have zero means and the same variance?

8.13.1 Find $R_{zz}(\tau)$ for Problem 8.12.4.

8.13.2 Let $x(t)$ be a stationary random process with an autocorrelation function which is periodic with a period of 2 sec. Over the interval $(-1, 1)$ this autocorrelation function is $R_{xx}(\tau) = (1 - |\tau|)$. Find and sketch the power spectral density of $x(t)$.

8.13.3 White noise with a (two-sided) power spectral density of $\eta/2$ watts per Hz is applied to a RC low-pass filter. Find (a) the power spectral density of the output; (b) the autocorrelation function of the output. (c) Determine the quantity $\tau_0 = \int_0^\infty R(\tau)d\tau/R(0)$, called the "effective correlation time," of the filter output.

8.13.4 A binary communication system transmits equally probable 1's and 0's using the signal A over $(0, T_b)$ for 1 and 0 for 0.

a) Determine the power spectral density of the transmitted waveform.
b) Determine the power spectral density of this binary transmission in the presence of the multipath described in Problem 3.10.2. (Assume $\tau < T_b$.)

★ 8.13.5 A random binary source generating equally probable 1's and 0's with levels A, $-A$ over $(0, T_b)$ is filtered by the transversal filter in Problem 3.10.1. Determine the power spectral density of the output waveform by (a) first finding the auto-correlation function, and then taking a Fourier transform; and (b) first finding the system transfer function, and then multiplying the input power spectral density by the magnitude-squared transfer function.

★ 8.13.6 A binary communication system transmits equally probable 1's and 0's using the waveform $f_1(t) = \cos(\pi t/T_b)$ over $(-T_b/2, T_b/2)$ for 1 and $f_2(t) = -f_1(t)$ for 0.

a) Compute the power spectral density of the transmitted waveform. [*Hint*: see Problem 3.6.5.]
b) Repeat if $f_2(t) = 0$.

c★ 8.14.1 a) Run a 256-point power spectral density program for the uniform random number generator in your computer library; plot over 0–100.
b) Perform frequency smoothing by taking each point in the power spectral density and replacing it with the arithmetic average of that point and one point to either side. Repeat this smoothing operation three times and plot the spectral density.

c★ 8.14.2 Repeat Problem 8.14.1 for the random number generator of Drill Problem 8.7.1.

CHAPTER 9

INFORMATION AND DIGITAL TRANSMISSION

Equipped with some knowledge of probability, we now return to a discussion of pulse code modulation (PCM). Of particular importance is the performance of PCM in the presence of noise and how it compares not only with other modulation systems but to the ideal communication system whose performance is only restricted by band-limited noise.

9.1 A MEASURE OF INFORMATION

The purpose of a communication system is, in the broadest sense, the transmission of information from one point in space and time to another. In preceding chapters we have described several ways of accomplishing this goal by use of electrical signals. We have not, however, really specified what we mean by the term *information* even though we have referred to it and such related topics as messages, bandwidth, etc. In this and the two succeeding sections we shall briefly develop basic ideas about what information is, how it can be measured, and how these ideas relate to bandwidth, signal-to-noise ratio, etc. A development of these ideas becomes necessary in order to make meaningful comparisons of PCM with analog and analog pulse-modulation systems in the presence of noise.

The amount of information about an event is closely related to its probability of occurrence. Messages containing knowledge of a high probability of occurrence (i.e., those indicating very little uncertainty in the outcome) convey relatively little information. In contrast, those messages containing knowledge of a low probability of occurrence convey relatively large amounts of information. Thus a measure of the information received from the knowledge of occurrence of an event is inversely related to the probability of its occurrence.

To formulate a mathematical equation following the above reasoning, we note that if an event is certain (i.e., one occurring with probability one), it conveys zero information. In the other extreme, if an event is impossible (i.e., one occurring with probability zero), its occurrence conveys an infinite amount of information. We also desire our measure of information to be real-valued and monotonic, and to be additive for events which are statistically independent. Based on these considerations, the measure of information associated with an event A occurring with probability P_A is defined as[†]

$$I_A = \log \frac{1}{P_A} . \tag{9.1}$$

The choice of base of the logarithm in Eq. (9.1) has purposely been left undefined at this time and will be considered in the following discussion.

Consider two equiprobable events, A and B, and a binary system to transmit the knowledge of their occurrence. A minimum of one binary pulse or signal is required, and this basic quantum unit of information is called a *bit*. The case of four equiprobable events requires four distinct binary pulse patterns, or two bits, while eight equiprobable events require three bits, etc. In general, any one of n equiprobable messages then contains $\log_2 n$ bits of information. Because we have assumed all n messages to be equiprobable, the probability of occurrence of each one is $P_i = 1/n$ and the information associated with each message is then

$$I_i = \log_2 n, \tag{9.2}$$

$$I_i = \log_2 \frac{1}{P_i} \qquad \text{bits.}$$

Equation (9.2) confirms what we had in Eq. (9.1) and supplies the units in terms of the base 2 logarithm. Even if the number of possibilities is not a power of two, we can use Eq. (9.2) to express the information in bits, although the answer will have a fractional value. For numerical calculations it is often convenient to make use of the base 10 logarithm and then use the conversion,

$$\log_2 N = (\log_{10} N)(\log_2 10) = (\log_{10} N)/\log_{10} 2,$$

$$\log_2 N \cong 3.322 \log_{10} N. \tag{9.3}$$

The above results define our measure of information for the somewhat special case in which all messages are equally likely. To generalize, we define an *average information,* which is called the *entropy H,* of each message by taking the expected value of Eq. (9.2),[‡]

$$H \triangleq I_{\text{avg}} = E\{I_i\},$$

[†] We have not proved that Eq. (9.1) follows from these conditions but have only stated it. Proof can be found in references on the subject (see, for example, R. Ash, *Information Theory,* New York: Wiley, 1965, Ch. 1).

[‡] The name "entropy" and its symbol H are borrowed from a similar equation in statistical mechanics.

or

$$H = \sum_{i=1}^{n} P_i \log_2 \frac{1}{P_i} \quad \text{bits.} \tag{9.4}$$

Thus each event or message is assigned an average information content even though the information for each event may fluctuate considerably.

Now we encounter a problem in terminology. Previously we have always referred to binary pulses as bits but now we are defining the bit as a measure of information content. For equiprobable events, the two definitions coincide but this is admittedly a special case. One way to avoid this problem is by making use of the contraction *binit* for *binary digit* to distinguish it from the bit (which then is reserved for a measure of information content). In fact, we can easily show that the bit is always equal to or less than the binit because from probability theory we can write

$$E\{I_i\} \leq I_i|_{\max},$$

or,

$$H \leq I_i|_{\max}. \tag{9.5}$$

The equality holds if all events or messages are equiprobable [this can be shown by using Eqs. (9.2) and (9.4) in Eq. (9.5)]. In other words, if all messages are equiprobable, then one binit is equal to one bit. This is an appealing approach but the use of the term *bit* is so firmly entrenched in both senses that we bow to convention and use only the term bit for both meanings. Thus the term bit is used *both* as a binary digit and as a measure of information and we must rely on the context to determine which meaning is applicable. The saving fact is that both definitions coincide for the equally likely case.

Example 9.1.1 The four symbols A, B, C, D occur with probabilities 1/2, 1/4, 1/8, 1/8, respectively. Compute the information in the three-symbol message $X = BDA$, assuming that the symbols are statistically independent.†

Solution. Because the symbols are independent, the measure of information is additive and we can write

$$I_X = \log_2 4 + \log_2 8 + \log_2 2,$$

$$I_X = 2 + 3 + 1,$$

$$I_X = 6 \text{ bits.}$$

---

† In construction of languages, one is constrained by certain rules of spelling and grammar which limit the choice in successive symbols. Therefore the choice of symbols is no longer completely independent. This can be accounted for by use of conditional probabilities and hence conditional entropy, topics which are interesting in their own right but are not covered here.

Example 9.1.2   Determine and graph the entropy (i.e., average information) of the binary code in which the probability of occurrence of the two symbols is p and $q = (1 - p)$.

Solution.   Using Eq. (9.4), we have

$$H = \sum_{i=1}^{2} P_i \log_2(1/P_i),$$

$$H = p \log_2(1/p) + (1 - p) \log_2[1/(1 - p)].$$

A plot of H vs. p is shown in Fig. 9.1. Note that the maximum entropy is one bit/symbol and occurs for the equiprobable case (i.e., $p = q = 1/2$)—cf. Eq. (9.5).

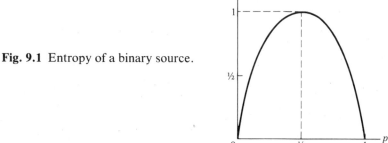

Fig. 9.1 Entropy of a binary source.

Drill Problem 9.1.1   Determine the average information (entropy) per symbol for (a) the four-symbol code in Example 9.1.1; (b) the case in which four symbols are equiprobable.

Answer.   (a) 1.75 bits; (b) 2 bits.

Drill Problem 9.1.2   An efficient binary code for Example 9.1.1 is $A = 0$, $B = 10$, $C = 110$, $D = 111$. By weighting each with the appropriate probability of occurrence, determine (a) the probability of occurrence of 0's in the code; (b) the average information per codeword.

Answer.   (a) 1/2; (b) 1.75 bits.

The above concepts can be extended to the case of continuous messages, but the mathematics become quite involved and abstract. We shall prefer instead to use the discrete approach and employ the sampling theorem and quantization to reduce continuous analog messages to discrete ones.

In electrical communication systems we are particularly concerned with the rate of information transmission. We define the information-transmission rate R by

$$R = rH \text{ bits per second,} \tag{9.6}$$

where r is the symbol rate (corresponding to the symbols over which H is determined).

Although it is not generally within the realm of a communication system design to specify message probabilities (these are usually either known or can be approximated by measurements), a major concern is the maximization of the information-transmission rate of the system. This will, in fact, form the subject of the next section.

Drill Problem 9.1.3 A limit on the information-transmission rate R in the presence of thermal noise and assuming no bandwidth restrictions can be expressed as†

$$R_{\max} = \frac{1.44 P_r}{kT_{eq}} \text{ bits per second,}$$

where P_r is the received power in watts, k is the Boltzmann constant, and T_{eq} is the equivalent receiver noise temperature in Kelvin. Calculate the minimum transmitter power required to send a standard black-and-white television picture with an information content of 10^6 bits from a deep-space probe within a time interval of one hour if the attenuation losses are 200 dB and $T_{eq} = 50$ K.

Answer. 13.3 watts.

We conclude this section with a word of caution. The development of the concepts of information and maximum information-transmission rate have proceeded strictly from an engineering viewpoint. Thus the message content has been judged from its quality of uncertainty, reducing the transmission problem to a highly systematic basis. However, the validity of this approach may be open to question when we attempt to apply it to problems involving human behavior. The problem is not so much the concept of a measure of information but how it relates to human value and decision processes and what criteria are involved. For example, the artistic and aesthetic senses are not necessarily best served (if at all) by the criterion of a most efficient transfer of information. Failure to recognize some of these and other related factors has resulted in not a few engineers striving to milk the very last ounce of performance out of a communication system using information theory only to find that no one could really use the output data in the first place!

---

† This relation is developed in Example 9.2.3.

9.2   CHANNEL CAPACITY

Previously we have referred to the link between transmitter and receiver, including noise sources, as the transmission medium. In information theory it is convenient to treat it as a more abstract and general noisy filter (which may be nonlinear, time-varying, etc.) called the *channel*. The limiting rate of information transmission through a channel is called the *channel capacity*.

Importance of the concept of channel capacity stems from a theorem first stated and proved by C. E. Shannon. Shannon's theorem states that if the entropy rate R is equal to or less than the channel capacity C, then there exists a coding technique which enables transmission over the channel with an arbitrarily small frequency of errors, or

$$R \leq C. \tag{9.7}$$

This restriction holds even in the presence of noise in the channel. A converse to this theorem states that it is not possible to transmit messages without errors if $R > C$. Thus the channel capacity is defined as the maximum rate of reliable information transmission through the channel.

Now consider a source with an available alphabet of α discrete messages. We assume that each message sent can be identified at the receiver; therefore this case is often called the "discrete noiseless channel." The maximum entropy of the source is $\log_2 \alpha$ bits, and if T is the transmission time of each message, the channel capacity is†

$$C = \frac{1}{T} \log_2 \alpha \text{ bits per second}. \tag{9.8}$$

To attain this maximum, the messages must be equiprobable and statistically independent. These conditions form a basis for the coding of the information to be transmitted over the channel.

In the presence of noise, the capacity of the discrete channel decreases as a result of the errors made in transmission. This may be accounted for by subtracting the entropy rate of the symbols which were erroneously detected from the channel capacity of the discrete noiseless channel.

Example 9.2.1   A binary source sends α equiprobable and identifiable messages using the binary symbols 0, 1 at a symbol rate r over a time T. However, as a result of noise, a 0 may be mistaken for a 1 and vice versa. For the case known as the "binary symmetric channel," the probabilities of making these two types

---

† For unequal message transmission times,

$$C = \lim_{T \to \infty} \frac{1}{T} \log_2 \alpha(T).$$

of errors are equal and are designated here by p. Determine the resulting channel capacity.

Solution.   The number of equiprobable messages is $\alpha = 2^{rT}$. Using this in Eq. (9.8), we have

$$C = \frac{1}{T} \log_2 2^{rT} = r.$$

In the presence of noise, errors will be made at the entropy rate R_e which, from the results of Example 9.1.2, is

$$R_e = rH_e = rp \log_2(1/p) + r(1 - p) \log_2[1/(1 - p)].$$

The channel capacity is then

$$C = r - R_e$$

$$= r[1 + p \log_2 p + (1 - p) \log_2(1 - p)].$$

Note that for $p = 0.5$, $C = 0$ and no information can be transmitted. For $p = 0.001$, $C = 0.989r$ so that C is about 989 bps for r of 1000 symbols per second.

In making comparisons between various types of communication systems, it is convenient to consider a channel which is described in terms of bandwidth and signal-to-noise ratio. Another theorem in information theory, known as the Hartley-Shannon theorem, states that the channel capacity of a white band-limited gaussian channel is

$$C = B \log_2(1 + S/N) \text{ bits per second}, \qquad (9.9)$$

where B is the channel bandwidth and S/N is the mean-square signal-to-noise ratio. Although restricted to the gaussian case, this result is of widespread importance to communication systems because many channels can be modeled by the gaussian channel. It is applicable to both discrete and continuous channels.

A proof of the Hartley-Shannon theorem is not easy and is quite involved. We shall follow an intuitive explanation based on the above reasoning for the discrete channel.

Consider a signal which is band-limited to B Hz. Using the sampling theorem, we can describe this signal by its samples taken at the rate

$$r = 2B. \qquad (9.10)$$

How many levels shall we use to describe each sample in the presence of noise? A plausible choice is one in which each discernible level is spaced by one standard deviation of the noise. For zero-mean noise with power N, this gives level spacings of $\sigma_n = \sqrt{N}$. Using the rms value of the observed signal-plus-noise, $y(t)$, for the range of possible signals, we can formulate the number of discernible levels M as

$$M = \sigma_y/\sigma_n = \frac{\sqrt{S + N}}{\sqrt{N}} = \sqrt{1 + S/N}. \tag{9.11}$$

The channel capacity over $(0, T)$ is, from Eq. (9.8),

$$C = \frac{1}{T} \log_2 M^{rT} = r \log_2 M. \tag{9.12}$$

Use of Eqs. (9.10) and (9.11) in Eq. (9.12) gives

$$C = 2B \log_2 \sqrt{1 + S/N}$$

$$= B \log_2(1 + S/N).$$

An application of this result reveals that if $S/N = 1$, then $C = B$; in other words, the channel capacity in bps is equal to the channel bandwidth in Hz. However, if $S/N = 3$, then $C = 2B$, if $S/N = 7$, then $C = 3B$, etc. Thus the channel capacity increases as the S/N increases. Alternatively, for a fixed channel capacity, the bandwidth B can be reduced in exchange for an increase in signal-to-noise ratio S/N. In the ideal case, the results indicate that this trade-off is approximately exponential.

Example 9.2.2   A black-and-white television picture may be considered as composed of approximately 3×10^5 picture elements. Assume that each picture element is equiprobable among 10 distinguishable brightness levels. Thirty (30) picture frames are transmitted per second. Calculate the minimum bandwidth required to transmit the video signal assuming that a 30-dB signal-to-noise ratio is necessary for satisfactory picture reproduction.

Solution

Information per picture element $= \log_2 10 = 3.32$ bits.
Information per picture frame $= (3.32)(3 \times 10^5) \doteq 9.96 \times 10^5$ bits.
Information rate $R = (30)(9.96 \times 10^5) = 29.9 \times 10^6$ bps.

Because R must be less than or equal to C, we let $R = C = 29.9 \times 10^6$ and Eq. (9.9) gives

$$B_{min} = \frac{C}{\log_2(1 + S/N)},$$

$$B_{min} = \frac{29.9 \times 10^6}{(3.32)(3.0004)} \approx 3 \text{ MHz}.$$

As noted in Appendix D, commercial television transmissions actually use a video bandwidth of 4 MHz.

Drill Problem 9.2.1   What is the minimum time required for the facsimile transmission of one picture over a standard telephone circuit? There are about 2.25 $\times 10^6$ picture elements to be transmitted and 12 brightness levels are to be used for good reproduction. Assume all brightness levels are equiprobable. The telephone circuit has a 3-kHz bandwidth and a 30-dB signal-to-noise ratio (these are typical parameters).

Answer.   4.5 min.

Although the potential signal-to-noise trade-off with increased bandwidth is exponential in PCM, its performance in conveying information still falls short of that predicted by the Hartley-Shannon law. One reason for this is that the Hartley-Shannon law assumed no limitation on the number of pulse levels except that constrained by the signal-to-noise ratio of the channel. Thus such a system would use more levels as the signal-to-noise ratio improved. In contrast, practical PCM systems make use of a (fixed) finite number of pulse levels and their maximum information transmission is limited by this fact.

A second reason is that PCM, like the other wideband communication systems we have examined, suffers from a threshold effect. Therefore, an increase in the minimum signal-to-noise, on the order of 6–8 dB, is required for significant signal-to-noise improvement. We shall consider this threshold effect in a later section when the probability of error for PCM is discussed.

Example 9.2.3   Starting with the Hartley-Shannon law, derive a relation for the channel capacity in the presence of white noise as the bandwidth is increased without limit.

Solution.   For white noise with power spectral density $\eta/2$, the Hartley-Shannon law is

$$C = B \log_2 \left(1 + \frac{S}{\eta B} \right)$$

$$= \frac{S}{\eta} \left(\frac{\eta B}{S} \right) \log_2 \left(1 + \frac{S}{\eta B} \right),$$

so that we wish to investigate the limit,

$$\lim_{B \to \infty} \left[\frac{S}{\eta} \log_2 \left(1 + \frac{S}{\eta B} \right)^{\eta B/S} \right].$$

However, we can also write

$$\lim_{x \to 0} (1 + x)^{1/x} = e,$$

and, identifying x as $x = S/(\eta B)$, we have

$$\lim_{B \to \infty} C = \frac{S}{\eta} \log_2 e,$$

or

$$\lim_{B \to \infty} C = 1.44 \frac{S}{\eta} = \frac{S}{0.69\eta}.$$

This is an interesting result because it gives the maximum information-transmission rate possible for a system of given power but no bandwidth limitations. This result is useful in the design of minimum-power systems, such as satellite and space probe communication systems, for which the bandwidth is not specified or is a secondary consideration. The power spectral density can be specified in terms of equivalent noise temperature by $\eta = kT_{eq}$.

9.3 IDEAL DEMODULATOR DETECTION GAIN

The Hartley-Shannon law provides us with a measure of the maximum information-transmission rate over a band-limited channel in the presence of white gaussian noise. But it also provides us with the relationship between bandwidth and signal-to-noise ratio for an optimum system. And because it holds for continuous as well as discrete channels, we can apply the results to both to make comparisons between types of systems.

Consider a message signal which is band-limited to f_m Hz having an information content of I bits. Before transmission of this information we may choose to use a modulator (or encoder). If performed correctly, this process of modulation (encoding) will not alter the information content of the message but is designed to combat the degrading effects of the channel. Let the transmitted bandwidth be B Hz and assume that the information-transmission rate R is matched to the channel so that $R = C$, where C is the channel capacity. At the receiver, the transmitted signal is received within the bandwidth B in the presence of white gaussian noise, so we can write

$$C = B \log_2(1 + S_i/N_i). \tag{9.13}$$

Here S_i/N_i designates the signal-to-noise ratio at the input of the receiver.

In the receiver, the modulated signal is applied to the input of an ideal demodulator (see Fig. 9.2). There is no loss of information in the ideal demodulator and therefore the information-transmission rate at the output is identical to that at the input (assuming no time scaling). Thus we can also write

$$C = f_m \log_2(1 + S_o/N_o), \tag{9.14}$$

where S_o/N_o is the signal-to-noise ratio at the receiver output.

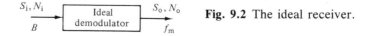

Fig. 9.2 The ideal receiver.

Setting the right-hand sides of Eqs. (9.13) and (9.14) equal, we have

$$B \log_2(1 + S_i/N_i) = f_m \log_2(1 + S_o/N_o),$$

or

$$(1 + S_o/N_o) = (1 + S_i/N_i)^{B/f_m}. \tag{9.15}$$

For S_i/N_i, $S_o/N_o \gg 1$, we obtain

$$\frac{S_o}{N_o} \approx \left(\frac{S_i}{N_i}\right)^{B/f_m}. \tag{9.16}$$

We conclude that in the optimum system the receiver output signal-to-noise ratio increases *exponentially* with the transmitted bandwidth B. In contrast, recall that for angle modulation (FM, PM) and pulse-timing modulation (PPM, PWM) the signal-to-noise (power) ratio increases proportional to the square of the bandwidth. Although superior in this respect to AM and PAM, which offer no signal-to-noise vs. bandwidth advantage, such systems are far from optimum when $B \gg f_m$. As we shall see later, the signal-to-noise improvement in PCM systems also falls short of the optimum trade-offs.

9.4 QUANTIZATION NOISE

In PCM, the information to be transmitted is contained in the code. Noise is added in the transmission and this may cause errors in recognition of the code symbols sent. But if the signal pulses are received above a certain signal-to-noise threshold (which we will discuss later in this chapter), the average error rate can be kept very low.

However, in order to encode a continuous signal we must first quantize it into a finite number of discrete amplitude levels. Once quantized, the instantaneous values of the continuous signal can never be reconstructed exactly. The random errors introduced are called *quantization noise*. Quantization noise, in contrast to transmission noise, is artificially generated by the quantization process prior to being transmitted. It can be reduced to any desired degree by choosing the number and distribution of the quantizing levels.

To calculate the mean-square quantization noise, we assume equal amplitude increments between levels. Specifically, let the input signal be quantized into n levels, each spaced by an amplitude increment, a. (For binary PCM, n is chosen to be a power of 2.) If we assume that the signal amplitude is bipolar and there is no dc level to offset the amplitudes, a reasonable distribution of the quantizing levels is $\pm a/2$, $\pm 3a/2$, $\pm 5a/2$, ... $\pm (n - 1)a/2$ (see Fig. 9.3).

Assuming that the number of signal levels is even and that all signal levels are equiprobable, we find the probability density function of the levels is

$$p(x) = \sum_{\substack{i=-n \\ \text{odd}}}^{n} \left(\frac{1}{n}\right) \delta(x - ia/2). \tag{9.17}$$

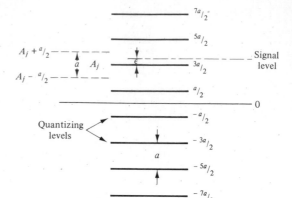

Fig. 9.3 Quantization uncertainty for continuous input.

The mean-square signal after quantization is

$$S = \int_{-\infty}^{\infty} x^2 p(x)\, dx$$

$$= 2 \int_{0}^{\infty} x^2 p(x)\, dx$$

$$= \frac{2}{n} \{(a/2)^2 + (3a/2)^2 + \cdots + [(n-1)a/2]^2\}$$

$$= \frac{2}{n}\, \frac{n(n-1)(n+1)}{6}\, (a/2)^2$$

$$S = \frac{n^2 - 1}{12}\, a^2. \tag{9.18}$$

The same method can be applied if not all levels are equiprobable, but the result corresponding to Eq. (9.18) cannot be written in as compact a form.

In the quantization process, each sample of the continuous input signal is approximated to the nearest allowed level. This is shown in Fig. 9.3, where the input signal level is ϵ units from the nearest quantizing level A_j. At the eventual receiver output, the quantized level A_j could have been due to any signal amplitude in the range $A_j - a/2$ to $A_j + a/2$. This uncertainty could just as well have been due to additive noise and therefore we treat it as an additive noise source.

The quantizing error ϵ is the difference between the signal level and the nearest allowed quantizing level, as shown in Fig. 9.3. Lacking any information to the contrary, we assume that all values of ϵ are equally likely anywhere in the range $-a/2 \le \epsilon < a/2$ so that it can be described by the uniform probability density function,

$$p(\epsilon) = \begin{cases} \dfrac{1}{a} & -a/2 \leq \epsilon < a/2, \\ 0 & \text{elsewhere.} \end{cases} \tag{9.19}$$

Actually, the result is not highly dependent on this assumption as long as the probability density function is not strongly peaked somewhere in the interval (see, for example, Drill Problem 9.4.2). The mean-square quantization noise now follows readily:

$$\overline{\epsilon^2} = \int_{-\infty}^{\infty} \epsilon^2 p(\epsilon) \, d\epsilon$$

$$= \int_{-a/2}^{a/2} \frac{1}{a} \epsilon^2 \, d\epsilon$$

$$\overline{\epsilon^2} = \frac{a^2}{12}. \tag{9.20}$$

Because the peak signal excursion (range) is $na/2$, the peak signal-to-noise (power) ratio is

$$\left(\frac{S}{N}\right)_{\text{pk qnt}} = \frac{(na/2)^2}{a^2/12} = 3n^2. \tag{9.21}$$

Expressed in decibels, this becomes

$$\left[\left(\frac{S}{N}\right)_{\text{pk qnt}}\right]_{\text{dB}} = 4.8 + 20 \log_{10} n. \tag{9.22a}$$

For a binary code, we have $n = 2^m$ so that, for the binary case, Eq. (9.22a) becomes

$$\left[\left(\frac{S}{N}\right)_{\text{pk qnt}}\right]_{\text{dB}} = 4.8 + 6m \qquad \text{(binary system).} \tag{9.22b}$$

Thus the peak-signal–to–rms-quantization-noise ratio increases by 6 dB for every additional bit used in a binary system.

For the case in which all signal levels are equiprobable, the mean-square signal is given by Eq. (9.18); dividing this by Eq. (9.20), we get

$$\left(\frac{S}{N}\right)_{\text{qnt}} = n^2 - 1 \approx n^2. \tag{9.23}$$

The performance quality, on a mean-square basis, increases as the square of the number of levels used and differs only by a constant from the peak S/N relation given in Eq. (9.21). In both cases, the signal-to-quantization-noise ratio increases about 6 dB for each additional bit used. However it is difficult to assign a criterion for performance in general. For speech, experimental results indicate that 8 to 16 levels are adequate for intelligibility, $n = 128$ to $n = 256$ (i.e., $m = 7$ to $m = 8$) is standard for voice telephony, while as many as 4096 ($m = 12$) levels are used for music-quality audio.†

The above calculation assumed that all signal levels were equiprobable. However, this may not be the case (and often is not in practice). This results in some increments in signal amplitude having a poorer signal-to-quantization noise than others. For example, experiment indicates that typical instantaneous-speech amplitudes are less than one-fourth the rms value for 50% of the time.‡ Thus in the encoding of speech the small-amplitude signals, which occur most of the time, have relatively large quantization noise resulting in a rather poor overall signal-to-quantization-noise ratio.

A remedy for this situation is to use a nonlinear (i.e., nonuniform) quantizer which has tapered quantization steps (e.g., see Fig. 7.23). A method often used in practice to achieve an equivalent effect is to weight the input signal before quantizing with a linear (uniform) quantizer. An inverse weighting is applied at the receiver so that the overall transmission is distortionless.

The input-output characteristic of a nonlinear predistortion system which could be used for signals such as speech is shown in Fig. 9.4. The effect of the nonlinear characteristic is to compress the larger signal amplitudes and therefore

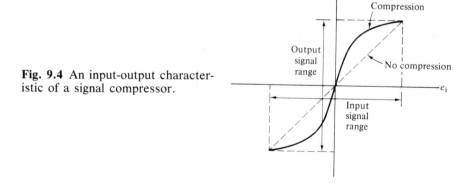

Fig. 9.4 An input-output characteristic of a signal compressor.

---

† The passenger entertainment systems in the Boeing 747 and Lockheed LM1011 aircraft use 16-channel 12-bit PCM time-multiplexed onto a single wire pair. Fourteen of these form a 7-channel stereo system.

‡ J. A. Betts, *Signal Processing, Modulation and Noise,* New York: American Elsevier Publishing Co., 1970, Ch. 7.

such a system is called a *compressor*.† The inverse operation at the receiver is called an *expander*. The combination of a compressor and an expander is called a *compander*. Companding systems are often synthesized using active circuits whose gains are controlled by biased diode networks.‡ By this means, one can approximate the required nonlinear characteristic with a piecewise linear approximation.

Example 9.4.1 Consider a continuous input signal having the following probability density function:

$$p(x) = \begin{cases} 1 - |x| & -1 \le x \le +1, \\ 0 & \text{elsewhere.} \end{cases}$$

a) Determine the quantizer step size and levels if a linear (uniform) 8-level ($n = 8$) quantizer is used.

b) Determine the nonlinear quantizer step levels required to make the quantized signal levels equiprobable. Also plot the required compressor characteristic to precede a linear quantizer.

Solution

a) The net signal range is 2 so that the level step size, a, is $a = 2/n = \frac{2}{8} = \frac{1}{4}$. The quantizer levels are then set at $\pm\frac{1}{8}$, $+\frac{3}{8}$, $\pm\frac{5}{8}$, $\pm\frac{7}{8}$, and the quantizer round-off boundaries are set at 0, $\pm\frac{1}{4}$, $\pm\frac{1}{2}$, $+\frac{3}{4}$, ± 1.

b) As a result of the symmetry, we consider the case only for $x > 0$. To make all signal levels equiprobable we set the quantizer step levels such that the area under $p(x)$ is divided equally (see Fig. 9.5a). Designating the quantizer steps by a, b, c, d, respectively, we have

$$\int_0^a (1 - x)\, dx = \int_d^1 (1 - x)\, dx = \tfrac{1}{16},$$

and

$$\int_a^b (1 - x)\, dx = \int_b^c (1 - x)\, dx = \int_c^d (1 - x)\, dx = \tfrac{1}{8}.$$

---

† This type of signal compression is sometimes used in AM systems for voice communications to effectively obtain the effects of a higher-modulation index without exceeding the peak-modulation limit. A logarithmic taper is generally preferred for speech applications.

‡ In PCM telephone systems in the U.S., the companding law used prior to transmission is $e_o = [\log (1 + 255e_i)]/[\log (256)]$, where both e_i, e_o are normalized to a $(0, 1)$ range. This is approximated using 15 linear segments in the gain characteristic of an amplifier.

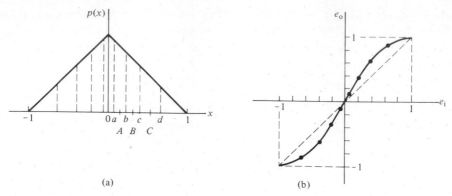

Fig. 9.5 The probability density function and compressor characteristic for Example 9.4.1.

Solving in succession, we get the quantizer steps,

$$a = 1 - (\sqrt{7/2})/2 = 0.065,$$

$$b = 1 - (\sqrt{5/2})/2 = 0.209,$$

$$c = 1 - (\sqrt{3/2})/2 = 0.388,$$

$$d = 1 - (\sqrt{1/2})/2 = 0.646.$$

The corresponding compressor characteristic to precede a linear (uniform) quantizer to achieve the same results can be found by plotting the points $(0.065, 0.125)$, $(0.209, 0.375)$, etc. This is shown in Fig. 9.5(b). Similarly, the quantizer round-off boundaries are 0, $\pm A$, $\pm B$, $\pm C$, ± 1, where A, B, C are determined from

$$\int_0^A (1 - x)\, dx = \int_A^B (1 - x)\, dx = \int_B^C (1 - x)\, dx = \int_C^1 (1 - x)\, dx = \tfrac{1}{8},$$

or

$$A = 1 - \sqrt{3}/2 = 0.134,$$

$$B = 1 - \sqrt{2}/2 = 0.293,$$

$$C = 1 - 1/2 = 0.500.$$

Drill Problem 9.4.1   Compressor characteristics determined by a mapping to a uniform probability density, as illustrated in Example 9.4.1, may tend to compress too much if there are large differences in $p(x)$ from one region to another. This can be corrected by using the area under $[p(x)]^{1/3}$ instead of $p(x)$ between increments.† Recompute the quantizer points for Example 9.4.1 using this method (in this case, there are no large differences between adjacent regions so the results may be expected to be similar to those in Example 9.4.1).

Answer.   0.063, 0.194, 0.333, 0.481.

Drill Problem 9.4.2   (a) Calculate the mean-square quantization noise for the signal described by the pdf of Example 9.4.1 if there were only two quantization levels (± 0.5). Do not assume that the pdf of the quantization error is constant over each level. (b) Calculate the mean-square quantization noise using Eq. (9.20), and compare.

Answer. $\frac{1}{12}$ (both cases).

The method of choosing quantization steps suggested in Example 9.4.1 is relatively easy to apply. It does not, however, guarantee optimum results even though the results may be satisfactory for many applications. A more fundamental method is to choose the quantization points in such a way as to minimize the quantization error in a mean-square sense. A solution to this problem can be found in a manner similar to that used for the orthogonal signal representations in Section 2.5.‡ However, the limits of integration are dependent upon the quantization points, and an iterative numerical procedure must be used to find a solution (cf. Problem 9.4.6).

9.5 PROBABILITY OF ERROR IN TRANSMISSION

We now devote some attention to the details of transmitting digitally encoded messages and begin by considering the use of the type of digital signals which are intended for direct transmission over a line. These are called *baseband* signals. The case of modulated pulses intended for transmission via radio-frequency propagation is considered in Chapter 10.

In a binary PCM system, the two pulse levels may be represented as an on-off signal—i.e., by an amplitude A and an amplitude 0 as the two possible levels—or by equal-amplitude pulses of opposite polarities—i.e., by amplitudes $-A/2$ and $+A/2$. The latter is known as a *polar* binary signal. The on-off signal results in a nonzero average level which is dependent on the relative proportions of ones and zeros being transmitted. In a random sequence of on-off pulses the slight variation in average level gives rise to a line spectrum in addition to the

---

† See, for example, J. J. Spilker, Jr., *Digital Communications by Satellite*, Englewood Cliffs, NJ: Prentice-Hall, Inc., 1977, p. 45.

‡ See, for example, F. Haber, *An Introduction to Information and Communication Theory* (Advances in Modern Engineering Series, vol. 4), Reading, Mass.: Addison-Wesley, 1974, Ch. 6.

continuous frequency spectrum of the information. The spectral component at the bit rate can be used for receiver synchronization. On the other hand, the polar signal representation requires less transmitted power for a specified error rate in the presence of noise. Its use also makes setting the receiver decision threshold easier if the relative frequencies of occurrence of both binary signals are equal. Both of these topics will be discussed later in this section.

A block diagram of an overall system for baseband communication is shown in Fig. 9.6. Because the bandwidth of the transmission line is severely limited by wire capacitance and inductance, line equalization is usually employed at regular intervals along the line. This counteracts the capacitive and inductive effects and widens the usable bandwidth of the line. Even with equalization, however, the transmission frequency response does not closely approximate that of an ideal low-pass filter; and the methods of pulse shaping discussed in Section 7.3 are employed to minimize the intersymbol interference (ISI).

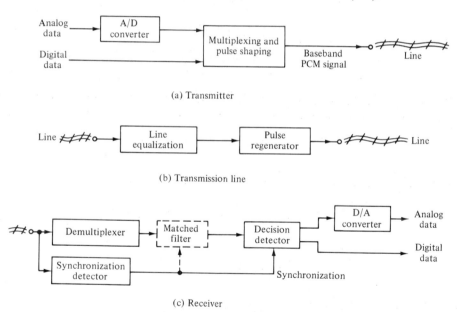

Fig. 9.6 A baseband PCM communication system.

Note that in using the methods for pulse shaping to minimize ISI, we are more concerned with combating the bandwidth limitations of the transmission line than with the effects of transmission noise. The matched filter, which was discussed in Chapter 7, can be employed to maximize the signal-to-noise ratio in the receiver. However, the matched filter dictates a different type of pulse-waveform design. In the design of radio systems handling digital signals, both of these types of requirements need to be considered and appropriate compro-

45

mises can be made. In line systems the signal-to-noise ratio is generally sufficiently high so that the matched filter may be deleted. Instead, attention is
concentrated more on widening the effective bandwidth using equalization to
gain higher data rates and on minimizing the intersymbol interference by appropriate waveform shaping.

We begin our discussion of the probability of error by not using a matched
filter receiver. Assume that the binary levels at the receiver are 0 and A; i.e.,
the received signal $f(t)$ is either at the level 0 or at the level A. The observed
waveform $y(t)$ at the receiver also contains additive noise $n(t)$ so that

$$y(t) = f(t) + n(t). \tag{9.24}$$

At a given time $t = t_1$, the possible receiver inputs are either

$$y(t_1) = A + n(t_1) \quad \text{(signal present)}, \tag{9.25}$$

or

$$y(t_1) = n(t_1) \quad \text{(signal absent)}. \tag{9.26}$$

The correct decision whether $f(t)$ is present or absent could easily be made from
Eqs. (9.25) and (9.26) were it not for the fact that $n(t_1)$ is random. Therefore we
use probability to describe the performance. Realizing that in general there is
a nonzero probability that the noise amplitude can exceed that of the signal, we
prefer to speak of a *probability of error* in making the correct decision.

To calculate the probability of error, we first make a rule for the decision.
Let the decision threshold be designated by μ. At the time $t = t_1$, we use the
following decision rule:

$$\begin{cases} y(t_1) > \mu: & \text{``signal present''} \\ y(t_1) < \mu: & \text{``signal absent''} \\ y(t_1) = \mu: & \text{``guess''}\dagger \end{cases}. \tag{9.27}$$

The next step in determining the optimum-decision threshold value μ will depend
on the shape of the probability density function of $n(t_1)$.

We assume that the noise $n(t_1)$ is gaussian-distributed with zero mean value
and a mean-square value of:

$$\overline{n^2(t_1)} = \sigma^2. \tag{9.28}$$

Using Eqs. (9.28) and (8.50), we find that the probability density function of the
input when there is no signal present, designated here by $p_0(y)$, is

$$p_0(y) = \frac{1}{\sqrt{2\pi}\sigma} e^{-y^2/(2\sigma^2)}. \tag{9.29}$$

---

† It will soon be evident that, except in highly unusual cases, this alternative occurs with
probability zero so it is not of major concern to us.

The probability density function of the input when the signal $f(t)$ is present, designated here by $p_1(y)$, is also gaussian but it has a mean value equal to A [see Eq. (9.25)],

$$p_1(y) = \frac{1}{\sqrt{2\pi}\sigma} e^{-(y-A)^2/(2\sigma^2)}. \tag{9.30}$$

These probability density functions are shown in Fig. 9.7.

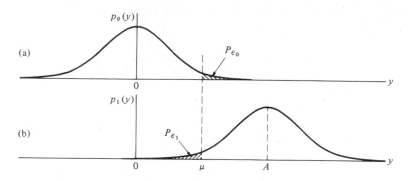

Fig. 9.7 Probability density functions for the binary decision case.

The decision threshold is μ, and we see from Fig. 9.7 that there are instances when $y > \mu$ even though the signal is absent. This type of error is called a "false alarm" because the receiver decides that the signal is present even though it is not present. The probability of false alarm, designated here by P_{ϵ_0}, is given by the shaded area in Fig. 9.7(a). We can write this in equation form as

$$P_{\epsilon_0} = \int_{\mu}^{\infty} \frac{1}{\sqrt{2\pi}\sigma} e^{-y^2/(2\sigma^2)} \, dy. \tag{9.31}$$

In a similar manner, we see from Fig. 9.7(b) that there are instances when $y < \mu$ even though the signal is present. This type of error is called a "false dismissal" because the receiver decides that the signal is absent even though it is present. The probability of false dismissal, designated here by P_{ϵ_1}, is given by the shaded area in Fig. 9.7(b), or,

$$P_{\epsilon_1} = \int_{-\infty}^{\mu} \frac{1}{\sqrt{2\pi}\sigma} e^{-(y-A)^2/(2\sigma^2)} \, dy. \tag{9.32}$$

Although we can continue to speak of the probability of false-alarm errors and the probability of false-dismissal errors, it is more convenient to define a net probability of error to incorporate both types. For convenience, let us associate the binary digit 1 with the condition "signal present" and the binary digit

0 with the condition "signal absent" (we could just as well reverse the definition; the order is arbitrary). Now let P_0, P_1 be the source digit probabilities of zeros and ones, respectively; that is, P_0 is the probability that a binary 0 is sent. The usual case is for $P_0 = P_1 = \frac{1}{2}$ (although they can be different as long as the sum is one—either a 0 or a 1 must be sent). In terms of these definitions, we define the net *probability of error*, P_ϵ, as:

$$P_\epsilon = P_0 P_{\epsilon_0} + P_1 P_{\epsilon_1}. \tag{9.33}$$

If we assume the binary levels (zeros and ones) are equiprobable (that is, $P_0 = P_1 = \frac{1}{2}$), the net probability of error is given by one-half the sum of the two shaded areas in Fig. 9.7:

$$P_\epsilon = \tfrac{1}{2}(P_{\epsilon_0} + P_{\epsilon_1}). \tag{9.34}$$

It is obvious from Fig. 9.7 that the sum of these two areas will be minimum if

$$\mu = A/2. \tag{9.35}$$

This sets the decision threshold μ for a minimum probability of error. (Note that the value of μ for minimum probability of error is not so obvious if $P_0 \neq P_1$.) With this decision threshold the two shaded areas shown in Fig. 9.7 are equal as a result of symmetry, and we can write

$$P_\epsilon = \int_\mu^\infty \frac{1}{\sqrt{2\pi}\sigma} e^{-y^2/(2\sigma^2)}\, dy. \tag{9.36}$$

With the change of variable $z = y/\sigma$, this becomes

$$P_\epsilon = \int_{\mu/\sigma}^\infty \frac{1}{\sqrt{2\pi}} e^{-z^2/2}\, dz. \tag{9.37}$$

Referring to Eq. (8.59), we recognize this as the complementary error function so that Eq. (9.37) can be rewritten as

$$P_\epsilon = \mathrm{Erfc}\,(\mu/\sigma). \tag{9.38}$$

Inserting the value of the decision threshold $\mu = A/2$, we obtain

$$P_\epsilon = \mathrm{Erfc}\,[A/(2\sigma)]. \tag{9.39}$$

A graph of P_ϵ as a function of $A/(2\sigma)$ is shown in Fig. 9.8.

These results can be put into more familiar units for the on-off case by letting the average signal power S be $S = (\frac{1}{2})(0)^2 + (\frac{1}{2})(A)^2 = A^2/2$. Also, we have that the average noise power is: $N = \sigma^2$ so that Eq. (9.39) can be rewritten for the on-off case as

$$P_\epsilon = \mathrm{Erfc}\,\sqrt{\frac{S}{2N}} \qquad \text{(on-off binary).} \tag{9.40}$$

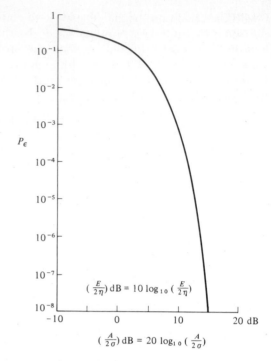

Fig. 9.8 Probability of error versus $A/(2\sigma)$.

$$\left(\tfrac{E}{2\eta}\right) dB = 10 \log_{10} \left(\tfrac{E}{2\eta}\right)$$

$$\left(\tfrac{A}{2\sigma}\right) dB = 20 \log_{10} \left(\tfrac{A}{2\sigma}\right)$$

The polar binary case follows in the same way except that $S = (A/2)^2$ so that

$$P_\epsilon = \mathrm{Erfc} \sqrt{\frac{S}{N}} \qquad \text{(polar binary).} \qquad (9.41)$$

Therefore the average transmitted power for the on-off binary signal must be twice that of the polar binary signal to achieve the same net probability of error. The polar binary signal also has an advantage in that the optimum-decision threshold is at zero for $P_0 = P_1$, whereas the receiver for the on-off binary signal must adjust the decision threshold to be one-half the received signal amplitude.

The preceding equations can be rewritten explicitly in terms of the conditional probabilities

$$p_0(y) = p_{Y|0}(y) \qquad (9.42)$$

and

$$p_1(y) = p_{Y|1}(y), \qquad (9.43)$$

and the additional observation that

$$p_Y(y) = P_0 p_{Y|0}(y) + P_1 p_{Y|1}(y). \qquad (9.44)$$

When written in this form, the probability density functions shown in Fig. 9.8, properly scaled by P_0, P_1, can be plotted together on one set of axes. If the appropriate areas in $p_Y(y)$ are identified as $P_0 P_{\epsilon_0}$, $P_1 P_{\epsilon_1}$, Eqs. (9.35)–(9.39) follow. This is illustrated in Fig. 9.9. It is often advantageous to use only one set of axes for a given set of conditions. The reader is cautioned, however, that the source digit probabilities P_0, P_1 must be included, as given in Eq. (9.44), before the two probability densities are combined.

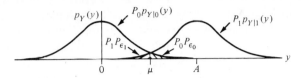

Fig. 9.9 Conditional probability density functions for the binary case.

The probability of error is dependent on the signal-to-noise ratio at a given time. Because the matched filter discussed in Section 7.9 is the best linear time-invariant system to maximize the peak-signal–to–rms-noise ratio in the presence of white noise, we now consider this case.

Consider the on-off signal $f(t)$ over $(0, T)$ in the presence of white noise $n(t)$. Letting $y(t)$ be the output of the matched filter, we have

$$y(t) = f_o(t) + n_o(t), \tag{9.45}$$

where $f_o(t)$, $n_o(t)$ are the matched-filter outputs due to input signal and noise, respectively. The matched filter is designed to maximize the signal-to-noise ratio at the time $t = t_m$ and we shall assume that $t_m = T$. From Eq. (7.49), the output of the matched filter at $t_m = T$ is E, the energy in $f(t)$. Therefore Eq. (9.45) for $f(t)$ present becomes

$$y(T) = E + n_o(T) \qquad \text{(signal present).} \tag{9.46}$$

On the other hand, if $f(t)$ is absent, the matched-filter output is

$$y(T) = n_o(T) \qquad \text{(signal absent).} \tag{9.47}$$

A comparison of Eqs. (9.46) and (9.47) with Eqs. (9.25) and (9.26) shows that we have already solved this problem with the exception that A is now replaced by E. Also, assuming that the output noise is gaussian-distributed with zero mean, we have, from Eq. (7.50),†

$$\sigma^2 = \overline{n_o^2(t)} = \frac{E\eta}{2}, \tag{9.48}$$

---

† Because the matched filter is a linear time-invariant system, then $n_i(t)$ is also gaussian-distributed, although we shall not attempt to prove this.

where $\eta/2$ is the (two-sided) power spectral density of the white noise input to the matched filter. Substituting E for A and $\sqrt{E\eta/2}$ for σ in Eq. (9.39) gives the probability of error for matched-filter detection as

$$P_\epsilon = \text{Erfc} \sqrt{\frac{E}{2\eta}}. \tag{9.49}$$

This is also indicated in Fig. 9.8. From this graph, we observe that the net probability of error begins to decrease very rapidly with increases in signal energy beyond $E/(2\eta) \approx 10$ dB. Above this level, the probability of error decreases approximately by an order of magnitude for each dB increase in signal energy.

We have derived the result expressed in Eq. (9.49) for the case of an on-off signal in which all of the signal energy E is used to convey one binary state (say, a one). This result also holds for more general binary signals if E is interpreted as the total energy used for the two binary states (i.e., for a one and a zero). In fact, we can interpret the term $E/2$ as the average energy per binary level so that

$$P_\epsilon = \text{Erfc} \sqrt{\frac{E_{\text{avg}}}{\eta}} \qquad \text{(on-off binary)}. \tag{9.50}$$

The polar binary case follows in the same way except that the output of the matched filter is

$$y(T) = \pm E + n_o(T). \tag{9.51}$$

Substituting $2E$ for A and $\sqrt{E\eta/2}$ for σ in Eq. (9.39) gives

$$P_\epsilon = \text{Erfc} \sqrt{\frac{2E}{\eta}} \qquad \text{(polar binary)}. \tag{9.52}$$

The choice of a matched filter receiver, as used here, is warranted for systems in which additive noise is a more dominant design factor than ISI. If ISI is the more dominant factor, then the overall frequency transfer characteristic of the transmitter filter, channel, and receiver filter should be chosen to minimize the ISI. If the channel characteristics are not completely known, or may vary, an adjustment is usually made at the receiver to correct the overall transfer function to yield minimum ISI. Such an adjustment is called equalization, the subject of a later section. If the effects of both ISI and additive noise must be considered, then the transmitter and receiver filters each should have magnitude frequency transfer functions proportional to the square root of the desired spectral density at the output of the receiver filter and inversely proportional to the square root of the frequency transfer function of the channel.†

---

† See, for example, K. S. Shanmugam, *Digital and Analog Communication Systems,* New York: John Wiley & Sons, 1979, Ch. 10; and R. W. Lucky, F. Salz, and E. J. Weldon, Jr., *Principles of Data Communication,* New York: McGraw-Hill, 1968, Ch. 5.

Example 9.5.1   The concept of a net probability of error can be extended to the multilevel (*M*-ary) detection problem, and for gaussian noise the calculation is straightforward. Consider the polar trinary waveform in additive gaussian noise whose probability density functions are shown in Fig. 9.10, representing the trinary digits 0, 1, and 2. Calculate the net probability of error if the decision thresholds are set at $\pm A/4$ and the source probabilities of the digits are equal.

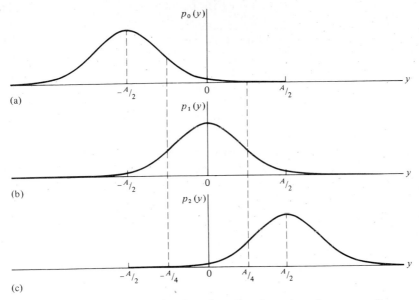

Fig. 9.10 Probability density functions for the polar trinary waveform.

Solution.   Referring to Fig. 9.10, we can write

$$P_{\epsilon_0} = \int_{-A/4}^{\infty} \frac{1}{\sqrt{2\pi}\sigma} e^{-(y+A/2)^2/2\sigma^2} \, dy = \text{Erfc} \left[A/(4\sigma) \right].$$

In a similar manner, we have

$$P_{\epsilon_2} = \text{Erfc} \left[A/(4\sigma) \right] \qquad \text{and} \qquad P_{\epsilon_1} = 2 \, \text{Erfc} \left[A/(4\sigma) \right].$$

We are given that $P_0 = P_1 = P_2 = 1/3$ so that

$$P_{\epsilon} = \tfrac{4}{3} \text{Erfc} \left[A/(4\sigma) \right].$$

Example 9.5.2   Assuming that errors are made in mistaking adjacent levels only, calculate the net probability of error for the *n*-level quantizer illustrated in Fig. 9.3 for the case of additive gaussian noise with zero mean and variance $\sigma^2 = N$.

Solution.   Referring to the result of Example 9.5.1, we note that the factor 4/3 arose because there were $2(n - 1)$ areas to be counted in determining the probability of error and each was weighted by $1/n$ because all were equiprobable. Generalizing the result of Example 9.5.1 and letting $a = A/2$, we can write

$$P_\epsilon = \frac{2(n-1)}{n} \text{Erfc} \, [a/(2\sigma)].$$

To write this in terms of S/N, we use Eq. (9.18) in the above to give

$$P_\epsilon = 2\left(1 - \frac{1}{n}\right) \text{Erfc} \left(\sqrt{\frac{3}{n^2-1} \cdot \frac{S}{N}}\right).$$

Drill Problem 9.5.1   An on-off binary system uses the signal levels 0 and A for transmission in the presence of gaussian noise. The probability of 1's and 0's is equal. Let the signal amplitude A be K times the standard deviation of the noise. (a) Calculate the net probability of error if $K = 10$. (b) Determine the required value of K to give a net probability of error of 10^{-5}.

Answer.   (a) 2.87×10^{-7}; (b) 8.53.

Drill Problem 9.5.2   Assign the equiprobable binary digits 0, 1 to the probability density functions shown in Fig. 9.10(a), (c), reducing the polar trinary case to the polar binary case. With the decision threshold set at zero, repeat the calculations of Drill Problem 9.5.1 for the polar binary case.

Answer.   Same as those of Drill Problem 9.5.1.

9.6   *S/N* PERFORMANCE OF PCM

We now examine the S/N performance of PCM in the presence of white gaussian-distributed noise (e.g., thermal noise). To simplify things, we make some assumptions which are usually valid in typical operating conditions. First we assume that the probability of error, P_ϵ, is small enough that the probability of more than one bit error in a code word is negligibly small. For example, if $P_\epsilon = 10^{-4}$ and a 10-bit code word ($m = 10$) is used, the probability of more than one error is on the order of (using the Poisson approximation) $1 - P_0 - P_1 = 1 - (1.001) \exp(-0.001) = 5 \times 10^{-7}$.

There are m binary digits to describe $n = 2^m$ levels and a is the amplitude-quantizing increment. We assume that the code words formed by the m binary digits are in order of the numerical significance of the code word. Using a linear quantization rule, an error which occurs in the least significant bit corresponds to an amplitude error a, an error in the next higher significant bit corresponds to an error $2a$, etc. The mean-square output noise arising from thermal transmission noise is then

$$\overline{(\Delta m)^2} = \frac{1}{m}[(a)^2 + (2a)^2 + (4a)^2 + (8a)^2 + \cdots + (2^{m-1}a)^2]. \quad (9.53)$$

The sum of the geometric progression in Eq. (9.53) is

$$\overline{(\Delta m)^2} = \frac{2^{2m} - 1}{3m} a^2. \tag{9.54}$$

The probability that a given bit is in error is P_ϵ and therefore the mean-square thermal noise per code word is

$$\overline{n_{th}^2(t)} = mP_\epsilon\overline{(\Delta m)^2} = P_\epsilon(2^{2m} - 1)a^2/3. \tag{9.55}$$

Recalling Eqs. (9.18) and (9.20) for the mean-square signal and quantization noise present and using Eq. (9.55), we have

$$\frac{S_o}{N_o} = \frac{\overline{s^2(t)}}{\overline{n_{qnt}^2(t)} + \overline{n_{th}^2(t)}}, \tag{9.56}$$

$$\frac{S_o}{N_o} = \frac{2^{2m} - 1}{1 + 4P_\epsilon(2^{2m} - 1)} = \frac{n^2 - 1}{1 + 4P_\epsilon(n^2 - 1)}. \tag{9.57}$$

For the polar binary case, we use Eq. (9.41) in (9.57) to give

$$\frac{S_o}{N_o} = \frac{n^2 - 1}{1 + 4(n^2 - 1)\,\text{Erfc}\,(\sqrt{S_i/N_i})} \tag{9.58}$$

where S_i/N_i is measured at the receiver input. A plot of Eq. (9.58) is shown in Fig. 9.11. There are two distinct modes of performance, one limited by the

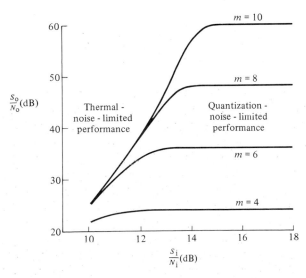

Fig. 9.11 Performance of PCM in the presence of noise.

(thermal) transmission noise and the other limited by the quantization noise. PCM systems usually are operated in the region where quantization begins to be the predominant factor.

One way to obtain a fairly good qualitative indication of the performance of an operating PCM system is to display the bit stream on an oscilloscope. The time base of the oscilloscope is set to trigger on the bit rate with a sweep duration of a few bit intervals. For no additive noise or bandwidth limitations, the oscilloscope display would appear like Fig. 9.12(a). Figure 9.12(b) illustrates the effects of a limited bandwidth, and some random noise is added to obtain Fig. 9.12(c). The resulting pattern suggests an eye and is called an *eye pattern*. For a given system, the eye closes as more noise is added.

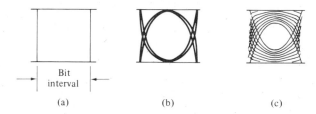

Bit
interval

(a)   (b)   (c)

Fig. 9.12 Eye patterns for PCM: (a) wideband, no noise; (b) bandwidth limitations, no noise; (c) bandwidth limitations with additive noise.

The best time to sample the received waveform is when the eye opening is largest. The presence of ISI also causes the eye to close, and the relative amount of closure at the best sampling time gives an indication of the degradation caused by ISI. For example, the eye closing by 20% due to the presence of ISI represents an equivalent S/N degradation of $-20 \log_{10} (1 - 0.20) = 1.94$ dB. Thus such a system would require an additional 1.94 dB of signal-to-noise ratio above that required for a specified error rate and zero ISI.

9.7   DELTA MODULATION AND DPCM

Up to this point, we have described PCM systems as those with fairly straight-forward digital codes for the transmission of information. There are some alternative PCM systems that have features distinctive enough to warrant special notation. We briefly describe two such types of systems here.

In the transmission of messages having repeated sample values (i.e., an average value), the repeated transmission represents a waste of communication capability because there is little information content in the repeated values. One way to improve this situation is to send only the digitally encoded differences between successive sample values. This is known as *differential pulse-code modulation*, abbreviated as DPCM. For example, DPCM is advantageous in the transmission of sampled picture information because an appreciable portion of the code assigned to each level merely describes the average background level. Thus a picture which has been quantized to 6 bits (i.e., 64 brightness levels per picture element) can be transmitted with comparable fidelity using 4-bit DPCM. Block diagrams of a DPCM modulator and demodulator are shown in Fig. 9.13.

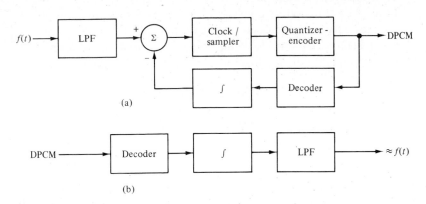

Fig. 9.13 Differential pulse code modulation: (a) modulator; (b) demodulator.

Disadvantages of DPCM systems include the fact that if an error is made, an incorrect bias is retained until corrected. Also, such systems suffer from a possible rate-of-rise overloading as a result of the differencing and truncation operations. For instance, if two adjacent picture samples in the above example would differ by more than ± 7 levels (one bit in the 4-bit code is reserved for sign), the system could only send a ± 7, and the resulting overload would cause an error in the reconstruction. This problem can be alleviated by preceding the encoder with a linear filter (say, an RC filter) to limit the maximum rate of rise or fall in the waveform. (This also, of course, decreases the resolution of the reconstructed picture.) It is possible to detect the overload condition and vary the filtering when this condition arises (i.e., nonlinear filtering). Going still farther, one can make the procedure adaptive and dependent on the input data.

Viewed from a wider perspective, the purpose of the feedback path in the DPCM encoder is to perform a prediction of the next sample value. The reasoning behind this is that if the system can do a good job of predicting the next sample value, then there is not as much information needed for the encoder to send (i.e., the encoder has removed some of the redundancy in the data). Considerable research has centered on improving the prediction algorithms. Linear prediction systems using tapped delay lines with adjustable tap gains have been used with success to remove redundancy from the data prior to transmission, but at the expense of some added complexity.

An encoding system which is particularly simple to implement results when a DPCM system is used with only a one-bit output, that bit being used for the sign of the sample difference. This PCM system is known as *delta modulation*, abbreviated either as DM or ΔM. DM systems have an advantage in that the electronic circuitry required for modulation at the transmitter and demodulation at the receiver is substantially simpler than that required for other PCM systems.

A block diagram of a delta-modulation system for line applications is shown in Fig. 9.14(a). The transmitter consists of a comparator/modulator whose input is the difference between its accumulated output and the input signal $f(t)$. The modulator is triggered by the clock pulses. If the input-signal amplitude $f(t)$ is greater than the accumulated output at the clock time, a positive output pulse results. Conversely, if $f(t)$ is less than the feedback signal, a negative output pulse is generated. Figure 9.14(b) shows that the output of the accumulator in the transmitter is a stepped approximation to the continuous input signal $f(t)$. Therefore a similar accumulator at the receiver, smoothed by a low-pass filter,

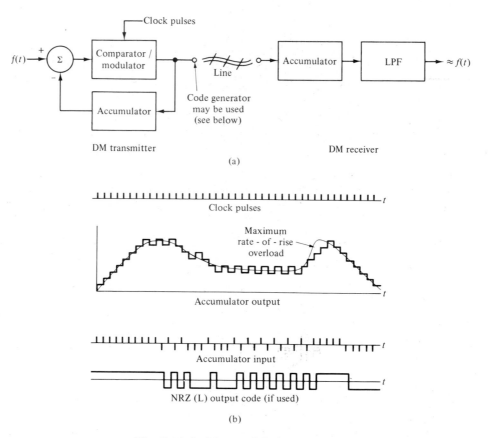

Fig. 9.14 A delta-modulation system.

will produce an output approximating the original signal $f(t)$. The DM system shown in Fig. 9.14 is intended for line applications where the signal-to-noise ratio is high. For higher noise conditions, the accumulator in the receiver can be preceded by a matched filter and decision threshold.

DM, like DPCM, is subject to rate-of-rise overload problems whenever the input changes too rapidly for the stepped waveform to follow it (see Fig. 9.14b). As a result of the accumulator, it eventually "catches up" and no long-term bias error results if the input is noise-free.

We observe from Fig. 9.14 that if the input-signal level remains constant, the reconstructed DM waveform exhibits a hunting behavior known as *idling noise*. This idling noise is a square wave at one-half the clock rate. If the clock rate is much greater than twice the highest frequency in the input signal, most of the idling noise can be filtered out at the receiver.

An estimate of the rate-of-rise limit, or *slope-overload*, condition for DM may be obtained quite easily for sinusoidal modulation. Let the input be $f(t) = b \cos \omega_m t$ so that

$$\left| \frac{df}{dt} \right|_{\max} = b2\pi f_m. \tag{9.59}$$

If the step size used in the DM system is a, then the maximum rate of rise which can be handled is $a/T = af_s$ so that

$$af_s \geq b2\pi f_m,$$

$$f_s \geq \frac{2\pi f_m}{a/b}. \tag{9.60}$$

If the spectral weighting is uniform, Eq. (9.60) may be applied to band-limited signals by letting f_m be the highest frequency. If the spectral weighting is not uniform but decreases for higher frequencies, Eq. (9.60) may be used by replacing f_m with an equivalent frequency f_0. For voice signals in telephone applications it is found that $f_0 \approx 800$ Hz can be used satisfactorily.

Small step size is desirable to accurately reproduce the input waveform. However, small step size must be accompanied by a fast clock rate to avoid slope overload.

We shall assume that the quantization noise in DM is uniformly distributed over $(-a, a)$ so that the mean-square quantization error is $a^2/3$. To obtain a measure of the quantization noise, we also assume that its power spectral density is flat up to $f_s = 1/T$. Filtering this noise to a bandwidth f_m (where $f_m < f_s$), we have

$$\overline{n_{\text{qnt}}^2(t)} = \frac{a^2}{3} \frac{f_m}{f_s}. \tag{9.61}$$

The mean-square value of the signal is $\overline{f^2(t)}$ so that the signal-to-quantization noise ratio for DM is

$$\frac{S}{N} = \frac{3f_s}{a^2 f_m} \overline{f^2(t)}.$$

(9.62)

Because $\overline{f^2(t)} = b^2/2$ for sinusoidal signals, using Eq. (9.60) to eliminate a, and assuming a flat input-signal spectrum, we can rewrite Eq. (9.62) as

$$\frac{S}{N} = \frac{3}{8\pi^2} \left(\frac{f_s}{f_m}\right)^3.$$

(9.63)

Thus the signal-to-quantization noise ratio increases by $10 \log_{10} 2^3 = 9$ dB for every octave increase in the sampling frequency. This improvement, however, is realized only within the approximations made and cannot be expected to increase indefinitely with higher and higher sampling rates. Finally, if we use $f_s = 2B$, Eq. (9.63) becomes

$$\frac{S}{N} = \frac{3}{\pi^2} \left(\frac{B}{f_m}\right)^3.$$

(9.64)

From Eq. (9.64) we conclude that the S/N performance of DM is superior to angle modulation (for example, FM) and pulse-timing modulation (for example, PPM) but falls short of the exponential characteristic of PCM. This is more noticeable when the number of levels used in PCM becomes large.

 More extensive analyses and experimental tests have been made using various classes of signals in DM systems. A typical result for the net signal-to-noise ratio is shown in Fig. 9.15. Considerable research has produced adaptive DM

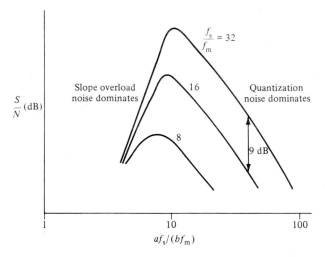

Fig. 9.15 Typical output signal-to-noise ratio characteristics for delta modulation systems.

methods that seek to minimize the effects of slope overload and idling noise at the expense of increased complexity. In these methods, the step size is varied automatically in some predetermined manner and in accordance with the time-varying characteristics of the input signal.

Delta modulation is used primarily for speech transmissions in telephone and telemetry systems. It has been found that PCM is preferable for high-quality speech transmissions, whereas DM is easier to implement and yields transmissions of acceptable quality. One interesting application is to use a DM encoder followed by an up-down binary counter (i.e., a counter that increments on positive pulses and decrements on negative pulses). With appropriate reset provisions, this combination can serve as a comparatively simple A/D converter for telephone and telemetry applications.

Drill Problem 9.7.1   Starting with Eq. (9.62), develop an expression for the S/N performance of DM for typical voice transmissions and calculate for the case $f_0 = 800$ Hz; $f_m = 3.3$ kHz; $f_s = 32$ kHz.

Answer.   27.7 dB.

★ 9.8   ERROR ANALYSIS OF PCM REPEATERS

All communication channels cause some attenuation and distortion of the signals that pass through them. To ensure the satisfactory reception of signals at their final destination, repeaters are often provided at appropriate spacings along the transmission route. These provide the amplification and correction of channel distortion (i.e., equalization) necessary to enable satisfactory signal recognition at the receiver.

In analog repeaters, the signal at the end of each segment of the transmission path is equalized and amplified before being sent out again on the next segment of the transmission path. Repeater spacings along the transmission path are governed by path losses and the dynamic range of the repeater. If the repeater itself is not to be a primary source of distortion, its dynamic range is limited by possible nonlinear amplification characteristics at the higher levels.

Net repeater gain is limited also by possible feedback from output to input. In modulated analog repeater systems (e.g., FM) with antennas, feedback can be a primary consideration. To minimize feedback, repeaters may use separate highly directive antennas for reception and transmission. They also may shift the carrier frequency slightly and filter out the original carrier frequency before transmission in order to obtain better isolation between input and output. (Systems with this operation are called *duplexers*.) Note that in the analog repeater the additive noise effects of all preceding transmission paths (and repeaters) are amplified, as well as the signal, before the composite signal-plus-noise is sent on its way.

A significant advantage for using all-digital transmission is that digital signals lend themselves to periodic conditioning and reshaping. Thus each repeater can be used to regenerate the pulse waveforms at each sample time. This type of repeater is called a *regenerative* repeater. Our primary interest in this section is to make a comparison of relative error rates of systems using analog repeaters and regenerative repeaters in digital transmission. Baseband transmission is assumed, and we ignore the possible modulation and demodulation processes required to transmit the signal through a channel.

Assume that the binary signals are NRZ-polar and that the additive noise present has zero mean value and variance σ^2. Let the signal plus noise at the input to the first analog repeater at a given sample time be

$$y_1 = \pm \frac{A}{2} + n_1.$$

This signal is amplified to offset the attenuation and sent on to the second repeater. At the input to the second repeater, we have

$$y_2 = \pm \frac{A}{2} + n_1 + n_2.$$

This continues until ultimately, after m transmission paths, the signal plus noise at the input to the receiver is

$$y = \pm \frac{A}{2} + n_1 + n_2 + \cdots + n_m,$$

$$y = \pm \frac{A}{2} + n. \tag{9.65}$$

We assume that the additive noise terms are statistically independent so that (cf. Example 8.9.1)

$$\sigma_n^2 = \sigma_1^2 + \sigma_2^2 + \cdots + \sigma_m^2.$$

In addition, we assume that the amplifiers in the analog repeaters are all identical and that amplifier noise dominates the other noise sources. Under these conditions, we have

$$\sigma_n^2 = m\sigma^2.$$

If, in addition, we assume that $n(t)$ is gaussian-distributed, then the probability of error is exactly the same as that calculated in Section 9.5 with σ^2 replaced by $m\sigma^2$:

$$P_\epsilon = \int_0^\infty \frac{1}{\sqrt{2\pi m}\ \sigma} e^{-(y + A/2)^2/(2m\sigma^2)}\ dy,$$

$$P_\epsilon = \text{Erfc}\ [A/(2\sqrt{m}\ \sigma)]. \tag{9.66}$$

Next we consider a regenerative repeater. Assuming identical repeaters and gaussian noise, the probability of error at each repeater is

$$p = \text{Erfc}\,[A/(2\sigma)]. \tag{9.67}$$

The probability of making i errors in m trials is given by the binomial distribution [cf. Eq. (8.46)]

$$P_i = \binom{m}{i} p^i (1 - p)^{m-i}. \tag{9.68}$$

However, an error is made at the receiver only if an *odd* number of incorrect decisions is made along the total transmission path. The net probability of error is obtained by summing over all values of i, giving

$$P_\epsilon = \sum_{\substack{i=1 \\ i \text{ odd}}}^{m} \binom{m}{i} p^i (1 - p)^{m-i} \tag{9.69}$$

$$= mp(1 - p)^{m-1} + \frac{m(m-1)(m-2)}{3!} p^3 (1 - p)^{m-3} + \dots$$

$$\approx mp, \tag{9.70}$$

where the approximation applies if $p \ll 1$ and $mp \ll 1$. Using Eq. (9.67) in Eq. (9.70), we have

$$P_\epsilon \approx m\,\text{Erfc}\,[A/(2\sigma)]. \tag{9.71}$$

We conclude that, within the approximations given, P_ϵ increases linearly with m for the use of regenerative repeaters.

A comparison of Eqs. (9.66) and (9.71) for several different error rates is shown in Fig. 9.16. Note that the relative advantage in the use of regenerative

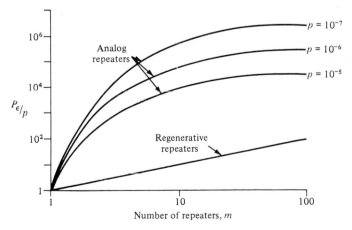

Fig. 9.16 Net probability of error as a function of the number of repeaters.

repeaters becomes more pronounced as p, the error rate for one repeater, decreases.

★ 9.9   POWER SPECTRAL DENSITIES OF PCM SIGNALS

The power spectral density of the transmitted signal in a PCM system depends on the signaling format used. Major types of PCM signaling formats are discussed in Chapter 7 (see Fig. 7.30). Choice of signaling format is influenced by the transmission characteristics. Because these are often specified in terms of frequency, we shall examine the spectral densities of the PCM signaling formats that are commonly used.

For our purposes here, we assume that 0's and 1's are generated with equal probability by a source and are then encoded into one of the PCM signaling formats shown in Fig. 7.30. The pulse waveform $f_1(t)$ is used to signal a 0, the pulse waveform $f_2(t)$ a 1. The bit interval is $(0, T_b)$, and in each case the pulse waveform is limited to an interval of time equal to T_b. We rely on the results of Examples 8.13.2–8.13.4 to determine the power spectral densities within the above restrictions. In all cases, the power spectral densities are normalized to a one-ohm resistance.

9.9.1   Nonreturn-to-Zero (NRZ)

In NRZ, the output level is at a specified constant value $(\pm A)$ during the full bit interval. For equally probable 1's and 0's, the result of Example 8.13.2 for the transmitted signal $y(t)$ is

$$S_y(\omega) = A^2 T_b \, \text{Sa}^2(\omega T_b/2). \tag{9.72}$$

This is shown in Fig. 9.17. The $\text{Sa}^2(x)$ function occurs often in digital data systems. For this power spectral density, approximately 77% of the power is contained in the frequency components $(-\pi/2, \pi/2)$, 90% in $(-\pi, \pi)$ and 94.8% in $(-2\pi, 2\pi)$.

Random NRZ data do not have discrete spectral components. Therefore either an external clock or nonlinear processing in the receiver is necessary for bit rate acquisition.

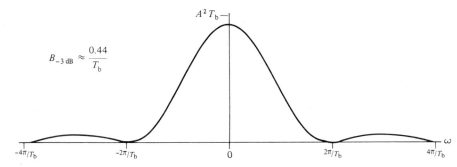

Fig. 9.17 Power spectral density of random data (NRZ signaling format).

9.9.2 Return-to-Zero (RZ)

In this signaling format, we have

$$f_1(t) = 0,$$

$$f_2(t) = \begin{cases} A & 0 < t \le \dfrac{T_b}{2} \\[2mm] 0 & \dfrac{T_b}{2} < t \le T_b \end{cases}.$$

Using the result of Example 8.13.4 for equally probable zeros and ones, we can write

$$S_y(\omega) = \frac{A^2 T_b}{16} \operatorname{Sa}^2(\omega T_b/4) + \frac{\pi A^2}{8} \sum_{n=-\infty}^{\infty} \operatorname{Sa}^2(n\pi/2)\,\delta(\omega - 2\pi n/T_b). \qquad (9.73)$$

This is shown in Fig. 9.18.

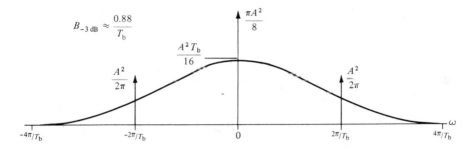

Fig. 9.18 Power spectral density of random data (RZ signaling format).

The first zero crossings of the RZ power spectral density are at twice the frequency as those for the NRZ case. In addition to the continuous spectral density, random RZ data have a nonzero average value and discrete lines at odd multiples of the bit rate. Although it requires twice the bandwidth of NRZ, the presence of the discrete lines permits simple synchronization for the RZ signaling format.

9.9.3 Split-Phase (Manchester Code)

In this signaling format, we have

$$f_1(t) = \begin{cases} -A & 0 < t \le \dfrac{T_b}{2} \\[2mm] A & \dfrac{T_b}{2} < t \le T_b \end{cases}$$

$$f_2(t) = -f_1(t).$$

Using the result of Example 8.13.4 for equally probable zeros and ones, we can write

$$S_y(\omega) = A^2 T_b \, \mathrm{Sa}^2(\omega T_b/4) \sin^2 (\omega T_b/4). \tag{9.74}$$

This is shown in Fig. 9.19.

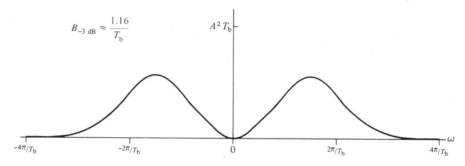

Fig. 9.19 Power spectral density of random data (Split-Phase or Manchester code signaling format).

In contrast to that for both RZ and NRZ, the power spectral density for the split-phase signaling format has low relative weighting at low frequencies (and zero at zero frequency). The power spectral density is maximum at a frequency (in Hz) of $0.743/T_b$ and is down by 3 dB at a frequency of $1.16/T_b$. There are no discrete lines, and bandwidth occupancy is similar to that for RZ data at the higher frequencies.

9.9.4   Bipolar Return-to-Zero (BRZ)

The BRZ signaling format uses three amplitude levels: $A, 0, -A$ and is sometimes classified as a pseudo-ternary signal. For Alternate Mark Inversion, we have

$$f_1(t) = 0,$$

$$
f_2(t) =
\begin{cases}
\left.
\begin{cases}
A & 0 < t \le \dfrac{T_b}{2} \\[2mm]
0 & \dfrac{T_b}{2} < t \le T_b
\end{cases}
\right\} \text{first 1,} \\[8mm]
\left.
\begin{cases}
-A & 0 < t \le \dfrac{T_b}{2} \\[2mm]
0 & \dfrac{T_b}{2} < t \le T_b
\end{cases}
\right\} \text{next 1,}
\end{cases}
$$

etc.

This case can be modeled as a random binary generator whose output is filtered by a one-stage transversal filter to obtain a ternary output (see Example 8.13.3). For convenience here we assume that the output can take any of the levels: $-1, 0, +1$. Assuming that $P(0) = \frac{1}{2}$, $P(-1) = P(+1) = \frac{1}{4}$,

$$R_{cc}(0) = \sum_{i=-1}^{1} i^2 P(c_k = i) = \frac{1}{2}.$$

Correlation between successive output symbols for the Alternate Mark Inversion case can be evaluated with the aid of Table 9.1 to give

$$R_{cc}(1) = \sum_{i=-1}^{1} \sum_{j=-1}^{1} ij P(c_k = i, c_{k+1} = j)$$

$$= -\frac{1}{4}.$$

Table 9.1   Enumeration of Joint Probabilities of Successive Symbols for BRZ (Alternate Mark Inversion)

Binary gen. output		Transversal filter output		Probability
b_k	b_{k+1}	c_k	c_{k+1}	
0	0	0	0	$\frac{1}{4}$
0	1	0	1	$\frac{1}{4}$
1	0	1	0	$\frac{1}{4}$
1	1	1	-1	$\frac{1}{4}$

Because the filter has only one stage ($N = 1$) (see Example 8.13.3),

$$R_{cc}(k) = 0 \qquad \text{for } k > 1.$$

Using the result of Example 8.13.3 and designating the power spectral density of the filter input by $S_x(\omega)$, we have

$$S_y(\omega) = S_x(\omega)[\tfrac{1}{2} - \tfrac{1}{2} \cos(\omega T_b)]$$

$$= S_x(\omega) \sin^2(\omega T_b/2).$$

Using the result of Example 8.13.2 and scaling the output to the range $(-A, A)$, we have

$$S_y(\omega) = \frac{A^2 T_b}{4} \, \text{Sa}^2(\omega T_b/4) \sin^2(\omega T_b/2). \tag{9.75}$$

This is shown in Fig. 9.20.

There are no discrete lines in the power spectral density of the BRZ (AMI) signaling format. Bandwidth occupancy is that of NRZ at the higher frequencies.

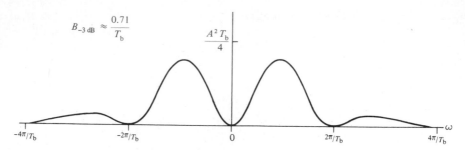

Fig. 9.20 Power spectral density of random data (BRZ or Alternate Mark Inversion signaling format).

9.9.5 Delay Modulation (Miller Code)

Derivation of the power spectral density of the delay modulation signaling format is beyond the level of this book.† A graph of the power spectral density is shown in Fig. 9.21. The major spectral weighting is in those frequency components that are greater than zero frequency but less than one-half the symbol rate ($1/T_b$). This makes delay modulation attractive for magnetic tape recording and similar applications that are severely restricted in bandwidth and require relatively low power at very low frequencies.

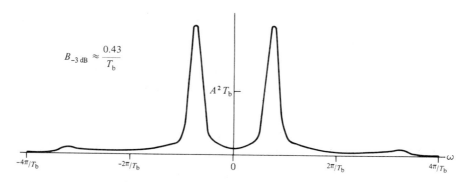

Fig. 9.21 Power spectral density of random data (delay modulation or Miller code signaling format).

Drill Problem 9.9.1 Determine the power spectral density of the RB signaling format (see Fig. 7.30) in which the signal levels are $-A$, 0, A and 0's and 1's are equiprobable.

Answer. $S_y(\omega) = (A^2 T_b/4)\, \text{Sa}^2\,(\omega T_b/4)$.

---

† Some results are given in W. C. Lindsay and M. K. Simon, *Telecommunication Systems Engineering*, Englewood Cliffs: Prentice-Hall, Inc., 1973, Ch. 1.

9.10   PARTIAL-RESPONSE SIGNALING

In Chapter 7, we found that transmission rates up to $2B$ symbols per second could be attained within a bandwidth of B hertz. To avoid intersymbol interference (ISI), however, we assumed the use of an ideal low-pass filter to attain this rate. Introduction of raised-cosine spectral shaping to avoid ISI while providing for a more realizable filter characteristic resulted in transmission rates that decreased from $2B$ symbols per second to B symbols per second.

An alternative is to purposely use controlled amounts of ISI by combining a number of successive binary pulses prior to transmission. Because these are combined in a known way, the receiver can isolate the correct data stream again. This approach is called *partial-response signaling* or, alternatively, *correlative coding*. Basically, a partial-response system generates an ℓ-level signal from a binary input data stream x_k using a linear weighted superposition of the x_k. A transversal filter to accomplish this is shown in Fig. 9.22. The superposition memory extends over $(N + 1)$ bit periods and is determined by the number N of delay elements (shift registers) used. The rate of the input data stream is T^{-1}. A low-pass filter (LPF) restricts the bandwidth of the resultant signal to the Nyquist frequency $(2T)^{-1}$ and the ℓ-level output is designated as y_k.

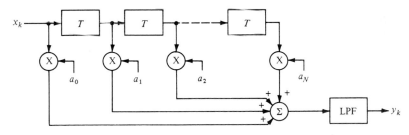

Fig. 9.22 Transversal filter for generation of partial-response signaling.

Partial-response signaling permits control of the spectral properties of the signal to make efficient use of available bandwidth and to tailor these properties to the channel. It offers the capability of increasing the transmission rate above B symbols per second but does demand some additional required transmitted power. In addition, known redundancy in the received signal can be used for error detection.

We shall examine two specific cases of partial-response signaling.† In the

---

† For additional coverage, see A. Lender, "Correlative Level Coding for Binary Data Transmission," *IEEE Spectrum*, vol. 3, no. 2 (February 1966): 104–115; E. R. Kretzmer, "Generalization of a Technique for Binary Data Communication," *IEEE Trans. Commun. Technol.*; COM–14 (February 1966): 67–68; S. Pasupathy, "Correlative Coding: A Bandwidth-Efficient Signaling Scheme," *IEEE Commun. Soc. Mag.*, vol. 15, no. 4 (July 1977): 4–11.

first case, two successive binary input pulses are added so that

$$y_k = x_k + x_{k-1}. \tag{9.76}$$

This is termed *duobinary* signaling. A transversal filter ($N = 1$) to accomplish this operation is shown in Fig. 9.23(a). The resulting frequency transfer function is

$$H(\omega) = (1 + e^{-j\omega T})_{\text{LP}} \tag{9.77}$$

$$H(\omega) = \begin{cases} 2e^{-j\omega T/2} \cos \dfrac{\omega T}{2} & |\omega| \leq \dfrac{\pi}{T} \\ 0 & \text{elsewhere} \end{cases}. \tag{9.78}$$

The corresponding impulse response is

$$h(t) = \frac{4 \cos [\pi(t - T/2)/T]}{\pi T[1 - 4(t - T/2)^2/T^2]} \tag{9.79}$$

and this is shown in Fig. 9.23(c). From this result, we conclude that duobinary signaling introduces ISI, but it is controlled in such a way that the interference comes only from the immediately preceding symbol. Note that the operation of

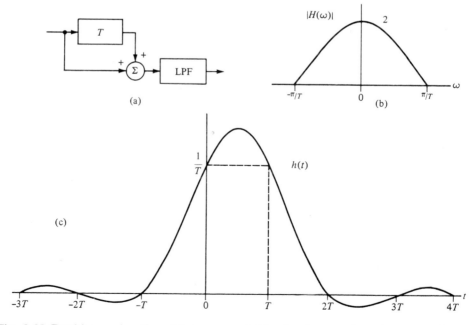

Fig. 9.23 Duobinary signaling: (a) transversal filter implementation; (b) magnitude of frequency transfer function; (c) unit impulse response.

adding two successive symbols has provided an effective spectral shaping to the transmitted signal.†

Suppose that the binary data stream x_k uses the levels 0, 1 for convenience; then the y_k will assume the levels 0, 1, 2. Also, let the transmitter use a polar signal representation y_k. Without additive noise or attenuation, the received signal, Y_k, will take on one of the three possible levels $-A$, 0, A. For a given amount of additive noise, it was shown in Section 9.5 that discrimination between logic levels at the receiver is proportional to the amplitude separation between levels. Therefore it follows that for a given error rate a duobinary system must transmit more power than a binary system for the same amount of additive noise present. However, the duobinary system permits the use of a higher transmission rate and also furnishes the proper spectral weighting to accomplish it.

The role of the receiver is to solve the equation [cf. Eq. (9.76)]

$$\tilde{x}_k = y_k - \tilde{x}_{k-1}, \tag{9.80}$$

where $\tilde{x}_k$ is the decoded binary data stream. Because $\tilde{x}_k$ is decoded according to the decoded value of $\tilde{x}_{k-1}$, decoding errors tend to propagate in the duobinary system described above. A method proposed by A. Lender eliminates this tendency by precoding the input binary data stream before transmission. In this precoding procedure for duobinary systems, the input binary data stream x_k is converted to another binary stream b_k according to the rule

$$b_k = x_k \oplus b_{k-1} \tag{9.81}$$

where, as before, the symbol $\oplus$ represents modulo-2 (EXCLUSIVE-OR) addition.

The binary stream b_k is applied to the input of the duobinary filtering operation described by Eq. (9.76), and yields

$$y_k = b_k + b_{k-1} = (x_k \oplus b_{k-1}) + b_{k-1}. \tag{9.82}$$

Examination of Eq. (9.82) shows that if $x_k = 1$, then $y_k = 1$ *regardless* of the binary value of b_{k-1}. If $x_k = 0$, then we have $y_k = 0$ or $y_k = 2$. These values are listed in Table 9.2. Thus the y_k can be decoded to find the x_k following the rule

$$\tilde{x}_k = y_k \bmod 2. \tag{9.83}$$

A similar situation occurs if the polar signal representation $Y_k = -A$, 0, A is used. If $x_k = 0$, then $b_k = b_{k-1}$ [see Eq. (9.81)], and Y_k will be $-A$ or A. If $x_k = 1$, then b_k will be the complement of b_{k-1}, and Y_k will be 0. The corresponding decoding rule is

$$\tilde{x}_k = \begin{cases} 0 & \text{if } Y_k = \pm A \\ 1 & \text{if } Y_k = 0 \end{cases}. \tag{9.84}$$

---

† A comparison of Fig. 9.23(c) and Fig. 7.9(c)—with $B_x = f_m = (2T)^{-1}$—is instructive here.

Table 9.2   Decoding of Precoded Duobinary Signals

x_k	b_{k-1}	y_k	Y_k
1	0	1	0
1	1	1	0
0	0	0	$-A$
0	1	2	A

In either case, precoding has enabled the receiver to make each binary decision based only on the current received sample; and, therefore, the possibility of error propagation has been eliminated.

The frequency response of the channel in many transmission systems (e.g., commercial telephone systems) does not extend down to zero frequency. It is advantageous in such systems to modify the duobinary signaling to obtain zero spectral weighting at zero frequency and yet retain the desirable property of allowing signaling up to the Nyquist rate. In one method, called *modified duobinary* signaling, binary pulses spaced by two sample periods are subtracted, giving [see Fig. 9.24(a)]

$$y_k = x_k - x_{k-2}. \tag{9.85}$$

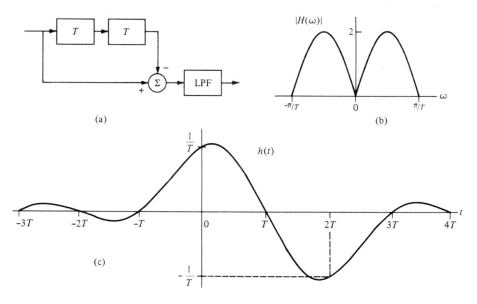

(a)

(b)

(c)

Fig. 9.24 Modified duobinary signaling: (a) transversal filter implementation; (b) magnitude of frequency transfer function; (c) unit impulse response.

The resulting frequency transfer function is

$$H(\omega) = (1 - e^{-j2\omega T})_{\mathrm{LP}} \tag{9.86}$$

$$H(\omega) = \begin{cases} 2je^{-j\omega T} \sin \omega T & |\omega| \leq \dfrac{\pi}{T} \\ 0 & \text{elsewhere} \end{cases} \tag{9.87}$$

The corresponding impulse response is

$$h(t) = (1/T)\{\mathrm{Sa}(\pi t/T) - \mathrm{Sa}\,[\pi(t - 2T)/T]\} \tag{9.88}$$

and this is shown in Fig. 9.24(c). From this result, we conclude that modified duobinary signaling introduces controlled ISI from only one symbol removed by two sample periods. Note that the relatively steep transition in $h(t)$ at $t = T$ makes this signaling method more sensitive to small timing errors.

Again, error propagation becomes a problem unless precoding is used. Precoding used for the modified duobinary case is

$$b_k = x_k \oplus b_{k-2}. \tag{9.89}$$

The binary stream b_k is applied to the modified duobinary operation described by Eq. (9.85), yielding

$$y_k = b_k - b_{k-2} = (x_k \oplus b_{k-2}) - b_{k-2}. \tag{9.90}$$

The various possibilities using Eq. (9.90) are listed in Table 9.3. From this table, we see that the y_k can be decoded to find the x_k if we follow the rule

$$\tilde{x}_k = |y_k| = y_k \bmod\text{-}2. \tag{9.91}$$

Similarly, the receiver rule for modified duobinary signaling that corresponds to Eq. (9.84) for the duobinary case is

$$\tilde{x}_k = \begin{cases} 1 & \text{if } Y_k = \pm A \\ 0 & \text{if } Y_k = 0 \end{cases} . \tag{9.92}$$

Again, error propagation is eliminated because each binary decision depends only on the current received value.

Table 9.3 Decoding of Precoded Modified Duobinary Signals

x_k	b_{k-2}	y_k	Y_k
1	0	1	A
1	1	-1	$-A$
0	0	0	0
0	1	0	0

Table 9.4   Classes of Partial-Response Signaling

Class	Generating Equation	Magnitude Frequency Transfer Function, $|H(\omega)|$
1	$y_k = x_k + x_{k-1}$	$2 \cos \omega T/2$
2	$y_k = x_k + 2x_{k-1} + x_{k-2}$	$4 \cos^2 \omega T/2$
3	$y_k = 2x_k + x_{k-1} - x_{k-2}$	$[(2 + \cos \omega T - \cos 2\omega T)^2 + (\sin \omega T - \sin 2\omega T)^2]^{1/2}$
4	$y_k = x_k - x_{k-2}$	$2 \sin \omega T$
5	$y_k = x_k - 2x_{k-2} + x_{k-4}$	$4 \sin^2 \omega T$

Both duobinary and modified duobinary are cases of a wider selection of partial-response signaling methods. These and some others, often referred to by classes of partial response, are listed in Table 9.4. Note that duobinary and modified duobinary are referred to as Class 1 and Class 4 signaling, respectively.

Drill Problem 9.10.1   Determine the transmitted data stream corresponding to the input sequence x_k = 0 1 1 1 0 0 0 1 0 1 1 0 0, using (a) precoded duobinary signaling, and (b) precoded modified-duobinary signaling. Assume that the precoder b_k is initialized with 1's and that the transmitter levels are designated as $+$, 0, $-$. Also verify that the receiver decoding rule yields the x_k.

Answer.   (a) $+0\ 0\ 0 - - - 0 + 0\ 0 + +$; (b) $0 - - - + 0\ 0\ 0 - 0 + + 0\ 0$.

Drill Problem 9.10.2   Show that the sum of the weight coefficients in the transversal filter in Fig. 9.22 must be zero for a spectral null at zero frequency.

9.11   EQUALIZATION

In preceding sections, steps were described for the improvement of spectral weightings of a transmission channel to minimize intersymbol interference (ISI). In many systems, however, either departures from ideal characteristics are not known or they may vary. For example, the characteristics of a telephone channel may vary as a function of a particular connection and line used, as well as possible variations with time. It is advantageous in such systems to include a filter that can be adjusted to compensate for the nonideal characteristics of the transmission channel. Because their purpose is to compensate for imperfect channel transmission characteristics, these filters are called *equalizers*. Although equalizers could be included in the transmitter, receiver, or channel, for the purposes of this discussion we include them in the receiver.

It is difficult to construct equalizers that will compensate exactly for all channel imperfections. Various criteria can be proposed for equalizer designs.

We shall examine several types of equalizers which attempt to minimize ISI. However, we must keep in mind that additive noise is also a problem. Because cost is proportional to the number of adjustable parameters in an equalizer, the available number of adjustable parameters must be divided carefully between matched filtering (to minimize the effects of additive noise) and equalization (to minimize ISI). For example, in a high-noise environment with small ISI it might be best to concentrate on matched filtering. On the other hand, in a low-noise environment, such as a telephone channel, the most important consideration is the minimization of ISI.

Because of its versatility and relative ease in implementation, the transversal filter is a popular equalizer design. Also, the transversal filter design lends itself to the compensation of channels for which the time dispersion of pulse signals is only over a small number of symbol intervals. Because this is typical of many data transmission channels (including telephone), the transversal filter is the only equalizer implementation we discuss here. The transversal equalizer consists of a tapped delay line, as shown in Fig. 9.25. The delay line is tapped at intervals of T seconds, where T is the symbol interval of the input pulse stream.† The output of each tap is weighted by a variable gain factor a_n, and the weighted outputs are added and sampled to form the output. For convenience, we assume here that there are $(2N + 1)$ taps with the corresponding weights $a_{-N}, \ldots, a_0, \ldots, a_N$, as indicated in Fig. 9.25.

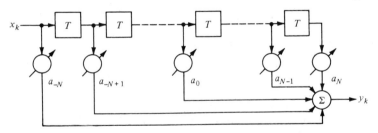

Fig. 9.25 Block diagram of a transversal equalizer.

The problem of minimizing ISI is simplified by considering only those signals at correct sample times. Thus we represent the input to the transversal equalizer by $x(kT) = x_k$ and the corresponding output by $y(kT) = y_k$. For zero ISI, we require that

$$y_k = \begin{cases} 1 & \text{for } k = 0 \\ 0 & \text{elsewhere} \end{cases} \tag{9.93}$$

† The transversal filter can be used for matched filtering also, by spacing the taps on the delay line at Nyquist intervals rather than at symbol intervals.

The output y_k can be expressed in terms of the input x_k and tap weights a_n as

$$y_k = \sum_{n=-N}^{N} a_n x_{k-n}. \tag{9.94}$$

Equation (9.94) yields $2N + 1$ independent equations in terms of the a_n. This limits us to $2N + 1$ constraints, and therefore Eq. (9.93) must be modified to

$$y_k = \begin{cases} 1 & k = 0 \\ 0 & k = \pm 1, \pm 2, \ldots, \pm N \end{cases}. \tag{9.95}$$

Thus the transversal equalizer can force the output to go to zero at N sample points to either side of the desired peak output (which has been normalized to unity for convenience). For this reason, this is called a *zero-forcing* equalizer. There is no guarantee that the output will be zero at T-second intervals beyond $\pm N$, where $2N + 1$ is the number of taps used. However, it can be shown that this equalizer is optimal in the sense that it minimizes the peak ISI.†

We can combine Eqs. (9.94) and (9.95) to exhibit the $2N + 1$ equations that can be solved for the tap weights, a_n:

$$\begin{vmatrix} x_0 & x_{-1} & \cdots x_{-N} & \cdots x_{-2N-1} & x_{-2N} \\ x_1 & x_0 & \cdots x_{-N+1} & \cdots x_{-2N} & x_{-2N+1} \\ \vdots & \vdots & & \vdots & \vdots \\ x_N & x_{N-1} & \cdots x_0 & \cdots x_{-N-1} & x_{-N} \\ \vdots & \vdots & & \vdots & \\ x_{2N-1} & x_{2N-2} & \cdots x_{N-1} & \cdots x_{-2} & x_{-1} \\ x_{2N} & x_{2N-1} & \cdots x_N & \cdots x_{-1} & x_0 \end{vmatrix} \begin{vmatrix} a_{-N} \\ a_{-N+1} \\ \vdots \\ a_0 \\ \vdots \\ a_{N-1} \\ a_N \end{vmatrix} = \begin{vmatrix} 0 \\ 0 \\ \vdots \\ 1 \\ \vdots \\ 0 \\ 0 \end{vmatrix} \tag{9.96}$$

Computation of required tap weights using Eq. (9.96) is illustrated in the following example.

Example 9.11.1   Determine the tap weights of a three-tap, zero-forcing equalizer for the input shown in Fig. 9.26(a), where $x_{-2} = 0.0$, $x_{-1} = 0.2$, $x_0 = 1.0$, $x_1 = -0.3$, $x_2 = 0.1$, $x_k = 0$ for $|k| > 2$.

Solution.   Writing the $2N + 1 = 3$ equations, the three tap weights a_{-1}, a_0, a_1 are solutions of the following set of simultaneous equations:

$$\begin{aligned} a_{-1} + 0.2a_0 &= 0 \\ -0.3a_{-1} + a_0 + 0.2a_1 &= 1 \\ 0.1a_{-1} - 0.3a_0 + a_1 &= 0. \end{aligned}$$

---

† R. W. Lucky, J. Salz, and E. J. Weldon, Jr., *Principles of Data Communication*, N.Y.: McGraw-Hill, 1968, Ch. 6.

Fig. 9.26 (a) Input pulse waveform and (b) output (equalized) pulse waveform for the three-tap, zero-forcing, transversal equalizer in Example 9.11.1.

Solving, we obtain the required tap weights

$$a_{-1} = -0.1779$$
$$a_0 = 0.8897$$
$$a_1 = 0.2847.$$

Using this three-tap equalizer, we compute the values of the equalized pulse from Eq. (9.94) to find

$$
\begin{array}{ll}
y_{-3} = 0.0 & y_1 = 0.0 \\
y_{-2} = -0.0356 & y_2 = 0.0153 \\
y_{-1} = 0.0 & y_3 = 0.0285 \\
y_0 = 1.0 & y_4 = 0.0.
\end{array}
$$

The resulting equalized pulse waveform is shown in Fig. 9.26(b). Note that this pulse has the desired zeros to either side of the peak, but some ISI has been introduced at sample points further from the peak. Except in special cases, the finite-N, zero-forcing equalizer cannot completely eliminate ISI. As the number of filter taps become large, however, the ISI can be made small if pulse-dispersion effects caused by the channel do not extend over very many symbol intervals.

Drill Problem 9.11.1   Let the input sample values of a given pulse be $x_0 = 1.0$, $|x_{-1}| \ll 1$, $|x_1| \ll 1$, $x_k = 0$ for $|k| > 1$. For this case, develop a convenient approximation for the tap weights of a three-tap, zero-forcing transversal equalizer.

Answer. $a_n \approx (-1)^n x_n; \ n = -1, 0, 1.$

Adjustment of tap weights of the transversal equalizer involves the solution of a set of $2N + 1$ simultaneous equations. Systems have been developed, by using iterative techniques, to adjust the weights automatically. Generally, these adjustments can be divided into two categories. The *preset equalizer* makes use of a special sequence of pulses prior to or during pauses in the transmission of data to determine and adjust the tap weights. In contrast, the tap weights in the *adaptive equalizer* are adjusted during the transmission of data by using the data signal itself. Hybrid systems make use of both of these approaches.

We discuss two methods of equalizer adjustment. The first method is better suited to preset equalizers, whereas the second method is applicable to both the preset and the adaptive equalizer. In the first method of adjustment, the preset equalizer uses widely separated test pulses to set the tap weights prior to data transmission. A simplified block diagram of a three-tap preset equalizer is shown in Fig. 9.27. Upon reception of a test pulse, a binary decision is made in the level slicer at each sample time, and these decision bits are loaded into a shift register, one at a time. The threshold in the level slicer is zero for all sampled values, except that the peak is detected by a peak detector and compared against $+1$ (or whatever reference level is desired). At the end of each test pulse, the binary contents of the shift register are used as correction signals to the respective tap weights in the transversal filter. Because each error signal is simply a binary value (i.e., ± 1), each tap weight is adjusted by a fixed increment, $\pm \Delta$, in each iteration. This iterative "training" algorithm continues, ideally, until all tap weights converge to their optimum values.

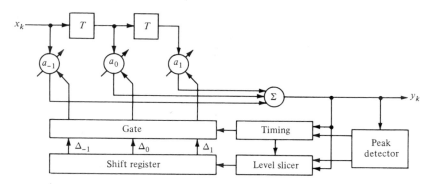

Fig. 9.27 Example of a three-tap preset transversal equalizer.

The accuracy to which the tap weights are set is proportional to Δ in the fixed-increment adjustment procedure. Equalizer settling time, on the other hand, is inversely proportional to Δ, so that the increment size must be chosen judiciously. The effects of additive noise can be decreased by averaging the generated error signals over several test pulses, but doing so increases the settling time required for convergence within a given accuracy constraint. Refinements to decrease the required settling time include use of the variable-increment adjustment of tap weights and the use of a pseudorandom noise (PN) pulse sequence instead of widely separated pulses in the training phase. A disadvantage of the preset equalizer is that once the training period is completed, the stability of the tap weights must be such that they can be held within the system's desired range of accuracy until the next training period occurs.

The second method for adjustment of the tap weights in a transversal equalizer is to minimize the mean-square error between the output sequence and a known (or desired) sequence. Recall that in the zero-forcing equalizer the tap weights are adjusted to yield N 0's to either side of the peak in the equalized output. In contrast, this second method seeks to minimize the mean-square error over all taps. In practice, the results of both methods are very similar when channel distortions are relatively small. However, they do yield different algorithms for the implementation of automatic equalizers.

To investigate a mean-square minimization criterion, let $c_k = c(kT)$ be a given pulse sequence and let $y_k = y(kT)$ be the output of a transversal filter. Defining an error $e(kT)$ as the difference between these two, we have

$$e(kT) = y(kT) - c(kT). \tag{9.97}$$

The mean-square error for K samples is

$$\overline{e^2} = \frac{1}{K} \sum_{k=1}^{K} [y(kT) - c(kT)]^2. \tag{9.98}$$

A necessary condition on a given tap weight a_n when adjusted to its optimal setting is

$$\frac{\partial \overline{e^2}}{\partial a_n} = 0. \tag{9.99}$$

Using Eq. (9.98) in Eq. (9.99), we get

$$\frac{\partial \overline{e^2}}{\partial a_n} = \frac{2}{K} \sum_{k=1}^{K} [y(kT) - c(kT)] \frac{\partial y(kT)}{\partial a_n}. \tag{9.100}$$

Then using Eq. (9.94) in Eq. (9.100), we get

$$\frac{\partial \overline{e^2}}{\partial a_n} = \frac{2}{K} \sum_{k=1}^{K} [y(kT) - c(kT)] x(kT - nT). \tag{9.101}$$

We rewrite Eq. (9.101) as

$$\frac{\partial \overline{e^2}}{\partial a_n} = 2R_{ex}(nT), \tag{9.102}$$

where R_{ex} is the deterministic crosscorrelation of the output error sequence $e(kT)$ and of the input sequence $x(kT)$. Combining Eqs. (9.99) and (9.102), we have

$$R_{ex}(nT) = 0 \qquad \text{for } n = 0, \pm 1, \cdots, \pm N. \tag{9.103}$$

Therefore the tap weights are optimum, in a minimum mean-square error sense, when the crosscorrelation between the output error sequence $e(kT)$ and the input sequence $x(kT)$ is zero for all integer multiples of the tap-delay increment T.

Now let the input sequence $x(kT)$ be equal to the sequence $c(kT)$; then Eq. (9.101) becomes

$$\frac{\partial \overline{e^2}}{\partial a_n} = \frac{2}{K} \sum_{k=1}^{K} [y(kT) - c(kT)]c(kT - nT) \tag{9.104}$$

$$= 2R_{yc}(nT) - 2R_c(nT). \tag{9.105}$$

The procedure is that the n-th tap weight, a_n, is to be adjusted so that $\partial \overline{e^2}/\partial a_n$ goes to zero. Noting that $R_c(0) = \overline{c^2}$, we form an estimator of the n-th tap weight as follows:†

$$\hat{\epsilon}_n = \frac{1}{K\overline{c^2}} \sum_{k=1}^{K} [y(kT) - c(kT)]c(kT - nT). \tag{9.106}$$

Because $y(kT)$ depends upon a_n, this adjustment is an iterative procedure. Also, the estimator must be checked for random inputs. It can be shown that this estimator is an unbiased estimator of the tap weight error if‡

$$R_c(nT) = 0 \qquad \text{for } n \neq 0. \tag{9.107}$$

Equations (9.106) and (9.107) suggest that an efficient means of adjusting a preset equalizer is the use of the minimum mean-square error criterion. Instead of transmitting a sequence of widely separated pulses, a pseudorandom noise (PN) sequence is sent—which takes care of satisfying Eq. (9.107). At the receiver a maximal-length codeword generator can be used to generate the PN sequence. Then Eq. (9.106) can be used to generate an error signal for each tap weight. Use of the PN sequence in this manner, instead of widely separated test pulses, decreases the equalizer settling time required for a given tap-weight accuracy.

In contrast to preset equalizers, the error signals in adaptive equalizers are estimated continuously during the transmission of data. Thus adaptive equalizers can adapt to slowly varying changes during data transmissions and do not require long training periods.

In the presence of data transmission, the random bit sequence needed to adjust the tap weights in an adaptive equalizer can be derived from the binary decisions that the receiver makes on the data sequence. Such a learning procedure is called *decision-directed* because the receiver learns by employing its own decisions. A three-tap adaptive equalizer using this approach is shown in Fig. 9.28. Decision-directed adaptive equalizers do not establish initial equalization easily. Once the correct initial equalization is acquired, however, the error estimates are accurate and the equalization loop tracks changes in channel characteristics unless the changes are faster than the equalizer's time constants.

A hybrid system—a combination of a preset equalizer and an adaptive equal-

---

† *Ibid.*
‡ *Ibid.*

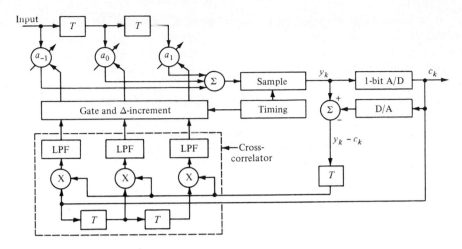

Fig. 9.28 Example of a three-tap adaptive transversal equalizer.

izer—may be used to gain the best overall performance. In such a system, a PN sequence is sent (and also generated at the receiver) to enable the system to acquire the correct initial equalization values in a preset mode of operation. When the initial equalization is reasonably good, the system switches to an adaptive mode and data transmission commences. The system may also provide a reset mode in which data transmission pauses temporarily to permit the sending of a PN sequence to verify the correctness of equalizer settings.

Equalizers based on the principles in the preceding discussion do a good job of minimizing ISI in channels where additive noise is not of major concern. They are essential parts of modern high-speed modems for the transmission of digital data using telephone channels.

9.12   *M*-ARY SIGNALING

In the PCM communication systems that we have considered up to this point, one of only two possible signals is transmitted during each *T*-second signaling interval. We call a signal waveform that might be sent in a *T*-second interval a *symbol*. In a binary transmission system, only two symbols are used (e.g., $\pm A$). The symbol rate, often called the *baud rate*, is $(1/T)$ symbols per second. For example, in a 300-baud system, symbols are sent at a rate of 300 symbols per second.

In *M*-ary signaling, one of *M* possible symbols is transmitted during each *T*-second interval. In baseband transmission systems, each symbol corresponds to one of *M* distinct levels. However, *M*-ary transmissions are not restricted to amplitude-level variations. We could, for example, vary the amplitude, frequency, and phase of a sine wave to form the various symbols. The subject of digital modulation is introduced and discussed in Chapter 10.

It is convenient to envision all sampling and multiplexing operations being carried out by binary signals and then a composite bit stream being applied to a M-ary encoder prior to transmission. This also turns out to be the usual case for operating systems. For example, in a quaternary ($M = 4$) transmission we could assign the binary groupings 00, 01, 10, and 11 to levels A, B, C, and D, respectively. Each of possible output levels A, B, C, D is transmitted within a T-second interval.

If a channel bandwidth of B Hz is available for transmission, the Nyquist rate is $1/T = 2B$ symbols per second. In M-ary signaling, each symbol represents $\log_2 M$ bits of information and may be uniquely coded into $\ell = \log_2 M$ levels ($M = 2^\ell$). It follows that the equivalent bit rate is $2B\ell$ bps. Thus the bandwidth required for M-ary signaling with a fixed information rate is inversely proportional to ℓ. For example, let B_2 be the bandwidth required for a binary PCM system; then $B = B_2/\ell$ is the bandwidth required for a M-ary system to transmit information at the same rate as a binary system.

Because the bandwidth required for transmission is proportional to the baud rate $1/T$, but the information transmitted is proportional to the bit rate ($\log_2 M)/T$, we see that M-ary systems provide a means for increasing the rate-of-information transmission within a given bandwidth. However, this increased information rate comes at the expense of added transmitter power and increased system complexity. Moreover, intersymbol interference is also a problem in M-ary signaling systems, and the bandwidths required for operating these systems are greater than the theoretical minimums discussed previously. Mathematical results are difficult to obtain for the effects of ISI in M-ary systems, however, and these effects are often studied using system simulation studies and computer programs to generate multilevel eye diagrams.†

In an analysis of M-ary signaling, we assume that the spacing of amplitude levels is uniform and that the M amplitude levels are centered at zero. For convenience, we assume also that M is an even integer (the usual case is that M is a power of two, or $M = 2^\ell$, where ℓ is an integer). Designating the spacing between adjacent levels by A, we find the various levels are

$$A_j = \pm\frac{A}{2}, \pm\frac{3A}{2}, \pm\frac{5A}{2}, \ldots, \pm\frac{(M-1)A}{2}. \tag{9.108}$$

Assuming that all M levels are equiprobable, we obtain the average signal power [cf. Eq. (9.18)]

$$S = \frac{2}{M}\left\{\left(\frac{A}{2}\right)^2 + \left(\frac{3A}{2}\right)^2 + \cdots + \left[\frac{(M-1)A}{2}\right]^2\right\}$$

$$S = \frac{M^2 - 1}{3}\left(\frac{A}{2}\right)^2. \tag{9.109}$$

---

† See, for example, K. Feher, *Digital Communications: Microwave Applications*, Englewood Cliffs, NJ: Prentice-Hall, 1980.

Therefore, for a given level spacing, the transmitted power increases approximately in proportion to the square of the number of levels.

Next, we assume that $M = 4$ (i.e., a quaternary system) and that the signaling levels are $-3A/2$, $-A/2$, $A/2$, $3A/2$. For convenience, we let the corresponding symbols be labeled A, B, C, D, respectively. We also let the additive noise present be gaussian-distributed with zero mean value and variance σ^2. The probability density function for the case in which all four symbols are equiprobable is shown in Fig. 9.29. Because the symbols have been chosen to be equiprobable, the optimum threshold levels are midway between levels. Noting that there are six equal areas in Fig. 9.29 to be included in the probability-of-error calculation, we have

$$P_\epsilon = (6)(\tfrac{1}{4}) \int_A^\infty \frac{1}{\sqrt{2\pi}\,\sigma} e^{-[y-(A/2)]^2/(2\sigma^2)}\, dy. \tag{9.110}$$

With the change of variable, $z = [y - (A/2)]/\sigma$, this expression becomes

$$P_\epsilon = \frac{3}{2} \int_{A/(2\sigma)}^\infty \frac{1}{\sqrt{2\pi}\,\sigma} e^{-z^2/2}\, dz \tag{9.111}$$

$$P_\epsilon = \frac{3}{2} \operatorname{Erfc} \frac{A}{2\sigma}. \tag{9.112}$$

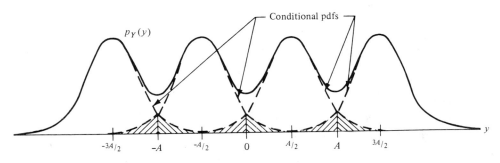

Fig. 9.29 Probability density function for signal plus noise in an *M*-ary system, $M = 4$.

A comparison of Eq. (9.112) with Eq. (9.39) for the binary case reveals that the two expressions are identical except for the factor of 3/2. This factor arose from counting the areas indicated in Fig. 9.29 and multiplying this number by $1/M$ (because all symbols are equiprobable). Noting that in general there are $2(M - 1)$ areas to be counted, we see that Eq. (9.110) can be extended to the *M*-ary case to give

$$P_\epsilon = \frac{2(M - 1)}{M} \operatorname{Erfc} \frac{A}{2\sigma}. \tag{9.113}$$

Therefore the probability of error increases slowly with an increase in number of levels.

A third comparison of interest is that of bandwidth. Designating the bandwidth of a binary system B_2, and the corresponding bit interval T_b, we have

$$B_2 = \frac{1}{2T_b} \tag{9.114}$$

and

$$B = \frac{B_2}{\log_2 M} = \frac{B_2}{\ell}. \tag{9.115}$$

From Eqs. (9.109), (9.113), and (9.115), we conclude that, in M-ary signaling at a given information rate, the required bandwidth is inversely proportional to $\log_2 M$, the transmitted power is proportional to $M^2 - 1$, and the probability of error varies over a maximum of one to two times that of an equivalent binary system. These dependencies are illustrated in the following example.

Example 9.12.1   Compare the required transmitter power and minimum bandwidth for those M-ary systems for which $1 \leq \ell \leq 5$. Assume additive white gaussian noise with two-sided power spectral density $\eta/2$ W/Hz and a probability of symbol error $P_\epsilon = 10^{-4}$. Normalize all results to the minimum bandwidth required for an equivalent binary system B_2.

Solution.   Noting that $M = 2^\ell$ and $\sigma^2 = \eta B$, we see that the equations that must be solved are

$$B = \frac{B_2}{\ell}$$

$$10^{-4} = \frac{2(M - 1)}{M} \, \mathrm{Erfc} \, \frac{A}{2\sigma}$$

$$S = \frac{M^2 - 1}{3} \left(\frac{A}{2\sigma}\right)^2 \eta B.$$

Use of the equations above and Appendix G gives the results that are listed in Table 9.5. Note that in this range the average power required rises by about 4–5 dB for every twofold increase in the number of levels used.

If the input to the M-ary encoder is binary, it is more meaningful to use for comparison the bit-error rather than the symbol-error probability. While there are no unique relationships between the bit-error probability and the M-ary symbol-error probability, we can obtain relationships for two special cases.

In the first case we assume that whenever the receiver is in error for a

Table 9.5   Comparison of *M*-ary Signaling for $P_\epsilon = 10^{-4}$

ℓ	M	B/B_2	$A/(2\sigma)$	$[S/(\eta B_2)]$ dB
1	2	1	3.72	11.4
2	4	$\frac{1}{2}$	3.82	15.6
3	8	$\frac{1}{3}$	3.86	20.2
4	16	$\frac{1}{4}$	3.88	25.0
5	32	$\frac{1}{5}$	3.88	30.1

M-ary symbol, the output is equally likely to be any one of the $M - 1$ erroneous ℓ-bit code words. Using a standard binary code (cf. Table 9.6), we find that the sum of the Hamming distances from one ℓ-bit word to all of the others in the code is $\ell 2^{\ell-1}$. The total number of bits in all of the erroneous words is $(M - 1)\ell = (2^\ell - 1)\ell$. If we assume all are equally likely, the average probability of bit error, P_{be}, is

$$P_{\text{be}} = \frac{\ell 2^{\ell-1}}{(2^\ell - 1)\ell} P_\epsilon$$

$$P_{\text{be}} = \frac{2^\ell}{2(2^\ell - 1)} P_\epsilon. \tag{9.116}$$

Note that

$$\frac{1}{2} \le \frac{P_{\text{be}}}{P_\epsilon} \le \frac{2}{3}$$

for this case.

One difficulty with the standard binary code is that the Hamming distance varies (from 1 to ℓ) between one adjacent level and another. In *M*-ary signaling, we would prefer to use a code in which only one binary digit varied as the amplitude changed from one adjacent level to the next. In other words, the code words for adjacent levels should be separated by a Hamming distance of one. The Gray code shown in Table 9.6 is an example of such a code.†

In the second case of the relationship between bit-error and symbol-error probability, we assume that the input to the *M*-ary encoder is coded so that the binary codewords for adjacent amplitude levels differ in only one binary digit (as in the Gray code). We assume also that the symbol probability of error is

---

† The binary digits in the Gray code, g_k, can be found from those in the corresponding binary code b_k from

$$g_k = \begin{cases} b_k & k = 1 \\ b_k \oplus b_{k-1} & k > 1 \end{cases}.$$

Table 9.6   Four-Bit Binary Code and Four-Bit Gray Code

Decimal No.	Level	Binary Code		Gray Code	
0	15A/2	0000		0000	
1	13A/2	0001		0001	
2	11A/2	0010		0011	
3	9A/2	0011		0010	
4	7A/2	0100		0110	
5	5A/2	0101		0111	
6	3A/2	0110		0101	Reverse-order
7	A/2	0111	Image, except	0100	image, except
			for first bit.		for first (sign)
8	−A/2	1000		1100	bit.
9	−3A/2	1001		1101	
10	−5A/2	1010		1111	
11	−7A/2	1011		1110	
12	−9A/2	1100		1010	
13	−11A/2	1101		1011	
14	−13A/2	1110		1001	
15	−15A/2	1111		1000	

low, so that it is very unlikely that an error of more than one level will be made. Within these assumptions, we have

$$P_{be} = \frac{1}{\ell} P_\epsilon \qquad \text{(Gray level coding)}. \qquad (9.117)$$

Equations (9.116) and (9.117) can be used to give some approximate answers for comparison purposes.

Drill Problem 9.12.1   Repeat Example 9.12.1, except use $P_{be} = 10^{-4}$ and a Gray code.

Answer.   See Table 9.7.

Table 9.7   Comparison of M-ary Signaling for $P_{be} = 10^{-4}$ and Gray Code

ℓ	M	B/B_2	$A/(2\sigma)$	$[S/(\eta B_2)]$ dB
1	2	1	3.72	11.4
2	4	$\frac{1}{2}$	3.65	15.2
3	8	$\frac{1}{3}$	3.58	19.5
4	16	$\frac{1}{4}$	3.52	24.2
5	32	$\frac{1}{5}$	3.47	29.2

★ 9.13   CODING FOR RELIABLE COMMUNICATION

In binary pulse code modulation, each sample of the signal is represented by a codeword of, say, k bits. These bits are transmitted and the role of the receiver is to recognize each codeword in order to reconstruct the samples. However, errors may occur in transmission as a result of noise. One way to improve the reliability of communication in the presence of noise is to increase the signal-to-noise ratio. An alternative is to add extra bits, at the expense of an increase in bandwidth, to detect and possibly even to correct the errors. This section is intended to introduce the interested reader to the latter approach. We shall restrict our attention to the binary case.

Suppose that we wish to transmit a signal which has been quantized to 16 equiprobable levels. In this case each codeword will consist of $k = 4$ bits. To each codeword we add r bits, which will be used to check for errors, and possibly to correct them. These added bits are redundant because they bear no additional information about the signal. Each codeword is then composed of n bits in the following manner:

$$k = \text{number of message digits per word};$$
$$r = \text{number of check digits per word}; \tag{9.118}$$
$$n = k + r = \text{total number of digits per word}.$$

Such a code is referred to as a (n, k) code. The code rate efficiency is defined as k/n and is an indicator of the information rate relative to the bit rate in the code.

9.13.1 Algebraic Codes†

One type of code can be formed by adding extra binary digits at the end of each codeword for error detection—and, possibly, error correction. These extra bits are called parity-check bits.

A general codeword for a parity-check code can be written in the form

$$a_1 a_2 a_3 \cdots a_k c_1 c_2 \cdots c_r$$

where a_i is the ith bit of the message code word and c_j is the jth parity-check bit. The check bits are chosen to satisfy the $r = n - k$ linear equations:

$$0 = h_{11} a_1 \oplus h_{12} a_2 \oplus \cdots \oplus h_{1k} a_k \oplus c_1$$
$$0 = h_{21} a_1 \oplus h_{22} a_2 \oplus \cdots \oplus h_{2k} a_k \oplus c_2$$
$$\vdots$$
$$0 = h_{r1} a_1 \oplus h_{r2} a_2 \oplus \cdots \oplus h_{rk} a_k \oplus c_r . \tag{9.119}$$

---

† This section requires some knowledge of matrix notation.

These equations can be expressed more conveniently in terms of an $n \times 1$ column matrix $[T]$ representing the codeword

$$[T] = \begin{bmatrix} a_1 \\ a_2 \\ \vdots \\ a_k \\ c_1 \\ \vdots \\ c_r \end{bmatrix}$$

(9.120)

and a rectangular $r \times n$ parity-check matrix $[H]$,

$$[H] = \begin{bmatrix} h_{11} & h_{12} & \cdots & h_{1k} & 1 & 0 & \cdots & 0 \\ h_{21} & h_{22} & \cdots & h_{2k} & 0 & 1 & \cdots & 0 \\ \vdots & \vdots & & \vdots & & & & \vdots \\ h_{r1} & h_{r2} & \cdots & h_{rk} & 0 & 0 & \cdots & 1 \end{bmatrix}$$

(9.121)

so that Eq. (9.119) can be written as:

$$[H][T] = 0.$$

(9.122)

Now let the received codeword be $[R]$, which may or may not be equal to $[T]$. If $[H][R] = 0$, we know that $[R]$ is a codeword—and most likely the transmitted codeword. On the other hand, if $[H][R] \neq 0$, then $[R]$ is not a codeword and at least one error has been made.

The codeword $[R]$ may be written in terms of the transmitted codeword by introducing an $n \times 1$ error matrix $[E]$:

$$[R] = [T] \oplus [E].$$

(9.123)

If $[E]$ contains all zeros, no error has been made.

In order to correct the errors we need to determine $[E]$. As a first step we determine a matrix $[S]$—called the *syndrome*—from the received codeword and the parity-check matrix:

$$[S] = [H][R]$$

$$= [H][T] \oplus [H][E]$$

$$= [H][E].$$

(9.124)

The syndrome $[S]$ has dimensions $r \times 1$ and can be any of 2^r sequences (including the all-zero sequence). The error sequence $[E]$ can be any of $2^n = 2^{r+k}$ sequences. Therefore the syndrome cannot unambiguously determine the error sequence. The task of the decoder is to select one of several error sequences which are associated with a given syndrome. Ordinarily, this selection is based on a minimum-distance criterion. If there is a single error, the error matrix $[E]$ will have one nonzero entry and the multiplication of Eq. (9.124) will result in an exhibition of the corresponding column of $[H]$. The position of the error in the codeword is then known by recognition of the appropriate column in $[H]$. This is illustrated in the examples below.

Hamming codes are those parity-check codes for which the columns of $[H]$ consist of all the distinct nonzero r sequences of binary numbers. Thus a Hamming code has as many parity-check matrix columns as there are single-error sequences; i.e.,

$$r + k = n = 2^r - 1. \tag{9.125}$$

Hamming codes are capable of correcting all single-error sequences, and every possible received sequence is either a codeword or is a distance of one from a codeword.

The codes which we have discussed are examples of *block* codes. In an (n, k) block code, the $r = n - k$ redundant bits are combined with the k message bits to give an n-digit codeword. Each n-digit codeword is encoded and decoded independently from any other codeword; n is called the *block length* of the codeword.

Example 9.13.1 A given (7, 4) Hamming code has the following parity-check matrix:

$$[H] = \begin{bmatrix} 1 & 1 & 1 & 0 & 1 & 0 & 0 \\ 1 & 1 & 0 & 1 & 0 & 1 & 0 \\ 1 & 0 & 1 & 1 & 0 & 0 & 1 \end{bmatrix}$$

a) Determine the codeword for the message code 0011.

b) If the received codeword is 1000010, determine if an error has been made. If it has, find the correct codeword.

Solution

a) The codeword is 0011 $c_1 c_2 c_3$ and we can write (all operations are modulo–2)

$$[H][T] = \begin{bmatrix} 1 + c_1 \\ 1 + c_2 \\ 0 + c_3 \end{bmatrix}.$$

Using Eq. (9.122), we find that the codeword is 0011110.

b) Computing the syndrome, we have:

$$[S] = [H][R] = \begin{bmatrix} 1 \\ 0 \\ 1 \end{bmatrix}.$$

Because the syndrome is the third column of the parity-check matrix, the third position of the received codeword is in error and the corrected codeword is 1010010.

Example 9.13.2   The columns of the parity-check matrix for the Hamming code may be chosen in any order as long as each is distinct and no column has all zeros. Repeat Example 9.13.1 for the (7,4) Hamming code which is chosen so that the columns of the parity-check matrix are the binary representations of the successive integers 1, 2, 3, . . . , n.

Solution.   a) The parity-check matrix is found by writing the binary represen-tations of 1, 2, . . . , 7 in successive columns so that:

$$[H] = \begin{bmatrix} 0 & 0 & 0 & 1 & 1 & 1 & 1 \\ 0 & 1 & 1 & 0 & 0 & 1 & 1 \\ 1 & 0 & 1 & 0 & 1 & 0 & 1 \end{bmatrix}.$$

Therefore a general codeword is $c_3 c_2 a_1 c_1 a_2 a_3 a_4$. The parity equations are

$$[H][T] = \begin{bmatrix} c_1 + 0 \\ c_2 + 0 \\ c_3 + 1 \end{bmatrix} = 0,$$

so that the codeword is 1000011.
b) Computing the syndrome, we have:

$$[S] = [H][R] = \begin{bmatrix} 1 \\ 1 \\ 1 \end{bmatrix}.$$

Thus the error is in the seventh position, and the decoded codeword is 1000011. Note that for a single error the syndrome is the binary representation of the position in the codeword which is in error.

9.13.2   Convolutional Codes

In contrast to block codes, the parity-check information in convolutional codes is distributed over a span of message symbols, called the *constraint span* of the code. In this way, long streams of message bits can be encoded continuously without the necessity of grouping them into blocks. This is accomplished by using shift registers whose outputs are combined in a preset manner to give certain constraints within the encoded bit stream.

A shift register is a cascaded series of one-stage binary memories. The binary state of each memory is transferred to the next at the command of a clock which is synchronized to the input bit stream. Thus the output of a four-stage shift register is the input delayed by four clock intervals, and at any one time this shift register will contain the four successive most recent bits in the input data stream.

A convolutional code of constraint span K can be generated by combining the outputs of K shift registers with v modulo-2 adders. At each clock time the outputs of the v adders are sampled by a commutator. Thus v output symbols are generated for each input symbol, giving a code of rate $1/v$.

An example of a convolutional encoder for $K = 4$, $v = 3$ is shown in Fig. 9.30. The binary state is available at the output of each stage until it is shifted into the next stage. The equations for these three adders are

$$v_1 = S_1,$$

$$v_2 = S_1 \oplus S_2 \oplus S_3 \oplus S_4,$$

$$v_3 = S_1 \oplus S_3 \oplus S_4. \tag{9.126}$$

Note that not all stages are connected to all adders. The connections influence the code which is generated and their selection is beyond the level of this discussion.

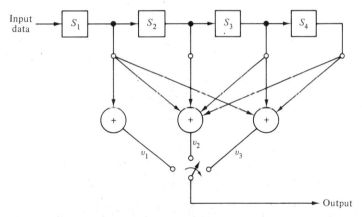

Fig. 9.30 Example of a rate $\tfrac{1}{3}$ convolutional encoder.

We demonstrate the convolutional encoding by assuming that the input data stream begins with 10110 and that all shift registers are initially at zero. The first message digit to enter is 1 and the corresponding output is 111 (if the first message digit were 0, the output would be 000). The second message digit is 0 and the corresponding output is 010 (if the second message digit were a 1, the output would be 101). Note, however, that if the first message digit had been 0, then the output would have been 000 if the second digit were 0, or 111 if the second digit were 1. Therefore the generated code depends on previous message digits within the constraint span of the code.

Next we consider the decoder problem. At each increment in the message code, the three-bit output code in the above example yields 8 possible choices.

Some of these choices can be ruled out immediately as impossible for a given encoder. The remaining ones can be examined in reference to the previous digits within the constraint span of the code. In the absence of noise, this procedure is quite straightforward and involves making decisions after each group of $v = 3$ input digits. This procedure involving successive branches can be diagrammed as a "code tree." A code tree for the convolutional encoder in Fig. 9.30 is shown in Reference 2.

In the presence of noise, the optimum procedure is similar. It examines the possible code choices in terms of the conditional probabilities between each choice within the constraint span of the code. A decoding algorithm using this approach, known as the Viterbi algorithm,[†] is popular for codes of short constraint spans. Other decoding algorithms have also been devised. However, this section is intended as a brief introduction to the subject and we shall not pursue the topic further here.

Drill Problem 9.13.1   Determine the first 18 output bits in the convolutional code for the message 110011. Use the encoder shown in Fig. 9.30 and assume that all shift registers are initially at zero.

Answer.   111101001000100101.

9.14 SUMMARY

Communication is the transmission of information from one point in space and time to another. Information conveyed by a knowledge of the occurrence of an event is proportional to the logarithm of the inverse of the probability of its occurrence. The binary unit of information is the bit. The information rate is the product of the average information (entropy) per symbol and the symbol rate.

The filter-like characteristics of the propagation medium and transmitting and receiving stages can be described by a mathematical model known as the channel. The limiting rate of information transmission through a noisy channel is called the channel capacity. The Hartley-Shannon law gives the channel capacity C in the presence of band-limited white gaussian noise as $C = B \log_2 (1 + S/N)$. Use of this result shows that the optimum trade-off between the signal-to-noise ratio, S/N, and bandwidth B is exponential.

Quantizing a signal introduces noise. This quantization noise is inversely proportional to the number of quantizing levels. A nonlinear characteristic known as companding is used to make all given signal levels more equiprobable in order to minimize the effects of quantization noise.

Direct transmission of a low-frequency signal is called baseband communi-

---

[†] See, for example, G. D. Forney, "The Viterbi Algorithm," *Proceedings IEEE*, **61** (March 1973): 268–273.

cation. Two types of binary baseband PCM are the on-off signal and the polar signal. The polar baseband PCM signal is more efficient and uses one-half as much average power for a given error probability. Differential PCM (DPCM) is used for possible bandwidth savings; delta modulation (DM) uses simpler circuitry to send and receive coded transmissions.

Overall S/N performance in PCM systems is limited by both transmission (thermal) noise and quantization noise. Usually, PCM systems are designed so that the latter is the controlling factor.

In long-distance transmission, signals often are equalized and amplified by repeaters spaced at distances determined by expected signal distortions and attenuations. Effects of transmission noise are compounded in successive repeaters in analog systems, whereas these effects increase approximately in proportion to the number of repeaters in digital systems.

A random binary waveform's power spectral density is of the form $(\sin x/x)^2$. Various coding formats have some effect on this form. For PCM signals having transitions predominantly toward a given polarity in direct relationship to the clock rate, the power spectral density will contain discrete frequency components.

Partial-response signaling uses controlled amounts of intersymbol interference to transmit data at increased rates in nonideal band-limited channels. This increased rate is at the expense of some additional transmitted power, and precoding is necessary to prevent undesired error propagation.

Filters adjusted to compensate for channel imperfections are called equalizers. Transversal filters are popular equalizer designs because they permit adjustment of parameters either prior to data transmission (preset types) or during data transmission (adaptive types).

Transmission at increased information rates can be achieved using one of more than two possible levels within each symbol duration. In M-ary signaling, one of M levels is sent and, thus, each symbol represents $\log_2 M$ bits of information. Because bandwidth is inversely proportional to symbol duration, such systems can transmit at increased data rates within a given bandwidth by increasing M. This is accomplished at the expense of an increase in transmitted power and complexity of equipment.

In error-control coding, redundancy is introduced by adding extra bits to the transmitted message, making it possible to detect and, possibly, to correct errors that may have occurred. Potential increases in system performance are at the expense of bandwidth. Algebraic codes are examples of block codes in which redundancy is introduced as parity-check digits in each codeword. In contrast, the parity-check digits in convolutional codes are continuously interleaved in the coded bit stream.

Selected References for Further Reading

1. W. R. Bennett and J. R. Davey. *Data Transmission*. New York: McGraw-Hill, 1965. Digital communication techniques with particular emphasis on telephone systems.

2. H. Taub and D. L. Schilling. *Principles of Communication Systems*. New York: McGraw-Hill, 1971.
   Chapters 6, 12, and 13 contain good discussions of quantization, delta modulation, and block codes.

3. A. B. Carlson. *Communication Systems: An Introduction to Signals and Noise in Electrical Communication*, Second Ed. New York: McGraw-Hill, 1975.
   Chapter 4 covers baseband transmission; Chapter 9 has a good treatment of information theory.

4. K. S. Shanmugam. *Digital and Analog Communication Systems*. New York: John Wiley & Sons, 1979.
   Chapter 5 has a good treatment of baseband data transmission; various coding techniques are discussed briefly in Chapter 9.

5. R. W. Lucky, J. Salz and E. J. Weldon, Jr. *Principles of Data Communication*. New York: McGraw-Hill, 1968.
   Chapters 4–6 cover baseband data transmission and equalization with particular emphasis on telephone systems.

6. S. Lin. *An Introduction to Error-Correcting Codes*. Englewood Cliffs, NJ: Prentice-Hall, 1970.
   Well-presented coverage of various methods of error-correcting codes.

7. K. W. Cattermole. *Principles of Pulse Code Modulation*. London: Iliffe (New York: American Elsevier Pub. Co.), 1969.
   A treatment of PCM with particular emphasis on quantization noise effects.

8. R. Steele. *Delta Modulation Systems* (Halsted Press). New York: John Wiley & Sons, 1975.
   Discusses various forms of delta modulation and their S/N performance; comparisons with PCM; advanced.

9. D. J. Sakrison. *Communication Theory: Transmission of Waveforms and Digital Information*. New York: John Wiley & Sons, 1968.
   An advanced treatment of information theory and signal theory for digital transmission.

10. R. G. Gallager. *Information Theory and Reliable Communication*. New York: John Wiley, 1968.
    An advanced treatment of coding is covered in Chapter 6.

Problems

9.1.1 A standard kilowatt-hour meter for residential use has five dials, each numbered 0–9. Calculate the amount of information (in bits) required to make a meter reading.

9.1.2 A black-and-white television picture may be considered as composed of approximately 3×10^5 picture elements, each having 10 brightness levels. Suppose one attempts to describe such a picture orally using 1000 words out of a vocabulary of 100,000 words (not very realistic but enough to illustrate the idea).

    a) Determine the amount of information required to describe the picture.

    b) Assuming equiprobable words, comment on the validity of the old adage: "one picture is worth one thousand words."

9.1.3 A given code has six messages with probabilities 1/2, 1/4, 1/8, 1/16, 1/32, 1/32. Determine the entropy of the code.

9.1.4 The entropy of average English text is 4.16 bits with a letter frequency (per 1000):

E	124	R	60	F	23	V	9
T	96	H	51	M	22	K	5
A	81	L	40	W	20	Q	2
O	79	D	37	Y	19	X	2
N	72	C	32	B	16	J	1
I	72	U	31	G	16	Z	1
S	66	P	23				

a) What letter conveys the maximum information?
b) What letter conveys the minimum information?
c) If John asks you to guess the name of the object he is thinking of and gives the first letter, which is a more helpful clue, "e" or "z"? Why?
d) What would be the entropy of English text if all letters were equiprobable?

9.1.5 Using the table in Problem 9.1.4, determine the entropy of English text if all the vowels were deleted from the alphabet.

9.1.6 A given code consists of 271 words of which 15 occur with probability $(16)^{-1}$ and the remaining 256 each occur with probability $(4096)^{-1}$. If 100 code words are transmitted each second, what is the rate of information transmission?

9.1.7 A given code is composed of dots and dashes. Assume that a dash is twice as long as a dot (this includes allowances for spaces) and has one-half the probability of occurrence.

a) Calculate the information in a dot and in a dash.
b) Calculate the entropy in the code.
c) If a dot interval is 10 msec, what is the average rate of information transmission?

9.2.1 A standard voice-grade telephone line has a bandwidth of 3 kHz.

a) If $S/N = 30$ dB, what is the channel capacity?
b) In practice, the maximum data rate on such a line is 4800 bits/sec. What is the minimum theoretical S/N required to maintain this rate?
c) Determine the minimum amount of time required to transmit the meter information of Problem 9.1.1 over a standard telephone line ($B = 3$ kHz, $S/N = 30$ dB).

9.2.2 Investigate the relative channel-capacity trade-offs between bandwidth and S/N ratio for S/N of (a) -30 dB; (b) 0 dB; (c) 30 dB.

9.2.3 a) Plot channel capacity C versus B for $S/\eta = $ constant for a white gaussian channel with (two-sided) spectral density $\eta/2$.
b) If $B = 0.1$ MHz and $S/\eta = 1$ kHz, determine the channel capacity C.

9.3.1 Suppose that the information in the last four digits of a telephone number is to be transmitted digitally. How many pulses are necessary if each pulse has (a) two levels? (b) four levels? (c) eight levels? (d) Discuss the advantages and disadvantages, relative to the binary case, of using pulses with more than two levels. [Consider required bandwidth, S/N, etc.]

9.3.2 Plot a graph of S_o/N_o (in dB) versus S_i/N_i (in dB) using Eq. (9.15); also indicate the approximation given in Eq. (9.16).

9.4.1 A signal voltage which is band-limited to 3 kHz and amplitude-limited to a peak-to-peak amplitude of 2 V is converted to a binary PCM code using 256 evenly spaced levels.

a) Calculate the minimum bandwidth required.
b) Calculate the peak-signal–to–rms-quantization-noise ratio.

9.4.2 a) A logarithmic companding law often used for speech (e.g., telephone) is

$$e_o = \frac{\log(1 + \mu e_i)}{\log(1 + \mu)},$$

where e_i, e_o are normalized to $(0, 1)$ and μ is a constant. (Odd symmetry is used for negative arguments.) Plot the compressor characteristic for $\mu = 10$, 100, and 255. (PCM telephone systems in the U.S. use μ-law companding with $\mu = 255$.)

b) Another logarithmic compression used for telephone service in Europe is the A-law defined by

$$e_o = \begin{cases} \dfrac{Ae_i}{1 + \log(A)} & 0 \le e_i \le \dfrac{1}{A} \\ \dfrac{1 + \log(Ae_i)}{1 + \log(A)} & \dfrac{1}{A} \le e_i \le 1 \end{cases}.$$

Plot the compressor characteristic for $A = 100$, and compare with the $\mu = 255$ characteristic in (a).

9.4.3 A given signal has a probability density

$$p(x) = \begin{cases} K[1 - (x/4)^2] & -4 < x < 4, \\ 0 & \text{elsewhere.} \end{cases}$$

a) Determine the mean-square quantization noise for four equally spaced quantization levels over $(-4, 4)$. Do not assume that the distribution is constant within each interval.

b) Compare your answer to (a) with that assuming a uniform distribution.

9.4.4 Determine the four quantization levels required in Problem 9.4.3 to yield a uniform signal amplitude distribution.

9.4.5 The noise generated in a photomultiplier tube (PMT) has zero mean value and variance proportional to the signal level. What compressor and expander characteristics should be used to give a uniform distribution of PMT noise in signal transmission for equiprobable signal levels?

c **9.4.6** Quantization levels can be chosen to minimize the error on a mean-square basis as follows. If b_k is the kth quantizing level and a_k, a_{k+1} are the adjacent round-off boundaries, then the quantizing error for n quantizing levels is:

$$\overline{\epsilon^2} = \sum_{k=1}^{n} \int_{a_k}^{a_{k+1}} (x - b_k)^2 p(x)\, dx.$$

This can be solved for a minimum by choosing:

$$a_k = (b_k + b_{k-1})/2$$

$$b_k = \frac{\int_{a_k}^{a_{k+1}} xp(x)\, dx}{\int_{a_k}^{a_{k+1}} p(x)\, dx}.$$

Because the end-points (a_1, a_n) are known, the quantizing levels (b_k) can be found from these relations. Solve Example 9.4.1 using these equations.

9.5.1 A binary PCM waveform uses the Manchester code representation for 0, 1, as shown in Fig. P–9.5.1. Find an expression for the average probability of error for equiprobable 1's and 0's, assuming the use of a matched-filter detector in the presence of additive white gaussian noise with zero mean and two-sided power spectral density $S_n(\omega) = \eta/2$ W/Hz.

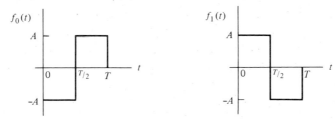

Fig. P–9.5.1

9.5.2 A received signal is either $+1$ V or -1 V over $(0, T)$. Additive white gaussian noise is also present with a (two-sided) spectral density of 1 mW/Hz across 50 Ω. If the composite signal-plus-noise is processed by a matched filter, determine the minimum time T for a probability of error less than 10^{-4}.

9.5.3 An on-off binary system uses the pulse waveform shown in Fig. P–9.5.3 for a one and absence of a pulse for a zero. Let $A = 1$ V and $T = 30$ μsec. Additive white gaussian noise is also present with a spectral density of 1 μW/Hz across one ohm.

a) Find the optimum receiver.
b) Determine the probability of error for equiprobable ones and zeros.

Fig. P–9.5.3

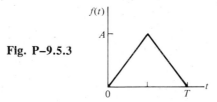

9.5.4 Repeat Problem 9.5.3 for a binary system which sends $[-f(t)]$ for zeros and transmits the same average power.

9.6.1 Let a PCM threshold be defined where S_o/N_o decreases by 1 dB from the quantization-noise-limited performance of Fig. 9.11. Determine the S_i/N_i threshold values for $m = 4, 6, 8,$ and 10.

9.6.2 A PCM signal is passed through a channel such that the normalized response of the channel to one pulse of width T is

$$g(t) = [1 + \cos \pi(t - T)/T]/2 \qquad 0 < t < 2T.$$

If the bit duration of the PCM signal is T, sketch the eye patterns for the following: (a) an alternating 1010 sequence; (b) an alternating 11001100 sequence.

9.6.3 A polar-NRZ system suffering from ISI has sampled values at the receiver of

$$y(t_k) = \pm A + n(t_k) + z(t_k),$$

where $n(t_k)$ is additive noise and $z(t_k)$ represents the ISI. The pdf of $z(t_k)$ is given as

$$p_Z(z) = \tfrac{1}{4}\delta(z + \alpha) + \tfrac{1}{2}\delta(z) + \tfrac{1}{4}\delta(z - \alpha).$$

a) Assuming zero-mean gaussian noise $n(t)$ with variance σ^2, obtain an expression for the net probability of error in terms of A, α, and σ for equiprobable 1's and 0's.

b) For $A/\sigma = 4.0$, compute the equivalent degradation (in dB) in $S/N = A^2/\sigma^2$ for $\alpha/A = 0.10$, and $\alpha/A = 0.20$. Interpret your results in terms of an eye diagram closure.

9.7.1 A given DM system operates with a sampling frequency $f_s = 1/T$ and a fixed step size a. If the input to the system is $f(t) = kt$ for $t > 0$, determine (a) the value of k above which a slope overload will occur; (b) the value of k which minimizes the mean-square quantization noise.

c 9.7.2 Write a computer program for the DM system shown in Fig. 9.14(a). Let the input be $f(t) = \sin 2\pi t$ and use sample spacings of 0.05 sec. Run your program and plot both $f(t)$ and the accumulator output for a step size of (a) 0.10; (b) 0.20; (c) 0.40.

c 9.7.3 Convert the DM system of Problem 9.7.2 into a two-bit DPCM system by adding the following statement at the appropriate point: "If the magnitude of the difference is greater than one step size, double the step size." Repeat Problem 9.7.2 for this system with sample spacings of 0.10 sec.

★ 9.8.1 Modify the results of Section 9.8, assuming the gains of all repeaters are not identical.

★ 9.8.2 NRZ polar binary signals are transmitted over a transmission path using nine regenerative repeaters. Assume additive gaussian noise with zero mean value and a probability of error of 10^{-7} for each repeater. Compute (a) the net probability of error; (b) the net probability of error if analog repeaters are used; (c) the additional transmitted power, in dB, required in each analog repeater to give the same net probability of error as a system in which regenerative repeaters are used.

★ 9.9.1 Determine the power spectral density of the following transmissions for equiprobable 1's and 0's: (a) the binary system in Problem 9.5.4; (b) the binary system in Problem 9.5.3.

★ 9.9.2 Suppose that a certain binary system transmits the waveform shown in Fig. P–9.5.3 for a 1 and a fixed amplitude $-A/2$ over $(0, T)$ for a 0. Assume equiprobable 1's and 0's.

a) Do you expect any impulses in the power spectral density of the transmitted data stream? Explain.

b) Determine the power spectral density.

★ 9.9.3 Determine the power spectral density of the waveform

$$\phi(t) = \begin{cases} A \sin \omega_c t & \text{for a 1} \\ \gamma A \sin \omega_c t & \text{for a 0} \end{cases} \text{over } (0, T),$$

where $0 \leq \gamma < 1$.

9.10.1 Sketch a block diagram of the transversal filter needed to implement the partial-response signaling codes described by

$$y_k = x_k - 2x_{k-2} + x_{k-4}.$$

Also determine the magnitude of its frequency transfer function. [*Hint:* Examples 3.6.4 and 3.6.2 may be helpful.] Check your answer with Table 9.4.

9.10.2 Use the sequence of binary digits shown in the first line of Fig. 7.30.

a) Determine the corresponding sequence of duobinary and modified-duobinary digits y_k. Let the levels be represented as 0, 1, 2 and -1, 0, 1, respectively.

b) Decode the y_k found in (a), assuming no knowledge of the input sequence (x_k) except that the first two input bits can be assumed to be 1's.

c) Repeat (b), assuming that an error occurs so that the first $y_k = 0$ occurring beyond the first three digits is interpreted by the receiver to be 1. Trace the effect of this error on the decoding of the signal.

d) Repeat (a) and (c) using precoding. Assume that the precoder is initialized with 1's.

9.10.3 Investigate the required receiver rule for use of the precoding operation $b_k = [(x_k + b_{k-1})\text{mod-3}]$ in duobinary signaling. For a binary input data stream x_k, how many levels are required for y_k?

9.10.4 Let the three levels in a trinary digital system be designated as two extreme levels and one intermediate level. Verify by three examples that the following possible error-detection rule holds:† (a) the values of two successive bits at the extreme levels in a duobinary system must differ if the number of intervening bits at the center level is odd; (b) they must have the same value if the number of intervening bits at the center level is even (zero is included as an even number).

9.11.1 In the presence of multipath interference, replicas of the transmitted signal arrive at the receiver with various attenuations and delays. In this problem, assume that there is only one predominant source of multipath. If the transmitted signal in such a system is $x(t)$, then the (noise-free) received signal $y(t)$ is

$$y(t) = x(t) + \alpha x(t - \tau_m),$$

where α is the attenuation of the multipath component and τ_m is its delay. Design

† M. Schwartz, *Information Transmission, Modulation, and Noise*, Third Ed., New York: McGraw-Hill, 1980, 198–199.

a transversal equalizer (cf. Fig. 9.25) for this particular problem, assuming that $T = \tau_m$ and $\alpha \ll 1$. [*Hint:* Work with frequency transfer functions.]

9.11.2 Repeat Example 9.11.1, with the exception that $x_{-2} = 0.1$.

9.11.3 Show that under the conditions of Drill Problem 9.11.1 and $|x_{-1}| = |x_1|$, the three-tap transversal equalizer provides (a) amplitude equalization if $x_{-1} = x_1$, and (b) phase equalization if $x_{-1} = -x_1$.

9.12.1 Repeat Example 9.12.1 for $P_e = 10^{-5}$, and compare your results with those in Table 9.5.

9.12.2 The received waveform in a trinary PCM system in the absence of noise has the following pdf:

$$p_Y(y) = \tfrac{1}{4}\delta(y + 5\,\text{mV}) + \tfrac{1}{2}\delta(y) + \tfrac{1}{4}\delta(y - 5\,\text{mV}).$$

Additive noise is present also with zero mean and standard deviation of 1 mV.

a) Determine the optimum threshold settings for decoding.
b) Find the resulting net probability of error.

9.12.3 Design a M-ary signaling system to transmit 9600 bps over a 3 kHz channel, and calculate the probability of error. Assume additive white gaussian noise with two-sided power spectral density of 0.5×10^{-7} W/Hz and an absolute maximum signal power level of 0 dB.

★ **9.13.1** The parity-check matrix for a repetition code with code symbols 000 and 111 is

$$[H] = \begin{bmatrix} 1 & 1 & 0 \\ 1 & 0 & 1 \end{bmatrix}.$$

a) Show that Eq. (9.122) is satisfied.
b) Demonstrate by way of an example that this repeated code will correct single errors. Does this conclusion agree with Eq. (7.38)?

★ **9.13.2** Explain why a column of all zeros is not allowed in the parity-check matrix $[H]$.

★ **9.13.3** Find all possible codewords in the parity-check code which has the parity-check matrix of Example 9.13.1.

★ **9.13.4** A given convolutional encoder is constructed using shift registers and modulo–2 adders. The equations for the adder outputs are

$$v_1 = S_1 \oplus S_2 \oplus S_3, \quad v_2 = S_1, \quad v_3 = S_1 \oplus S_2.$$

Repeat Drill Problem 9.13.1 for this encoder.

CHAPTER 10

DIGITAL MODULATION

In the preceding chapter, we covered various aspects of the digital transmission of information at baseband frequencies. For digital communication systems employing bandpass channels, it becomes advantageous to modulate a carrier signal with the digital data stream prior to transmission. Three basic forms of digital modulation, corresponding to AM, FM, and PM, are known as *amplitude-shift keying* (ASK), *frequency-shift keying* (FSK), and *phase-shift keying* (PSK). In this chapter, we consider each of these modulation methods briefly from the perspective of their probability-of-error advantage and bandwidth efficiency. Emphasis in the early part of the chapter is on binary systems. Variants of the three basic modulation methods, some using more than two possible signaling states, are described later in the chapter. The chapter closes with two sections that discuss more generalized signaling representations and considerations.

10.1 AMPLITUDE-SHIFT KEYING (ASK)

In amplitude-shift keying, the amplitude of a high-frequency carrier signal is switched between two or more values in response to the PCM code. For the binary case, the usual choice is *on-off keying* (sometimes abbreviated as OOK). The resultant amplitude-modulated waveform consists of RF pulses, called *marks,* representing binary 1, and *spaces* representing binary 0. An ASK waveform is shown in Fig. 10.1 for a given PCM code (the same code used in Fig. 7.32). As in AM, the baseband bandwidth is doubled in ASK.

The ASK waveform for one pulse (i.e., a binary 1) can be written as:

$$\phi(t) = \begin{cases} A \sin \omega_c t & 0 < t \le T, \\ 0 & \text{otherwise.} \end{cases} \tag{10.1}$$

The impulse response of the matched filter for optimum detection of this ASK waveform in the presence of white noise is, within an arbitrary constant,

$$h(t) = \phi(T - t).$$

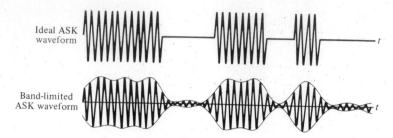

Fig. 10.1 Binary ASK waveforms.

The matched-filter output for the (noiseless) input $\phi(t)$ is

$$y(t) = \phi(t) \circledast h(t)$$

$$= \int_{-\infty}^{\infty} \phi(\tau)\phi(T - t + \tau)\, d\tau$$

$$= r_\phi(T - t), \tag{10.2}$$

where $r_\phi(t)$ is the time-autocorrelation function for the finite-energy signal $\phi(t)$ [cf. Eq. (4.43)]. The optimum decision time is for $t = T$, so that

$$y(T) = r_\phi(0) = E. \tag{10.3}$$

A sketch of the matched-filter output is shown in Fig. 10.2. Using Eq. (10.1), we find the signal energy is

$$E = \int_0^T A^2 \sin^2 \omega_c t\, dt = A^2 T/2. \tag{10.4}$$

The receiver must make a decision at $t = T$ based on the two possibilities $y(T) = n_o(T)$ and $y(T) = E + n_o(T)$. For equal source probabilities of ones and

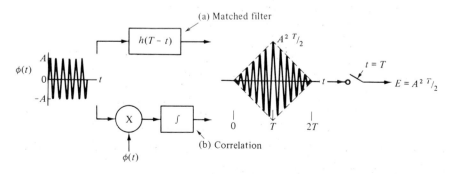

Fig. 10.2 Matched-filter detection of ASK waveforms.

zeros and noise which has a symmetrical probability density function, the optimum-decision threshold is set at $E/2$. Thus the net probability-of-error computation reduces to that of the on-off baseband system. For gaussian-distributed noise, we have found that this gives [(cf. Eq. (9.49)]

$$P_\epsilon = \text{Erfc} \sqrt{\frac{E}{2\eta}}. \tag{10.5}$$

For purposes of comparison with other systems, we express the probability of error in terms of the average signal energy per bit, $E_{\text{avg}} = ST$ so that Eq. (10.5) can be rewritten as

$$P_\epsilon = \text{Erfc} \sqrt{\frac{E_{\text{avg}}}{\eta}}. \tag{10.6}$$

The average signal power is $S = (1/2)(A^2/2)$; as before, $N = \eta B$, and if we assume Nyquist sampling, $B = 1/(2T)$† so that we can rewrite Eq. (10.5) in terms of the average signal-to-noise ratio:

$$P_\epsilon = \text{Erfc} \sqrt{\frac{S}{2N}}. \tag{10.7}$$

From this result, we conclude that the ASK system is equivalent to an on-off baseband system in terms of the average signal-to-noise ratio required for a given probability of error. Note, however, that the *peak* signal-to-noise ratio required is increased by a factor of two (i.e., by 3 dB).

The matched-filter detection of ASK is essentially a synchronous detection, as shown in Fig. 10.2(b). Envelope detection can be used and is much simpler to implement but the mathematics needed to derive the net probability of error is more involved and is not covered here. For equiprobable 1's and 0's, the result is‡

$$P_\epsilon = \frac{1}{2} \exp \frac{-E}{4\eta} + \frac{1}{2} \text{Erfc} \sqrt{\frac{E}{2\eta}}. \tag{10.8}$$

It turns out that for $P_\epsilon < 10^{-4}$, the required signal-to-noise ratios are high enough that there is only about a 1-dB penalty (or less) for the use of envelope detection.

The power spectral density of ASK is centered at ω_c and has an identical shape to that of the corresponding on-off-keyed baseband signal. Because the bandwidth has been doubled in the modulation process, the theoretical maximum

---

† The bandwidth B used here is referred to baseband; the bandpass bandwidth is doubled, but after synchronous detection it can be reduced again, yielding a detection gain of two [cf. Eq. (5.73)].

‡ See, for example, R. E. Ziemer and W. H. Tranter, *Principles of Communication Systems, Modulation, and Noise.* Boston: Houghton Mifflin, 1976, 326–330.

bandwidth efficiency is 1 bps/Hz. Operating systems typically use two to three times this amount of bandwidth.

10.2   FREQUENCY-SHIFT KEYING (FSK)

In frequency-shift keying, the instantaneous frequency of the carrier signal is switched between two (or more) values in response to the PCM code. Figure 10.3(a) shows an idealized FSK signal corresponding to the binary PCM code of Fig. 7.32. This suggests that we can consider the FSK waveform as composed of two ASK waveforms of differing carrier frequencies, as shown in Fig. 10.3(b).

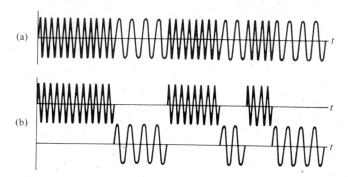

Fig. 10.3 (a) An idealized FSK waveform and (b) its decomposition into two ASK waveforms.

Thus to convey either of the binary symbols, we have a choice of the two waveforms:

$$\phi_1(t) = \begin{cases} A \sin m\omega_0 t & 0 < t \le T, \\ 0 & \text{elsewhere,} \end{cases} \tag{10.9a}$$

$$\phi_2(t) = \begin{cases} A \sin n\omega_0 t & 0 < t \le T, \\ 0 & \text{elsewhere.} \end{cases} \tag{10.9b}$$

The two received signal waveforms are now different so we use two matched filters, one for each waveform. Two possible matched-filter receivers for FSK are shown in Fig. 10.4.

The average energy per binary digit is

$$E = \int_0^T A^2 \sin^2 m\omega_0 t \, dt = A^2 T/2. \tag{10.10}$$

If one signaling frequency is present in the absence of noise, we assume that the one matched-filter output is zero and the other output is at E.[†] Conversely,

---

† This implies orthogonality over $(0,T)$; Problem 10.2.2.

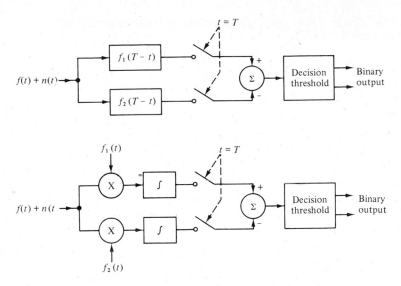

Fig. 10.4 Matched-filter detection of FSK waveforms.

if the second signaling frequency is present, the first matched-filter output is zero and as a result of the subtraction the net output is at $-E$. This is illustrated in Fig. 10.5.

This would appear to be analogous to the polar-baseband case were it not for the fact that the noise voltages are subtracted at the outputs of the two matched filters. If we assume that the two matched-filter frequency responses do not overlap, the output-noise voltages are statistically independent and hence add on a power (mean-square) basis. If the two filter bandwidths are the same (the usual case), we can then just double the variance from $\sigma_n^2 = \eta E/2$ to $\sigma_n^2 = \eta E$ (cf. Fig. 10.5). The remainder of the analysis proceeds in the same

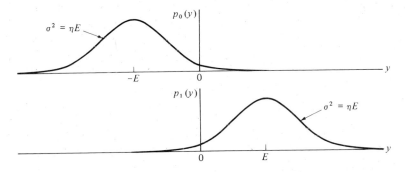

Fig. 10.5 Probability density functions for binary FSK waveforms.

manner as before. For gaussian-distributed noise and equiprobable ones and zeros, we get

$$P_\epsilon = \int_0^\infty \frac{1}{\sqrt{2\pi\eta E}} e^{-(y+E)^2/(2\eta E)} \, dy$$

$$= \text{Erfc} \sqrt{\frac{E}{\eta}}, \tag{10.11}$$

where E is the (average) energy per binary digit. Therefore we conclude that, on an average bit energy-to-noise basis, the net probability of error for FSK is the same as that for ASK. On the other hand, for the same peak-power requirements FSK has a 3-dB advantage over ASK because the latter is off approximately one-half of the time.

As noted before, matched-filter detection is really a synchronous detection. This requires both frequency and phase synchronization for two oscillators, one at each signaling frequency. One way to implement this is to use phase-locked loops, as shown in Fig. 10.6(a). The frequency range of each loop is restricted and the low-pass filter is narrow enough so that the voltage-controlled oscillators

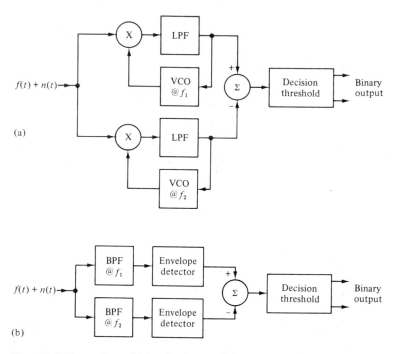

Fig. 10.6 Examples of (a) coherent and (b) noncoherent FSK detection systems.

(VCO) will not change frequency appreciably during a pause. A split-phase or Manchester code is well suited to this type of operation. Often one phase-locked loop is used and adjusted so that it will follow the input frequency.

A popular way to implement the FSK receiver is to use only the magnitude response of the matched filters, yielding the noncoherent FSK receiver shown in Fig. 10.6(b). The analysis of the envelope detector is omitted here, except for the result†

$$P_\epsilon = \frac{1}{2} \exp \frac{-E}{2\eta}.$$
(10.12)

Noncoherent FSK results in a S/N penalty of no more than 1 dB over that for orthogonal FSK with coherent detection for error rates of interest. It is a popular choice in operating systems. With this approach, the frequency spacing—to prevent significant overlap of the passbands of the two filters—must be at least $2 \Delta fT \geq 1$, where $2 \Delta f$ is the frequency difference between the two frequencies used and T is the symbol duration. Another possible approach is the use of a discriminator to convert the frequency variations to amplitude variations, followed by envelope detection. This second approach removes the constraint inherent in the first on $2 \Delta fT$, but yields slightly poorer performance. Still another approach is the use of a zero-crossing detector, which also yields a slightly poorer performance.

The overall bandwidth of the FSK transmission depends on the frequency separation used. To see this, we refer both signaling frequencies to a center frequency so that the modulated signal can be written as

$$\phi(t) = A \sin \left\{ \int_0^t [\omega_c + (\Delta\omega)p(t)] \, dt \right\},$$
(10.13)

where $2 \Delta\omega = (m - n)\omega_0$ [cf. Eq. (10.9)] and $p(t)$ is a binary switching function with possible states of ± 1 over $(0, T)$. For the case of alternating ones and zeros, $p(t)$ is a symmetrical square wave. The spectrum for this case is found in Example 6.3.3 and is shown in Fig. 10.7 for $2 \Delta fT \gg 1$. It can easily be

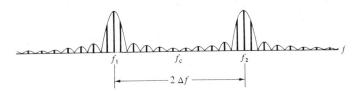

Fig. 10.7 Magnitude spectrum of a periodic FSK waveform (positive frequencies).

---

† A. B. Carlson, *Communication Systems: An Introduction to Signals and Noise in Electrical Communication.* New York: McGraw-Hill, 1975, 397.

shown that the minimum frequency separation is one-half cycle per bit interval between the two signaling frequencies in order for $\phi_1(t)$, $\phi_2(t)$ to be orthogonal over $(0, T)$.

What is the optimum frequency separation for the detection of FSK? The low-pass difference in synchronous detector outputs is proportional to

$$\int_0^T (\sin n\omega_1 t - \sin m\omega_1 t) \sin n\omega_1 t \, dt$$

$$\approx \frac{T}{2}\left[1 - \frac{\sin(n - m)\omega_1 T}{(n - m)\omega_1 T}\right] \text{ for } (n - m) \ll n, m. \quad (10.14)$$

For bandpass systems in which $\omega_c \gg \Delta\omega$ and $\omega_c T \gg 1$, Eq. (10.14) can be rewritten and incorporated into Eq. (10.11) as

$$P_\epsilon = \text{Erfc}\sqrt{[1 - \text{Sa}(2\,\Delta\omega T)]E/\eta}. \quad (10.15)$$

The factor within the brackets in Eq. (10.15) is maximum for $(2\,\Delta\omega T) \approx (3\pi)/2$, so that the frequency separation $2\,\Delta f$ for minimum probability of error requires about three-fourths of a cycle difference within one signaling interval T (i.e., $(2\,\Delta fT) \approx \frac{3}{4}$). Using this value, we find the factor within the brackets yields $1 + 2/(3\pi) = 1.21$, or,

$$P_\epsilon = \text{Erfc}\sqrt{1.21\frac{E}{\eta}}. \quad (10.16)$$

Coherent FSK systems may take advantage of this extra 0.8 dB in effective S/N, but the orthogonality condition $\Delta fT = m$ (where m is an integer), is usually assumed for other systems, giving the result expressed in Eq. (10.11).

Derivation of the FSK power spectral density for random waveform inputs is rather involved.[†] However, some general trends in the results are as follows. For low values of $2\,\Delta fT$, the power spectral density has a single peak centered at the carrier frequency f_c and decreases smoothly to either side of the peak. As $2\,\Delta fT$ is increased, the central peak in the power spectral density decreases and peaks begin to develop near the deviation frequencies $f_c \pm \Delta f$. For still larger values of $2\,\Delta fT$, the FSK power spectral density tends toward two separately identifiable spectral groupings centered at $f_c \pm \Delta f$, as illustrated by the line spectrum in Fig. 10.7. The power spectral density is continuous for random binary waveform inputs, with the exception that it will contain impulses if $2\,\Delta fT = m$, with m an integer. For the particular choice $2\,\Delta fT = 1$, 50% of the total power in the FSK waveform is in line components at the two transmitted fre-

---

[†] See, for example, R. W. Lucky, J. Salz, and E. J. Weldon, Jr., *Principles of Data Communications*. New York: McGraw-Hill, 1968, Ch. 8.

quencies, and this choice is generally avoided to prevent possible interchannel interference. Systems intended primarily for the more inexpensive noncoherent receivers use $2 \Delta f T > 1$, whereas FSK systems intended primarily for coherent detection often use choices in the range $\frac{1}{2} < 2 \Delta f T < 1$ to gain some S/N advantage and minimize the bandwidth required.

It might prove tempting to try to approximate the required bandwidth for FSK, using the approximations discovered in frequency modulation. Recall that in analog modulation the bandwidth of the modulated signal is not less than the double-sided bandwidth of the modulating signal and the bandwidth of a frequency-modulated signal is equal to or greater than the bandwidth of an amplitude-modulated signal. Neither of these conclusions is necessarily valid in digital modulation, however; and therefore some caution must be used in approximating the required bandwidth for FSK. For $2 \Delta f T > 1$, a rough approximation of the bandwidth can be obtained using Carson's rule with $f_m = 1/T$. For $2 \Delta f T < 1$, however, the required bandwidth, although it will always be greater than $2 \Delta f$, may be less than the two-sided bandwidth of the modulating signal. These are predetection bandwidths. A post-detection bandwidth can be narrowed to a bandwidth on the order of $1/T$ (as assumed in the matched-filter receiver).

Because it is relatively efficient in terms of peak power requirements and also relatively simple to implement, FSK is almost universally used for low-speed modems. For binary signaling up to 300 bps over commercial telephone channels, the usual choice for the transmit frequencies is 1070, 1270 Hz and 2025, 2225 Hz. Use of two-frequency pairs permits the use of full-duplex operation over one telephone channel. For modems operating at 1200 bps over commercial telephone channels, the transmit frequencies are 1200, 2200 Hz, and such modems can only transmit in the half-duplex mode. FSK modems intended for data transmissions up to 1800 bps are sometimes used, but they generally require conditioned telephone lines.

Example 10.2.1 NRZ binary data is transmitted at 300 bps through a telephone channel using FSK with transmit frequencies of 2025, 2225 Hz.

   a) Assuming an 800-Hz bandwidth centered on the carrier, calculate the minimum probability of error if the average signal-to-noise ratio is 8 dB;

   b) Repeat for $S/N = 7$ dB.

Solution

   a) Here we have $f_c = 2125$ Hz, $\Delta f = 200$ Hz, and $T = 1/300$ sec. Because $\omega_c T \gg 1$ and $\omega_c \gg \Delta \omega$, we can use Eq. (10.15); the bracketed term in this equation gives

$$1 - \text{Sa}\left(2\pi \frac{200}{300}\right) = 1.21.$$

Also, we have

$$\frac{S}{800\eta} = 10^{0.8},$$

so that

$$P_\epsilon = \text{Erfc} \sqrt{1.21 \frac{ST}{\eta}}$$

$$P_\epsilon = \text{Erfc} (4.51) = 3.26 \times 10^{-6}.$$

b) Changing S/N to $10^{0.7}$, we find the calculation above gives

$$P_\epsilon = \text{Erfc} (4.02) = 2.93 \times 10^{-5}.$$

Comparing the results of (a) and (b) shows that in the assumed range the probability of error changes by about one order of magnitude for a 1-dB change in S/N.

Using two oscillators and switching between them in response to the binary input is easy to visualize, but more popular in operating systems is the application of a polar binary waveform to the input of a voltage-controlled oscillator (VCO), causing the frequency of the oscillator output to vary in response to the input. At the switching times, both methods have sharp, random-magnitude phase transitions, which can be controlled by filtering and amplitude-limiting. An alternative in the latter method is to band-limit the input to the VCO.

Note that FSK is not true frequency modulation and, in fact, even has more similarities to AM (e.g., see Fig. 10.3). Therefore it does not provide the wideband noise reduction usually associated with FM. Any noise reduction attained arises from the PCM encoding, not from the frequency-shift keying.

10.3   PHASE-SHIFT KEYING (PSK)

Even though a symmetrical distribution about zero was achieved for the FSK operating characteristic, we were not able to obtain the superior probability-of-error performance of the polar-baseband system. It is instructive to pause and to consider the overall binary PCM detection problem to gain some insight into why this is true.

The optimum receiver for binary PCM need only make a decision on the difference between two given alternatives based on an observation over a finite time interval. We therefore return to investigate a matched filter for the detection of the difference between two signals. First, we let

$$g(t) = f_1(t) - f_2(t) \qquad 0 < t \le T, \tag{10.17}$$

where $f_1(t)$ and $f_2(t)$ are the two signals chosen to convey the binary information. Using Eq. (7.46), we find that the corresponding peak signal-to-noise ratio at the output of the matched filter is

$$\frac{|g_o(T)|^2}{n_o^2(t)} = \frac{1}{\pi\eta} \int_{-\infty}^{\infty} |G(\omega)|^2 \, d\omega. \tag{10.18}$$

From Parseval's theorem, we have

$$\frac{1}{2\pi} \int_{-\infty}^{\infty} |G(\omega)|^2 \, d\omega = \int_{-\infty}^{\infty} |g(t)|^2 \, dt. \tag{10.19}$$

Using Eqs. (10.17) and (10.19) in Eq. (10.18) and restricting our attention to the real-valued case, we obtain

$$\frac{g_o^2(T)}{n_o^2(t)} = \frac{2}{\eta} \int_0^T [f_1(t) - f_2(t)]^2 \, dt$$

$$= \frac{2}{\eta} \left[\int_0^T f_1^2(t) \, dt + \int_0^T f_2^2(t) \, dt - 2 \int_0^T f_1(t) f_2(t) \, dt \right]. \tag{10.20}$$

The first two integrals in Eq. (10.20) represent the energy in $f_1(t)$ and $f_2(t)$. We now constrain these energies to be equal so that

$$\int_0^T f_1^2(t) \, dt = \int_0^T f_2^2(t) \, dt \triangleq E. \tag{10.21}$$

From Eq. (10.20), the maximum peak signal-to-noise ratio for a given signal energy is then obtained for the condition

$$f_2(t) = -f_1(t). \tag{10.22}$$

The optimal class of signals for which Eqs. (10.21) and (10.22) are valid is called *antipodal*; i.e., the two signals denoting the two possible information symbols have exactly the same shape but opposite polarity. For linear time-invariant channels corrupted only by additive white gaussian noise, antipodal signals are optimal in the sense of requiring minimum E/η for a specified probability of error.

Recall that Eq. (10.22) does not hold for FSK [Eq. (10.21) does hold] although it did hold for the polar-baseband case. Hence we did not obtain the optimal performance from FSK. On the other hand, it will soon be evident that both Eqs. (10.21) and (10.22) do hold for the case of phase-shift keying (PSK).

In phase-shift keying, the phase of the carrier signal is switched between two (or more) values in response to the PCM code. For binary PCM, a 180° phase shift is a convenient choice because it simplifies the modulator design and hence is often used. This particular choice is commonly known as *phase-reversal keying* (PRK). The PRK waveform can be written as

$$\phi_1(t) = A \sin \omega_c t, \qquad \phi_2(t) = -A \sin \omega_c t. \qquad (10.23)$$

Note that Eqs. (10.21) and (10.22) are satisfied for this type of modulation. A PRK waveform is shown in Fig. 10.8 (also cf. Fig. 7.32).

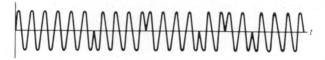

Fig. 10.8 A PRK waveform.

Because our choice of signal design obeys Eq. (10.22) we require only one reference function in the correlation detector, as shown in Fig. 10.9. Also, because our choice of signal design obeys Eq. (10.21), the decision threshold is at zero.

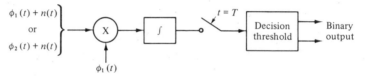

Fig. 10.9 A correlation detector for antipodal signals.

Referring to the correlation detector shown in Fig. 10.9, if $[\phi_1(t) + n(t)]$ is present at the input, the output at $t = T$ is $y(T) = E + n_o(T)$. On the other hand, if $[\phi_2(t) + n(t)]$ is present at the input, the output at $t = T$ is $y(T) = -E + n_o(T)$. The variance of the noise is [cf. Eq. (7.50)] $\overline{n_o^2(T)} = \eta E/2$ and the corresponding probability density functions are shown in Fig. 10.10.

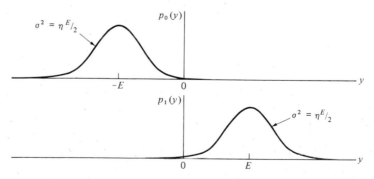

Fig. 10.10 Probability density functions for PRK.

As is evident from Fig. 10.10, the optimum-decision threshold is at zero so
that the net probability of error (for equiprobable ones and zeros) is

$$P_\epsilon = \int_0^\infty \frac{1}{\sqrt{\pi\eta E}} e^{-(y+E)^2/(\eta E)} \, dy,$$

$$P_\epsilon = \mathrm{Erfc}\sqrt{\frac{2E}{\eta}}. \tag{10.24}$$

Comparison of Eq. (10.24) with Eq. (9.52) shows that PRK has the same probability-of-error performance as polar-baseband systems. Comparison of Eq.
(10.24) with Eq. (10.11) shows that to attain a given error probability the average
power required for FSK and ASK is twice that required for PRK.

A more general signal representation for binary PSK (BPSK) is

$$\phi(t) = A \sin [\omega_c t + \Delta\theta \, p(t)], \tag{10.25}$$

where $\Delta\theta$ is the peak phase deviation and $p(t)$ is a binary switching function
with possible states of ± 1. Also, it is convenient for us to define a modulation
index, m, for BPSK as

$$m = \cos \Delta\theta, \tag{10.26}$$

where $0 \leq m \leq 1$. Expanding Eq. (10.25) using a trigonometric identity, we get

$$\phi(t) = A \sin \omega_c t \cos [p(t) \cos^{-1} m] + A \cos \omega_c t \sin [p(t) \cos^{-1} m].$$

Using the fact that $\cos (\pm\cos^{-1} m) = m$, and $\sin (\pm\cos^{-1} m) = \pm\sqrt{1 - m^2}$,
we have

$$\phi(t) = mA \sin \omega_c t + p(t) \sqrt{1 - m^2} A \cos \omega_c t. \tag{10.27}$$

The first term in Eq. (10.27) is the carrier component, and the second term is
the modulation component.

The average power in the carrier component of the BPSK waveform is
$m^2A^2/2$ and the power in the modulation component is $(1 - m^2)A^2/2$. Therefore
the carrier component has the fraction m^2 of the total power in the modulated
signal. It follows that the carrier component is zero in a PRK waveform (i.e.,
one for which $\Delta\theta = \pi/2$) (*see also* Drill Problem 6.5.1).

To determine the probability of error for BPSK, we use the correlation
detector (see Fig. 10.9). Using Eq. (10.25), we find the two possible signal outputs
at $t = T$ are

$$\int_0^T A^2 \sin (\omega_c t \pm \Delta\theta) \cos \omega_c t \, dt = \pm\tfrac{1}{2}A^2 T \sin \Delta\theta. \tag{10.28}$$

The probability-of-error computation follows in the same manner as for the PRK
case [cf. Eq. (10.24)], except that the probability density functions are centered

at $\pm E \sin \Delta\theta$:

$$P_\epsilon = \int_0^\infty \frac{1}{\sqrt{\pi\eta E}} e^{-(y+E\sin\Delta\theta)^2/(\eta E)} \, dy,$$

$$P_\epsilon = \text{Erfc} \sqrt{(2E \sin^2 \Delta\theta)/\eta},$$

$$P_\epsilon = \text{Erfc} \sqrt{2E(1 - m^2)/\eta}. \tag{10.29}$$

Thus the effect of allocating the fraction m^2 of the total transmitted power to the carrier component is to degrade P_ϵ by an equivalent S/N loss of $10 \log_{10} (1 - m^2)$ dB.

The superior performance of PSK comes with the disadvantage of a need for synchronous detection because the information is in the phase. An advantage in retaining a carrier component in the PSK waveform is that it can be used for receiver synchronization. For example, a phase-locked loop (PLL) can be used to demodulate BPSK if a sufficient carrier component is present. This is done, however, at the expense of a degradation in P_ϵ unless more transmitter power is allocated. A reduction of $\Delta\theta$ from 90° to 63° would result in a 1-dB S/N penalty for fixed P_ϵ while providing $m^2 = 21\%$ of the total power in the carrier component for carrier synchronization.

The power spectral density of PRK is centered around ω_c and has an identical shape to that of the double-sideband modulating spectral density. For NRZ modulation, this power spectral density has the $(\sin x/x)^2$ shape [cf. Sec. 9.9]

$$S_\phi(\omega) = \tfrac{1}{2}A^2 T \, \text{Sa}^2[(\omega + \omega_c)T/2] + \tfrac{1}{2}A^2 T \, \text{Sa}^2[(\omega - \omega_c)T/2]. \tag{10.30}$$

Even though the power spectral density of the randomly modulated PRK signal is highest around the carrier, there is no discrete spectral line (or impulse) at the carrier frequency. Thus PRK is really a double-sideband, suppressed-carrier modulation technique. In fact, the binary phase modulator can be implemented simply as a balanced mixer with a polar binary input.

For BPSK with $\Delta\theta < \pi/2$, there is a carrier component [cf. Eq. (10.27)], and therefore the spectral density has a discrete spectral line (impulse) at the carrier frequency. In this case the spectral density is analogous to that for double-sideband with carrier, although the carrier component need not be large compared to the sidebands. The theoretical bandwidth efficiency of BPSK systems is 1 bps/Hz. For NRZ modulation, the first-nulls bandwidth is $2/T$, which is commonly used in operating systems designs.

Example 10.3.1 Using PRK modulation, the GOES (Geostationary Orbiting Experimental Satellite) series of satellites transmit quantized, meteorological cloud-picture data at a rate of 1.75 Mbps. Assuming $\eta = 1.26 \times 10^{-20}$ W/Hz (corresponding to a net receiving system noise temperature of 229 K) and total path and system losses including antenna gains of 144 dB (cf. Table 4.1), calculate the minimum satellite transmitter power required for $P_\epsilon = 10^{-7}$.

Solution.   Using Eq. (10.24), we require

$$\text{Erfc}\sqrt{\frac{2ST}{\eta}} = 10^{-7}$$

or

$$\frac{2ST}{\eta} = (5.2)^2.$$

Multiplying by the expected losses, we have

$$2S = (5.2)^2(1.26 \times 10^{-20})(1.75 \times 10^6)(10^{14.4})$$

$$S = 75 \text{ W.}$$

Note that the system bandwidth does not enter explicitly into this computation. However, we must remember that our results assume the use of a matched-filter receiver. The bandwidth of such a receiver is matched to that of the signal. If desired, of course, the probability-of-error results for BPSK systems can be rewritten in terms of S/N by noting that $E = ST$ and $N = \eta B$ so that

$$\frac{E}{\eta} = \left(\frac{S}{N}\right)BT,$$

where T is the unit bit duration and B is the receiver noise bandwidth. In most practical systems the receiver noise bandwidth is larger than the double-sided Nyquist bandwidth, so that $BT > 1$.

Drill Problem 10.3.1   Suppose a phase error is present in the correlation detector for a PRK system so that the reference carrier signal is $A \cos(\omega_c t + \psi)$, with ψ the phase error. (a) Derive an expression for P_ϵ corresponding to Eq. (10.24) for this case. (b) Estimate the phase error ψ required to reduce P_ϵ from 10^{-5} to 10^{-4}.

Answer.   (a) $\text{Erfc}[\sqrt{(2E\cos^2\psi)/\eta}]$;   (b) 29.4°.

Several methods have been proposed for generating a reference carrier signal from a received PRK waveform. One possible solution is to first square the input PRK waveform. The phase of the resulting double-frequency term is determined regardless whether the phase is 0 or $\pm\pi$ radians. A frequency divider is used to obtain the desired carrier reference, as shown in Fig. 10.11. The frequency divider may be mechanized using a phase-locked loop and a divider in the feedback loop. The squaring loop tends to increase the noise near the desired

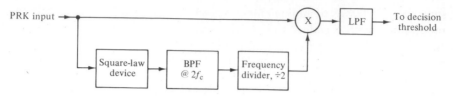

Fig. 10.11 A PRK detection system.

double-frequency carrier component so that a very narrow BPF is required.† For this reason, phase-locked loops (PLL) or crystal-controlled phase-locked loops are often used.

Choice of NRZ modulation is a popular one for PRK systems. If NRZ is used, however, a symbol timing recovery system must be used to regenerate the symbol clock frequency. A symbol timing recovery circuit can be constructed in a manner similar to that described for carrier recovery. The first step in the symbol timing recovery loop is applying the bit stream to a nonlinear circuit (e.g., square-law), as shown in Fig. 10.12. A discrete spectral line is present at the symbol rate at the output of the nonlinear circuit and can be filtered with a BPF or PLL. The recovered symbol clock signal with an integrate-and-dump threshold circuit can be used to demodulate the bit stream.

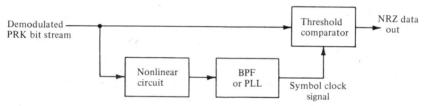

Fig. 10.12 A symbol timing recovery system for NRZ modulation.

Unfortunately, it is impossible to filter out all of the undesired noise power about the spectral line of the desired symbol clock reference. Those undesired spectral components close to the desired clock frequency are a frequent cause of timing jitter in the recovered symbol clock signal. Timing jitter results in performance degradation because the sampling is not done at the optimum times (i.e., at the maximum eye opening of the demodulated bit stream). Timing jitter tends to accumulate from the regenerative repeaters and is a fairly serious problem in long-haul systems, often placing an upper bound on the total distance of

---

† The reader will profit from sketching the power spectral density of a carrier in the presence of band-limited white noise and then taking a convolution to see how the noise increases near the double-frequency carrier term.

transmission. Circuits called "dejitterizer" circuits can be used to help maintain as low a timing jitter as possible. As a rule of thumb, the maximum permissible timing jitter should not exceed 30% of the symbol duration.[†]

Another method of carrier recovery for PRK, called a *Costas loop,* uses both in-phase and quadrature phase detectors to keep a VCO centered at the suppressed-carrier frequency. A Costas loop is shown in Fig. 10.13. For an input of the form $p(t) \cos (\omega_0 t + \theta)$, where $p(t) = \pm 1$, the loop will track θ while remaining insensitive to the sign of $p(t)$.

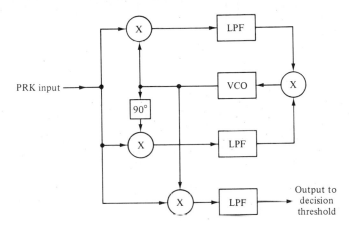

Fig. 10.13 The Costas loop.

Both of these methods have a phase ambiguity of $\pm\pi$ radians which can be solved by sending a known preamble code to establish the identity of the initial state. An alternative is to use a modified form of PSK called *differentially encoded PSK* (DE-PSK). With DE-PSK, the information is conveyed via transitions in carrier phase (e.g., no transition for a space and a 180° transition for a mark). Because a bit decision error on the current bit will induce another error on the subsequent bit, the performance of DE-PSK is slightly inferior to that of coherent PSK.[‡] The above methods can also be extended to the detection of multilevel PSK.

Another way to get around the synchronization problem is to use a modification of PSK known as *differential PSK* (DPSK). In DPSK, the information is encoded using the differences between bits in two successive bit intervals, as illustrated in Fig. 10.14(a). A differential binary sequence is generated from the

† K. Feher, *Digital Modulation Techniques in an Interference Environment,* EMC Encyclopedia, Vol. 9. Gainesville, VA: Don White Consultants, 1977, Ch. 4.

‡ See, for example, W. C. Lindsey and M. K. Simon, *Telecommunication Systems Engineering.* Englewood Cliffs, N.J.: Prentice-Hall, 1973, p. 252.

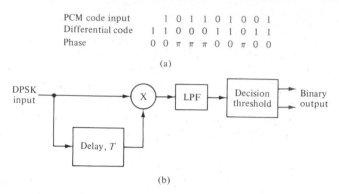

PCM code input	1 0 1 1 0 1 0 0 1
Differential code	1 1 0 0 0 1 1 0 1 1
Phase	0 0 π π π 0 0 π 0 0

(a)

(b)

Fig. 10.14 DPSK and its detection.[†]

input binary message at the transmitter. This sequence has one extra starting digit, which is arbitrary, and which is assumed to be a one here. Succeeding digits in the differential encoding are determined by the rule that there is no change in the output state if a 1 is present. There is a change in output state if a 0 is present.

A system for the detection of DPSK is shown in Fig. 10.14(b). The phase of the previous digit serves as the reference signal. If the phases are the same, a plus output results; if they differ, a minus output results. Checking back in Fig. 10.14(a), we see that this exactly decodes into the original message. A disadvantage of DPSK is that the signaling speed is fixed by the delay used. Also, because a bit determination is made on the basis of the signal received in two successive bit intervals, there is a tendency for bit errors to occur in pairs. The probability of error of DPSK is[‡]

$$P_\epsilon = \frac{1}{2} \exp \frac{-E}{\eta} . \qquad (10.31)$$

When compared to PSK systems, those systems using DPSK suffer a signal power penalty of 1 dB (or less) for $P_\epsilon < 10^{-4}$.

10.4   COMPARISON OF BINARY DIGITAL MODULATION SYSTEMS

The net probability-of-error performances of ASK, FSK, and PSK systems, as presented in the preceding sections, are plotted in Fig. 10.15 over a range of error rates of interest in typical operating systems. These results are graphed versus the bit energy E divided by η, the (one-sided) noise power spectral density.

---

† Multiplication here is defined as an equivalence (EXCLUSIVE-NOR logic) operation.
‡ *Ibid.*, p. 248.

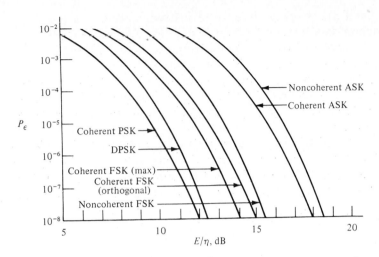

Fig. 10.15 Error probabilities for binary digital modulation systems.

Thus the same peak power is assumed for all systems in this comparison. (If the comparison were made in terms of average power requirements, then the ASK schemes would be displaced to the left by 3 dB in the graph and would require about the same amount of power as the FSK schemes.)

Figure 10.15 shows that coherent PSK signaling requires the least amount of power of any binary digital modulation method. Requiring the next least amount of power is DPSK, followed by coherent FSK, noncoherent FSK, coherent ASK, and noncoherent ASK. Whereas differences between some plots are only on the order of 1–2 dB, recall that, in the ranges of practical interest, a change in signal power of 1 dB will result in a corresponding approximate order-of-magnitude change in P_ϵ.

The transmitters for ASK systems are very easy to build and have an advantage in that there is no power transmitted when there is no data being sent (if OOK is used). Such systems find some applications in short-range miniature telemetry systems. Receivers for noncoherent ASK systems are easy to build. The difference in performance between coherent and noncoherent detection is slight compared to the increase in complexity required so that coherent detection of OOK is generally not used. A disadvantage of ASK is that the decision threshold in the receiver must be adjusted with changes in received signal levels. These adjustments are normally made with an automatic gain control.

FSK systems, in contrast to ASK, operate symmetrically about a zero decision-threshold level regardless of carrier signal strength. Besides a possible increase in required frequency stability, there is little difference in the complexity of FSK and PSK transmitters over those for ASK. Receiver complexity depends primarily upon whether a coherent or noncoherent modulation method is used.

Noncoherent FSK is relatively easy to instrument and is a popular choice for low-to-medium data-transmission rates such as teletype. FSK transmissions intended for noncoherent demodulation require more bandwidth for a given bit rate than either ASK or PSK. Bandwidths of FSK transmissions intended for coherent demodulation can be made as small as desired by controlling Δf, but cases for $2 \Delta f T < \frac{1}{2}$ demand an S/N penalty. Bandwidths of FSK transmissions intended for coherent demodulation are typically about equal to or slightly greater than those required for ASK or PSK.†

As we have noted, PSK systems are superior to both ASK and FSK systems in that they require less transmitted power for a given error probability. However, synchronous detection is required, and carrier recovery systems are more difficult (and therefore more expensive) to build. DPSK systems are often a good compromise that sacrifices some error performance but permits a more economical receiver. The three most widely used digital modulation methods for communication systems are PSK, DPSK, and noncoherent FSK.

Not one of the digital modulation methods we have described is particularly efficient in terms of the bandwidth used. Although we discuss this topic in greater detail in later sections of this chapter, we note here that one method we have not mentioned is the use of baseband signal shaping with a linear modulation (such as VSB). This method provides a means of modulation that is more conserving of bandwidth than any of the preceding methods.

The final choice of which type of digital modulation is chosen depends on trade-offs between performance, costs, bandwidth, etc. In addition, propagation distortions (e.g., delay distortion), fading, nonwhite-noise spectra, nongaussian noise (e.g., interference), etc., may affect these choices.

10.5 QUADRATURE AM (QAM) AND QUATERNARY PSK (QPSK)

In the binary digital modulation systems we have considered so far, only one of two possible signals can be transmitted during each signaling interval. Such digital modulation systems have a theoretical bandwidth efficiency of 1 bps/Hz. In many applications, a transmission system is more cost effective if, within a given bandwidth, more bits per second can be transmitted. This leads to a consideration of M-ary digital modulation methods in which one of M possible signals are transmitted during each signaling interval.

Our first approach to increasing spectral efficiency is to use the principle of quadrature multiplexing (cf. Example 5.1.1) in which two modulated signals are combined in phase quadrature. A basic system to accomplish this is shown in Fig. 10.16. This is called *quadrature AM* (QAM).

---

† A problem in attempting to be more specific than this is that the definition of bandwidth used makes a significant difference; see, for example, F. Amoroso, "The Bandwidth of Digital Data Signals," *IEEE Commun. Mag.*, vol. 18, no. 6, 13–24.

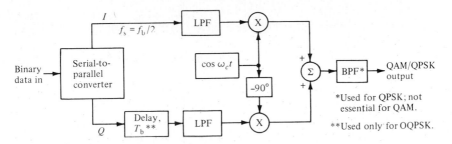

Fig. 10.16 Block diagram of a QAM/QPSK modulator.

In the QAM system of Fig. 10.16, the serial-to-parallel converter accepts the incoming binary data stream at a bit rate of $f_b = 1/T_b$ bps, and supplies two parallel binary data streams at $f_s = 1/T_s$ bps, where $T_s = 2T_b$. Here f_s is the symbol rate or *baud rate*, and f_b is the overall system bit rate. Following the serial-to-parallel conversion, low-pass filters are used to restrict the bandwidth and provide the desired spectral shaping. The I and Q data signals are modulated using DSB-SC modulation; the I signal, using the in-phase carrier reference; and the Q signal, using the quadrature carrier reference. The modulated I and Q signals are added to form the resultant QAM signal.

Because the QAM signal is the result of the linear addition of two quadrature DSB-SC signals, its spectral density is that of the DSB-SC signals. Thus the QAM power spectral density for a random NRZ input data stream is in the $(\sin x/x)^2$ form given by (for positive frequencies)

$$S_x(\omega) = C\left[\frac{\sin (\omega - \omega_c) T_s/2}{(\omega - \omega_c) T_s/2}\right]^2, \tag{10.32}$$

where C is a constant proportional to the average transmitted QAM power. Because $f_s = f_b/2$, the bandwidth efficiency of QAM is 2 bps/Hz.

A consideration of the phasor diagrams, as shown in Fig. 10.17, reveals that QAM can be viewed as phase modulation if the I and Q signals have identical magnitudes. Such systems are popularly known as *quadrature PSK* (QPSK) systems. As indicated in Fig. 10.17(d), Gray coding is used so that adjacent signal states differ by only one bit.

In QPSK, all of the information is conveyed by the phase, and a constant envelope is desirable. The low-pass filters of the modulator diagram in Fig. 10.16 are used merely to restrict the bandwidth of the data stream to less than one-half the carrier frequency. The major spectral shaping in QPSK systems is accomplished by a bandpass filter after the summation of the I and Q signals. The relative attenuations or gains in the I and Q circuits are kept as equal as possible. Thus a difference between quaternary QAM and QPSK is that the QAM systems employ premodulation (low-pass) filtering for spectral shaping,

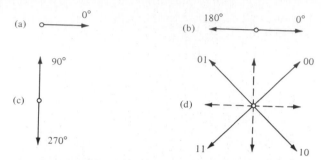

Fig. 10.17 Phasor diagrams of QAM/QPSK modulation: (a) unmodulated carrier; (b) in-phase (I) modulated signal; (c) quadrature (Q) modulated signal; (d) composite QAM/QPSK signal (Gray code values indicated).

whereas QPSK systems use post-modulation (bandpass) filtering and attempt to maintain a constant envelope in the modulated waveform. Theoretically, both quaternary QAM and QPSK systems have identical power spectral densities and probabilities of error. As a result of these similarities and the relative popularity of QPSK systems, the term QPSK is frequently used also for QAM systems.†

In QPSK, one of four possible signal waveforms is transmitted during each symbol interval T_s. These waveforms are as follows:

$$s_1(t) = A \cos \omega_c t,$$

$$s_2(t) = -A \sin \omega_c t,$$

$$s_3(t) = -A \cos \omega_c t,$$

$$s_4(t) = A \sin \omega_c t.$$

These waveforms correspond to phase shifts of 0°, 90°, 180°, and 270°, as shown in the phasor diagram, Fig. 10.17. Because the transmitted signal for a QPSK system can be viewed as the sum of two BPSK signals in quadrature, it is reasonable to expect that demodulation and detection would involve two BPSK receivers, or correlators, in parallel, one for each quadrature carrier. The block diagram of such a system is shown in Fig. 10.18.

Referring to Fig. 10.18, we find that the possible outputs of the first correlator in the absence of noise are

$$y_1(T_s) = \int_0^{T_s} [\pm A \cos \omega_c t][A \cos (\omega_c t + \pi/4)] \, dt, \qquad (10.33)$$

$$y_1(T_s) = \frac{\pm A^2 T_s}{2\sqrt{2}} = \frac{\pm E_s}{\sqrt{2}}, \qquad (10.34)$$

---

† The term *unbalanced QPSK* is sometimes used to describe the special case in which the in-phase and quadrature data rates or powers (or both) are unequal.

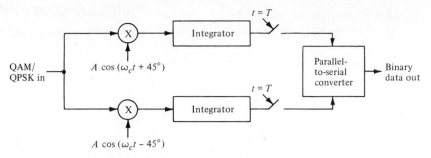

Fig. 10.18 Simplified block diagram of a QAM/QPSK demodulator.

where E_s is the energy per symbol. In a similar manner, the possible outputs of the second correlator are

$$y_2(T_s) = \frac{\pm E_s}{\sqrt{2}}.$$

The probability of error in the output of the first correlator when additive white gaussian noise is present is calculated in the same manner as that for PRK [cf. Eq. (10.24)], except that the probability density functions are centered at $\pm E_s/\sqrt{2}$, or

$$P_{\epsilon_1} = \int_0^\infty \frac{1}{\sqrt{\pi \eta E}} e^{-(y + E_s/\sqrt{2})^2/\eta E_s} \, dy,$$

$$P_{\epsilon_1} = \mathrm{Erfc}\sqrt{\frac{E_s}{\eta}}. \tag{10.35}$$

Similarly, the probability of error for the output of the second correlator is

$$P_{\epsilon_2} = \mathrm{Erfc}\sqrt{\frac{E_s}{\eta}}. \tag{10.36}$$

The probability that the QPSK receiver will correctly identify the transmitted signal is equal to the product of the probabilities that both correlators have yielded correct results, or†

$$P_c = (1 - P_{\epsilon_1})(1 - P_{\epsilon_2}). \tag{10.37}$$

Because $P_{\epsilon_1} = P_{\epsilon_2}$, we can write

$$P_c = 1 - 2P_{\epsilon_1} + P_{\epsilon_1}^2, \tag{10.38}$$

$$P_c \approx 1 - 2P_{\epsilon_1} \qquad \text{for } P_{\epsilon_1} \ll 1. \tag{10.39}$$

---

† This simplification assumes that the noise at the output of the two correlators is un-correlated. This assumption is valid if the period of integration T extends over one or more complete cycles of the carrier waveform.

The probability of error of the QPSK system is

$$P_\epsilon = 1 - P_c \tag{10.40}$$

$$P_\epsilon \approx 2P_{\epsilon_1} = 2\,\mathrm{Erfc}\sqrt{\frac{E_s}{\eta}}. \tag{10.41}$$

Comparing Eq. (10.41) with Eq. (10.24), we conclude that the probability of error for QPSK is greater than that for BPSK (PRK). If, however, we account for the fact that twice as much data is being transmitted, the probability of error performance of both systems is the same. (Note that E_s in Eq. (10.41) is the energy per symbol; Eq. (10.33) could be rephrased in terms of energy per bit by using $T_s = 2T_b$.) Because some uncertainty can arise in the use of E for both energy per symbol and energy per bit, systems calculations often refer specifically to the *bit error rate*, BER. This is illustrated in Example 10.5.1.

An objective in QPSK signaling is to maintain a constant waveform envelope, letting the information be conveyed by the phase. Phase transitions in each frequency mixer (cf. Fig. 10.16) occur at the symbol interval period and are smoothed by the bandpass filter. In practice, the realizability constraints of the filter cause envelope fluctuations during the filtering of these phase transitions. Envelope fluctuations are not of major concern in linear channels. However, many QPSK systems designed for satellite and terrestrial use operate with nonlinear output stages. The nonlinear amplifier reduces the envelope fluctuations at the expense of spreading the spectrum, which defeats the purpose of the bandpass filter and can cause unacceptable interference in adjacent frequency bands.

To decrease envelope fluctuations after filtering, some QPSK systems use a one-bit delay (T_b) in the quadrature data stream before modulation, as shown in Fig. 10.16. With this delay, the phase transitions at the two frequency mixers are separated by $T_b = T_s/2$ seconds. Such systems are called offset-keyed QPSK (or OQPSK) systems.[†] Consideration of a phasor diagram (e.g., see Fig. 10.17) shows that the maximum phase transition at any point in the OQPSK waveform is $\pm 90°$ (although these transitions can occur twice as often as in QPSK waveforms). As a result of the reduced magnitude of the phase transitions, the composite OQPSK signal has smaller envelope fluctuations following the bandpass filter. Another approach is to derive the carrier frequency from a multiple of the symbol timing clock so that the phase transition times can be controlled. This approach offers minimal envelope fluctuations and spectral spreading at some increase in equipment complexity.

Carrier recovery circuits must be used in QPSK demodulators. Like that for BPSK, one method quadruples the incoming QPSK waveform and then divides by four to derive the carrier (see Problem 10.5.3). A second method is the Costas

---

† OQPSK is also sometimes referred to as staggered QPSK.

loop shown in Fig. 10.19. A third method, often called a decision feedback loop, also uses both in-phase and quadrature feedback loops like the Costas loop. However, the bit decisions are made using, for example, such devices as integrate-and-dump filters before the feedback. Systems that use this third method are capable of good carrier recovery, but their acquisition time may be relatively long. In a fourth method, the modulated signal is first demodulated, then remodulated, and finally phase-compared to a delayed replica of the modulated signal. This method offers an advantage in much shorter acquisition times. In all of the feedback-type methods of carrier recovery, an error signal is generated that can be used to correct a VCO in phase and frequency, and then serve as the coherent reference. Other variations of these methods are described in the literature.†

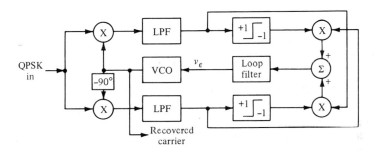

Fig. 10.19 Costas loop for QPSK carrier recovery.

In summary, QAM and QPSK systems have bandwidth efficiencies up to 2 bps/Hz. They offer attractive compromises between increased data rates with good performance characteristics (acceptable power versus bandwidth tradeoffs) and relative ease of implementation. They are widely used in practice for medium-speed data transmission.

Example 10.5.1 A given QPSK system is designed to transmit binary data at 44 Mbps in the presence of additive white gaussian noise with one-sided power spectral density $\eta = 1.67 \times 10^{-20}$ W/Hz. Assume Gray coding, a transmitter average power of 9 dB, and an expected net system loss of 120 dB.

a) Compute the bit error rate (BER) for this system.

b) Compare this BER with that of an equivalent BPSK (PRK) system that handles the same data rate.

---

† See, for example, W. C. Lindsey, *Synchronization Systems in Communications and Control.* Englewood Cliffs, NJ: Prentice-Hall, 1972; *see also* J. J. Stiffler, *Theory of Synchronous Communications.* Englewood Cliffs, NJ: Prentice-Hall, 1971.

Solution

a) Using Eqs. (9.117) and (10.41), we find the bit error rate for the QPSK system using Gray coding is

$$BER = Erfc \sqrt{\frac{ST_s}{\eta}}$$

$$= Erfc \sqrt{\frac{7.94 \times 10^{-12}}{(1.67 \times 10^{-20})(22 \times 10^6)}}$$

$$= Erfc\ 4.65 = 1.66 \times 10^{-6}.$$

b) Using Eq. (10.24), the bit error rate for the BPSK system is

$$BER = Erfc \sqrt{\frac{2ST_b}{\eta}}$$

$$= Erfc \sqrt{\frac{2(7.94 \times 10^{-12})}{(1.67 \times 10^{-20})(44 \times 10^6)}}$$

$$= Erfc\ 4.65 = 1.66 \times 10^{-6}.$$

We conclude that the BER for both systems is identical if Gray coding is used for the QPSK system (the usual case). The minimum theoretical bandwidth of the QPSK system is 22 MHz, whereas that for the BPSK system is 44 MHz.

10.6 CONTINUOUS-PHASE FSK (CPFSK) AND MINIMUM-SHIFT KEYING (MSK)[†]

The primary objective of spectrally efficient modulation methods is to maximize the bandwidth efficiency as measured in bps/Hz. In the preceding section, for example, we found QPSK offers a theoretical bandwidth efficiency of 2 bps/Hz. A second objective is to minimize those spectral components that fall outside the minimum (Nyquist) bandwidth. In a linear channel, this can be accomplished by filtering. In nonlinear channels, OQPSK can be used to reduce the out-of-band spectral components after filtering. Away from the carrier frequency, the power spectral densities of QPSK and OQPSK decrease at a rate proportional to ω^{-2}. With filtering, the rate of fall-off can be increased; but then the waveform envelope may not be constant, and any nonlinearities in the channel tend to restore the initial spectral fall-off characteristics.

[†] S. Pasupathy, "Minimum Shift Keying: A Spectrally Efficient Modulation," *IEEE Commun. Soc. Mag.*, 19 (July 1979): 14–22; *see also* S. A. Gronemeyer and A. L. McBride, "MSK and Offset QPSK Modulation," *IEEE Trans. Commun.*, COM-24 (August 1976): 809–820.

In this section, we investigate a modulation method that offers the bandwidth efficiency of QPSK and OQPSK, but whose power spectral density decreases more rapidly beyond the minimum bandwidth. This method is a version of frequency-shift keying discussed in Section 10.2.

We recall that in the transmission of information, the phase of FSK was not exploited, other than to provide possible synchronization of the receiver to the transmitter. Now we show how the phase can be utilized fully to decrease the out-of-band spectral content. To do this, we control the phase in such a way as to avoid any discontinuities in the modulated signal while using the frequency shift to convey the information. This type of digital modulation is called *Continuous Phase FSK* (CPFSK). It turns out that the power spectral density of CPFSK decreases at a rate proportional to at least ω^{-4} away from the carrier frequency. In addition, CPFSK offers the possibility of a theoretical bandwidth efficiency of 2 bps/Hz. These advantages are gained at the expense of some increase in equipment complexity for modulation and detection.

We define the CPFSK waveform by

$$\phi(t) = A \cos [\omega_c t + \gamma(t)] \tag{10.42}$$

where the phase with respect to carrier, $\gamma(t)$, is a continuous function of time. Rewriting Eq. (10.42) in terms of the two frequencies ω_1, ω_2 that are used to represent the binary symbols 0, 1, respectively, we have

$$\phi(t) = A \cos [\omega_c t \pm \Delta\omega t + \gamma(0)], \qquad 0 < t \le T_b , \tag{10.43}$$

where

$$\omega_c = \frac{\omega_1 + \omega_2}{2} , \tag{10.44}$$

$$\Delta\omega = \frac{\omega_2 - \omega_1}{2} . \tag{10.45}$$

Comparing Eqs. (10.42) and (10.43), we find that the phase is a linear function of time in the interval $0 < t \le T_b$, or

$$\gamma(t) = \pm\Delta\omega t + \gamma(0), \qquad 0 < t \le T_b . \tag{10.46}$$

The initial phase, $\gamma(0)$, depends on the past history of the modulation process and is to be chosen in such a way as to avoid any discontinuities.

In Section 10.2 we stated that the condition for orthogonal keying is that $2 \Delta\omega T_b = n\pi$. Since our goal here is spectral efficiency, we choose the minimum condition

$$\Delta\omega T_b = \frac{\pi}{2} \tag{10.47}$$

or

$$2 \Delta f T_b = \frac{1}{2} . \tag{10.48}$$

Because this is the minimum frequency spacing between ω_1 and ω_2 that allows the two FSK signal waveforms to be orthogonal to each other, this particular choice of CPFSK is called *Minimum Shift Keying* (MSK). MSK is the only specific case of CPFSK that we shall consider further in this section. Note that the criterion of Eq. (10.48) requires a frequency separation between f_1 and f_2 such that there is a one-half cycle difference in one bit interval.

Using Eq. (10.47), we find that Eq. (10.46) can be rewritten for MSK as

$$\gamma(t) = \pm\frac{\pi}{2T_b}t + \gamma(0), \qquad 0 < t \le T_b . \tag{10.49}$$

Choosing $\gamma(0) = 0$ for convenience here [recall that $\gamma(0)$ depends on the past history of the modulation process], the possible values of $\gamma(t)$ for $t > 0$ are shown as a phase trellis diagram in Fig. 10.20 for several successive bit intervals.

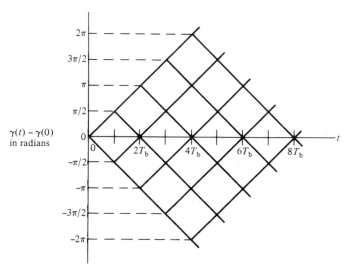

Fig. 10.20 Excess phase trellis for MSK.

Each path from left to right through the phase trellis corresponds to a specific binary input sequence. Figure 10.20 illustrates that over each bit interval the phase of the MSK waveform is advanced or retarded exactly 90°, depending upon whether the data for that interval is 0 or 1, with respect to the carrier phase in the preceding bit interval. Because all phase shifts are modulo 2π, the phase can take on only the two values $\pm\pi/2$ at odd multiples of T_b and only the values $0, \pi$ at even multiples of T_b.

Next we rewrite Eq. (10.49) to extend it over more than one bit interval

$$\gamma(t) = \frac{\pi t}{2T_b}p_k + \gamma_k , \tag{10.50}$$

where

p_k = a binary switching function with possible states of ± 1 representing the binary input data; $k = 1, 2, \ldots,$

$\gamma_k = \gamma[(k - 1)T_b]$ = excess phase at the beginning of the k-th bit interval required to maintain a continuous waveform.

In addition, we can write the recursive phase constraint

$$\gamma_k = \gamma_{k-1} + \frac{\pi}{2} p_k . \tag{10.51}$$

Using Eq. (10.50) in Eq. (10.42), we find that the MSK waveform can be expressed as

$$\phi_{MSK}(t) = A \cos \left[\omega_c t + \frac{\pi t}{2T_b} p_k + \gamma_k \right]. \tag{10.52}$$

Using a trigonometric identity, we can express Eq. (10.52) in terms of in-phase and quadrature components as

$$\phi_{MSK}(t) = A \left[\cos \left(\frac{\pi t}{2T_b} p_k + \gamma_k \right) \cos \omega_c t - \sin \left(\frac{\pi t}{2T_b} p_k + \gamma_k \right) \sin \omega_c t \right]. \tag{10.53}$$

As in QPSK signaling, we group the input binary stream into data pairs (e.g., assigning all even-indexed sample values to the in-phase component and odd-indexed sample values to the quadrature component). In doing this, we note that for successive values of either of the component values, the excess phase γ_k will always increment by 0, π, modulo 2π. The phases assigned to one component will differ from the other by $\pm \pi/2$ [see Eq. (10.51)] which, except for sign, can be taken care of by a delay of T_b. Therefore our analogy is more closely related to OQPSK signaling. The sign of the needed $\pm \pi/2$ shift is to be chosen to maintain phase continuity.

Using a trigonometric identity and noting that $\sin \gamma_k = 0$ in both the in-phase and quadrature components, we get

$$\phi_{MSK}(t) = A \left\{ \cos \gamma_k \cos \frac{\pi t}{2T_b} \cos \omega_c t - p_k \cos \gamma_k \sin \frac{\pi t}{2T_b} \sin \omega_c t \right\}, \tag{10.54}$$

$$\phi_{MSK}(t) = A \{ I(t) \cos \omega_c t - Q(t) \sin \omega_c t \}. \tag{10.55}$$

Consider this representation of the MSK waveform as being composed of two quadrature data channels, $a_I(t)$ and $a_Q(t)$, of an OQPSK signaling system

$$\phi_{MSK}(t) = A \left\{ a_I(t) \cos \frac{\pi t}{2T_b} \cos \omega_c t - a_Q(t) \sin \frac{\pi t}{2T_b} \sin \omega_c t \right\}. \tag{10.56}$$

Now MSK can be thought of as a special case of OQPSK with sinusoidal pulse weighting (rather than rectangular weighting). Comparing Eqs. (10.54) and (10.56), we have

$$a_I(t) = \cos \gamma_k,$$

$$a_Q(t) = p_k \cos \gamma_k,$$

and recalling that $\gamma_k = 0, \pi$, modulo 2π, we have

$$p_k = a_I(t)\, a_Q(t). \tag{10.57}$$

The quadrature data channels and sinusoidal weightings can be sketched using Eq. (10.57) and the requirement of phase continuity.

Various components of the MSK signal are illustrated in Fig. 10.21 for an input binary sequence 1 0 0 1 0 0 1. The even-indexed sample values in Fig. 10.21(a) are indicated by ± 1, held constant over two bit periods ($T_s = 2T_b$), and weighted by $\cos (\pi t/2T_b)$. The odd-indexed sample values, offset by one bit period with respect to the even-indexed sample values, held for two bit periods and weighted by $\sin (\pi t/2T_b)$, are shown in Fig. 10.21(c). The modulated in-phase and quadrature carrier terms are shown in Fig. 10.21(b), (d), respectively. Subtracting these two waveforms [see Eq. (10.56)] yields the MSK waveform shown

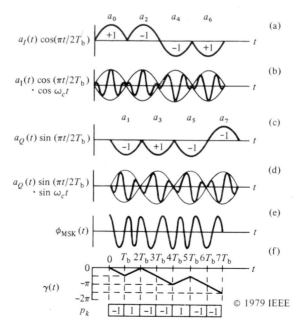

Fig. 10.21 MSK waveforms.†

---

† © 1979 IEEE. Reprinted, with permission, from "Minimum Shift Keying: A Spectrally Efficient Modulation," by S. Pasupathy from *IEEE Communications Magazine*, Vol. 19, no. 4 (July 1979), 18.

in Fig. 10.21(e). The excess phase trellis diagram for this MSK waveform is shown in Fig. 10.21(f). Also, the binary values of the p_k are indicated [these can be found by using Eq. (10.57)].

Because MSK signaling is essentially equivalent to OQPSK, the probability-of-error performance of MSK using matched-filter detection in the presence of additive white gaussian noise is the same as that for QPSK and OQPSK [see Eq. (10.41)]. From this we conclude that detection of MSK as two orthogonal binary channels provides a 3-dB E/η advantage over detection of orthogonal FSK. Note that the integration time for the matched-filter detection for each of the quadrature signals is $2T_b$ seconds. The bandwidth efficiency of MSK is 2 bps/Hz, the same as for QPSK and OQPSK signaling.

The power spectral density for QPSK and OQPSK signaling is

$$S_x(\omega) = 2A^2T_b[\mathrm{Sa}(\omega'T_b)]^2 \tag{10.58}$$

and the power spectral density for MSK is

$$S_x(\omega) = \frac{16A^2T_b}{\pi^2}\left[\frac{\cos\omega'T_b}{1 - (2\omega'T_b/\pi)^2}\right]^2 \tag{10.59}$$

where ω' is the radian frequency measured from the carrier frequency. Graphs of Eqs. (10.58), (10.59) are shown in Fig. 10.22. From these graphs, we observe that MSK has a wider main lobe (the first null is at $0.75/T_b$) than QPSK and OQPSK (the first null is at $0.50/T_b$) but the power spectral density of MSK has lower sidelobes than QPSK and OQPSK at frequencies more remote from the carrier frequency. These observations are supported in the comparison of various measures of bandwidth listed in Table 10.1.

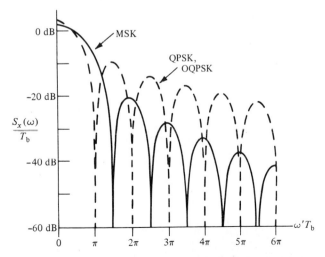

Fig. 10.22 Power spectral densities of QPSK, OQPSK, and MSK waveforms.

Table 10.1 Bandwidth Comparison of MSK and QPSK Signaling

	Bandwidths		
Modulation Type	-3 dB	-50 dB	99% Power Containment
QPSK, OQPSK	$0.44/T_b$	$100/T_b$	$10.3/T_b$
MSK	$0.59/T_b$	$8.18/T_b$	$1.17/T_b$

The modulation/demodulation circuits for MSK are not a great deal more difficult to implement than those for OQPSK. A block diagram of a MSK modulator is shown in Fig. 10.23 (see Example 10.6.1). The matched-filter MSK receiver is similar to that used for OQPSK with the exception that the reference signals for the synchronous detection of the in-phase and quadrature components are cos $(\pi t/2T_b)$ cos $\omega_c t$, and sin $(\pi t/2T_b)$ sin $\omega_c t$, respectively. A carrier recovery system for MSK is described in Problem 10.6.3.

In summary, MSK can be considered either as an OQPSK signal with sinusoidal pulse weighting or as a CPFSK signal with a frequency separation (2 Δf) equal to one-half the bit rate. The MSK waveform has a continuous phase at the bit transition times and has advantages of a constant envelope and a power spectral density that decreases at a rate proportional to ω^{-4} away from the carrier frequency. The rate of spectral fall-off can be improved by using other weightings in the in-phase and quadrature components, for example, Sa$(2\pi t/T_b)$, but only at the expense of a wider main lobe in the power spectral density.

CPFSK signaling may be generalized to include other choices of $\Delta\omega$ and the use of longer integration times before a decision is made in detection. However, potential improvements in performance are small and gained at the expense of added system complexity. In general, MSK offers a compromise for digital systems in which both bandwidth conservation and the use of amplitude-saturating (nonlinear) transmitters are important requirements.

Example 10.6.1 Show that the system illustrated in Fig. 10.23 is capable of generating an MSK waveform.

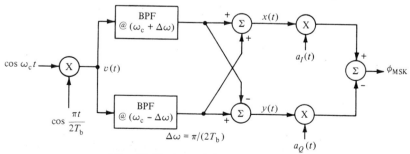

Fig. 10.23 A modulator for MSK.

Solution.   At the output of the first frequency mixer, we have the DSB-SC signal

$$v(t) = \tfrac{1}{2}\cos(\omega_c - \Delta\omega)t + \tfrac{1}{2}\cos(\omega_c + \Delta\omega)t,$$

where $\Delta\omega = \pi/(2T_b)$. Each BPF (assume ideal) attenuates one sideband, and the filter outputs are combined to give

$$
\begin{aligned}
x(t) &= \tfrac{1}{2}\cos(\omega_c - \Delta\omega)t + \tfrac{1}{2}\cos(\omega_c + \Delta\omega)t, \\
&= \cos\Delta\omega t \cos\omega_c t; \\
y(t) &= \tfrac{1}{2}\cos(\omega_c - \Delta\omega)t - \tfrac{1}{2}\cos(\omega_c + \Delta\omega)t, \\
&= \sin\Delta\omega t \sin\omega_c t.
\end{aligned}
$$

Modulating by the data streams $a_I(t)$ and $a_Q(t)$, respectively, we obtain the desired MSK signal [cf. Eq. (10.56)],

$$\phi_{\text{MSK}}(t) = a_I(t)\cos\Delta\omega t \cos\omega_c t - a_Q(t)\sin\Delta\omega t \sin\omega_c t.$$

Drill Problem 10.6.1   What relationship should be maintained between the carrier frequency f_c and the bit rate f_b in a MSK system if (a) the net phase is continuous at the bit transitions; (b) the bit transitions in both the in-phase and the quadrature components are to occur at the zero crossings of the carrier?

Answer.   (a) $f_c = nf_b/4$, n an integer; (b) cannot occur.

★ 10.7   *M*-ARY ORTHOGONAL FSK

Two basic resources in communication systems are transmitted power and channel bandwidth. In terms of channel capacity, they are related by the Hartley-Shannon theorem. One of these resources may be more precious than the other in a particular situation, and therefore many channels can be classified as either primarily power-limited or band-limited. Preceding sections have emphasized spectrally efficient methods of digital modulation that find application in band-limited channels. In this section we turn our attention to *M*-ary orthogonal signaling, which may be more useful in power-limited channels to conserve transmitted power at the expense of added bandwidth and equipment complexity.

In a series of *M*-ary signals, we transmit each one in the interval $(0, T_s)$, and we are interested in uniquely identifying each symbol at the receiver. To do so, we make use of a set of *M* orthogonal signals, $\phi_n(t)$. These signals are all taken to be of equal energy E_s and satisfy the orthogonality condition [cf. Eq. (2.21)]

$$\int_0^{T_s} \phi_n(t)\phi_m(t)\, dt = \begin{cases} 0 & n \neq m \\ E_s & n = m \end{cases}. \qquad (10.60)$$

Although many different sets of orthogonal functions could be chosen, we are interested particularly in frequency-shift keying and therefore choose the set of sinusoids

$$\phi_n(t) = A \cos \omega_n t \qquad 0 < t \le T_s. \tag{10.61}$$

Using Eq. (10.61) in Eq. (10.60), we require a minimum frequency separation between the signals $\phi_n(t)$ of

$$(\omega_m - \omega_n)_{\min} = \frac{\pi}{T_s}. \tag{10.62}$$

Using this result, we find the minimum net bandwidth needed in signaling using the $\phi_n(t)$ is

$$B_{\min} \approx \frac{M}{2T_s}, \tag{10.63}$$

where

$$T_s = T_b \log_2 M. \tag{10.64}$$

The optimum receiver for this orthogonal signal set consists of a bank of M matched filters as shown in Fig. 10.24. (It is interesting to compare this with Figs. 2.5 and 7.43.) At the times $t = kT_s$, we find which matched-filter output is the largest, and that output symbol is taken as the correct one for a symbol interval. In the presence of noise, some errors in choice will occur. If the noise is additive white gaussian noise with zero mean and with power spectral density $S_n(\omega) = \eta/2$, we can show that the probability of being correct is[†]

$$P_c = \frac{1}{\pi^{M/2}} \int_{-\infty}^{\infty} e^{-z^2} \left(\int_{-\infty}^{z + \sqrt{E_s/\eta}} e^{-y^2} \, dy \right)^{M-1} dz. \tag{10.65}$$

Then the probability of error can be found from

$$P_\epsilon = 1 - P_c. \tag{10.66}$$

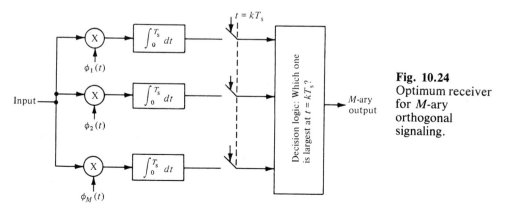

Fig. 10.24
Optimum receiver for M-ary orthogonal signaling.

[†] W. C. Lindsey and M. K. Simon, *Telecommunication Systems Engineering*. Englewood Cliffs, NJ: Prentice-Hall, 1973, Ch. 5.

The integration in Eq. (10.65) must be evaluated numerically in terms of E_s/η.† Some results are graphed in Fig. 10.25 versus $E_s/(\eta \log_2 M)$.‡ Using both $E_s = ST_s$, where S is the average signal power ($A^2/2$), and Eq. (10.64), we can write

$$\frac{E_s}{\eta \log_2 M} = \frac{ST_b}{\eta}. \tag{10.67}$$

Also, some values of $E_s/(\eta \log_2 M)$ for a given probability of error are given in Table 10.2.

From the results of Table 10.2 and Fig. 10.25, we conclude that the average power requirements for fixed η, T_b, and P_ϵ in M-ary orthogonal signaling decrease as M is increased. However, the required bandwidth increases in proportion to

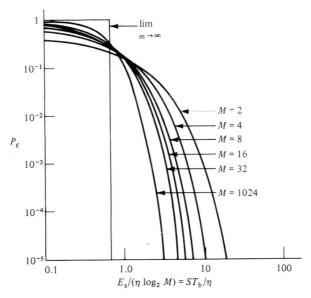

Fig. 10.25 Probability of error for M-ary orthogonal signaling.

Table 10.2 *S/N* Requirements of *M*-ary Orthogonal Signaling for Fixed Error Rates

log$_2$ *M*	*M*	$E_s/(\eta \log_2 M)$	
		$P_\epsilon = 10^{-4}$	$P_\epsilon = 10^{-5}$
1	2	13.8	18.2
2	4	7.94	10.1
3	8	5.82	7.29
4	16	4.72	5.83
5	32	4.04	4.94
6	64	3.58	4.34
7	128	3.25	3.91
8	256	3.00	3.58
9	512	2.81	3.33
10	1024	2.65	3.12
∞	∞	0.69	0.69

M [Eq. (10.63)], and larger values of *M* tend to increase the complexity of the transmitting and receiving equipment. Therefore *M*-ary orthogonal signaling conserves power at the expense of bandwidth and equipment complexity.

A more surprising result is that as $M \to \infty$, the *M*-ary signaling method is capable of transmitting information error-free $(P_\epsilon \to 0)$ if $E_s/(\eta \log_2 M) > 1/(\log_2 e)$. In contrast, the probability of error reduces to a certainty $(P_\epsilon \to 1)$ if

$$\frac{E_s}{\eta \log_2 M} < \frac{1}{\log_2 e} ;$$

in this condition, the system fails completely in any attempted transmission of information. Therefore the limiting bit rate f_b as $M \to \infty$ is, using Eq. (10.67),

$$\lim_{M \to \infty} f_b = \frac{S}{\eta} \log_2 e. \tag{10.68}$$

Because the bandwidth is proportional to *M*, Eq. (10.68) is exactly what was obtained in Example 9.2.3 for the channel capacity of an ideally coded system operating in a channel with no bandwidth limitation. Thus we have succeeded in demonstrating a signaling method which, at least in theory, is capable of approaching the limit of the Shannon-Hartley theory!

Finally, we note that the basic principles of orthogonal signaling are applicable to a much wider choice of orthogonal signals than sinusoids. Our particular interest here is in the sinusoids as an example of FSK modulation. In a broader sense, however, *M*-ary orthogonal signaling is a method of encoding information and can include more generalized sets of orthogonal signals (see, for example, those in Problem 2.5.3).

Example 10.7.1 Determine the savings in required transmitter power, assuming 16-tone orthogonal FSK in place of QPSK signaling in Example 10.5.1 if both systems are operated at $P_\epsilon = 10^{-5}$.

Solution. The average power required in the $M = 16$ FSK system is (using Table 10.2)

$$\frac{S_1 T_b}{\eta} = 5.83.$$

The average power of the QPSK system is

$$\frac{2 S_2 T_b}{\eta} = 18.2.$$

Taking the ratio, we get

$$\frac{S_1}{S_2} = 2 \frac{5.83}{18.2} = 0.641 \quad (-1.93 \text{ dB}).$$

The minimum bandwidth of the FSK system is, using Eqs. (10.63) and (10.64), $B \approx 2f_b = 88$ MHz. Also, the 16-tone FSK system uses 16 matched filters (correlators) whereas the QPSK system uses four. Therefore the orthogonal FSK signaling uses about 2 dB less average power than QPSK at the expense of added bandwidth and receiver complexity.

Drill Problem 10.7.1 Derive an expression for the maximum bandwidth efficiency of M-ary orthogonal FSK signaling.

Answer. $\dfrac{2 \log_2 M}{M}$ bps/Hz.

10.8 *M*-ARY PSK

There is an increasing interest in multi-level modulation methods for digital communication systems that are required to handle high data rates within fixed bandwidth constraints. In this section we investigate the use of M-ary PSK modulation for this purpose. Other methods will be investigated in a later section.

A convenient set of signals for M-ary PSK is the set

$$\phi_i(t) = A \cos(\omega_c t + \theta_i), \qquad 0 < t \le T_s, \qquad (10.69)$$

where the M possible phase angles θ_i are chosen as

$$\theta_i = 0, \quad \frac{2\pi}{M}, \quad \frac{4\pi}{M}, \quad \cdots, \quad \frac{2(M-1)\pi}{M}. \qquad (10.70)$$

The one-sided power spectral density of M-ary PSK, for a random binary input with equiprobable 1's and 0's at a bit rate $f_b = 1/T_b$, is

$$S_\phi(\omega) = A^2 T_s \, \text{Sa}^2 \left[(\omega - \omega_c) \frac{T_s}{2} \right] \tag{10.71}$$

where T_s is the unit symbol duration given by

$$T_s = T_b \log_2 M. \tag{10.72}$$

Because the minimum required bandwidth is $B = f_s$, the potential bandwidth efficiency of M-ary PSK is

$$\frac{f_b}{B} = \log_2 M \quad \text{bps/Hz}. \tag{10.73}$$

For example, $M = 8$ PSK signaling systems can transmit data at bandwidth efficiencies up to 3 bps/Hz.

A phasor diagram for M-ary PSK, $M = 8$, is shown in Fig. 10.26(a). All signals in the set have equal energy E_s over a symbol interval $(0, T_s)$, and each signal is demodulated correctly at the receiver if the phase is within $\pm \pi/M$ radians of the correct phase θ_i at the sample time. Thus the receiver decision thresholds are centered in phase angle between the θ_i, as shown in Fig. 10.26(a). No information is contained in the signal amplitude, and the input amplitude is assumed to be limited to a fixed level. All possible signal states (if a noise-free case is assumed) lie at equidistant spacings in a circular pattern centered at the origin, as illustrated for the $M = 8$ case in Fig. 10.26(b). The type of signal-state diagram shown in Fig. 10.26(b) is called a *signal constellation*. Signal constellations for M-ary PSK have circular symmetry, and the usual choice of M is a power of two.

In the presence of noise, a calculation of the probability of error involves a computation that the received phase lies outside the angular segment $(-\pi/M) \le$

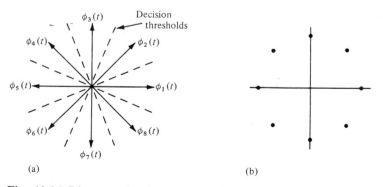

(a)   (b)

Fig. 10.26 Diagrams for M-ary PSK, $M = 8$: (a) phasor diagram; (b) signal state constellation.

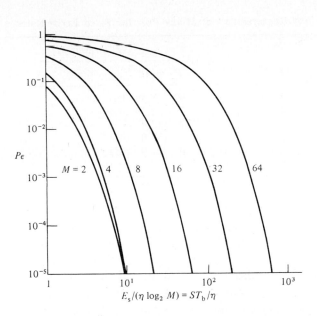

Fig. 10.27 Probability of error of *M*-ary PSK.

$\theta_i < (\pi/M)$ at the sample time. This is rather difficult to compute exactly.[†] Some numerical results versus $E_s/(\eta \log_2 M)$[‡] are graphed in Fig. 10.27. Also, some values of $E_s/(\eta \log_2 M)$ for a given probability of error are given in Table 10.3. For $P_\epsilon < 10^{-3}$, the probability of error of *M*-ary PSK can be approximated by[§]

$$P_\epsilon \approx 2 \, \text{Erfc} \sqrt{\frac{2E_s}{\eta} \sin^2 \frac{\pi}{M}}, \qquad M > 2. \tag{10.74}$$

This approximation improves for fixed M, as E_s/η increases. It should be noted that these results are for the *symbol* probability of error. The bit probability of error, P_{be}, can be found by modifying the preceding results according to methods described in Section 9.12. In particular, if a Gray code is used (the usual case),

$$P_{be} \cong P_\epsilon/\log_2 M \tag{10.75}$$

if P_ϵ is acceptably small [see Eq. (9.117)].

[†] See, for example, S. Stein and J. J. Jones, *Modern Communication Principles*. New York: McGraw-Hill, 1967, Ch. 14.

[‡] Tables of P_ϵ are given in W. C. Lindsey and M. K. Simon, *Telecommunication Systems Engineering*. Englewood Cliffs, NJ: Prentice-Hall, 1973, 232.

[§] W. C. Lindsey and M. K. Simon, *op. cit.; see also* J. R. Pierce and E. C. Posner, *Introduction to Communication Science and Systems*. New York: Plenum Press, 1980, 238.

Table 10.3 *S/N* Requirements of *M*-ary PSK for Fixed Error Rates

log₂ M	M	$E_s/(\eta \log_2 M)$	
		$P_\epsilon = 10^{-4}$	$P_\epsilon = 10^{-5}$
1	2	6.92 (8.40 dB)	9.10 (9.59 dB)
2	4	7.57 (8.79 dB)	9.75 (9.89 dB)
3	8	17.2 (12.4 dB)	22.2 (13.5 dB)
4	16	49.7 (17.0 dB)	64.1 (18.1 dB)
5	32	158.0 (22.0 dB)	203.0 (23.1 dB)
6	64	523.0 (27.2 dB)	673.0 (28.3 dB)

Values shown in Table 10.3 indicate that QPSK ($M = 4$) offers a good trade-off between power and bandwidth, requiring very modest increases (0.3–0.4 dB) in transmitted power for a potential doubling of bandwidth efficiency over that of coherent PSK ($M = 2$). For this reason, QPSK is widely used in practice for medium data rate transmissions in band-limited channels, even in some situations in which transmitted power is also a major concern (e.g., high data rate satellite transmissions). For higher data rate transmissions in band-limited channels, the choice $M = 8$ is often used. While this choice offers a potential bandwidth efficiency of 3 bps/Hz, it exacts a *S/N* penalty of almost 4 dB (see Table 10.3) over that required for coherent PSK ($M = 2$) at error rates between 10^{-4} and 10^{-5}. Choices of $M > 8$ are seldom used in *M*-ary PSK signaling as a result of the excessive additional power requirements.

When compared to BPSK signaling, the use of *M*-ary PSK requires more complex equipment for signal generation and detection, including that for carrier recovery. The requirement for recovery of the carrier for coherent detection can be alleviated by using a comparison between the phase of two successive symbols. This is the same concept as that described in Section 10.3 for $M = 2$ DPSK signaling and is called *M-ary differential PSK*. A derivation of the probability of error for *M*-ary differential PSK is rather involved; an approximation for large signal-to-noise ratios is[†]

$$P_\epsilon \approx 2\mathrm{Erfc}\sqrt{\frac{2E_s}{\eta}\sin^2\frac{\pi}{\sqrt{2M}}}. \qquad (10.76)$$

A comparison of Eqs. (10.74) and (10.76) shows that, within the approximations, differential detection increases the power requirements by a factor of approximately

$$\Gamma = \frac{\sin^2 \pi/M}{\sin^2 (\pi/\sqrt{2M})}. \qquad (10.77)$$

[†] W. C. Lindsey and M. K. Simon, *op. cit.*, 248; tables of P_ϵ using numerical integration procedures are also given.

Using this approximation and $M = 4$, we find the increase in power requirement for differential detection is about 2.5 dB. In some applications, savings in equipment complexity may be worth this moderate increase in power. Note that Eq. (10.77) approaches 3 dB for $M \geq 8$.

Example 10.8.1   Compute the bit error rate (BER) of a $M = 8$ PSK system operating under the conditions described in Example 10.5.1.

Solution.   Using Eqs. (10.74), (10.75), we have

$$P_{be} = \frac{2}{\log_2 M} \text{Erfc} \sqrt{\frac{2ST_s}{\eta} \sin^2 \frac{\pi}{8}}$$

$$P_{be} = \frac{2}{3} \text{Erfc} \sqrt{\frac{2(7.94 \times 10^{-12}) \sin^2 (\pi/8)}{(1.67 \times 10^{-20})(14.67 \times 10^6)}}$$

$$P_{be} = \frac{2}{3} \text{Erfc}(3.08) = 6.91 \times 10^{-4}.$$

Drill Problem 10.8.1   How many decibels must the transmitter power be increased in the $M = 8$ PSK system in Example 10.8.1 to give the same BER as the QPSK system described in Example 10.5.1?

Answer.   3.4 dB.

10.9   AMPLITUDE-PHASE KEYING (APK)

The need for increases in data transmission rates in band-limited channels leads us to a consideration of digital communication systems in which both amplitude and phase modulation are used. This general category of digital modulation is called *amplitude-phase keying* (APK). APK offers the advantage of less required power than PSK for a given probability of error and alphabet size M, but at the cost of increased equipment complexity and a sensitivity to possible nonlinearities in the channel.

     Two possible signal-state constellations for 16-ary APK systems are shown in Fig. 10.28. Note that each involves a combination of amplitude and phase modulation. The constellation shown in Fig. 10.28(a) uses three amplitudes, two with four phases and one with eight phases. In contrast, the constellation shown in Fig. 10.28(b) uses four amplitudes and four phases. Both are examples of signal constellations used in 9600 bps modem designs for conditioned telephone channels over the frequency range 300–3000 Hz with 1650 Hz used as the carrier frequency.† The constellation shown in Fig. 10.28(a) can be generated by inserting

---

† See, for example, E. R. Kretzmer, "The Evolution of Techniques for Data Communication over Voiceband Channels," *IEEE Commun. Soc. Mag.*, vol. 16 (January 1978), 10–14.

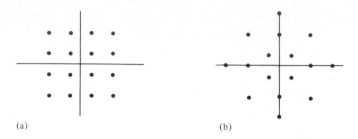

(a)   (b)

Fig. 10.28 Two examples of signal-state constellations for 16-ary APK signaling.

two-to-four-level converters in each of the I and Q signal paths before the balanced modulators and low-pass filters in a QPSK modulator (see Fig. 10.16). Because the two-to-four-level converters reduce the symbol rate by a factor of two and the serial-to-parallel converter reduces the symbol rate by a factor of two, it follows that the potential bandwidth efficiency of such a modulator is 4 bps/Hz.

In general, APK can be generated using binary-to-$L = \sqrt{M}$-level converters in each of the I and Q signal paths in a QPSK modulator. In this sense, APK can be considered also to be M-ary QAM. The block diagram of a corresponding M-ary APK demodulator using in-phase and quadrature detection is shown in Fig. 10.29. The L-level A/D flash encoders each consist of $(L - 1)$ comparators, which are set at the various threshold levels. The comparator outputs are sampled at the correct sampling time to determine the logic states of the comparators. These states are applied to a logic circuit that determines the output bit stream of the I and Q signal paths.

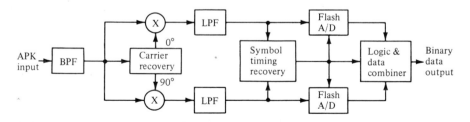

Fig. 10.29 Block diagram of an APK demodulator.

In band-limited channels with phase distortion and multipath distortion, the demodulator often is preceded by an adaptive equalizer operating at an intermediate frequency (IF). Operation of these equalizers is similar to those described in Section 9.11 for baseband operation, with the exception that equalization is made for both the I and Q signal components. Modern high-speed modems intended for band-limited channels with distortion (e.g., a telephone channel) often consist of an *M*-ary APK modulator/demodulator with adaptive equalization.

Another possible means of generation, which corresponds to the signal-state constellation shown in Fig. 10.28(a), is to first use a serial-to-parallel converter to divide the incoming data stream into two bit streams, each at half the rate. Then each of these bit streams can be applied to a QPSK modulator, as shown in Fig. 10.30(a). The outputs of the two QPSK modulators are added to yield a 16-ary APK system, as demonstrated by the phasor addition in Fig. 10.30(b). The theory of the corresponding demodulator is more complex and is described in the literature.†

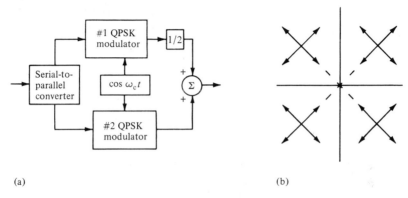

(a)   (b)

Fig. 10.30 Generation of 16-ary APK using the superposed QPSK method: (a) modulator diagram; (b) superimposed phasor diagrams.

The symbol probability-of-error performance of several PSK and APK systems is shown in Fig. 10.31. Note that 8-ary APK systems can require about 1 dB less power than 8-ary PSK for the same probability of error, while 16-ary APK systems can require about 3.5 dB less power than 16-ary PSK. Therefore a definite advantage develops for use of APK for $M > 8$. The bandwidth efficiency of *M*-ary APK systems is the same as that of *M*-ary PSK systems.

Another interesting approach in APK is to make use of the partial-response signaling techniques, which are described in Section 9.10, in both the *I* and *Q* signal paths of a QPSK (QAM) modulator. This technique is called *quadrature partial response* (QPR) signaling. Recall that a partial-response system generates an ℓ-level baseband signal data stream x_k using a linear weighted superposition of the binary input. For the specific case of duobinary (also known as Class 1) or modified duobinary (also known as Class 4) signaling, three possible output levels result. Thus if both the *I* and *Q* signal paths of a QPSK (QAM) modulator

---

† See K. Miyauchi, S. Saki, H. Ishio, "New Technique for Generating and Detecting Multi-level Signal Formats," *IEEE Trans. Comm.*, COM-24 (February 1976): 263–267.

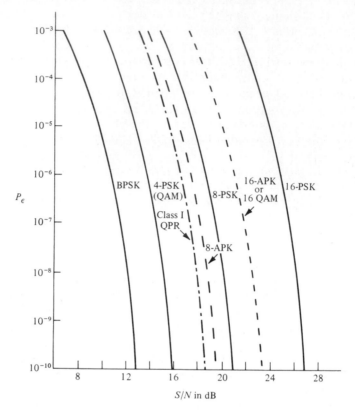

Fig. 10.31 Probability-of-error performance of M-ary PSK, QAM, QPR, and M-ary APK coherent systems. The mean-square S/N is specified in the double-sided Nyquist bandwidth that equals the symbol-rate bandwidth.†

use duobinary signaling, the resulting output constellation will have nine signal states, as shown in Fig. 10.32. An advantage of QPR systems is that they have some S/N advantage over comparable PSK systems (see Fig. 10.31) and that they are relatively easy to implement.

In general, problems of carrier recovery for M-ary APK systems are similar to those for M-ary PSK systems. An exception is that the $\times M$ frequency multiplier loop does not lend itself to applications in which the signal-state constellation does not have circular symmetry. One alternative is to use a Costas loop (see Fig. 10.19) with $L = \sqrt{M}$-level converters in each of the I and Q signal paths. Other methods are described in the literature.

---

† From K. Feher, *Digital Communications: Microwave Applications*, © 1981, p. 71. Reprinted by permission of Prentice-Hall, Inc., Englewood Cliffs, New Jersey.

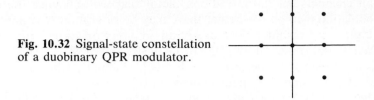

Fig. 10.32 Signal-state constellation of a duobinary QPR modulator.

Although *M*-ary PSK and FSK are very popular and widely used digital modulation methods for current data transmission systems, the APK and QPR methods have been receiving much attention recently. The ever-increasing requirements for data transmission through band-limited channels are expected to make these—and versions of them—popular design choices in new operating systems.

10.10 COMPARISON OF DIGITAL MODULATION SYSTEMS

Choice of digital modulation methods is dependent primarily upon bandwidth efficiency (in bps/Hz), error performance (P_ϵ versus S/N) and equipment complexity (cost). Required transmitted power and equipment complexity generally increase with improved bandwidth efficiency, as we have emphasized in the preceding sections. Results for bandwidth efficiencies and error performances were given in preceding sections, and representative digital modulation methods are ranked according to their inherent equipment complexity in Fig. 10.33. However, a specific modulation choice may be dependent on other factors, and it is the purpose of this section to delineate some of these other factors.

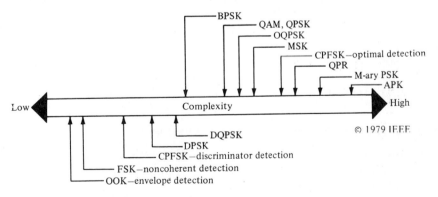

Fig. 10.33 Relative complexity of representative modulation schemes.†

---

† © 1979 IEEE. Reprinted, with permission, from "A Comparison of Modulation Techniques for Digital Radio," J. D. Oetting, *IEEE Trans. on Comm.*, COM-27 (December 1979): 1757.

Not all channels are linear and one factor that will influence the choice of digital modulation method is whether there may be saturation-type nonlinearities present (e.g., those resulting from use of an amplifier operated with a nonlinear gain characteristic). FSK, BPSK, DPSK, OQPSK, and MSK systems are notably good in maintaining their performance in such channels, whereas *M*-ary PSK and *M*-ary APK systems are poor choices in this respect.

Another type of channel impairment is *delay distortion*. If the magnitude response of a linear time-invariant filter does not vary with frequency, then the slope of the phase-frequency characteristic can be identified with time delay (cf. Section 3.12). For nonideal filters, it is convenient to define the delay as the slope. Any departures of the derivative of the phase-frequency characteristic from this constant are defined as delay distortion. (Appendix E presents some delay-distortion data for commercial telephone channels.) OOK and BPSK maintain good performance in the presence of linear delay distortion, whereas the error performances of QPR and of coherent biorthogonal signaling methods (i.e., QAM, QPSK, OQPSK, MSK) are degraded significantly. For quadratic-type delay distortion, FSK is a good choice, whereas DQPSK suffers severe degradation.

In the presence of fading, the error performances of OOK and the coherent biorthogonal systems do not degrade as rapidly as the others, whereas FSK and the differentially coherent systems are poorer than average. BPSK, DPSK, FSK, VSB, and the coherent biorthogonal systems have above-average tolerance to interference from other signals, whereas OOK and *M*-ary PSK systems are below average in this respect. BPSK, VSB, and the coherent biorthogonal systems have above-average tolerance to ISI, whereas *M*-ary PSK is below average. For some of the newer methods (e.g., *M*-ary APK), not much data have been published to date and, therefore, not many comments can be made with respect to them.

Finally, we note that the performance characteristics in the preceding sections dealt with idealized cases. Examples of error performances and signaling speeds of some systems reported in the literature are given in Table 10.4. Observe that deratings of 1.0–1.1 dB in E_b/η and 80–90% in signaling speed (bandwidth efficiency) are typical.

★ 10.11 REPRESENTATION OF DIGITAL WAVEFORMS

The previous sections have described several specific methods of transmitting digital waveforms. Now we wish to address the more general problems of waveform design and subsequent detection of digital signals.

In digital communication systems there is a finite collection of distinguishable waveforms from which one is chosen at periodic intervals and transmitted. A convenient approach in the analysis of such systems is that of a signal space and vector space—a concept introduced in Chapter 2. In particular, we wish to express a given signal in terms of a set of orthonormal functions. A motivation

Table 10.4 Error Performance and Signaling Speeds of Representative Digital Modulation Methods, 10^{-4} BER†

Modulation Type	Ideal System E_b/η (dB)	Band-limited System E_b/η (dB)	B.W. Eff. (bps/Hz)
BPSK	8.4	9.4	0.8
DE-PSK	8.9	9.9	0.8
DPSK	9.3	10.6	0.8
OOK-coherent detection*	11.4	12.5	0.8
MSK	8.4	9.4	1.9
QAM	8.4	9.5	1.7
QPSK	8.4	9.9	1.9
MSK-differential encoding	9.4	10.4	1.9
DQPSK	10.7	11.8	1.8
8-ary PSK	11.8	12.8	2.6
QPR	10.7	11.7	2.25
16-ary APK	12.4	13.4	3.1
16-ary PSK	16.2	17.2	2.9

* Average E_b/η is used here so that the values shown for OOK are 3 dB below the peak values indicated in Fig. 10.16.

behind this is that there is a countable set of possible waveforms, and if the receiver is set to recognize one member of the set, then it will not respond to the other members in the set. We use an orthonormal set so that the energy in the waveform will not depend on which member of the set is used.

Let the interval for transmitting each digital waveform be $(0, T)$. The orthonormal basis functions are then [cf. Eq. (2.21)]‡

$$\int_0^T \phi_n(t)\phi_m(t)\, dt = \begin{cases} 1 & n = m, \\ 0 & n \neq m. \end{cases} \tag{10.78}$$

For convenience, we restrict our attention to the case of a set of two orthonormal functions; extensions to a larger orthonormal set are straightforward.

The signal waveforms $s_m(t)$ may be represented by the orthonormal set [cf. Eq. (2.22)],

$$s_m(t) = a_{m1}\phi_1(t) + a_{m2}\phi_2(t), \tag{10.79}$$

where the coefficients a_{mn} can be found from [cf. Eq. (2.26)]

$$a_{mn} = \int_0^T s_m(t)\phi_n(t)\, dt. \tag{10.80}$$

---

† J. D. Oetting, *op. cit.*, 1755.
‡ For simplicity in notation, we use only real-valued functions here.

Thus we can represent the two signals $s_1(t)$, $s_2(t)$ by

$$s_1(t) = a_{11}\phi_1(t) + a_{12}\phi_2(t),$$
$$s_2(t) = a_{21}\phi_1(t) + a_{22}\phi_2(t).$$
(10.81)

The energy in each waveform is [using Eqs. (10.78) and (10.79)]

$$\int_0^T s_1^2(t)\, dt = a_{11}^2 + a_{12}^2,$$

$$\int_0^T s_2^2(t)\, dt = a_{21}^2 + a_{22}^2.$$
(10.82)

In a similar way, the energy in the difference between signals is

$$\int_0^T [s_1(t) - s_2(t)]^2\, dt = (a_{11} - a_{21})^2 + (a_{12} - a_{22})^2.$$
(10.83)

These concepts can be extended to waveforms constructed using more than two orthonormal basis functions.

A geometrical interpretation can be given to these results by considering the orthonormal set $\phi_n(t)$ as an orthonormal set of basis vectors, $\boldsymbol{\phi}_n$. A given vector in this vector space can be expressed in terms of the basis vectors [cf. Eq. (2.18)] and the coefficients can be found in terms of the dot product between the given vector and the basis vector [cf. Eq. (2.19)]. In addition, Eqs. (10.82) and (10.83) show that distance (or length—called the "norm" of the space) in the vector space is analogous to the square root of the signal energy in the signal space.

The received signals have additive random noise present and this noise may also be represented in terms of the set of orthonormal basis functions. Because the noise is random, the coefficients in the orthonormal series representation of the noise are random. The coefficients are uncorrelated with each other. If the noise is white with zero mean value, it can be shown that the mean-square value of each coefficient is equal to the mean-square value of the noise.† Only those noise coefficients which use the same orthonormal functions as the signal waveforms are important—the remaining ones are noise-only terms and rejected by the receiver.

A geometric representation in two dimensions of received signal-plus-noise is shown in Fig. 10.34. Because the noise coefficients n_1, n_2 are uncorrelated and have equal variances, the effect of the noise is to spread out the possible positions of the received vectors over a region around the noiseless signal point. The location of the resultant can be described statistically; the dashed circles in Fig. 10.34 are used to outline equiprobable loci.

The noise distribution does not depend on the signal. Therefore the trans-

---

† See, for example, H. L. Van Trees, *Detection, Estimation and Linear Modulation Theory, Part I.* N.Y.: John Wiley & Sons, 1968, Ch. 3.

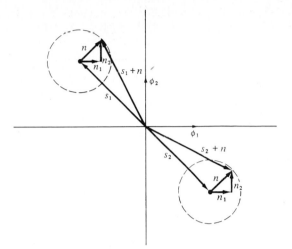

Fig. 10.34 Geometric representation of signal-plus-noise.

mitted signal states should be located as far apart as possible. If two waveforms are used (as shown in Fig. 10.34), the best placement within a fixed energy constraint is such that they are collinear and symmetrically opposite the origin; that is, $s_1(t) = -s_2(t)$. This type of waveform choice is called *antipodal* signaling. In contrast, an orthogonal waveform system provides maximum receiver output for the desired waveform while giving zero output for the others in the set. Thus antipodal signaling provides a lower error rate for a given average power.

We reconsider the ASK, FSK, BPSK, and QPSK digital waveforms which were used in previous sections but now in a more general context. Geometric representations of these are shown in Fig. 10.35 for a fixed average power (assuming equiprobable ones and zeros) and fixed noise power. The ASK waveforms are obtained by using one-sided collinear vectors (OOK is shown). The FSK waveforms are chosen to be orthogonal over $(0, T)$ while the BPSK waveforms are an example of antipodal signaling.

From Fig. 10.35, it is easy to see the relative probability-of-error advantage of BPSK over that of ASK and FSK. Besides the physical intuition it lends, the geometric representation also has the advantage that it can easily be extended to higher-dimensional waveforms.[†]

In QPSK, the two carriers used (i.e., $\cos \omega_c t$ and $\sin \omega_c t$) are orthogonal and, therefore, the two bit streams in the I and Q signal paths can be demodulated independently. For a given E/η, then, the error probabilities of coherently detected BPSK and QPSK are identical. Moreover, because staggering the bit

---

† For further reading on this topic, see E. Arthurs and H. Dym, "On the Optimum Detection of Digital Signals in the Presence of White Gaussian Noise—A Geometric Interpretation and a Study of Three Basic Data Transmission Systems," *IRE Trans. Commun. Systems*, CS-10 (December 1962): 336–372.

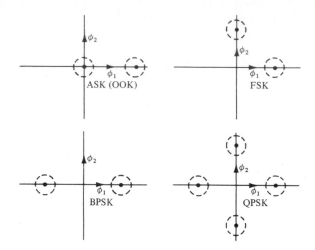

Fig. 10.35 Geometrical representation of ASK, FSK, BPSK, and QPSK.

streams does not alter the orthogonality of the carriers, OQPSK has the same error performance as BPSK and QPSK.

MSK also uses the modulation of two orthogonal carriers to achieve antipodal signaling of two binary data streams. It differs from OQPSK in that it uses weighting functions—$\cos(\pi t/2T_b)$ and $\sin(\pi t/2T_b)$—in the I and Q signal paths, but with no sacrifice in signal energy. Therefore MSK has the same error performance as BPSK, QPSK, and OQPSK when matched filtering is used to recover the data. However, if MSK were coherently detected as a FSK signal (with a bit decision time of T_b second), then the performance of MSK would be poorer than that of BPSK by 3 dB. The reason for this is that the distance between signal states has been reduced $1/\sqrt{2}$ of that for antipodal signaling.

In M-ary PSK, all signal states must lie in a circular pattern centered on the origin. This leaves little room for signal design choices. In contrast, the M signal states in M-ary APK can be placed in any pattern subject to the constraint of maximizing the distance of each signal state from all of the others to keep the error rate as low as possible. This must be achieved within a fixed average power constraint. Other important parameters are the minimum phase angle differences between signal states (a measure of immunity to phase jitter/delay distortion) and the ratio of peak-to-average power (a measure of immunity against nonlinear distortion).

★ 10.12   OPTIMUM DETECTION ALGORITHMS

In previous sections we have discussed the calculation of the probability of error for a given binary system. Now we wish to investigate the synthesis of an optimum signal processor in the receiver to yield a minimum probability of error.

Again we shall restrict our attention to the binary case even though these methods can be extended to more general cases.

To simplify the problem, we return to a consideration of on-off keying (OOK) in which the presence of a pulse signifies a one and the absence of a pulse signifies a zero. Thus either a one or a zero exists over $(0, T)$ and the receiver must make a decision as to which one. We begin by assuming that there is only one sample available on which to base this decision.

Let us label the two possible choices as H_0 and H_1; when a one is sent we call it H_1, and when a zero is sent we call it H_0. The binary transmission is in the presence of additive noise. For simplicity, we assume that the additive noise $n(t)$ is gaussian-distributed with zero mean. The two observed states over $(0, T)$ are [cf. Eqs. (9.25) and (9.26)]

$$y(t) = \begin{cases} s(t) + n(t) & \text{signal present,} \\ n(t) & \text{signal absent.} \end{cases} \tag{10.84}$$

At the end of the interval $(0, T)$ the receiver has the four following choices:

1. H_0 true, choose H_0;
2. H_0 true, choose H_1;
3. H_1 true, choose H_1;
4. H_1 true, choose H_0.

The first and third alternatives correspond to correct choices. In addition, two types of error may occur (cf. Fig. 9.7). In the second alternative the noise may be mistaken for signal when the signal is not present (P_{ϵ_0}) and in the fourth alternative the signal may be missed when it is present (P_{ϵ_1}). The overall (net) probability of error is then [cf. Eq. (9.33)]

$$P_\epsilon = P_0 P_{\epsilon_0} + P_1 P_{\epsilon_1}, \tag{10.85}$$

where P_1 is the probability of occurrence of a one (or signal in this case) and $P_0 = 1 - P_1$.

Both P_{ϵ_0} and P_{ϵ_1} are conditional probabilities because they depend on the signal being present or absent. To derive them, we write

$$p(y \mid 0) = p_n(y). \tag{10.86}$$

We assume that the noise is gaussian with zero mean so that

$$p_n(n) = \frac{1}{\sqrt{2\pi}\sigma_n} e^{-n^2/(2\sigma_n^2)}. \tag{10.87}$$

Using Eq. (10.87) in Eq. (10.86), we have

$$p(y \mid 0) = \frac{1}{\sqrt{2\pi}\sigma_n} e^{-y^2/(2\sigma_n^2)}. \tag{10.88}$$

Let us assume that the signal is a constant, A, over $(0, T)$ so that when the signal is present, $y(t) = A + n(t)$. This simply shifts the mean value of the noise from 0 to A (see Fig. 10.36) and

$$p(y \mid s) = p_n(y - A)$$

$$= \frac{1}{\sqrt{2\pi}\sigma_n} e^{-(y-A)^2/(2\sigma_n^2)} . \qquad (10.89)$$

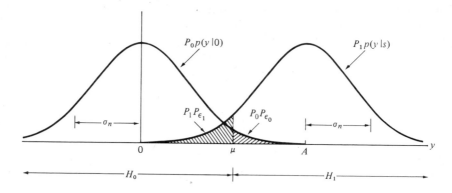

Fig. 10.36 Decision regions for the binary case.

From Fig. 10.36 we can write

$$P_{\epsilon_1} = \int_{H_0} p(y \mid s) \, dy, \qquad (10.90)$$

$$P_{\epsilon_0} = \int_{H_1} p(y \mid 0) \, dy, \qquad (10.91)$$

so that Eq. (10.85) becomes

$$P_{\epsilon} = P_1 \int_{H_0} p(y \mid s) \, dy + P_0 \int_{H_1} p(y \mid 0) \, dy. \qquad (10.92)$$

This is the equation we wish to minimize by choosing H_0, H_1 correctly.

We wish to minimize the probability of error by determining an algorithm (i.e., a rule) for deciding if a given observable, $y(t)$, should be interpreted as a one or a zero. Suppose we choose to develop a rule for H_1. We have

$$\int_{H_0 + H_1} p(y \mid s) \, dy = 1, \qquad (10.93)$$

so that Eq. (10.92) can be rewritten as

$$P_\epsilon = P_1\left[1 - \int_{H_1} p(y \mid s)\, dy\right] + P_0\int_{H_1} p(y \mid 0)\, dy$$

$$= P_1 + \int_{H_1} [P_0 p(y \mid 0) - P_1 p(y \mid s)]\, dy. \qquad (10.94)$$

We wish to make P_ϵ as small as possible. Because P_0, P_1 and both probability density functions in Eq. (10.94) are nonnegative, the best we can do is to make

$$P_1 p(y \mid s) > P_0 p(y \mid 0). \qquad (10.95)$$

Thus those values of y satisfying the inequality in Eq. (10.95) should be assigned to the decision H_1. By following the same reasoning, we conclude that those values of y which reverse the inequality should be assigned to H_0.

Equation (10.95) can be rewritten as

$$\ell(y) = \frac{p(y \mid s)}{p(y \mid 0)} > \frac{P_0}{P_1}, \qquad (10.96)$$

where $\ell(y)$ is called the *likelihood ratio*. The signal processing algorithm for a minimum probability of error can now be specified as

$$\ell(y) = \frac{p(y \mid s)}{p(y \mid 0)} \mathop{\gtrless}_{H_0}^{H_1} \frac{P_0}{P_1}. \qquad (10.97)$$

We have assumed in this algorithm that the two types of errors are equally costly to system performance. This is normally true in binary communication systems. If desired, allowance for errors of unequal cost can easily be made in the algorithm.

Because probability density functions are nonnegative and the logarithm function is monotonic, it is often convenient to take the natural logarithm of both sides of Eq. (10.97) to give

$$\ln \ell(y) \mathop{\gtrless}_{H_0}^{H_1} \ln \frac{P_0}{P_1}. \qquad (10.98)$$

The signal-processing algorithm specified in Eqs. (10.97) and (10.98) is called the *likelihood ratio test* (LRT).

Example 10.12.1 (a) Determine the LRT for on-off keying in the presence of additive gaussian noise with zero mean. (b) If in addition the noise is white, specify the LRT for the output of a matched filter.

Solution

a) Assume that the signal has an amplitude A when a one is sent and zero amplitude when a zero is sent. The conditional probabilities are given by

Eqs. (10.88) and (10.89). Using Eq. (10.98), the LRT is

$$\frac{(2Ay - A^2)}{(2\sigma_n^2)} \overset{H_1}{\underset{H_0}{\gtrless}} \ln\left(\frac{P_0}{P_1}\right),$$

or

$$y \overset{H_1}{\underset{H_0}{\gtrless}} \frac{A}{2}\left[1 + \frac{2\sigma_n^2}{A^2} \ln \frac{P_0}{P_1}\right].$$

Thus when $P_0 = P_1 = 0.5$, the signal processor sets the decision threshold at $A/2$, as we could expect [cf. Eq. (9.35)]. If P_1 differs from 0.5, the optimum decision threshold is shifted. The relative amount it is shifted is dependent on the natural logarithm of $(P_0/P_1)^2$ and inversely proportional to the peak signal-to-noise ratio.

b) The matched filter is a linear system so that the output noise is also gaussian-distributed with zero mean value and [cf. Eq. (9.48)] $\sigma_n^2 = E\eta/2$. The output amplitude with signal present is E so that the LRT of part (a) becomes

$$y \overset{H_1}{\underset{H_0}{\gtrless}} \frac{E}{2} + \frac{\eta}{2} \ln \frac{P_0}{P_1},$$

where E is the energy of the signal $s(t)$ over $(0, T)$ and $\eta/2$ is the (two-sided) power spectral density of the noise.

Drill Problem 10.12.1   Repeat Example 10.12.1 for the case of a polar binary signal with transmitted signals $-A/2$, $A/2$.

Answer.   (a) $y \gtrless (\sigma_n^2/A) \ln (P_0/P_1)$; (b) $y \gtrless (\eta/2) \ln (P_0/P_1)$.

The likelihood ratio test (LRT) was developed above for the case of one sample upon which to base a decision. However, the concept of the LRT is much more general than this and we now look at the more general case in which more than one sample is available over $(0, T)$.

The sampling theorem states that we can obtain $2BT$ independent samples of a deterministic signal band-limited to B Hz over $(0, T)$. For band-limited white noise, the noise samples are uncorrelated if they are spaced $1/(2B)$ seconds apart (see Drill Problem 8.13.1). For gaussian noise it follows that these noise samples are also independent. Therefore we shall assume that $N = 2BT$ independent samples of the observable, $y(t)$, are available.

Because the $N = 2BT$ samples are statistically independent, we can multiply their probability density functions to obtain the net probability density function. Thus the results obtained for the one-sample case are easily generalized to give a LRT for N samples:

$$\ell(y_1, y_2, \ldots, y_N) = \frac{\prod_{i=1}^{N} p(y_i|s)}{\prod_{i=1}^{N} p(y_i|0)} \overset{H_1}{\underset{H_0}{\gtrless}} \frac{P_0}{P_1} \tag{10.99}$$

where Π indicates multiplication. Taking the logarithm of both sides of Eq. (10.99), we get

$$\ln \ell(y_1, y_2, \ldots, y_N) \underset{H_0}{\overset{H_1}{\gtrless}} \ln \frac{P_0}{P_1}. \tag{10.100}$$

Note that Eqs. (10.99) and (10.100) are generalized versions of Eqs. (10.97) and (10.98).

Example 10.12.2   Assume that under H_1 the transmitter sends a constant level A and under H_0 it sends a zero level over $(0, T)$. Additive gaussian noise with zero mean and variance σ^2 is present and the observable signal-plus-noise is ideally band-limited to B Hz. Determine the LRT and the optimum signal processor.

Solution.   If we follow Eqs. (10.88) and (10.89) for the multiple-sample case, Eq. (10.99) is

$$\frac{\Pi_{i-1}^N (1/\sqrt{2\pi}\sigma)e^{-(y_i-A)^2/(2\sigma^2)}}{\Pi_{i=1}^N (1/\sqrt{2\pi}\sigma)e^{-y_i^2/(2\sigma^2)}} \underset{H_0}{\overset{H_1}{\gtrless}} \frac{P_0}{P_1}.$$

Cancelling common terms, using summations in the arguments of the exponentials for the products, and taking the natural logarithm, we have

$$\frac{1}{N} \sum_{i=1}^N y_i \underset{H_0}{\overset{H_1}{\gtrless}} \frac{A}{2} + \frac{2\sigma^2}{NA} \ln \frac{P_0}{P_1},$$

where $N = 2BT$. Therefore the optimum signal processor performs an operation known as the *sample mean*: $(1/N) \sum_{i=1}^N y_i$ on the observable signal-plus-noise in $(0, T)$ and compares it to a specified level. This result agrees with that obtained in Example 10.12.1(a) for the case where $N = 1$.

Drill Problem 10.12.2   Repeat Example 10.12.2 for the case of a polar binary signal with transmitted signals $-A/2$, $A/2$.

Answer

$$\frac{1}{N} \sum_{i=1}^N y_i \underset{H_0}{\overset{H_1}{\gtrless}} \frac{\sigma^2}{NA} \ln \frac{P_0}{P_1}.$$

The likelihood ratio test (LRT) is a very important one because it specifies the optimum signal processor algorithm in terms of the conditional statistics of the observable quantity. As illustrated in Example 10.12.2, it can be used to specify the required receiver operations (which are not required to be linear) to achieve this optimum. The criterion for optimality is a minimum probability of error for equal costs in making errors.

The next logical step is to try to find an optimum processor to estimate a parameter in the observed signal after its detection. This problem is an important one not only for digital communication systems but also for analog systems. For example, we would like to know what is the optimum signal processor to recover the variation of phase (or frequency) in a received signal. It turns out that the likelihood ratio plays a key role in finding the optimum processor for many of these problems. But we shall have to leave these (and other) very interesting topics for future work in communication systems.

10.13   SUMMARY

Digital modulation methods are used for transmission of PCM signals through bandpass channels. Binary modulation methods used are amplitude-shift keying (ASK), frequency-shift keying (FSK), and phase-shift keying (PSK). In these methods, the amplitude, frequency, or phase of a sinusoid is switched in response to the PCM input.

Using matched-filter detection in additive white gaussian noise, PSK requires up to 3 dB less average signal power than ASK or FSK for a given error probability. Coherent detection is required for PSK, whereas either coherent or noncoherent detection can be used for ASK and FSK. Differential PSK (DPSK) can be used to gain most of the advantages of PSK and yet avoid the requirement for coherent detection. For a given error performance, FSK requires about 3 dB less peak power than ASK, even though their average power requirements are similar. Popular choices for communication systems are PSK, DPSK, and noncoherent FSK.

Quadrature PSK (QPSK) uses quadrature multiplexing principles to offer twice the bit rate capability as that of binary PSK (BPSK) within the same bandwidth. Delay of either the in-phase (I) or quadrature (Q) signal by one bit period results in offset-keyed QPSK (OQPSK). An advantage of OQPSK is that it has a more constant envelope than QPSK after bandpass filtering. Both methods offer bandwidth efficiencies of 2 bps/Hz.

Minimum-shift keying (MSK) is an example of coherent FSK in which the phase in the modulated waveform is continuous and the frequency shift is such that there is a one-half cycle difference in one bit interval. MSK has signaling characteristics similar to those of OQPSK. A difference is that the power spectral density of MSK has a wider main lobe but decreases more rapidly farther away from the carrier frequency.

In M-ary PSK, the modulator generates constant-amplitude signals with phase spacings of $2\pi(M - 1)/M$ to transmit M-ary information. After bandpass filtering, such signals do not necessarily have constant envelope. For a fixed noise power, detection at low error rates requires substantial increases in signal power as M is increased. The 8-ary PSK systems are popular choices, but transmitted power requirements for $M > 8$ become excessive for many applications.

Amplitude-phase keying (APK) systems offer less required power for given

M than does M-ary PSK for a given probability of error. The generated signal envelope is not constant, and a linear channel is required. Both APK and M-ary PSK systems offer bandwidth efficiencies of $\log_2 M$ bps/Hz. Both are sensitive to channel distortions, and equalization is required for good transmission characteristics. In one particular type of APK signaling, called quadrature partial response (QPR), partial-response signaling is used in both the I and Q signals with quadrature amplitude modulation (QAM).

In contrast to the foregoing methods, M-ary orthogonal FSK offers savings in transmitted power at the expense of increased bandwidth for a given error rate. M-ary orthogonal FSK can be generalized to include other orthogonal signal sets. The performance of M-ary orthogonal signaling approaches that of the Hartley-Shannon result for channel capacity when $M \rightarrow \infty$.

Vector-space concepts may be used to give geometric interpretations to digital communication problems. The concept of a signal-state constellation is useful in M-ary signaling. Conditional probabilities may be used to form a likelihood ratio test. This test yields a signal processing algorithm that results in a minimum probability of error in detection of signals in noise.

Selected References for Further Reading

1. R. Techo. *Data Communications: An Introduction to Concepts and Design.* New York: Plenum Press, 1980.
   An introduction to modems, common-carrier services, multiplexing, and communication control procedures; very easy reading.

2. K. Feher. *Digital Modulation Techniques in an Interference Environment.* EMC Encyclopedia, Vol. 9. Gainesville, VA: Don White Consultants, Inc., 1977.
   A brief but easily read description of both practical and theoretical aspects of FSK, PSK, and APK systems.

3. M. Schwartz and L. Shaw. *Signal Processing.* New York: McGraw-Hill, 1975.
   Chapter 5 contains a well-written discussion of detection of signals in noise.

4. S. Haykin. *Communication Systems.* New York: John Wiley & Sons, 1978.
   Chapter 7 covers some digital modulation methods in the context of matched-filter detection and maximum likelihood estimation.

5. R. E. Ziemer and W. H. Tranter. *Principles of Communication Systems, Modulation, and Noise.* Boston: Houghton Mifflin, 1976.
   Chapters 7 and 8 offer an expanded treatment of many of the topics in this chapter with an emphasis on matched-filter detection and maximum likelihood estimation.

6. K. Feher. *Digital Communications: Microwave Applications.* Englewood Cliffs, NJ: Prentice-Hall, Inc., 1981.
   A good discussion of both theory and current practice of M-ary PSK, and APK methods are covered in Chapters 3, 6, 7, 8, with an emphasis on terrestial microwave systems.

7. R. W. Lucky, J. Salz, and E. J. Weldon, Jr. *Principles of Data Communication.* New York: McGraw-Hill, 1968.

Chapters 7, 8, 9 cover the topics of ASK, FSK, and PSK in some detail with an emphasis on telephone channels.

8. J. J. Spilker, Jr. *Digital Communications by Satellite*. Englewood Cliffs, NJ: Prentice-Hall, Inc., 1977.
   A fairly extensive discussion of generation, detection, and effects of distortion in BPSK and QPSK systems in Chapters 11, 12, 13, with an emphasis on satellite channels.

9. W. C. Lindsey and M. K. Simon. *Telecommunication Systems Engineering*. Englewood Cliffs, NJ: Prentice-Hall, Inc., 1973.
   An extensive treatment of phase-coherent systems.

Problems

10.1.1 Sketch a block diagram of the matched filter receiver for ASK, and show that it is an example of coherent detection.

10.1.2 A binary RZ code is transmitted using (OOK) amplitude-shift keying at a rate of 1 Mbps. Calculate the expected minimum probability of error if the ratio of the carrier power (when keyed on) to thermal noise temperature at the receiver input is 2×10^{-15} W/K.

10.1.3 A certain proposed binary ASK system is to utilize the following signals in the presence of white gaussian noise

$$\phi(t) = \begin{cases} A \sin \omega_c t & \text{for a 1} \\ \gamma A \sin \omega_c t & \text{for a 0} \end{cases} \text{over } (0, T),$$

where $0 \leq \gamma \leq 1$. Derive an expression for the minimum probability of error for this system assuming equiprobable 0's and 1's.

10.2.1 Sketch the (line) power spectrum of a FSK signal for an alternating sequence of 1's and 0's (cf. Example 6.3.3) for the following choices:

a) $2 \Delta fT = 2$;
b) $2 \Delta fT = 4$.

10.2.2 Let $\phi_1(t)$, $\phi_2(t)$ be two binary FSK signals over $(0, T)$ with a carrier frequency ω_c and a frequency difference of $2 \Delta\omega$, where $\Delta\omega \ll \omega_c$.

a) Show that $\phi_1(t)$, $\phi_2(t)$ are orthogonal over $(0, T)$ if $2 \Delta\omega T = n\pi$.
b) Plot $\log_{10} P_e$ versus $2 \Delta\omega T$ for $E/\eta = 16$ in Eq. (10.15).

10.2.3 Digital FSK transmissions are permitted in some high-frequency amateur radio bands using either $2 \Delta f = 170$ Hz ("narrowband") or $2 \Delta f = 850$ Hz ("wideband") with $2 \Delta fT = 3.75, 18.75$, respectively. Suppose that the rms (noise-free) signal strength is 10 μV across 50 Ω and that these transmissions are received in the presence of additive white gaussian noise with (two-sided) power spectral density $\eta/2 = 10^{-15}$ W/Hz.

a) Compute the theoretical minimum probability of error, using coherent demodulation.
b) Repeat (a), using noncoherent demodulation.
c) What are the relative advantages of each choice of $2 \Delta f$?

10.3.1 A coherent BPSK (PRK) system operates with $E/\eta = 8$ and probabilities of 0's and 1's equal to P_0, P_1, respectively.

  a) Find the net probability of error, P_e, for $P_1 = 0.4$, 0.5, and 0.6, assuming that the receiver threshold is set at zero.

  b) Find the additional signal power required, in dB, to maintain P_e for $P_1 = 0.5$ when $P_1 = 0.4$, 0.6.

  c) What is P_e for the optimum threshold settings?

10.3.2 Compare the average power requirements, expressed in dB, for BPSK (PRK), DPSK, and noncoherent FSK signaling at 300 bps, assuming $\eta/2 = 10^{-14}$ W/Hz and $P_e = 10^{-5}$.

10.3.3 A BPSK (PRK) system transmits data at 1 Mbps in the presence of additive white gaussian noise with $P_e = 10^{-5}$. Calculate the additional signal power required, in dB, to maintain this error rate for

  a) BPSK, $\Delta\theta = 60°$;

  b) BPSK, $\Delta\theta = 30°$;

  c) DPSK.

10.3.4 Evaluate the line spectrum of a BPSK signal for an input binary sequence that consists of an alternating sequence of 0's and 1's (cf. Drill Problem 6.5.1). Express your answer in terms of the modulation index m [cf. Eq. (10.26)]. Sketch the magnitude line spectrum for $\Delta\theta = 30°$ and $\Delta\theta = 90°$.

10.3.5 The bit stream 111011000110101 is to be transmitted using DPSK. Determine the encoded sequence.

10.3.6 Use the receiver in Fig. 10.14 to demodulate the DPSK signal obtained in Problem 10.3.5.

10.3.7 A given high-frequency transmitter is peak-power limited to 1 kW. Path losses to the receiver are estimated to be 138 dB and the receiver input noise power is 1 pW. Estimate the minimum probability of error using coherent (a) ASK(OOK); (b) PRK. Assume a random NRZ input with equiprobable 1's and 0's, and use a first-nulls bandwidth.

10.3.8 Evaluate the loop filter output signal in a phase-locked loop when the input signal is the BPSK signal in Eq. (10.25), the VCO output is

$$v_r(t) = \sin[\omega_c t + \epsilon(t)],$$

and those signal components at $2\omega_c$ are filtered out. In particular, show that when the loop is locked [i.e., $\epsilon(t) \to 0$], the output is proportional to the binary data signal $p(t)$ if $\Delta\theta < \pi/2$.

10.4.1 Rank the binary digital modulation systems covered, assuming that the primary criterion is (a) receiver simplicity, or (b) lowest peak power requirement for a given error rate (10^{-4}).

10.5.1 a) Sketch the power spectral density of the output of a 90 Mbps QPSK modulator with a random NRZ data input, assuming that no band-limiting filter has been used after modulation.

  b) Repeat for the case in which the band-limiting filter is an ideal minimum bandwidth filter.

10.5.2 Determine the probability of error per symbol for a 10-Mbps QPSK system when the measured S/N ratio at the receiver input is 10 dB. Assume additive white gaussian noise and a receiver noise bandwidth of 12 MHz.

10.5.3 The block diagram of a fourth-power loop used for QPSK carrier recovery is shown in Fig. P–10.5.3. Analyze the operation of this loop, assuming a noise-free input.

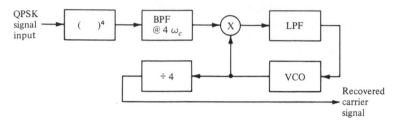

QPSK signal input

( )⁴ → BPF @ 4 ω_c → X → LPF

÷ 4 ← VCO

Recovered carrier signal

Fig. P–10.5.3

10.5.4 Sketch a complete block diagram of an OQPSK receiver for NRZ binary data, including both carrier recovery and symbol timing recovery subsystems.

10.5.5 Find $v_e(t)$ for the Costas loop in Fig. 10.19, and sketch $v_e(t)$ as a function of the error. Does this sketch have similarities to the error signal in a PLL? Explain.

10.6.1 Prepare a table comparing the null-to-null bandwidth and the -3-dB bandwidth for BPSK, QPSK, and MSK for a given bit rate $f_b = 1/T_b$.

10.6.2 Assuming that $\gamma(0) = 0$, sketch the following for the binary sequence 11000100: (a) the excess phase trellis path and MSK waveform; (b) the in-phase and quadrature data signals $a_I(t)$, $a_Q(t)$ and associated sinusoidal weighting functions.

10.6.3 The block diagram of a carrier recovery system for MSK is shown in Fig. P–10.6.3. Assuming a noise-free signal at the input, show that, except for a phase ambiguity of π radians, this system supplies the desired coherent reference signals for detection of MSK.

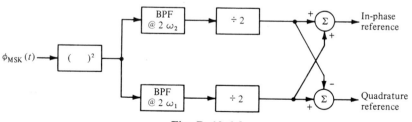

$\phi_{MSK}(t)$ → ( )² →

BPF @ 2 ω_2 → ÷ 2 → Σ (+ +) → In-phase reference

BPF @ 2 ω_1 → ÷ 2 → Σ (− +) → Quadrature reference

Fig. P–10.6.3

★10.7.1 A proposed receiving station for a space experiment requiring a data rate of 0.1 Mbps has a bandwidth of 1 MHz and a receiving antenna gain of 48 dB. The spacecraft antenna gain is 6 dB, and path losses (in dB) are $\alpha = 100 + 20 \log_{10} d$, where d is in miles. Assuming that the average transmitter power is 10 W and the power spectral density of the noise is $\eta/2 = 2 \times 10^{-21}$ W/Hz, estimate the maximum distance over which communication can be maintained with (a) $P_e = 10^{-5}$ and M-ary orthogonal FSK; (b) $P_e = 10^{-5}$ and QPSK.

★ **10.7.2** Using a change in the order of integration and a change of variables, show that Eq. (10.11) follows from Eqs. (10.65) and (10.66) for the case $M = 2$.

10.8.1 Show that, for a given average symbol error probability, an increase in M by a factor of two in M-ary PSK requires approximately a 6-dB increase in E_s/η for large M and small P_e. Is this also true for M-ary differential PSK?

10.8.2 A block diagram of a possible $M = 8$ PSK modulator is shown in Fig. P–10.8.2. The serial-to-parallel converter changes the input bit stream at rate f_b into three polar binary waveforms A, B, C, each at a rate $f_b/3$. The C waveform controls two amplifiers, each with two preset gains (the amplifiers are assumed to be identical). When $C = 1$, the gain of the upper amplifier is large and the gain of the lower amplifier is small; when $C = 0$, the converse is true. (a) Sketch a phasor diagram for this modulator when the amplifier gains are in the ratio $2 : 1$. (b) Determine the amplifier gain settings ratio for 8 equal modulation angles.

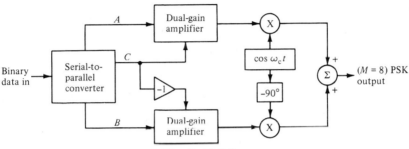

Fig. P–10.8.2

10.8.3 A telephone channel allows signal transmission in the frequency range 600–3,000 Hz. Assuming that the carrier frequency is 1800 Hz, show that (a) 2400 bps can be sent using four-phase PSK with raised-cosine shaping; and that (b) 4800 bps can be sent using eight-phase PSK with 50% sinusoidal roll-off. (c) What is the -6 dB bandwidth for each case?

10.9.1 Explain why there is a significant degradation in performance in an APK system when the modulated signal is passed through a nonlinear (saturation-type) amplifier. Consider both the case of a band-limited APK signal and one that is not band-limited.

10.9.2 Binary data at 9600 bps is transmitted over a conditioned telephone line using a bandwidth of 2400 Hz. Calculate the average S/N necessary to maintain $P_e = 10^{-3}$ in using (a) 16-ary PSK, and (b) 16-ary APK (see Fig. 10.31). Assume that the noise is additive white gaussian noise.

10.9.3 The symbol probability-of-error performance of M-ary APK systems in the presence of additive gaussian noise with zero mean and variance σ^2 for large S/N is[†]

$$P_e = \frac{1}{M} \frac{\sigma}{\sqrt{\pi}} \sum_{i=1}^{M} \sum_{\substack{j=1 \\ j \neq i}}^{M} \frac{\exp(-|s_i - s_j|^2/4\sigma^2)}{|s_i - s_j|},$$

---

† K. Feher, *Digital Modulation Techniques in an Interference Environment, EMC Encyclopedia*, Vol. 9. Gainesville, VA: Don White Consultants, Inc., 1977, Ch. 6.

where $|s_i - s_j|$ is the distance between signal states s_i, s_j in a signal-state constellation. Using this result, derive an expression for the probability of error (as a function of the average S/N) of the 4-ary signal-state constellation shown in Fig. P–10.9.3.

Fig. P–10.9.3

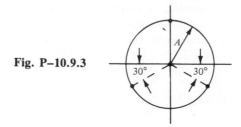

★**10.11.1** It is given that the probability of error for the binary PSK system in Fig. 9.31 is 10^{-4} with $P_0 = P_1 = 0.5$. Determine the corresponding probability of error geometrically for ASK(OOK) and FSK using the same average transmitted power.

★**10.11.2** A geometrical representation of a "bi-orthogonal" PSK coding system is shown in Fig. P–10.11.2. Derive the minimum probability of error of this system when operating in the presence of zero-mean additive gaussian noise with variance σ_n^2. Let the signal energy be E over $(0, T)$.

Fig. P–10.11.2

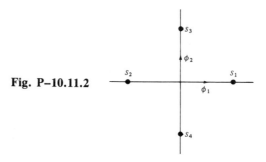

★**10.11.3** Repeat Problem 10.11.2, assuming the use of distance between adjacent signal states instead of the bi-orthogonal method for detection.

★**10.12.1** A bipolar signal of amplitude ± 1 V is received in the presence of additive gaussian noise with a variance of 0.1 V^2. Determine the optimum decision threshold if one sample of signal-plus-noise is taken and P_1 is (a) 0.5; (b) 0.3; (c) 0.7.

★**10.12.2** A single observation $y = s + n$ consists of a signal s which can be either 0 or 1 and additive noise with a pdf given by $p(n) = \exp(-2|n|)$. Determine the optimum detection threshold and corresponding probability of error if (a) $P_1 = P_0$; (b) $P_1 = 2P_0$.

★**10.12.3** In a given experiment, N observations are taken to determine which of the two possible hypotheses H_0, H_1 is correct. In both possibilities the observations are statistically independent zero-mean gaussian random variables. Under H_0 each observable y_i has a variance σ_0^2 and under H_1 each y_i has a variance σ_1^2. Determine the optimum signal-processing algorithm for detection with a minimum probability of error.

★ **10.12.4** A given signal $s(t)$ is to be detected in the presence of white gaussian noise based on N observations. Using the likelihood ratio test (LRT), show that for a fixed energy constraint the optimum receiver performs a cross-correlation. [*Hint*: Set up a decision based on the possible conditions $y = s + n$, $y = n$, write the pdf for the noise, and make a substitution of variables.]

APPENDIXES

APPENDIX A

SELECTED MATHEMATICAL TABLES

A.1 TRIGONOMETRIC IDENTITIES

$\sin (A \pm B) = \sin A \cos B \pm \cos A \sin B$

$\cos (A \pm B) = \cos A \cos B \mp \sin A \sin B$

$\sin A \sin B = \frac{1}{2}[\cos (A - B) - \cos (A + B)]$

$\cos A \cos B = \frac{1}{2}[\cos (A + B) + \cos (A - B)]$

$\sin A \cos B = \frac{1}{2}[\sin (A + B) + \sin (A - B)]$

$\sin 2A = 2 \sin A \cos A$

$\cos 2A = 2 \cos^2 A - 1 = 1 - 2 \sin^2 A = \cos^2 A - \sin^2 A$

$\sin^2 A = \frac{1}{2}(1 - \cos 2A)$

$\cos^2 A = \frac{1}{2}(1 + \cos 2A)$

$\sin A = \dfrac{1}{2j}(e^{jA} - e^{-jA})$

$\cos A = \frac{1}{2}(e^{jA} + e^{-jA})$

$e^{\pm jA} = \cos A \pm j \sin A$

A.2 COMPLEX FUNCTION IDENTITIES

$z = x + jy = \sqrt{x^2 + y^2}\, e^{j\tan^{-1}(y/x)}$

$z^* = x - jy = \sqrt{x^2 + y^2}\, e^{-j\tan^{-1}(y/x)}$

$|z|^2 = zz^* = x^2 + y^2$

$\mathcal{R}e\{z\} = \frac{1}{2}[z + z^*]$

$\mathcal{I}m\{z\} = \dfrac{1}{2j}[z - z^*]$

$\mathcal{R}e\{z_1 z_2\} = \mathcal{R}e\{z_1\}\mathcal{R}e\{z_2\} - \mathcal{I}m\{z_1\}\mathcal{I}m\{z_2\}$

$\mathcal{I}m\{z_1 z_2\} = \mathcal{R}e\{z_1\}\mathcal{I}m\{z_2\} + \mathcal{I}m\{z_1\}\mathcal{R}e\{z_2\}$

A.3 SERIES

MacLaurin:

$$f(x) = f(0) + f'(0)x + \frac{1}{2!} f''(0)x^2 + \frac{1}{3!} f'''(0)x^3 + \cdots.$$

Exponential:

$$e^x = 1 + x + \frac{1}{2!} x^2 + \frac{1}{3!} x^3 + \frac{1}{4!} x^4 + \cdots.$$

Trigonometric:

$$\sin x = x - \frac{1}{3!} x^3 + \frac{1}{5!} x^5 - \frac{1}{7!} x^7 + \cdots.$$

$$\cos x = 1 - \frac{1}{2!} x^2 + \frac{1}{4!} x^4 - \frac{1}{6!} x^6 + \cdots.$$

Binomial   (for $x^2 < 1$):

$$(1 \pm x)^n = 1 \pm nx + \frac{1}{2!} n(n-1)x^2 \pm \frac{1}{3!} n(n-1)(n-2)x^3 + \cdots.$$

$$(1 \pm x)^{-n} = 1 \mp nx + \frac{1}{2!} n(n+1)x^2 \mp \frac{1}{3!} n(n+1)(n+2)x^3 + \cdots.$$

A.4 INDEFINITE INTEGRALS

$$\int \sin ax\, dx = -\frac{1}{a} \cos ax$$

$$\int \cos ax\, dx = \frac{1}{a} \sin ax$$

$$\int \sin^2 ax\, dx = \frac{x}{2} - \frac{\sin 2ax}{4a}$$

$$\int \cos^2 ax\, dx = \frac{x}{2} + \frac{\sin 2ax}{4a}$$

$$\int \sin ax \cos ax\, dx = \frac{1}{2a} \sin^2(ax)$$

$$\int x \sin ax\, dx = \frac{1}{a^2} (\sin ax - ax \cos ax)$$

$$\int x \cos ax\, dx = \frac{1}{a^2} (\cos ax + ax \sin ax)$$

$$\int x^2 \sin ax\, dx = \frac{1}{a^3} (2ax \sin ax + 2 \cos ax - a^2 x^2 \cos ax)$$

$$\int x^2 \cos ax\, dx = \frac{1}{a^3} (2ax \cos ax - 2 \sin ax + a^2 x^2 \sin ax)$$

$$\int \sin ax \sin bx \, dx = \frac{\sin (a - b)x}{2(a - b)} - \frac{\sin (a + b)x}{2(a + b)} \qquad a^2 \neq b^2$$

$$\int \cos ax \cos bx \, dx = \frac{\sin (a - b)x}{2(a - b)} + \frac{\sin (a + b)x}{2(a + b)} \qquad a^2 \neq b^2$$

$$\int \sin ax \cos bx \, dx = -\frac{\cos (a - b)x}{2(a - b)} - \frac{\cos (a + b)x}{2(a + b)} \qquad a^2 \neq b^2$$

$$\int e^{ax} \, dx = \frac{1}{a} e^{ax}$$

$$\int xe^{ax} \, dx = \frac{1}{a^2} e^{ax}(ax - 1)$$

$$\int x^2 e^{ax} \, dx = \frac{1}{a^3} e^{ax}(a^2 x^2 - 2ax + 2)$$

$$\int e^{ax} \sin bx \, dx = \frac{1}{a^2 + b^2} e^{ax}(a \sin bx - b \cos bx)$$

$$\int e^{ax} \cos bx \, dx = \frac{1}{a^2 + b^2} e^{ax}(a \cos bx + b \sin bx)$$

$$\int \left[\frac{\sin ax}{x}\right]^2 dx = a \int \frac{\sin 2ax}{x} dx - \frac{\sin^2 ax}{x}$$

$$\int \frac{dx}{a^2 + b^2 x^2} = \frac{1}{ab} \tan^{-1}\left(\frac{bx}{a}\right)$$

$$\int \frac{x^2 \, dx}{a^2 + b^2 x^2} = \frac{x}{b^2} - \frac{a}{b^3} \tan^{-1}\left(\frac{bx}{a}\right)$$

$$\int \frac{dx}{(a^2 + b^2 x^2)^2} = \frac{x}{2a^2(a^2 + b^2 x^2)} + \frac{1}{2a^3 b} \tan^{-1}\left(\frac{bx}{a}\right)$$

$$\int \frac{x^2 \, dx}{(a^2 + b^2 x^2)^2} = \frac{-x}{2b^2 (a^2 + b^2 x^2)} + \frac{1}{2ab^3} \tan^{-1}\left(\frac{bx}{a}\right)$$

$$\int \frac{dx}{(a^2 + b^2 x^2)^3} = \frac{x}{4a^2(a^2 + b^2 x^2)^2} + \frac{3x}{8a^4(a^2 + b^2 x^2)} + \frac{3}{8a^5 b} \tan^{-1}\left(\frac{bx}{a}\right)$$

A.5 DEFINITE INTEGRALS

$$\int_0^\infty \frac{\sin ax}{x} \, dx = \begin{cases} \pi/2 & a > 0 \\ 0 & a = 0 \\ -\pi/2 & a < 0 \end{cases}$$

$$\int_0^x \frac{\sin u}{u} \, du \triangleq \text{Si}(x) \quad \text{(a tabulated integral as a function of } x)$$

$$\int_0^\infty \frac{\sin^2 ax}{x^2} \, dx = |a| \, \pi/2$$

$$\int_0^\infty e^{-ax^2}\, dx = \frac{1}{2}\sqrt{\pi/a}$$

$$\int_0^\infty xe^{-ax^2}\, dx = \frac{1}{2a}$$

$$\int_0^\infty x^2 e^{-ax^2}\, dx = \frac{1}{4a}\sqrt{\pi/a}$$

$$\int_0^\infty \frac{dx}{(x^2 + a^2)(x^2 + b^2)} = \frac{\pi}{2ab(a + b)} \qquad a > 0,\, b > 0$$

$$\int_0^\infty \frac{dx}{ax^4 + b} = \frac{\pi}{2\sqrt{2b}}\left(\frac{b}{a}\right)^{1/4} \qquad ab > 0$$

$$\int_0^\infty \frac{dx}{ax^6 + b} = \frac{\pi}{3b}\left(\frac{b}{a}\right)^{1/6} \qquad ab > 0$$

Suggestions for Further Reference

1. S. M. Selby, ed. *CRC Standard Mathematical Tables*, Fifteenth ed. Cleveland, Ohio: The Chemical Rubber Co., 1967.

2. H. B. Dwight. *Tables of Integrals and Other Mathematical Data*, Fourth ed. New York: Macmillan, 1961.

3. I. S. Gradshteyn and I. M. Ryzhik. *Table of Integrals, Series and Products*, Fourth ed. New York: Academic Press, 1965.

APPENDIX B

DECIBELS

The decibel is a logarithmic unit of measure used for comparing two power levels. Assigning P_r as the reference power level, the decibel (dB) is defined by the equation

$$(\text{dB}) \triangleq 10 \log_{10}(P/P_r). \tag{B.1}$$

Because the decibel is a logarithmic unit, operations of multiplication and division reduce to addition and subtraction, and powers and roots reduce to multiplication and division. Addition and subtraction, however, require conversion to numeric values. The sign of a logarithm changes when its argument is inverted. Given a power ratio expressed in decibels, the power ratio can be found from the inverse of Eq. (B.1):

$$P/P_r = 10^{(\text{dB})/10}. \tag{B.2}$$

Decibels are also used to indicate absolute power levels by adding a third letter to the notation. If the reference power level P_r is one watt, the power P is expressed in "decibels above one watt" (dBW) by

$$P_{\text{dBW}} = 10 \log_{10} P. \tag{B.3}$$

In a similar manner, if the reference power P_r is one milliwatt, the power P is expressed in "decibels above one milliwatt" (dBm).

The definition in Eq. (B.1) can be rewritten in terms of voltages V, V_r developed across resistances R, R_r in the following manner:

$$(\text{dB}) = 10 \log_{10}\left[\frac{V^2/R}{V_r^2/R_r}\right] \tag{B.4a}$$

$$= 20 \log_{10}(V/V_r) - 10 \log_{10}(R/R_r) \tag{B.4b}$$

For the case in which $R = R_r$, this becomes

$$(\text{dB}) = 20 \log_{10}(V/V_r). \tag{B.5}$$

If, in addition, the reference voltage level is taken as unity, we have

$$V_{\text{dB}} = 20 \log_{10} V. \tag{B.6}$$

Sometimes Eq. (B.6) is used as a definition of the decibel. This is valid only as long as the proper normalizing factors are considered whenever any comparison is made to power calculations using the definition of Eq. (B.1).

The predominant usage of the decibel in communication systems is for power ratios. Unless otherwise specified, Eqs. (B.1) and (B.2) are assumed to be the defining relations.

Drill Problem B.1   (a) Convert the following power ratios to dB: 4000, 0.003. (b) Convert the following to numeric values (in terms of power): 29.3 dB, −7 dBW, 27 dBm.

Answer.   (a) 36.0 dB, −25.2 dB; (b) 851, 0.200 W, 0.501 W.

BROADCAST FREQUENCY BANDS

Table C.1   Frequency Designations

VLF (very low frequencies)	3 Hz to	30 kHz
LF (low frequencies)	30 kHz to	300 kHz
MF (medium frequencies)	300 kHz to	3 MHz
HF (high frequencies)	3 MHz to	30 MHz
VHF (very high frequencies)	30 MHz to	300 MHz
UHF (ultrahigh frequencies)	300 MHz to	3000 MHz
SHF (superhigh frequencies)	3 GHz to	30 GHz
EHF (extra-high frequencies)	30 GHz to	300 GHz

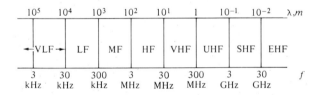

Table C.2   Radar Frequency Bands†

Band	Frequency range		Wavelength, cm
HF	3–30	MHz	10,000–1,000
VHF	30–300	MHz	1,000–100
UHF	300–1000	MHz	100–30
L	1.0–2.0	GHz	30–15
S	2.0–4.0	GHz	15–7.50
C	4.0–8.0	GHz	7.50–3.75
X	8.0–12.0	GHz	3.75–2.50
K_u	12.0–18.0	GHz	2.50–1.67
K	18.0–27.0	GHz	1.67–1.11
K_a	27.0–40.0	GHz	1.11–0.75
Millimeter	40–300	GHz	0.75–0.10

† IEEE Standards 521 and 686.

AM Band †

Stations are assigned carrier frequencies at 10 kHz intervals from 540 to 1600 kHz. The bandwidth of transmissions is nominally about 10 kHz. Stations in local proximity are usually assigned carrier frequencies which are separated by 30 kHz or more. Interference between transmissions is controlled by a combination of frequency allocation, transmitter power, transmitting antenna pattern, and possible night-time operating restrictions. Required carrier stability is ±20 Hz; licensed power output is for an unmodulated carrier.

FM Band †

Stations are assigned carrier frequencies at 200 kHz intervals from 88 MHz to 108 MHz. Transmissions received beyond the line-of-sight distance are weak and subject to fading. FM receivers tend to suppress weak signals in the presence of strong signals. These effects help to define the station broadcast area. Peak frequency deviation is 75 kHz. Minimum required carrier frequency stability is ±0.002% (that is, ±2 kHz at 100 MHz).

Table C.3 Permissable Transmitted Power Levels as Specified by FCC

AM Band:	Broadcast Class	Average Power, kW
	Local	0.1–1.0
	Regional	0.5–5.0
	Clear	0.25–50

FM Band: 0.25, 1, 3, 5, 10, 25, 50, 100 kW, depending on class of service (community size) and area coverage for license granted.

Television-video:	Channels	Effective Radiated Power (average), kW
	VHF (2–6)	100
	VHF (7–13)	316
	UHF	5000

Television-audio: 10% (−10 dB)–20% (−7 dB) of power in video carrier.

† For more information, see *Reference Data for Radio Engineers*, ITT, Sixth ed., Indianapolis: Howard W. Sams & Co., 1975, Ch. 30.

Table C.4   VHF Television Station Allocations

Channel number	Frequency span (MHz)	Video carrier (MHz)‡
1		Not used
2	54–60	55.25
3	60–66	61.25
4	66–72	67.25
5	76–82	77.25
6	82–88	83.25
		(Standard FM band 88–108)
7	174–180	175.25
8	180–186	181.25
9	186–192	187.25
10	192–198	193.25
11	198–204	199.25
12	204–210	205.25
13	210–216	211.25

‡ Audio carrier is 4.50 MHz higher than the video carrier.

Table C.5   UHF Television Station Allocations

Channel number	Frequency span (MHz)	Video carrier (MHz)†	Channel number	Frequency span (MHz)	Video carrier (MHz)†
14	470–476	471.25	42	638–644	639.25
15	476–482	477.25	43	644–650	645.25
16	482–488	483.25	44	650–656	651.25
17	488–494	489.25	45	656–662	657.25
18	494–500	495.25	46	662–668	663.25
19	500–506	501.25	47	668–674	669.25
20	506–512	507.25	48	674–680	675.25
21	512–518	513.25	49	680–686	681.25
22	518–524	519.25	50	686–692	687.25
23	524–530	525.25	51	692–698	693.25
24	530–536	531.25	52	698–704	699.25
25	536–542	537.25	53	704–710	705.25
26	542–548	543.25	54	710–716	711.25
27	548–554	549.25	55	716–722	717.25
28	554–560	555.25	56	722–728	723.25
29	560–566	561.25	57	728–734	729.25
30	566–572	567.25	58	734–740	735.25
31	572–578	573.25	59	740–746	741.25
32	578–584	579.25	60	746–752	747.25
33	584–590	585.25	61	752–758	753.25
34	590–596	591.25	62	758–764	759.25
35	596–602	597.25	63	764–770	765.25
36	602–608	603.25	64	770–776	771.25
37	608–614	609.25	65	776–782	777.25
38	614–620	615.25	66	782–788	783.25
39	620–626	621.25	67	788–794	789.25
40	626–632	627.25	68	794–800	795.25
41	632–638	633.25	69	800–806	801.25

† Audio carrier is 4.50 MHz higher than the video carrier.

Table C.6   Amateur Bands†

Band designation	Frequency allocation (MHz)
160 meter	1.800–2.000
80 meter	3.500–4.000
40 meter	7.000–7.300
20 meter	14.000–14.350
15 meter	21.000–21.450
10 meter	28.000–29.700
6 meter	50.0 –54.0
2 meter	144–148
	220–225
	420–450
	1,215–1,300
	2,300–2,450
	3,300–3,500
	5,650–5,925
	10,000–10,500
	24,000–24,500
	48,000–50,000
	71,000–76,000
	165,000–170,000
	240,000–250,000
	above 300,000

Notes

1. Maximum authorized power input to the final transmitter stage is 1 kW.
2. Maximum amplitude modulation allowed is 100%. Except for brief tests or adjustments, the transmitter may not emit a carrier wave on frequencies below 51 MHz unless modulated for the purpose of communication.
3. Single-sideband: lower sideband is customarily used on 80 m and 40 m, upper sideband on the higher frequencies.

† For more information, see *FCC Rules and Regulations* **VI,** Part 97, or *The Radio Amateur's Handbook*, American Radio Relay League, Newington, Conn., 06111.

Table C.7   Citizens Radio Service‡

These frequency allocations are for fixed and mobile stations intended for short-distance personal or business radiocommunications, radio signaling, and control of remote objects or devices by radio.

Classifications

Class station	Service	Band (MHz)	Maximum transmitter input power	Maximum average output power	Frequency tolerance, %
A	General	460–470	60 W	48 W	0.0005
B	General	460–470	5 W	4 W	0.5
C	Remote control	26.96–27.23	5 W	4 W	0.005
	Hobby control	72–76	5 W	4 W	0.005
D	Radio telephone	26.96–27.41	5 W	4 W	0.005

‡ For more information, see *FCC Rules and Regulations* **VI,** Part 95.

Frequency Allocations
Class A & B

Base station/ mobile (MHz)	Mobile only (MHz)
462.550	467.550
462.575	467.575
462.600	467.600
462.625	467.625
462.650	467.650
462.675	467.675
462.700	467.700
462.725	467.725

Class C (mobile only; may employ only amplitude tone modulation or on-off keying of the unmodulated carrier)

1. For control of remote objects or devices by radio:
   26.995 MHz
   27.045
   27.095
   27.145
   27.195
   27.255
2. For radio remote control of any model used for hobby purposes:
   72.16 MHz
   72.32
   72.96
3. For radio remote control of aircraft models only:
   72.08 MHz
   72.24
   72.40
   75.64

Class D
(mobile radio telephone use only)

Channel	Frequency (MHz)	Channel	Frequency (MHz)	Channel	Frequency (MHz)
1	26.965	15	27.135	28	27.285
2	26.975	16	27.155	29	27.295
3	26.985	17	27.165	30	27.305
4	27.005	18	27.175	31	27.315
5	27.015	19	27.185	32	27.325
6	27.025	20	27.205	33	27.335
7	27.035	21	27.215	34	27.345
8	27.055	22	27.225	35	27.355
9†	27.065	24	27.235	36	27.365
10	27.075	25	27.245	37	27.375
11	27.085	23	27.255	38	27.385
12	27.105	26	27.265	39	27.395
13	27.115	27	27.275	40	27.405
14	27.125				

† Channel 9 is reserved for emergency communications: (a) involving the immediate safety of life of individuals or the immediate protection of property; (b) necessary to render assistance to a motorist.

Table C.8   Civil Defense†
(Disaster Communications Service)

Channel	Assigned frequency (MHz)	Bandwidth (kHz)	Service
1	1.7505	1	radiotelegraph
2	1.7515	1	radiotelegraph
3	1.7525	1	radiotelegraph
4	1.7535	1	radiotelegraph
5	1.7545	1	radiotelegraph
6	1.7555	1	radiotelegraph
7	1.7565	1	radiotelegraph
8	1.7575	1	radiotelegraph
9	1.7615	7	radiotelephone
10	1.7685	7	radiotelephone
11	1.7755	7	radiotelephone
12	1.7825	7	radiotelephone
13	1.7895	7	radiotelephone
14	1.7965	7	radiotelephone

Channel 9 is reserved for "Scene of Disaster."

† For more information, see *FCC Rules and Regulations* **VI,** Part 99.

APPENDIX D

COMMERCIAL TELEVISION TRANSMISSIONS

Commercial television systems exhibit several aspects of signal design choices and compromises in communication systems. The emphasis here is on these signal design choices and not on the circuits. Television standards vary from country to country and the systems discussed here, unless otherwise stated, are those presently used in the United States.

D.1 B/W TELEVISION

In order to transmit two-dimensional (in this case, picture) information via a one-dimensional coordinate system (in this case, time), it is necessary to employ some type of scanning technique. In television, each picture is divided into 525 lines and is scanned line-by-line. The time for one complete scan of a picture has been chosen to be $\frac{1}{30}$ sec so that power line interference appears stationary and hence is much less objectionable in the picture display. Each complete picture is called a *frame*.

Even though the human eye interprets picture sequences shown at a rate of about 15 frames per sec or greater as continuous motion, there is some discernible flicker present until frame rates on the order of about 40 per sec are used.† To eliminate discernible flicker in the television picture, alternate lines are sent at a rate of 60 per sec. This is called *interlaced scanning* and is illustrated in Fig. D.1. The scan starts at the upper left and proceeds from left to right, retracing quickly after each scan. (The slower vertical scan rate produces a slight tilt to the lines.) This is repeated to the bottom of the display, where only a half-line is scanned. A total of 262.5 lines have now been scanned and this is called a *field*. The time to scan one field is $\frac{1}{60}$ sec. The required horizontal scan frequency is then $(262.5)(60) = 15,750$ Hz.

At the end of the field, the cathode ray tube beam retraces rapidly upward (while still zigzagging horizontally) until it reaches the top center of the display screen. The beam is biased off (blanked) during both horizontal and vertical retrace times so this is not visible in the display. Next the beam repeats the horizontal scan at the 15.75 kHz rate with a slow downward motion at the vertical scan rate of 60 Hz. It follows the dashed-line path shown

---

† Silent movie film is generally run at 16 frames per sec; sound movie film is run at 24 frames per sec. Theater projectors use a shutter which opens and closes at double the frame rate (i.e., at 48 per sec) to eliminate discernible flicker.

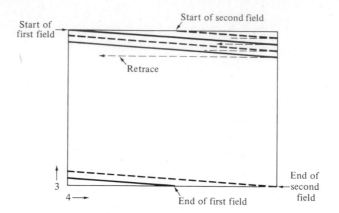

Fig. D.1   Interlaced scanning.

in Fig. D.1 until it reaches the lower right corner of the screen and the vertical retrace is again initiated. The two fields now total 525 lines, interlaced to form the complete picture (frame). In summary, the interlaced scanning method provides a 60 Hz field rate to eliminate flicker while maintaining a 30 Hz frame rate.

The line pattern used for each frame is called a *raster*. (The raster is all that is seen when no picture information is being received.) Although the size of the raster depends on the particular receiver (usually specified as a diagonal measurement), the relative dimensions of width-to-height—known as the aspect ratio—are standardized at 4:3.

Using the 4:3 aspect ratio, we can consider the complete picture as an array of 700×525 dots of varying intensity. Thus there are a maximum of $(700)(525)(30) = 11,025,000$ picture elements to be sent each second. (Actually, this is a pessimistic choice because about twenty-one lines are lost during the relatively long vertical retrace time.) Using the criterion that the system rise time must equal one picture element, a system bandwidth requirement ranging from *RC*-type response:

$$B \approx 0.35/t_r = (0.350)(11,025,000) = 3.86 \text{ MHz}$$

to a much sharper response:

$$B \approx 1/(2t_r) = (0.500)(11,025,000) = 5.51 \text{ MHz}$$

gives us some idea of the type of bandwidth required. In practice, a bandwidth of 4 MHz is assumed adequate.

Video information for television is transmitted using large-carrier amplitude modulation. Use of double-sideband would then require 8 MHz per channel for the video. However, channel allocations of 6-MHz width for experimental stations as early as 1936 have tended to restrict allowable bandwidths to within these limits. To circumvent these restrictions, a form of vestigial sideband transmission is used for the video information. A simplified spectral diagram of a television transmission is shown in Fig. D.2.

The lower sideband of the video transmission is attenuated below 0.75 MHz and is completely band-limited to 1.25 MHz below the video carrier. (Complete elimination of

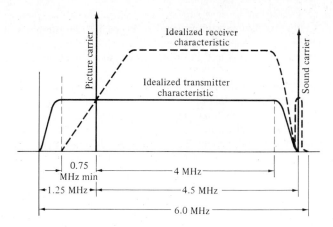

Fig. D.2   Simplified spectrum of a television channel.

one sideband at the transmitter would require much more elaborate and expensive filtering.) The receiver bandpass characteristic completes the vestigial sideband characteristic, as shown in Fig. D.2. The audio information is transmitted using FM, with a peak frequency deviation of 25 kHz, and is centered 4.5 MHz above the video carrier.

The amplitude modulation levels used for the video are shown in Fig. D.3. In the U.S., a negative modulation standard is used—i.e., less amplitude corresponds to a brighter scene while more amplitude corresponds to a darker scene. Because most pictures contain more white than black levels, somewhat greater power efficiency is possible with negative modulation than with positive modulation. Reference black is defined by 70% modulation and the minimum (white) modulation level is 12.5%, as shown in Fig. D.3.

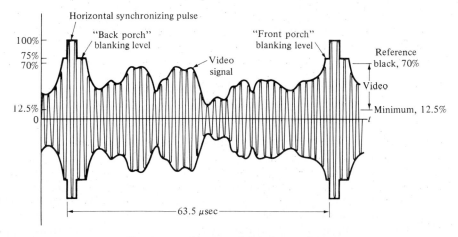

Fig. D.3   Video-modulated waveform.

Proper synchronization is necessary to reconstruct a stable picture. Synchronizing pulses are sent at the 100% modulation level to be easily identifiable and are sent at the beginning of each line scan. A blanking level sent at 75% modulation makes sure the CRT beam is cut off so that transients and the horizontal and vertical retraces are not visible in the display.

Vertical synchronization and the allowance for the one-half line scan is effected by generating a pulse train at the end of each field. This pulse train consists of a series of equalizing pulses and vertical synchronizing pulses generated at twice the horizontal scan rate (that is, at 31.5 kHz) as shown in Fig. D.4. The horizontal sweep oscillator is triggered by the leading edge of each pulse at the 31.5-kHz rate. (The fact that it can only sweep half the picture width during this time need not bother us because the CRT is blanked during the entire synchronization interval.) The equalizing pulses are kept narrow so that they will not trigger the vertical sweep oscillator. The vertical synchronization is composed of six consecutive wide pulses. The notches between these pulses are needed to maintain horizontal oscillator synchronization. These wide pulses are applied to an integrator to form a signal having sufficiently large amplitude to separate it from any other signals and to trigger the vertical sweep oscillator. More equalizing pulses and a few pulses at the 15.75-kHz horizontal scan rate follow. By the time the first video trace appears, all transient effects from synchronization have disappeared in the receiver. Note that the 31.5-kHz rate is necessary to cause the half-line horizontal trace variation for the interlaced scanning.

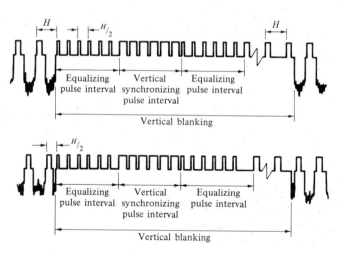

Fig. D.4   Vertical synchronization for each field.

Additional information can be transmitted by the commercial television signal in several ways without interfering with the normal picture and sound. These include use of the horizontal blanking interval, the vertical blanking interval, the audio channel (by using time or frequency multiplexing), and the video channel (by using time or frequency multiplexing). The first and last options are used in color broadcasting, as described in a later section. Multiplexing additional information in the audio channel is used by some subscription services. We devote some attention here to the recent interests in use of the vertical blanking interval.

As mentioned previously, the time taken for the vertical retrace interval is equivalent to twenty-one horizontal lines. The first nine lines are occupied by six equalizing pulses, followed by the vertical synchronizing pulse interval and another set of six equalizing pulses, as shown in Fig. D.4. This leaves lines numbered 10–21 available for other uses. (It is possible to see these lines by adjusting the vertical hold control of the television receiver until the bar separation between pictures rolls into view.)

Some of the lines in the vertical blanking interval have been assigned already. Lines 17, 18 carry vertical-interval test signal (VITS) transmissions. These video and chroma test signals are used to check the quality of network and station transmissions. Line 19 carries the vertical-interval reference (VIR) signals consisting of black, luminance, and chroma references. Some television receivers are equipped with automatic color tint and level circuitry using the VIR signals as a reference.

Line 21 is used by some stations for picture captioning to aid those viewers with hearing impairments. All of line 21 in the first field and one half of line 21 in the second field is available for this purpose. The NRZ data signal that is transmitted uses standard 7-bit-plus-parity ASCII coding. The decoder extracts the caption data from the vertical blanking interval and displays it near the bottom of the television screen.

Presently several experimental systems make use of the remaining lines in the vertical blanking interval for the transmission of calendar information (day, time, etc.), weather reports, and news. An idea of the communication potential of these lines can be obtained by assuming 62.5 μsec per line and a NRZ data format to give

$$(8 \text{ Mbps}) (62.5 \text{ } \mu\text{sec}) (60/\text{sec}) = 30 \text{ kbps/line}.$$

It is expected that such systems will gain in popularity in the near future as the cost of digital hardware decreases.

D.2 THEORY OF COLOR

Before we proceed directly to color television, it will be advantageous to briefly discuss some of the properties of color and color mixing. It is found that three primary colors are needed to make a color reproduction which is acceptable to human vision. By suitable mixing of these primary colors it is possible to approximate all colors found in nature. The three primary colors chosen are red, green, and blue. Other colors are synthesized by appropriate mixing of these primary colors. If we mix red and green light, for example, we obtain yellow.† In a similar manner, red and blue produce a pink-violet color known as magenta while green and blue produce a particular blue known as cyan. If we mix all three (in equal proportions), the result is known as white.

In television, it is customary to describe a picture in terms of its *luminance* (brightness) and *chrominance* (color). The chrominance information can in turn be described in terms of hue and saturation. To demonstrate the use of these terms, let us place the three primary colors at the corners of an equilateral triangle, as shown in Fig. D.5. From our discussion above on color mixing, it follows that yellow, magenta, and cyan will be located along the sides of this color (chrominance) triangle and white will be at the center.

Now suppose we describe this color information in terms of circular coordinates. This can be done by constructing a circle enclosing the color triangle and defining the colors by a radial magnitude and an angle. The *hue* (color) of the light then varies as the angle and

---

† All additions here are assumed to be noncoherent; i.e., addition on an energy basis.

the proportion of pure (i.e., saturated) color to white, called the *saturation*, varies as the radial distance from the center. Note that the saturation of a given color depends on its dilution with white light. Near the center a color is said to be highly desaturated (commonly called pale or pastel) while farther out it is said to be more saturated. However, all points along a line of fixed angle have the same hue.

The luminosity of light is a measure of its incident energy. This may be considered as a third dimension in Fig. D.5., perpendicular to the plane shown. The wavelength response

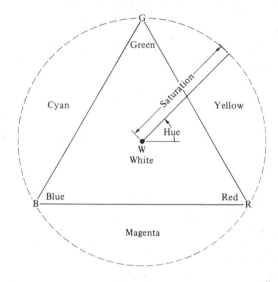

Fig. D.5 Color triangle showing hue and saturation.

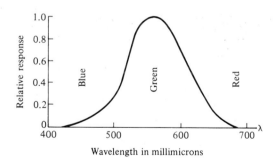

Fig. D.6 Response of the standard human eye.†

---

† This is the light-adapted response (photopic vision) corresponding to the brightness of normal daylight conditions. When adapted to darkness (scotopic vision) the peak response of the human eye shifts downward in wavelength by about 50 millimicrons. However, the eye loses some of its perception of color under scotopic conditions.

of detectors varies, so it is necessary to also consider the response of the detector. In the case of commercial television, of course, this is the human eye which is known to be unequally responsive to energy at different wavelengths. A composite of many measurements on the response of the human eye is shown in Fig. D.6. From this we conclude that the response of the human eye (i.e., luminosity as seen by the eye) is greatest at a wavelength of 550 millimicrons (a yellow–green color). The response drops off toward larger and shorter wavelengths, with a more rapid drop toward blue. In terms of the primary colors, then, the human eye interprets white light not as 33.3% of each but as about 59% green, 30% red, and 11% blue.

D.3   COLOR TELEVISION—THE NTSC SYSTEM

In the late 1940's and early 1950's several different color television systems were being proposed. Most were sequential systems, i.e., the color information was transmitted one color at a time in sequence. For example, one proposed system (CBS) which was approved by the FCC on a trial basis for a few years used a revolving color wheel so that one picture field was scanned in red, the next in green, etc. Such systems, while much simpler than ones in which all the color information is transmitted simultaneously, were found not to be readily compatible with existing B/W television methods and bandwidth restrictions.

In the early 1950's a committee—the National Television Systems Committee—was formed to try to resolve the problem of compatibility. The resulting NTSC system design was authorized by the FCC in 1953 and is the system presently used in the United States. Instead of transmitting the red, green, and blue signals sequentially, the NTSC system separates the information into luminance and chrominance signals which are transmitted simultaneously using frequency multiplexing. Most of the basic ideas of the NTSC system, with some refinements, have been incorporated into other color television systems used in other countries.

The NTSC system is designed to be compatible with the monochrome (i.e., B/W) system based on 525 lines and 60 fields per sec with interlaced scanning. A luminance component Y is formed by the summation of the three primary color components R, G, B. The resulting color signals are weighted in the same proportion as the human eye's response to the primary color frequencies so that

$$E_Y = 0.30E_R + 0.59E_G + 0.11E_B. \tag{D.1}$$

All of the information necessary for a monochrome picture is contained in this luminance signal.

In addition to the luminance signal, two color difference signals are transmitted. These color differences are formed by subtracting the luminance. Because the $(G - Y)$ component is smaller than the other two (and therefore subject to larger errors caused by noise), the $(R - Y)$ and $(B - Y)$ components are the two that are sent. These color difference signals are transmitted using quadrature multiplexing (DSB-SC) on a subcarrier placed ω_s rad/sec above the picture carrier. The chrominance (color difference) signal can then be written as

$$(E_R - E_Y) \cos \omega_s t + (E_B - E_Y) \sin \omega_s t. \tag{D.2}$$

The summation of Eqs. (D.1) and (D.2) forms the total video signal which is then modulated on the picture carrier using vestigial sideband techniques.

At the receiver, the following signals can be detected:†

$$\left\{\begin{array}{c} E_R - E_Y \\ E_B - E_Y \\ E_Y \end{array}\right\}. \tag{D.3}$$

Equation (D.1) can be rewritten as

$$(E_G - E_Y) = -\frac{0.30}{0.59}(E_R - E_Y) - \frac{0.11}{0.59}(E_B - E_Y). \tag{D.4}$$

Thus the third color difference signal $(E_G - E_Y)$ can be found by proper weighting and addition (known as "matrixing"). The signals E_Y, $(E_R - E_Y)$, $(E_G - E_Y)$, and $(E_B - E_Y)$ can then be applied to the appropriate control grids of the picture tube.

Now consider the values of the $(R - Y)$, $(B - Y)$, and Y components for six saturated colors. These are listed in Table D.1. The two color differences are transmitted in quadrature at the color subcarrier frequency ω_s and we can use phasor relationships. Writing the color differences in terms of magnitude and phase, we have

$$\sqrt{(R - Y)^2 + (B - Y)^2}\, \tan^{-1}[(R - Y)/(B - Y)]. \tag{D.5}$$

Note that the magnitude in Eq. (D.5) is a measure of color saturation and the phase is a measure of the hue. The total maximum instantaneous amplitude is proportional to the sum of the luminance component, Y, and the magnitude of Eq. (D.5). This quantity is listed in the last column of Table D.1.

Table D.1 Addition of Luminance and Chrominance Components for Six Saturated Colors

Color	R	G	B	Y	(R − Y)	(B − Y)	Magnitude/phase [using Eq. (D.5)]	Total maximum magnitude
Blue	0	0	1	0.11	−0.11	0.89	0.90 /353°	1.01
Red	1	0	0	0.30	0.70	−0.30	0.76 /113°	1.06
Magenta	1	0	1	0.41	0.59	0.59	0.83 / 45°	1.24
Green	0	1	0	0.59	−0.59	−0.59	0.83 /225°	1.42
Cyan	0	1	1	0.70	−0.70	0.30	0.76 /293°	1.46
Yellow	1	1	0	0.89	0.11	−0.89	0.90 /173°	1.78

For a monochrome transmission, we have $R = B = G$, and so $Y = 1.0$. The last column in Table D.1 shows, however, that the transmission of a fully saturated color requires larger instantaneous peaks, which may be as high as 78% over that required for

---

† It is possible to use detection along axes that are rotated by 33° (called I and Q axes) to gain some added bandwidth in the color difference signals. This will be discussed shortly. Modern receivers for home entertainment use generally employ the method outlined here and sacrifice the full color bandwidth (and hence definition) available.

monochrome. Because the peak modulation level is already 70% for monochrome, this is an intolerably large peak overload at the transmitter and it is necessary to reduce the amplitude of the color difference signals.

There are two color difference signals so we can introduce two independent scaling constants, a_1 and a_2, to reduce the peak modulation. For monochrome transmissions, $Y = 1.0$ and the maximum modulation index is 70%. A reasonable peak overload due to color differences is taken to be 33% (i.e., modulation index of 93%). Thus we can write the following equation for the maximum peak modulation level:

$$Y + \sqrt{a_1^2(R - Y)^2 + a_2^2(B - Y)^2} = 1.33. \tag{D.6}$$

From Table D.1, it can be seen that the two saturated colors resulting in the largest possible overloads are yellow and cyan. Substituting the values for yellow and cyan successively into Eq. (D.6) yields two equations in the two unknown scaling constants a_1, a_2:

$$0.89 + \sqrt{a_1^2(0.11)^2 + a_2^2(0.89)^2} = 1.33, \tag{D.7}$$

$$0.70 + \sqrt{a_1^2(0.70)^2 + a_2^2(0.30)^2} = 1.33. \tag{D.8}$$

Solving Eqs. (D.7) and (D.8) simultaneously yields $a_1 = 0.877$, $a_2 = 0.493$. The total video signal, including the corrected chrominance signal, which is used to amplitude modulate the transmitter carrier, is then†

$$E_T = E_Y + 0.493(E_R - E_Y) \cos \omega_s t + 0.877(E_B - E_Y) \sin \omega_s t. \tag{D.9}$$

The corrected chrominance values are listed in Table D.2. These results can also be portrayed by constructing a phasor diagram of the phase relationships at the color subcarrier frequency ω_s, as shown in Fig. D.7.

Table D.2   Phase Relationships for Corrected Chrominance Signal

Color	Components 0.877(R − Y)	0.493(B − Y)	Magnitude/phase [using Eq. (D.5)]
Blue	−0.0965	0.439	0.450 $\underline{/347°}$
Red	0.614	−0.148	0.631 $\underline{/103.5°}$
Magenta	0.517	0.291	0.594 $\underline{/60.7°}$
Green	−0.517	−0.291	0.594 $\underline{/240.7°}$
Cyan	−0.614	0.148	0.631 $\underline{/283.5°}$
Yellow	0.0965	−0.439	0.450 $\underline{/167°}$

A question that arises is how much bandwidth to allow for the chrominance signals. Fortunately the color resolution of the human eye is less than for black-and-white. In fact, studies of human vision have shown that it is not even the same for all color combinations. Finer chrominance detail can be resolved in orange and cyan than in green and magenta. To take advantage of this, the chrominance subcarrier reference can be advanced (at the

---

† At the receiver, the $(R - Y)$ component is weighted—relative to the luminance signal—by $1/0.877 = 1.14$, and the $(B - Y)$ component by $1/0.493 = 2.03$.

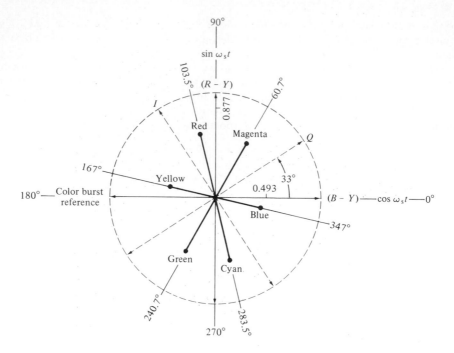

Fig. D.7 Phasor diagram of chrominance phase relationships.

transmitter) in Fig. D.7 by 33°. The new axes so formed are called the I and Q axes and are shown in Fig. D.7. For the Q signal, a comparatively narrow bandwidth of 0.5 MHz is found to be adequate. A much wider bandwidth is needed for the I signal. In order to accommodate the wider bandwidth, vestigial sideband modulation is used to restrict the I signal bandwidth to about 250 kHz above the subcarrier frequency. A simplified spectral diagram of a color television transmission is shown in Fig. D.8.

At the receiver, demodulation along the I and Q axes should be used to take advantage of the wider bandwidth of the I signal. However, this makes receiver design more difficult and generally this method of demodulation is not used. As noted earlier, it is simpler to demodulate along the $(R - Y)$ and $(B - Y)$ axes and then derive the $(G - Y)$ component by matrixing. The two demodulators use the same bandwidth so that there is some loss in color definition which is most noticeable in the orange and cyan hues (see Fig. D.7).

The next question that arises is that of the choice of the color subcarrier frequency. A subcarrier at a frequency that is too high would result in crowding the color sidebands too near the 4 MHz video bandwidth limit from the picture carrier. The possibility of heterodyne products between the audio carrier and the color-signal sidebands is also increased with a high subcarrier frequency. On the other hand, if the subcarrier frequency is too low, interference might result from heterodyne products between the main picture carrier and the color-signal sidebands. (In practice, the picture carrier is on the order of about 10 dB larger than the audio carrier, favoring a choice near the high end of the video

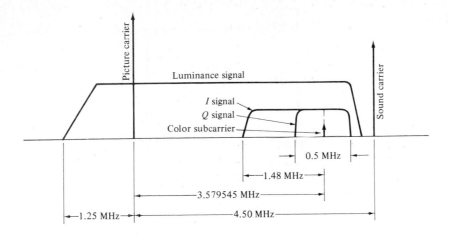

Fig. D.8   Color television spectrum showing I and Q spectra.

bandwidth.) The spectral components in a typical video signal are much stronger toward the lower frequencies so it is easier to insert the chrominance information toward the high end of the video bandwidth and remain compatible with monochrome receivers. A reasonable choice based on these considerations is in the vicinity of 3.5 MHz. To understand the choice of the specific frequency chosen (3.579545 MHz) requires a more careful consideration first of the spectrum of monochrome television transmissions.

A photograph of a spectrum analyzer display of a commercial television transmission is shown in Fig. D.9(a). An expanded portion of the spectrum near the picture carrier is shown in Fig. D.9(b). Note that as a result of the horizontal sweep sampling, the luminance spectral information tends to cluster about the harmonics of the horizontal sweep frequency. One can observe these very definite sidebands introduced by the sampling process at all multiples of the horizontal sweep frequency.† The reference or zeroeth harmonic is the main picture carrier.

The chrominance information is also sampled at the horizontal sweep rate but its reference is at the color subcarrier frequency. Therefore we can choose this subcarrier frequency in such a way that the spectra of the chrominance signals lie exactly between the spectra of the luminance signal. In this way it is possible to transmit both the luminance and the two chrominance signals within the same bandwidth as is normally used for monochrome television transmissions.

In order to position the chrominance sidebands between the luminance sidebands, it is necessary to choose the color subcarrier frequency at an integer-plus-a-half times the horizontal sweep frequency. In other words, the color subcarrier frequency must be an odd

---

† If the picture were absolutely stationary, the sidebands about each multiple of the 15.75 kHz horizontal sweep frequency should in turn have harmonic lines spaced at 60 Hz (the vertical sweep frequency). In practice, the picture information generally varies too much to resolve these lines.

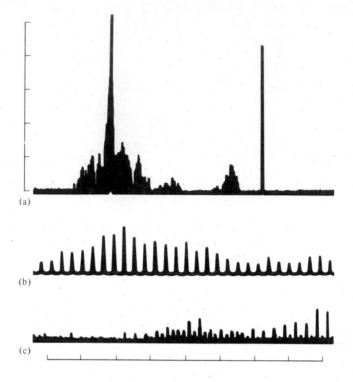

(a)

(b)

(c)

Fig. D.9 Spectrum analyzer display of a commercial television transmission: (a) 1 MHz per division; (b) 50 kHz per division near the picture carrier; (c) 50 kHz per division near the color subcarrier (vertical scale is 10 dB per division).

multiple of one-half the horizontal sweep rate. Another consideration is that we would like to have the sound carrier (at 4.5 MHz) positioned between the color sidebands to avoid possible low-frequency heterodyne effects. Taking the 571st harmonic, we get

$$\frac{15,750}{2}(571) = 4,496,625 \text{ Hz},$$

and for $n = 573$,

$$\frac{15,750}{2}(573) = 4,512,375 \text{ Hz}.$$

Thus the 4.50 MHz audio carrier is not centered, as desired, between the color sidebands. We have two possible choices to correct this situation. First, we could shift the audio carrier frequency up slightly to

$$\frac{4,496,625 + 4,512,375}{2} = 4,504,500 \text{ Hz}.$$

However, this 4.5 kHz shift in the audio FM spectrum would not be compatible with existing monochrome receivers.† The second choice is to lower the horizontal sweep frequency slightly. The required horizontal sweep frequency, f_h, must then satisfy the equation

$$\frac{4,500,000 + f_h/2}{573} = f_h/2, \tag{D.10}$$

from which

$$f_h = 2\,\frac{4,500,000}{572} = 15,734.266\ \text{Hz}. \tag{D.11}$$

This new horizontal sweep frequency for color transmission is within 0.1% of the monochrome value (15,750.) and is quite satisfactory with existing receiver designs. The corresponding new field frequency is:

$$\frac{2}{525}\,(15,734.266) = 59.94\ \text{Hz}. \tag{D.12}$$

Both the 59.94-Hz field rate and the 15,734.266-Hz horizontal sweep rate are sufficiently close to the 60-Hz and 15,750-Hz rates used in monochrome transmissions to permit synchronization in both types of receivers. However, the new rates are no longer in synchronism with the 60 Hz power line frequency. Thus any power line pickup in the video will appear to "roll" slowly through the picture vertically with about a 17-sec (that is, 1/.06) period when the receiver is tuned to a color broadcast. This is judged to be a minor disadvantage for the sake of compatibility and one which can be minimized with good design procedures.

The color subcarrier is chosen to be the 455th harmonic of one-half the horizontal sweep rate, or

$$(455)\left(\frac{15,734.266}{2}\right) = 3,579,545\ \text{Hz}. \tag{D.13}$$

Because quadrature multiplexing is used for the color information, it is essential that the synchronous detector in the receiver be exactly in phase and frequency with the transmitter. This is accomplished by sending eight cycles—known as the "color burst"—of the color subcarrier during the blanking interval following the horizontal sync pulse (known as the "back porch"). This is illustrated in Fig. D.10.

In the color television receiver the 8 cycles of the color subcarrier are selected by a delayed gating circuit triggered by the horizontal sync pulse. One method of synchronization is to compare the phase of the color burst with the phase of a local oscillator and correct it accordingly. An alternative approach is to use a very high-Q narrowband filter such that oscillations are started by the color burst and then will continue for the duration of one horizontal sweep. The color burst is not transmitted during the vertical retrace or during any black-and-white transmission. A "color killer" circuit in the color television receiver is used to detect the absence of the color burst and disables the color amplifiers so that possible color leakage is not displayed in the picture.

The stability of the color subcarrier frequency must be carefully controlled, particularly over the period of one horizontal line. Commercial broadcast station tolerances for the 3.579545-MHz frequency are ±0.0003%, or about ±10 Hz. The horizontal and

---

† A peak frequency deviation of 25 kHz is used in television audio transmissions and the shift of 4.5 kHz is not negligible.

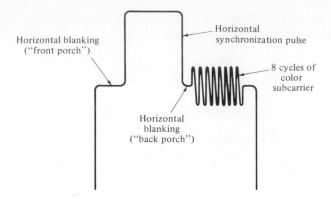

Fig. D.10   Color subcarrier burst.

vertical sweep frequencies at the transmitter are obtained by appropriate frequency dividers from the color subcarrier frequency.

Up to this point, we have tacitly assumed that the cathode ray tube in the receiver operates as a linear device so that the light intensity output is proportional to the amplitude of the input signal voltage. This is not true in practice and a more realistic relationship is

$$I = kE^\gamma, \tag{D.14}$$

where I represents the light intensity output, k is an arbitrary constant, E is the signal voltage, and γ (gamma) is a constant—about 2.2 to 2.5. Thus the light output increases more rapidly than the input signal voltage. If uncorrected, gradations in the highlights in the picture would be expanded while gradations in the shadows would be compressed. To compensate for this, a nonlinear amplifier which has an opposite transfer characteristic dependency is used at the transmitter to effect the needed gamma correction.

Gamma correction is quite straightforward for monochrome transmissions. An amplifier whose input/output transfer characteristic is of the form

$$E_0 = k_1 E_i^{1/\gamma} \tag{D.15}$$

is used to amplify the camera signal before modulation at the transmitter. For color transmissions, the three primary color signals are gamma-corrected, using Eq. (D.15), and then coded into the luminance and chrominance signals. This provides for the correct picture dependency in each color. The luminance signal is then given by

$$E_Y = 0.30E_R^{1/\gamma} + 0.59E_G^{1/\gamma} + 0.11E_B^{1/\gamma} \le (0.30E_R + 0.59E_G + 0.11E_B)^{1/\gamma}. \tag{D.16}$$

The equality in Eq. (D.16) holds when $E_R = E_G = E_B$; i.e., when a monochromatic (black-and-white) picture is being transmitted. For highly saturated colors, on the other hand, the luminance signal is smaller than it should be and yet the complete color information is present at the receiver. What is happening, of course, is that some of the luminance information for highly saturated colors is actually being sent in the chrominance signals when gamma correction is applied. This results in some lack of definition in scenes with highly saturated colors and some distortion on sudden changes in color as a result of a partial transmission of the luminance information through the narrower-band chrominance system. It is not judged to be a serious drawback in practice, however.

Because there is this cross-coupling effect between the luminance and chrominance channels when gamma correction is applied, it now makes a difference at which reference angle the color subcarrier burst is sent. Of the choices: 0, $\pm 90°$, 180°, the last one yields the least luminance cross-coupling and is chosen as the reference angle (see Fig. D.7).

In summary, the NTSC color television system produces color pictures of acceptable definition and hue to the human eye within an overall bandwidth allocation of 6 MHz.† It provides good compatibility with monochrome reception and it is reasonably simple to transmit and receive. It does have some disadvantages, however. One of the problems in the NTSC system is that an exact phase relationship must be maintained between transmitter and receiver for successful operation of the quadrature multiplexing. If the receiver oscillator is not in correct phase, or if differential phase errors occur in transmission (or in video recording), a distortion in the hue results. A second drawback is that the transmission of the I and Q chrominance signals with differing bandwidths results in some cross-coupling between color difference signals. This problem is normally avoided by demodulating along the $(R - Y)$ and $(B - Y)$ axes and using equal bandwidths for the two signals, sacrificing the potentially better color definition.

D.4 OTHER COLOR TELEVISION SYSTEMS

The NTSC color television system described above was the first system to be adopted for commercial broadcasting on a permanent basis. Other color systems have been devised since the NTSC system was placed in operation. Basically, these systems have many similarities to the NTSC system but they attempt to overcome some of its drawbacks. Two of these systems are described briefly here.

In the PAL (Phase Alternating Line) color television system, the color difference signals $(R - Y)$ and $(B - Y)$ each are transmitted within a bandwidth of 1 MHz using quadrature multiplexing. The $(R - Y)$ channel, however, is reversed in polarity on successive lines at the transmitter. The correct phase is restored by a synchronous polarity reversing switch in the receiver. By this means, effects of differential phase errors occurring in transmission (or in video recording) may be partially corrected by an averaging effect over several lines. (Recall from Fig. D.7 that the definition of the eye along the $(R - Y)$ axis is better than along the $(B - Y)$ axis.) The phase of the color burst alternates between 135° and 225° (that is, 180° $\pm$ 45°) to synchronize the polarity reversing switch. The required receiver design for the PAL system is therefore more complicated (and therefore more expensive), with some advantage in better performance. A simplified spectral diagram of a British color television transmission using a 625-line PAL system is shown in Fig. D.11.

Another color system is used in France—the SECAM (Séquential Couleur à Mémoire) system. In the SECAM system, the color difference signals $(R - Y)$ and $(B - Y)$ are sent separately on alternate lines, i.e., the $(R - Y)$ signal for one line is transmitted with the line while the $(B - Y)$ signal is discarded, and on the next line the

---

† For a discussion of some proposed improvements in the NTSC system, see R. Wilmotte, "TV Look-Ahead", *IEEE Spectrum* **13**, No. 2, (February 1976), 34–39.

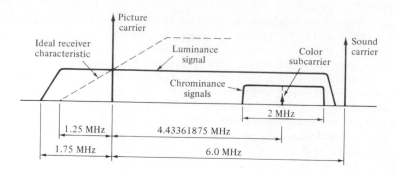

Fig. D.11 Simplified spectrum of the British 625-line PAL system.

$(B - Y)$ signal is transmitted while the $(R - Y)$ signal is discarded. Frequency modulation of the color subcarrier is used to transmit the color information.

In the SECAM receiver, a delay line and switch arrangement is used to allow a comparison of the previous line of color information with the present one. Thus the color resolution in the vertical dimension is reduced by one-half but the quadrature multiplexing and its possible cross-coupling between color signals is avoided. The use of FM for the color information makes the SECAM system immune—in contrast to the NTSC and PAL systems—to color distortion arising from amplitude variations. The major drawback of the SECAM system is that the use of FM for the color information tends to cause some video interference pattern effects in monochrome transmissions resulting in a poorer degree of compatibility.

Selected References for Further Reading

1. M. Mandl, *Principles of Electronic Communications*, Englewood Cliffs, N.J.: Prentice-Hall, 1973, Chapters 8, 9.

2. H. Stark and F. B. Tuteur, *Modern Electrical Communications: Theory and Systems*, Englewood Cliffs, N.J.: Prentice-Hall, 1979, Chapter 6.

3. C. N. Herrick, *Color Television: Theory and Servicing*, Reston, Va.: Reston Publishing Co., 1976.

4. H. V. Sims, *Principles of PAL Colour Television*, London: Iliffe Books Ltd., 1969.

5. G. N. Patchett, *Colour Television*, London: Norman Price (Publishers) Ltd., 1967.

6. R. H. Stafford, *Digital Television: Bandwidth Reduction and Communication Aspects*, Somerset, N.J.: Wiley-Interscience, 1980.

FACSIMILE TRANSMISSIONS

By "facsimile" we mean the transmission of two-dimensional imagery over commercial telephone systems. Examples include transmission of notices and invoices, weather forecast charts, satellite weather pictures, etc.

Facsimile transmission uses some of the same basic scanning techniques that are used in commercial television. Unlike television, it requires no vertical retrace, and relative allowances for horizontal retrace are negligible. As a result of some of the characteristics of telephone systems, however, additional problems may arise. We briefly discuss these here.

A basic voiceband telephone channel passes frequency components in the range 0.3–3 kHz with tolerable distortion to the human ear. When voiceband channels are used for data transmission, however, some problems may arise as a result of nonuniform magnitude frequency response and a phase delay which is not linear with frequency. To alleviate this, five types of "conditioned" lines are made available in the United States for data services. Specifications for lines maintained by AT & T are given in Table E.1.[†] (Envelope delay distortion will be defined later).

Most long-distance telephone systems use multiple SSB-SC channels which are frequency-multiplexed on a high-frequency (microwave) carrier. The subcarriers are not transmitted. The subcarrier which is reinserted for correct demodulation is not frequency or phase locked but is kept within ±5 Hz of the correct subcarrier frequency.

It is impractical to measure the phase-shift characteristic of a SSB channel because of the lack of phase lock. Therefore a narrowband sinusoidal AM waveform is sent through the channel and the difference between the phase shifts on the upper and lower sideband is detected as a function of the carrier frequency. For low modulating frequencies ($83\frac{1}{3}$ Hz is used by AT & T) this phase difference is approximately the derivative of the phase delay characteristic. The phase delay characteristic can be obtained, except for an arbitrary constant, by integration of the envelope delay. *Envelope delay distortion* is defined to be the maximum difference in the envelope delay over a given band of frequencies.

Delay distortion is quite tolerable for voice communications but it becomes a serious problem for pulse-type transmissions because not all frequency components arrive at the same time. For example, if a 1-kHz square wave were applied to a channel having

---

[†] "AT & T" is an abbreviation used for the American Telephone and Telegraph Company.

Table E.1   Some Typical Telephone Parameter Limits†

Channel conditioning	Attenuation distortion (frequency response) relative to 1 kHz		Envelope delay distortion relative to midband	
	Frequency range (Hz)	Variation (dB)	Frequency range (Hz)	Variation (μsec)
Basic	500–2500	−2 /+8	800–2600	1750
	300–3000	−3 /+12		
C1	1000–2400	−1 /+3	1000–2400	1000
	300–2700	−2 /+6	800–2600	1750
	300–3000	−3 /+12		
C2	500–2800	−1 /+3	1000–2600	500
	300–3000	−2 /+6	600–2600	300
			500–2800	3000
C3	500–2800	−0.5/+1.5	1000–2600	110
	300–3000	−0.8/+3	600–2600	300
			500–2800	650
C4	500–3000	−2 /+3	1000–2600	300
	300–3200	−2 /+6	800–2800	500
			600–3000	1500
			500–3000	3000
C5	500–2800	−0.5/+1.5	1000–2600	100
	300–3000	−1 /+3	600–2600	300
			500–2800	600

† From *Data Communications Using Voiceband Private Channels*, Bell System Pub. 41004, October 1973, © American Telephone and Telegraph Company 1973. Reprinted with permission.

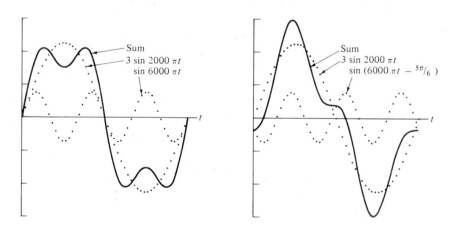

Fig. E.1   Input and output waveforms for a channel having delay distortion.

appreciable delay distortion, the fundamental and the third harmonic (all higher harmonics are severely attenuated) could undergo very different phase shifts. A typical resulting waveform is shown in Fig. E.1.

In order to make good use of the limited bandwidth available, analog facsimile transmissions are often amplitude-modulated, using a 2400-Hz carrier and a modulating bandwidth of 1600 Hz. Vestigial sideband filtering about the 2400-Hz carrier is then used to restrict the bandwidth to the range in which the delay distortion is tolerable. Line synchronization is sent by transmitting a low-level carrier (5% modulation) and then modulating the carrier with a step function to increase the modulation index to 100%. As a result of lack of phase coherence, the detected line synchronization can vary appreciably as shown in Fig. E.2 using the results of a computer model for the VSB channel. Note that in commercial television the carrier frequency is high enough that this is not a problem of concern.

As a result of the effects of delay distortion, facsimile systems are increasingly making use of digital encoding methods and frequency-shift (FSK) modulation. Some narrowband transmissions (e.g., EEG waveforms) are sent in analog form using frequency modulation (FM).

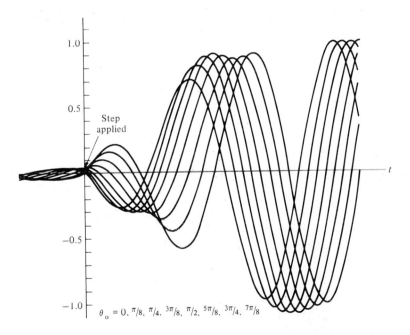

Fig. E.2 Facsimile line synchronization variation using VSB modulation without phase coherence.

SOME COMMERCIAL PREEMPHASIS/ DEEMPHASIS SYSTEMS

The use of passive high-frequency preemphasis/deemphasis has been practiced in audio recording and commercial FM broadcasting for years. Recently much more sophisticated preemphasis/deemphasis techniques have been applied successfully to the audio tape-recording field. Some are being proposed for FM broadcast use (the Dolby-B system is already used by many FM stations).

The three major types of new preemphasis/deemphasis systems in current use are described briefly. All three use a combination of filtering and dynamic-range compression to reduce the effects of the noise. Because the noise effects are most noticeable when the signal level is lowest, all three attempt to raise (boost) the low signal levels.

The Dolby-A system (Dolby Labs., London) divides the audio spectrum into four frequency bands and then applies a 10 to 15 dB boost to individual bands when the audio signal level falls below a preset level. This level is about 45 dB below a standard signal level known as the "Dolby level" and is set with reference to the flux level on the recording tape. The boosted signal is added to the original signal and then recorded. Assuming that the playback and record levels are properly set, the Dolby-A system can increase the net signal-to-noise (S/N) ratio 10 to 15 dB with no perceptible effect on the program material. Block diagrams of the Dolby-A system are shown in Figs. F.1 and F.2. A disadvantage of the Dolby system is that it has a limited dynamic range and cannot handle large orchestral peaks without some amplitude limiting ("clipping"). It is also fairly expensive for wide-spread use in domestic equipment.

A cheaper version, the Dolby-B system, is designed for domestic tape-recording equipment. To keep the costs down, it replaces splitting the audio spectrum into four bands

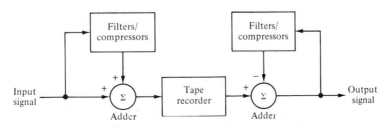

Fig. F.1 Block diagram of the Dolby system.

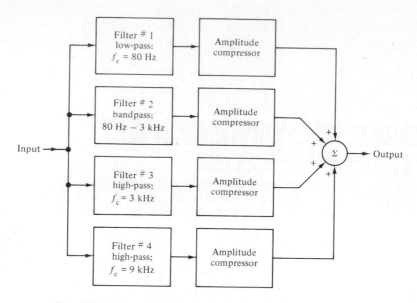

Fig. F.2   Filters/compressors for the Dolby-A system.

with a high-pass band above 600 Hz. A graph of the preemphasis characteristic of a Dolby-B system is shown in Fig. F.3. Its use in home-quality tape recorders is rapidly becoming popular with the availability of IC's. It is also being used for FM transmissions with a 25 μsec time-constant (6.4 kHz), high-pass band. This allows some S/N improvement while retaining compatibility with the 75 μsec fixed preemphasis used in FM transmissions.

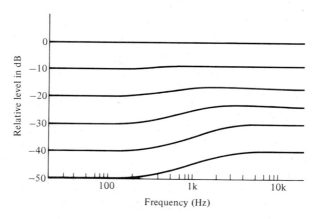

Fig. F.3   Preemphasis characteristic of the Dolby-B system.

The DBX system (DBX, Inc., Waltham, Mass.) treats the audio spectrum as a whole and uses high-frequency preemphasis above 1.6 kHz with a maximum high-frequency boost of 20 dB. A voltage-controlled amplifier with a wide dynamic range (130 dB) is used to amplify the signal. The gain of this amplifier is controlled by a rms detector. The overall effect is to detect the level of the signal and compress it by a 2:1 ratio. On playback, the signal is expanded to its original dynamic range and the high-frequency boost is removed. The DBX system offers 20–30 dB *S/N* improvement and can handle 26 dB signal peaks without distortion. A disadvantage of the DBX system is that the compression is governed by the largest input anywhere in the frequency band.

The Burwen system (Burwen Labs., Inc., Lexington, Mass.) is more elaborate and more expensive than either of the other two systems described above. It is similar in some respects to the DBX system but differs in that it uses precision peak signal detectors to control a wider compressor gain variation and uses low- as well as high-frequency preemphasis. The frequency-weighting preemphasis characteristics are dependent on the recording speed selected. The compression ratio is 3:1 and the net *S/N* improvement is approximately 50 dB. The Burwen system is expensive and its use is presently limited to high-quality studio use. A block diagram of the Burwen system is shown in Fig. F.4.

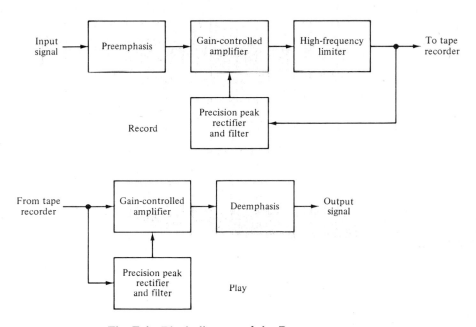

Fig. F.4 Block diagram of the Burwen system.

Selected References for Further Reading

1. R. S. Burwen. "A Dynamic Noise Filter," *Journal of the Audio Engineering Society* **19,** No. 2 (February 1971): 115.
2. R. M. Dolby. "An Audio Noise Reduction System." *Journal of the Audio Engineering Society* **15,** No. 4 (October 1967): 383.

APPENDIX G

THE COMPLEMENTARY ERROR FUNCTION

The gaussian probability density function with zero mean and unit variance is

$$p(x) = \frac{1}{\sqrt{2\pi}} e^{-x^2/2}. \tag{G.1}$$

The *error function*, Erf (x), is defined here as

$$\text{Erf}(x) = \int_{-\infty}^{x} \frac{1}{\sqrt{2\pi}} e^{-z^2/2} \, dz. \tag{G.2}$$

Note that Erf $(0) - \frac{1}{2}$ and Erf $(\infty) - 1$. The *complementary error function*, Erfc (x), is defined as†

$$\text{Erfc}(x) = 1 - \text{Erf}(x) = \int_{x}^{\infty} \frac{1}{\sqrt{2\pi}} e^{-z^2/2} \, dz. \tag{G.3}$$

Tabulated numerical values of Eq. (G.3) are given in Tables G.1 and G.2. For large values of x, this can be approximated by

$$\text{Erfc}(x) \approx \frac{1}{x\sqrt{2\pi}} \left[1 - \frac{1}{x^2} \right] e^{-x^2/2}. \tag{G.4}$$

The percentage error in the approximation is about -2% for $x = 3$, -1% for $x = 4$, and the approximation becomes increasingly better for larger values of x. An approximation with accuracy improved by about one order of magnitude over that of Eq. (G.4) is‡

$$\text{Erfc}(x) \approx \left[\frac{1}{(1 - a) x + a \sqrt{x^2 + b}} \right] \frac{1}{\sqrt{2\pi}} e^{-x^2/2}, \tag{G.5}$$

where $a = 1/\pi$ and $b = 2\pi$.

---

† Erfc (x), as used here, is also designated as $Q(x)$ in some texts.

‡ P. O. Börjesson and C. W. Sundberg, "Simple Approximations of the Error Function $Q(x)$ for Communications Applications," *IEEE Trans. Commun.*, COM-27 (March 1979): 639–643.

The error function used in statistics is usually defined as

$$\text{erf}(x) = \frac{2}{\sqrt{\pi}} \int_0^x e^{-t^2}\, dt, \tag{G.6}$$

and the corresponding complementary error function is

$$\text{erfc}(x) = 1 - \text{erf}(x). \tag{G.7}$$

Our particular choice of definition is more convenient for work in communication systems. It is because of the above differences that we use capital letters for our definitions.

Using a change of variables, it is easily seen that the relations between Eqs. (G.2) and (G.3) and Eqs. (G.6) and (G.7) are

$$\text{Erf}(x) = \frac{1}{2} + \frac{1}{2}\text{erf}\left(\frac{x}{\sqrt{2}}\right), \tag{G.8}$$

$$\text{Erfc}(x) = \frac{1}{2}\text{erfc}\left(\frac{x}{\sqrt{2}}\right), \tag{G.9}$$

or

$$\text{erf}(x) = 2\,\text{Erf}(\sqrt{2}\,x) - 1, \tag{G.10}$$

$$\text{erfc}(x) = 2\,\text{Erfc}(\sqrt{2}\,x). \tag{G.11}$$

Table G.1   Values of Erfc (x) vs. x.†

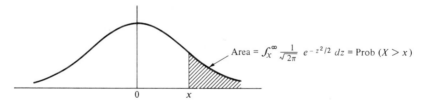

$$\text{Area} = \int_x^\infty \frac{1}{\sqrt{2\pi}}\, e^{-z^2/2}\, dz = \text{Prob}\,(X > x)$$

	0.00	0.01	0.02	0.03	0.04	0.05	0.06	0.07	0.08	0.09
0.0	.5000	.4960	.4920	.4880	.4840	.4801	.4761	.4721	.4681	.4641
0.1	.4602	.4562	.4522	.4483	.4443	.4404	.4364	.4325	.4286	.4247
0.2	.4207	.4168	.4129	.4090	.4052	.4013	.3974	.3936	.3897	.3859
0.3	.3821	.3783	.3745	.3707	.3669	.3632	.3594	.3557	.3520	.3483
0.4	.3446	.3409	.3372	.3336	.3300	.3264	.3228	.3192	.3156	.3121
0.5	.3085	.3050	.3015	.2981	.2946	.2912	.2877	.2843	.2810	.2776
0.6	.2743	.2709	.2676	.2643	.2611	.2578	.2546	.2514	.2483	.2451
0.7	.2420	.2389	.2358	.2327	.2296	.2266	.2236	.2206	.2177	.2148
0.8	.2119	.2090	.2061	.2033	.2005	.1977	.1949	.1922	.1894	.1867
0.9	.1841	.1814	.1788	.1762	.1736	.1711	.1685	.1660	.1635	.1611
1.0	.1587	.1562	.1539	.1515	.1492	.1469	.1446	.1423	.1401	.1379
1.1	.1357	.1335	.1314	.1292	.1271	.1251	.1230	.1210	.1190	.1170

† From J. S. Bendat and A. G. Piersol, *Random Data: Analysis and Measurement Procedures*, N.Y.: Wiley-Interscience, 1971, and D. B. Owen, *Handbook of Statistical Tables*, Reading, Mass.: Addison-Wesley Pub. Co., 1962, both by permission; courtesy of the U.S. Energy Research and Development Administration.

Table G.1   Values of Erfc (x) vs. x.

	0.00	0.01	0.02	0.03	0.04	0.05	0.06	0.07	0.08	0.09
1.2	.1151	.1131	.1112	.1093	.1075	.1056	.1038	.1020	.1003	.0985
1.3	.0968	.0951	.0934	.0918	.0901	.0885	.0869	.0853	.0838	.0823
1.4	.0808	.0793	.0778	.0764	.0749	.0735	.0721	.0708	.0694	.0681
1.5	.0668	.0655	.0643	.0630	.0618	.0606	.0594	.0582	.0571	.0559
1.6	.0548	.0537	.0526	.0516	.0505	.0495	.0485	.0475	.0465	.0455
1.7	.0446	.0436	.0427	.0418	.0409	.0401	.0392	.0384	.0375	.0367
1.8	.0359	.0351	.0344	.0336	.0329	.0322	.0314	.0307	.0301	.0294
1.9	.0287	.0281	.0274	.0268	.0262	.0256	.0250	.0244	.0239	.0233
2.0	.0228	.0222	.0217	.0212	.0207	.0202	.0197	.0192	.0188	.0183
2.1	.0179	.0174	.0170	.0166	.0162	.0158	.0154	.0150	.0146	.0143
2.2	.0139	.0136	.0132	.0129	.0125	.0122	.0119	.0116	.0113	.0110
2.3	.0107	.0104	.0102	.00990	.00964	.00939	.00914	.00889	.00866	.00842
2.4	.00820	.00798	.00776	.00755	.00734	.00714	.00695	.00676	.00657	.00639
2.5	.00621	.00604	.00587	.00570	.00554	.00539	.00523	.00508	.00494	.00480
2.6	.00466	.00453	.00440	.00427	.00415	.00402	.00391	.00379	.00368	.00357
2.7	.00347	.00336	.00326	.00317	.00307	.00298	.00289	.00280	.00272	.00264
2.8	.00256	.00248	.00240	.00233	.00226	.00219	.00212	.00205	.00199	.00193
2.9	.00187	.00181	.00175	.00169	.00164	.00159	.00154	.00149	.00144	.00139

Table G.2   Values of Erfc (x) for large x.

x	$\frac{10}{\log x}$	Erfc (x)	x	$\frac{10}{\log x}$	Erfc (x)	x	$\frac{10}{\log x}$	Erfc (x)
3.00	4.77	1.35E-03	4.00	6.02	3.17E-05	5.00	6.99	2.87E-07
3.05	4.84	1.14E-03	4.05	6.07	2.56E-05	5.05	7.03	2.21E-07
3.10	4.91	9.68E-04	4.10	6.13	2.07E-05	5.10	7.08	1.70E-07
3.15	4.98	8.16E-04	4.15	6.18	1.66E-05	5.15	7.12	1.30E-07
3.20	5.05	6.87E-04	4.20	6.23	1.33E-05	5.20	7.16	9.96E-08
3.25	5.12	5.77E-04	4.25	6.28	1.07E-05	5.25	7.20	7.61E-08
3.30	5.19	4.83E-04	4.30	6.33	8.54E-06	5.30	7.24	5.79E-08
3.35	5.25	4.04E-04	4.35	6.38	6.81E-06	5.35	7.28	4.40E-08
3.40	5.31	3.37E-04	4.40	6.43	5.41E-06	5.40	7.32	3.33E-08
3.45	5.38	2.80E-04	4.45	6.48	4.29E-06	5.45	7.36	2.52E-08
3.50	5.44	2.33E-04	4.50	6.53	3.40E-06	5.50	7.40	1.90E-08
3.55	5.50	1.93E-04	4.55	6.58	2.68E-06	5.55	7.44	1.43E-08
3.60	5.56	1.59E-04	4.60	6.63	2.11E-06	5.60	7.48	1.07E-08
3.65	5.62	1.31E-04	4.65	6.67	1.66E-06	5.65	7.52	8.03E-09
3.70	5.68	1.08E-04	4.70	6.72	1.30E-06	5.70	7.56	6.00E-09
3.75	5.74	8.84E-05	4.75	6.77	1.02E-06	5.75	7.60	4.47E-09
3.80	5.80	7.23E-05	4.80	6.81	7.93E-07	5.80	7.63	3.32E-09
3.85	5.85	5.91E-05	4.85	6.86	6.17E-07	5.85	7.67	2.46E-09
3.90	5.91	4.81E-05	4.90	6.90	4.79E-07	5.90	7.71	1.82E-09
3.95	5.97	3.91E-05	4.95	6.95	3.71E-07	5.95	7.75	1.34E-09

APPENDIX H

STEREO AM[†]

For more than fifty years in the United States interest has been expressed by various individuals and groups in the transmission of stereophonic (stereo) information using amplitude modulation (AM). In the late 1950s and early 1960s several groups (Philco, RCA, Westinghouse, and Kahn Communications) independently petitioned the Federal Communications Commission (FCC) to approve AM stereo broadcasting based on their respective systems. In 1962 the FCC ruled against all of these proposals, deciding instead to encourage the growth and expansion of stereo broadcasting using frequency modulation (FM). Since that time, public demand for stereo broadcasting has increased steadily and—more recently—has expanded into the automobile market.

As a result of the increasing interest in AM stereo, the National AM Stereophonic Radio Committee (NAMSRC) was formed in 1975, and this committee submitted a report on the subject to the FCC in 1977. Since 1977, the FCC has conducted evaluations and held hearings on AM stereo. Because no final decision has been reached on this matter at the time of this writing, this appendix summarizes briefly the five competing AM stereo options now before the Commission.

H.1 AM STEREO VERSUS FM STEREO

Before describing the various systems being proposed for AM stereo, we consider briefly what possible advantages AM stereo may have over FM stereo for certain situations. One advantage of wideband FM systems is the noise quieting that

---

[†] Because the material in this appendix depends on a knowledge of both amplitude modulation and angle modulation, we recommend that the reader cover Chapters 5 and 6 before attempting to read this appendix.

Several references to this appendix include the following: D. Mennie, "AM Stereo: Five Competing Options," *IEEE Spectrum*, June 1978, 24–31; R. Brownstein, "AM Stereo Getting Mixed Reviews," *Electronics*, July 6, 1978, 88–89; J. Deangelo, "AM Stereo: Soon on the Air?" *Popular Electronics*, December 1978, 59–64.

can be realized in demodulation, and the problem of an increasing noise power spectral density with frequency in FM detection is usually minimized with preemphasis/deemphasis. However, in FM stereo broadcasting, the stereo information is preemphasized prior to the DSB-SC modulation to avoid over-modulation of the transmitter. This results in a S/N degradation of 22 dB (Problem 6.10.4)—in FM stereo compared to FM mono—that partially offsets the FM noise quieting advantage.

Because the modulation in AM transmitters is limited by negative peaks in program material (to avoid overmodulation), many AM stations use audio compression circuits to compress the negative peaks in the program material before modulating the transmitter. In this manner the modulation on positive peaks can be increased (up to 125%), with the result that detected signal output in the receiver is louder in comparison to the noise level. This aids in the perceived signal-to-noise (S/N) ratio. However, this comparative gain is not so definitive because some FM stations also use audio compression to effectively increase loudness within the constraint of a fixed peak frequency deviation. Exceptions to this practice are those stations—many of which are classical music stations—that consider audio compression a type of undesired audio "distortion."

As a result of the higher frequencies used, the performance of FM systems tends to be influenced by multipath propagation. This is particularly noticeable in the rapid fading effects and varying signal intensities in the FM reception in vehicles moving in the vicinity of tall buildings or other structures. This is not as much of a problem in AM reception as long as the receiver is in an area of sufficient net signal strength and has a reasonably good automatic gain control. Also as a result of the frequencies used, FM has, basically, line-of-sight propagation, whereas AM has a greater potential range.

Finally, there is a question regarding adequate frequency response. In AM systems, the FCC requires a check of transmitter response for inputs from 100 to 7500 Hz (even though many AM receivers do not have good responses beyond 5 kHz). Some commercial AM stations have flat responses for inputs up to 10 kHz, and a few have good responses out to 15 kHz. Transmitted bandwidths are governed primarily by transmitter capabilities and possible cochannel interference from other stations in a particular geographical area. With the possibility of AM stereo, there is some concern that many stations will want to increase transmitter bandwidths to handle input frequencies in the 10–12 kHz range. Also, most of the AM stereo options under consideration at this time use some type of nonlinear modulation that creates extra sideband components which may fall outside the bandwidth allocated. One further complication is that 9-kHz spacings between AM station allocations instead of the present 10-kHz spacings are being considered in order to permit more stations in the AM band. These are unresolved problems.

Overall, it appears that AM stereo does hold some promise for improved stereo reception in mobile applications. It is estimated that 80% of the market for AM stereo receivers will be in automobile installations. In this type of application a slightly inferior frequency response and S/N, compared to FM stereo, may be acceptable in exchange for less fading and greater distance. Also, AM stereo receivers should be simpler, cheaper, and more stable than FM stereo receivers. With an increasing interest in stereo systems in automobiles, perhaps the time for AM stereo has arrived!

The five proposed AM stereo systems that attempt to meet these objectives have been submitted by Kahn Communications and the Hazeltine Corporation, Belar Electronics, Magnavox Company, Harris Corporation, and Motorola. In the following descriptions, each system is referred to by the company name.

H.2 THREE PROPOSED SYSTEMS (KAHN/HAZELTINE, BELAR, MAGNAVOX)

Desirable characteristics for AM stereo systems include compatibility with existing monophonic receivers, good stereo separation, good fidelity, minimal degradation (if any) in fringe-area reception, no increase in required bandwidth, adaptability to existing broadcast transmitter facilities, and low cost and complexity. As will be seen shortly, these characteristics are not as easy to obtain as they might seem at first glance.

H.2.1 Kahn/Hazeltine

In the Kahn/Hazeltine system, the upper and lower sidebands independently carry the right-channel (R) and left-channel (L) information, respectively. A mono receiver center-tuned to the carrier frequency adds the sidebands and detects the envelope, giving compatible (L + R) monophonic reception. Two mono AM receivers will detect stereo—with adequate but not good stereo separation—if the left receiver is tuned slightly below carrier and the right receiver is tuned slightly above carrier. A receiver with a single intermediate frequency (IF) has been developed for this system and gives better stereo reception. It uses envelope detection for the L + R signal and quadrature detection with phase-shifting networks for the L − R signal. These signals are then added and subtracted (this is called matrixing) to yield the L and R signals. During the early 1970's the Kahn system was used on an experimental basis at station WFBR in Baltimore, MD, and station XETRA in Tijuana, Mexico. The Hazeltine Corporation joined with Kahn in 1977 in support of this AM stereo system.

The transmitter in the Kahn/Hazeltine system uses the principles of compatible single-sideband (CSSB) modulation to obtain the independent sidebands. An L + R signal amplitude modulates the transmitter while the L − R signal is shifted by 90° (in practice, L + R is shifted −45° and L − R is shifted 45°) and then used to phase-modulate the transmitter. To give better stereo separation, a second-order correction signal is formed by squaring both the L and R signals and adding their difference to the L − R signal prior to the phase modulation. A 15-Hz tone can be added to the L − R signal, which, upon detection, is used for a stereo indicator light. This system has the potential for good stereo performance under varying propagation conditions (but at the expense of stereo separation in the double-IF version) and requires accurate phase shifting networks in the single-IF version.

H.2.2 Belar

A second AM stereo system is the AM/FM system proposed by Belar. The first proposal for this type of system for stereo was made by RCA in 1959 and was tested over station WRCA in New York City. Now known as the Belar system, this system uses narrowband FM to modulate $L - R$ information on the carrier. The peak frequency deviation is 1.25 kHz and preemphasis is used with a 100 μsec time constant. The $L + R$ signal is used to amplitude-modulate the carrier. A negative amplitude-modulation limit of 95% (5% residual carrier) is used to prevent excessive noise problems in the receiver limiter stages. The Belar proposal contains no provision for a stereo identification signal to operate a stereo indicator light in the receiver. Judging from their experience with FM stereo, however, manufacturers think it likely that potential users will want some type of indicator for stereo broadcasts.

H.2.3 Magnavox

A third AM stereo system is the AM/PM system proposed by Magnavox. In this system, an oscillator at the carrier frequency is frequency-modulated with a 5-Hz stereo identification tone using a peak frequency deviation of 20 Hz ($\beta = 4$). The $L - R$ signal is used to phase-modulate this carrier using a peak phase deviation of one radian ($\beta = 1$). The $L + R$ signal is used for the amplitude modulation signal to the transmitter. A simplified block diagram of the Magnavox system is shown in Fig. H.1.

Detection of the Magnavox AM/PM signal is implemented by using an envelope detector to obtain the $L + R$ signal and using an amplitude limiter and phase detector to obtain the $L - R$ signal. As in the Belar system, a negative modulation limit of 95% (5% residual carrier) is used to prevent excessive noise problems in the receiver limiter stages. The $L + R$ and $L - R$ signals are matrixed (i.e., added and subtracted) to obtain the L and R signals. Detection of the 5-Hz frequency modulation operates a stereo indicator light. Magnavox suggests the possibility of sending digital data (station identification, time, etc.) also using this 5-Hz tone capability.

Early in 1980, the Magnavox system was given preliminary acceptance for AM stereo broadcasting in the U.S. Shortly after that decision, however, the FCC was asked to reconsider its preliminary ruling and further tests, hearings and studies are still being conducted at this time.

Both the Belar and the Magnavox proposals use a combination of AM and angle modulation for stereo transmissions. Both are relatively easy and econom-

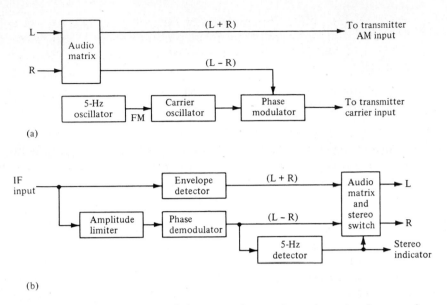

(a)

(b)

Fig. H.1 Simplified diagram of the transmitter and receiver signal processing for the proposed Magnavox AM stereo system.

ical to implement in the receiver as well as the transmitter. Both must maintain at least 5% residual carrier on negative peaks for acceptable angle demodulation. Also, the detected angle modulation could be expected to be noisy if the signal amplitude remains at this low level for any appreciable time interval.

H.3   TWO ADDITIONAL SYSTEMS (HARRIS AND MOTOROLA)

The remaining two proposed systems make use of modifications of quadrature multiplexing. Quadrature amplitude modulation (QAM) offers advantages of good performance in the presence of noise and the transmission and detection of two independent signals within the same bandwidth using a linear combination of the two signals. QAM is used successfully, for example, in the multiplexing of the chroma (color) information in commercial television broadcasting.

While QAM has much to offer for the transmission of stereo signals, a major problem is that of compatibility with envelope detection. Suppose we designate the left-channel signal as $\ell(t)$ and the right-channel signal as $r(t)$. The in-phase component $x(t)$ used to amplitude-modulate the carrier is

$$x(t) = 1 + \ell(t) + r(t) \tag{H.1}$$

and the quadrature signal is

$$y(t) = \ell(t) - r(t). \tag{H.2}$$

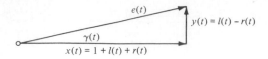

Fig. H.2 A phasor diagram of the in-phase and quadrature components in quadrature amplitude modulation (QAM) for AM stereo.

A phasor diagram of this result is shown in Fig. H.2. An in-phase detector will detect the AM signal (L + R); and a quadrature detector, the DSB-SC signal (L − R). However, an envelope detector will detect

$$e(t) = \sqrt{[1 + \ell(t) + r(t)]^2 + [\ell(t) - r(t)]^2}. \tag{H.3}$$

The presence of the L − R component in Eq. (H.3) may produce significant distortion in the envelope detector output if the L − R and L + R components are of comparable magnitude and if the modulation index is relatively high. Because these are realistic conditions, some modifications are necessary in quadrature multiplexing if compatibility with envelope detection is to be retained.

H.3.1   Harris

In the Harris AM stereo system, the L signal is applied to a balanced mixer whose reference is phase-shifted by −15 degrees with respect to carrier. This is shown in Fig. H.3(a). Similarly, the R signal is applied to a balanced mixer whose reference is phase-shifted by +15 degrees with respect to the carrier. The outputs of the two mixers are added together with the carrier, as indicated by the phasor diagrams in Fig. H.3(b).

The resulting signal envelope from the Harris AM stereo system is

$$e(t) = \sqrt{\{1 + [\ell(t) + r(t)] \cos 15°\}^2 + \{[\ell(t) - r(t)] \sin 15°\}^2}. \tag{H.4}$$

The reduction in the L − R component reduces the envelope distortion significantly at the penalty of some S/N loss. The signal power loss in the in-phase component is rather minor:

$$10 \log_{10} (\cos 15°)^2 = -0.30 \text{ dB},$$

but the power loss in the quadrature component is appreciable:

$$10 \log_{10} (\sin 15°)^2 = -11.7 \text{ dB}.$$

On the plus side, the Harris system is the only one of the current proposals that uses entirely linear modulation methods and, therefore, it is not as prone to the generation of harmonics outside a well-defined bandwidth as are the other four proposed systems. (Note that the second section of the transmitter diagram in Fig. H.3(a) is needed only to match into an existing AM transmitter; the signal

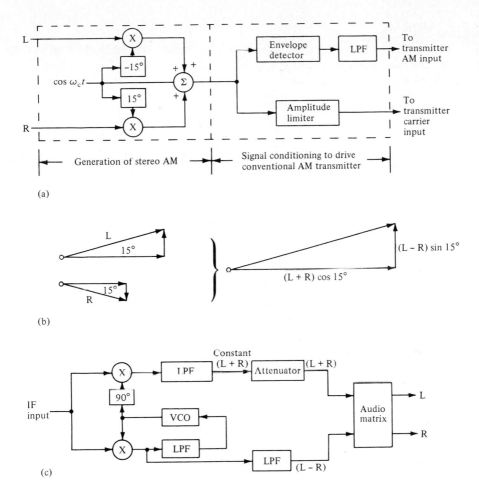

Fig. H.3 Simplified diagrams of the transmitter and receiver signal processing for the proposed Harris AM stereo system.

leaving the adder in the first section could be broadcast with a linear radio-frequency amplifier with the same results.)

While an envelope detector can do a good job of demodulating the mono signal from the Harris AM stereo system, a stereo receiver would have to have standard quadrature demultiplexing circuitry. A popular way to implement this circuitry is to use a phase-locked loop with both in-phase and quadrature mixers to demodulate the L + R and L − R signals. This is illustrated in Fig. H.3(c). For stereo identification, the Harris system transmits a 20–25 Hz tone at 9% modulation in the L − R channel only. This causes a negligible increase in envelope detector distortion and is relatively easy to separate in the stereo receiver by filtering.

H.3.2 Motorola

The Motorola system also uses quadrature multiplexing but retains a full 90° between the modulated in-phase and quadrature components. The resulting signal in terms of magnitude and phase is (see Fig. H.2):

$$\phi(t) = \sqrt{[1 + \ell(t) + r(t)]^2 + [\ell(t) - r(t)]^2} \cos [\omega_c t + \gamma(t)] \qquad (H.5)$$

where

$$\gamma(t) = \tan^{-1} \left[\frac{\ell(t) - r(t)}{1 + \ell(t) + r(t)} \right]. \qquad (H.6)$$

From Fig. H.2, note that

$$\cos \gamma(t) = \frac{1 + \ell(t) + r(t)}{\sqrt{[1 + \ell(t) + r(t)]^2 + [\ell(t) - r(t)]^2}}. \qquad (H.7)$$

Comparing Eq. (H.7) with Eq. (H.3), we find that the resulting envelope can be forced to be compatible with envelope detection by multiplying by $\cos \gamma(t)$,

$$e(t) \cos \gamma(t) = 1 + \ell(t) + r(t). \qquad (H.8)$$

Simplified diagrams of the Motorola AM stereo system are shown in Fig. H.4. After both the in-phase and quadrature components are modulated, they are added and amplitude-limited. This forms a phase-modulated carrier to the transmitter, and the L + R component is applied to the AM input to the transmitter. The resulting transmitted signal is the quadrature amplitude modulation signal described by Eq. (H.8) and is compatible with envelope detection for recovery of the mono signal. For stereo identification, a 25-Hz tone is added to the L − R component prior to modulation.

The burden of compatibility is placed on the stereo receiver in the Motorola system because the term $\cos \gamma(t)$ must be detected and factored out of the L − R component. A simplified receiver diagram is shown in Fig. H.4(b). A phase-locked loop follows the amplitude limiter; thus the in-phase and quadrature components of the phase-modulated carrier are generated. Synchronous detection with the quadrature component gives (L − R) $\cos \gamma$, and synchronous detection with the in-phase component gives $\cos \gamma$. These are applied to an analog divider to yield the L − R signal. The envelope detector output (L + R) and the divider output (L − R) are applied to an audio matrix that provides the desired stereo outputs.

H.4 CONCLUSION

The five AM stereo systems described here have all been field tested. Each system has its own advantages and disadvantages, and much of the weight of a decision concerning the implementation of one of them must be based on the

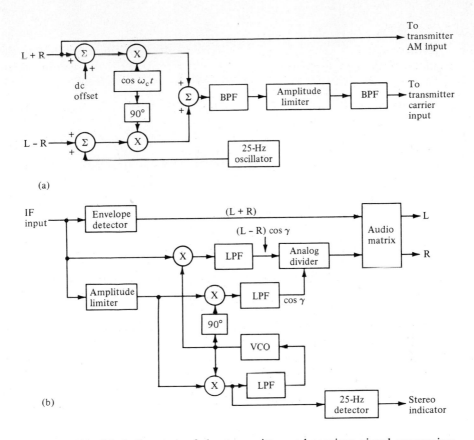

Fig. H.4 Simplified diagrams of the transmitter and receiver signal processing for the proposed Motorola AM stereo system.

cost efficiency trade-offs of various performance criteria. Among these are the increased bandwidth required, stereo separation, frequency response, and noise under various receiving conditions. Additional factors include not only mono and stereo receiver distortion under skywave-propagation and selective-fading conditions, but also distortion caused by both mistuning and restricted bandwidth conditions. Other considerations that may also be important are the potential reduction of the mono reception area, plans for system development through current and future transmitter and receiver designs, and the use of stereo indication. Trade-offs involving all of these are typical in the design of any new communication system. And which AM stereo system is chosen (if any) will result from a careful consideration of the relative costs—in both economic and performance terms—of each of the five competing systems.

APPENDIX I

A TABLE OF
BESSEL FUNCTIONS

Bessel Functions of the First Kind, $J_n(x)$

x	J_0	J_1	J_2	J_3	J_4	J_5	J_6	J_7	J_8	J_9	J_{10}
0.0	1.00										
.2	.99	.10									
.4	.96	.20	.02								
.6	.91	.29	.04								
.8	.85	.37	.08	.01							
1.0	.77	.44	.11	.02							
.2	.67	.50	.16	.03	.01⁻						
.4	.57	.54	.21	.05	.01⁻						
.6	.46	.57	.26	.07	.01						
.8	.34	.58	.31	.10	.02						
2.0	.22	.58	.35	.13	.03	.01⁻					
.2	.11	.56	.40	.16	.05	.01					
.4	.00	.52	.43	.20	.06	.02					
.6	−.10	.47	.46	.24	.08	.02	.01⁻				
.8	−.19	.41	.48	.27	.11	.03	.01⁻				
3.0	−.26	.34	.49	.31	.13	.04	.01				
.2	−.32	.26	.48	.34	.16	.06	.02				
.4	−.36	.18	.47	.37	.19	.07	.02	.01⁻			
.6	−.39	.10	.44	.40	.22	.09	.03	.01⁻			
.8	−.40	.01	.41	.42	.25	.11	.04	.01			
4.0	−.40	−.07	.36	.43	.28	.13	.05	.02			
.2	−.38	−.14	.31	.43	.31	.16	.06	.02	.01⁻		
.4	−.34	−.20	.25	.43	.34	.18	.08	.03	.01⁻		
.6	−.30	−.26	.18	.42	.36	.21	.09	.03	.01		
.8	−.24	−.30	.12	.40	.38	.23	.11	.04	.01		
5.0	−.18	−.33	.05	.36	.39	.26	.13	.05	.02	.01⁻	
.2	−.11	−.34	−.02	.33	.40	.29	.15	.07	.02	.01⁻	
.4	−.04	−.35	−.09	.28	.40	.31	.18	.08	.03	.01⁻	
.6	.03	−.33	−.15	.23	.39	.33	.20	.09	.04	.01	
.8	.09	−.31	−.20	.17	.38	.35	.22	.11	.05	.02	.01⁻

Bessel Functions of the First Kind, $J_n(x)$

x	J_0	J_1	J_2	J_3	J_4	J_5	J_6	J_7	J_8	J_9	J_{10}
6.0	.15	−.28	−.24	.11	.36	.36	.25	.13	.06	.02	.01⁻
.2	.20	−.23	−.28	.05	.33	.37	.27	.15	.07	.03	.01⁻
.4	.24	−.18	−.30	−.01	.29	.37	.29	.17	.08	.03	.01
.6	.27	−.12	−.31	−.06	.25	.37	.31	.19	.10	.04	.01
.8	.29	−.07	−.31	−.12	.21	.36	.33	.21	.11	.05	.02
7.0	.30	−.00	−.30	−.17	.16	.35	.34	.23	.13	.06	.02
.2	.30	.05	−.28	−.21	.11	.33	.35	.25	.15	.07	.03
.4	.28	.11	−.25	−.24	.05	.30	.35	.27	.16	.08	.04
.6	.25	.16	−.21	−.27	−.00	.27	.35	.29	.18	.10	.04
.8	.22	.20	−.16	−.29	−.06	.23	.35	.31	.20	.11	.05
8.0	.17	.23	−.11	−.29	−.11	.19	.34	.32	.22	.13	.06
.2	.12	.26	−.06	−.29	−.15	.14	.32	.33	.24	.14	.07
.4	.07	.27	−.00	−.27	−.19	.09	.30	.34	.26	.16	.08
.6	.01	.27	.05	−.25	−.22	.04	.27	.34	.28	.18	.10
.8	−.04	.26	.10	−.22	−.25	−.01	.24	.34	.29	.20	.11
9.0	−.09	.25	.14	−.18	−.27	−.06	.20	.33	.31	.21	.12
.2	−.14	.22	.18	−.14	−.27	−.10	.16	.31	.31	.23	.14
.4	−.18	.18	.22	−.09	−.27	−.14	.12	.30	.32	.25	.16
.6	−.21	.14	.24	−.04	−.26	−.18	.08	.27	.32	.27	.17
.8	−.23	.09	.25	.01	−.25	−.21	.03	.25	.32	.28	.19
10.0	−.25	.04	.25	.06	−.22	−.23	−.01	.22	.32	.29	.21

INDEX